MICROBIOLOGICAL CLASSIFICATION AND IDENTIFICATION

THE SOCIETY FOR APPLIED BACTERIOLOGY
SYMPOSIUM SERIES NO. 8

MICROBIOLOGICAL CLASSIFICATION AND IDENTIFICATION

Edited by

M. GOODFELLOW

AND

R. G. BOARD

1980

ACADEMIC PRESS

A Subsidiary of Harcourt Brace Jovanovich, Publishers

LONDON . NEW YORK . TORONTO . SYDNEY . SAN FRANCISCO

ACADEMIC PRESS INC. (LONDON) LTD
24–28 OVAL ROAD
LONDON NW1 7DX

U.S. Edition published by
ACADEMIC PRESS INC.
111 FIFTH AVENUE
NEW YORK, NEW YORK 10003

British Library Cataloguing in Publication Data

Impact of Modern Methods on Classification
and Identification (*Conference*), *University
of Birmingham, 1978*
Microbiological classification and identification.
– (Society for Applied Bacteriology. Symposium
series; no. 8).
1. Microbiology – Classification – Congresses
2. Micro-organisms – Identification – Congresses
I. Title   II. Goodfellow, Michael   III. Board,
Ronald, George   IV. Series
576'.01'2   QR12   80-40045

ISBN 0-12-289660-2

Typeset by Reproduction Drawings Ltd., Sutton, Surrey
Printed by Photolithography in Great Britain by
Whitstable Litho Ltd, Whitstable, Kent

Contributors

SHOSHANA BASCOMB, *Department of Bacteriology, St. Mary's Hospital Medical School, Wright–Fleming Institute, London W2 1PG, UK*

E. BILLING, *East Malling Research Station, East Malling, Maidstone, Kent ME19 6BJ, UK*

I. J. BOUSFIELD, *National Collection of Industrial Bacteria, Torry Research Station, P.O. Box 31, 135 Abbey Road, Aberdeen AB9 8DG, UK*

K. BØVRE, *Kaptein W. Wilhelmsen og Frues, Bakteriologiske Institutt, Riks-hospitalet, Oslo, Norway*

S. G. BRADLEY, *Department of Microbiology, MCV Station, P.O. Box 847, Virginia Commonwealth University, Richmond, Virginia 23298, USA*

J. S. CROWTHER, *Unilever Research, Colworth House, Sharnbrook, Bedford MK44 1LQ, UK*

C. M. E. GARRETT, *East Malling Research Station, East Malling, Maidstone, Kent ME19 6BJ, UK*

M. GOODFELLOW, *Department of Microbiology, The University, Newcastle-upon-Tyne NE1 7RU, UK*

H. HAMMANN, *Institut für Medizinische Mikrobiologie und Immunologie der Universität Bonn, Venusberg, 5300 Bonn 1, FDR*

C. R. HARWOOD, *Department of Microbiology, The Medical School, The University, Newcastle-upon-Tyne NE1 7RU, UK*

R. HOLBROOK, *Unilever Research, Colworth House, Sharnbrook, Bedford MK44 1LQ, UK*

C. W. JONES, *Department of Biochemistry, University of Leicester, Leicester LE1 7RH, UK*

DOROTHY JONES, *Department of Microbiology, School of Medicine and School of Biological Sciences, University of Leicester, Leicester LE1 7RH, UK*

R. M. KEDDIE, *Department of Microbiology, University of Reading, London Road, Reading RG1 5AQ, UK*

K. KERSTERS, *Laboratorium voor Microbiologie, K. L. Ledeganckstraat 35, B-9000 Gent, Belgium*

J. DE LEY, *Laboratorium voor Microbiologie, K. L. Ledeganckstraat 35, B-9000 Gent, Belgium*

D. E. MINNIKIN, *Department of Organic Chemistry, The University, Newcastle-upon-Tyne NE1 7RU, UK*

J. R. NORRIS, *Agricultural Research Council, Meat Research Institute, Langford, Bristol BS18 7DY*

M. J. SACKIN, *Department of Microbiology, School of Medicine and School of Biological Sciences, University of Leicester, Leicester LE1 7RH, UK*

H. N. SHAH, *The London Hospital Medical College, University of London, Turner Street, London E1 2AD, UK*

v

P. D. J. WEITZMAN, *Department of Biochemistry, University of Leicester, Adrian Building, University Road, Leicester LE1 7RH, UK*

E. M. H. WELLINGTON, *Department of Botany, University of Liverpool, P.O. Box 147, Liverpool L69 3BX, UK*

H. WERNER, *Institut für Medizinische Mikrobiologie und Immunologie der Universität Bonn, Venusberg, 5300 Bonn 1, FDR*

R. A. D. WILLIAMS, *The London Hospital Medical College, University of London, Turner Street, London E1 2AD, UK*

S. T. WILLIAMS, *Department of Botany, University of Liverpool, P.O. Box 147, Liverpool L69 3BX, UK*

Preface

A SYMPOSIUM on the topic "The Impact of Modern Methods on Classification and Identification" was held during the Summer Conference of the Society of Applied Bacteriology at the University of Birmingham in July 1978. The papers, which were given by leading specialists in their fields, are published in this, the eighth volume in the Symposium Series of the Society's publications.

In the early days of microbiology bacterial classifications were constructed primarily for the purposes of identification. These were based mainly on form and function and served to highlight the part played by bacteria in disease, food spoilage, organic matter turnover, the production of commercially useful compounds and the like. However, many of the features used in "classical" taxonomy were of little value in separating broad groups of bacteria whose taxonomy became confused and complex.

The past twenty years has witnessed a marked change in outlook in bacterial classification and identification. The application of methods such as numerical taxonomy, cell wall analysis, lipid analysis and nucleic acid pairing and base composition analyses is rapidly changing views on how bacteria should be classified and identified. Data of considerable taxonomic value are also being generated by the use of biochemical, genetical and physiological techniques and by the application of transmission and stereoscan microscopy. Many of the new approaches are quantitative, employing a variety of mathematical disciplines. The purpose of this book is to illustrate to readers the contribution modern taxonomic techniques have made on the classification and identification of some important groups of bacteria. The possible significance of plasmids in the processes of classification and identification is also discussed.

We wish to thank all the contributors for the efforts they took to produce contributions of quality. Indeed in some cases, the efforts to be comprehensive called for very extensive literature surveys and the labours involved have slightly delayed the publication of this volume in the Symposium Series. We wish also to thank Dr J. W. Hopton for the arrangement made for the Symposium at Birmingham.

M. GOODFELLOW
Department of Microbiology,
The University,
Newcastle upon Tyne NE1 7RU
April 1980

R. G. BOARD
School of Biological Sciences,
University of Bath,
Claverton Down,
Bath BA7 7AY

Contents

ABBREVIATED GENERA

This book includes reference to a wide range of common and less well recognized genera. In each chapter the names of genera are spelt out in full when first mentioned, but subsequently the four character alphabetic abbreviation, outlined by Leveritt *et al.* (1976) in the *International Journal of Systematic Bacteriology* (**26**, 443–445), has been adopted for consistency.

Introduction

J. R. NORRIS

Agricultural Research Council, Meat Research Institute,
Langford, Bristol, UK

THIS SYMPOSIUM of the Society for Applied Bacteriology comes, I believe, at a most opportune time. The last two decades have seen the development of new methods of characterizing and studying micro-organisms, some of which offer opportunities for the identification of micro-organisms by techniques which will be very much more rapid than those traditionally available. The time is ripe for someone to stand back from the technology and ask the question 'Where is all this getting us microbiologically?' The Society for Applied Bacteriology is to be congratulated on taking this step and I believe that the proceedings of this Symposium will become a landmark in the development of an important part of our science.

My task, as I see it, is to provide a background to the Symposium; a skeleton of ideas and concepts on which the main contributors will be able to put the flesh of specific examples. I would like to begin by looking briefly at the way in which our science has developed. It is often interesting to look at the origins of a science and to realize the way in which its development was determined by the individuals who were present at its birth and during its early years.

In microbiology we turn naturally to Pasteur as the acknowledged father of our subject. However, I do not personally believe that Pasteur had very much impact on the course of development of microbiology. Pasteur made a comment which I have always considered to be a significant one, telling us something about the man and about his attitude towards the science that he was creating. At the end of a lecture delivered late in his career, Pasteur was approached by an excited and enthusiastic young research worker who told him that he had made an important discovery; a bacterium which everyone thought was a coccus he had shown to be a short rod. Pasteur's comment was "I wish I could convey to you how little that information excites me". What he was saying was that he was interested in bacteria for what they did rather than for what they were. He was interested in the interaction between micro-organisms and their environment and other living things. I would suggest that Pasteur was not, in reality, a scientist as we would understand the term. He was, in fact, cast in the mould of the great naturalists of the eighteenth and nineteenth centuries.

It was another early microbiologist, Robert Koch, and the group of research workers he built around himself, that were to make the first scientific input to the subject. It was Koch who taught us to isolate, to purify, to count, to analyse; and he set the pattern for the next 100 years of microbiology. What he and his team did was quite remarkable. He gave to the early microbiologists techniques which were based on the ability of micro-organisms to grow in pure culture. Overnight the microbiologist could produce hundreds of thousands of copies of the cells he was studying. It was to be several decades before the chemist and physicist had at their disposal analytical techniques of comparable sensitivity.

The impetus given to microbiology by the possession of these powerful techniques was dramatic. I suspect that one had to live through the golden age of microbiology during the last few years of the last century and the first few of this to realize how quickly knowledge accumulated. Koch introduced nutrient gelatin in 1883. In 1886 de Bary in a series of lectures delivered in Strasbourg discussed Cohn's early classifications based almost entirely on morphology of bacteria and recorded the arguments which went on between Cohn and the monomorphists like Billroth and Nägeli in the 1870s and 1880s. The monomorphists believed that there was only one type of bacterium which occurred in many different forms. Yet by 1897 the monomorphists had all but gone. Fischer writing in his wonderful book "The Structure and Functions of Bacteria"* was able to tabulate the reactions of the anthrax bacillus, the typhoid bacillus, *Escherichia coli*, the cholera vibrio, *Bacillus subtilis* and *Pseudomonas aeruginosa* against sucrose, asparagine, tartrate, nitrate, ammonia and glycerin.

This represented an incredible rate of development and Koch undoubtedly did a tremendous service to microbiology. Perhaps he also did something of a disservice. By providing at such an early stage in the development of our science a range of powerful analytical techniques, he effectively focussed the attention of microbiologists on to pure cultures—man-made laboratory artefacts. I wonder whether he blocked, at that early stage, the development of our understanding of the natural history of micro-organisms. It is only in the last few years that we have begun again, to put together Koch's pure cultures to study their interactions with one another and investigate in a meaningful way the role of micro-organisms in their environment.

It may be helpful at this point for me to define some of the terms we shall be using during the Symposium:

Systematics is the scientific study of the diversity of organisms and their relationships. It includes:

---

*Alfred Fischer was Professor of Botany in the University of Leipzig when he wrote "Vorlesungen über Bakterien", an account of his course of lectures delivered to "students of biology, pharmacy, and agriculture, with here and there among them—as it were like a white raven!— a medical student". The English translation under the title "The Structure and Functions of Bacteria" was prepared by A. Coppen Jones. (Published in 1900 by the Clarendon Press.)

(a) *Classification*, the ordering of organisms into groups.

(b) *Taxonomy*, which is often used synonymously with classification but is sometimes used to mean the theory of classification.

(c) *Identification*, the assignment of unidentified organisms to a particular class in a previously made classification.

(d) *Nomenclature*, which deals with the assigning of the correct international scientific names to organisms.

(e) *Phylogeny*, which is the study of the evolutionary history of organisms.

The products of the activity of classification, or of taxonomy, are *classifications* or *taxonomies* (which are essentially the same thing) and the taxonomic groups so defined *taxa* (singular *taxon*). *Numerical taxonomy* is the grouping of organisms on the basis of numerical methods; it is sometimes called *taxometrics* or *numerical classification*.

The first quarter of the twentieth century saw many micro-organisms described and classifications grew up primarily for purposes of identification. These were based mainly on the shape of micro-organisms, their reaction to stains, the presence or absence of spores, movement, nutritional requirements, ability to utilize various sugars and so forth. In other words, they were based very largely on the behavioural properties of the cell. There are two good reasons for this; these characters were by far and away the easiest to study, we still lacked the sensitivity of analytical methods to enable us to analyse cell constituents in detail, and microbiologists were interested primarily in micro-organisms for what they did. The early classifications gave information relevant to the part played by micro-organisms in food spoilage, brewing, disease, cheese-making and so forth.

There was one exception to this behavioural basis for classification. In 1895 Pfeiffer showed that if the cholera vibrio was injected into a guinea-pig, serum taken later from the animal would immobilize and flocculate the vibrio. Furthermore, this was a specific reaction enabling the cholera vibrio to be differentiated from other micro-organisms and, although Fischer says "these facts have not been received without some scepticism", the use of specific antisera for purposes of identification had come to stay and to make a major impact on the development of our understanding of micro-organisms and their relationships to one another.

As the early years of the century passed, new techniques became available from chemistry and physics and the barriers of insensitivity began to fall away. Electron microscopy, chromatography, mass spectrometry enabled the microbiologist to ask more and different questions of organisms and their components and then in the mid 1950s came one of those mutational events which thrust understanding into a new era. Watson and Crick demonstrated the molecular basis of genetic information in the sequence of bases on the DNA molecule and

Level 1	The genome	DNA/DNA and DNA/RNA hybridization, G+C %, plasmids, phage, transfer of genetic material.
Level 2	Proteins	Amino-acid sequence, gel electrophoresis, serology.
Level 3	Cell components	Amino-acid pools, cell wall composition, lipid analysis, infra-red spectrophotometry, cytochromes, pyrolysis GLC and pyrolysis MS, bacteriocin, phage.
Level 4	Morphology and behaviour	Microscopic structure, motility, enzyme tests, physiology, nutritional requirements.

Fig. 1. Techniques for studying the genetic information in microbial cells at different levels of its expression.

opened a door to the understanding of microbial metabolism. Today every sixth-form biologist knows that the genetic information in a cell is carried in the form of a code on the DNA and that it determines, through the mediacy of RNA, the amino acid sequence in structural and enzymic proteins of the cell and hence its structure and its behaviour. We recognize that the genetic information present in a cell finds expression at several different levels (Fig. 1). At the first level it is expressed in the structure of the DNA itself, at the second level in the structure of protein molecules, at the third level in the chemical structure of cell components and products and at the fourth level in the morphology and behaviour of cells. With very few exceptions, the traditional classification which had evolved were concerned with information gleaned at the fourth level of genetic expression.

With the emergence of new techniques it became possible to ask questions at other levels and a new generation of characterization possibilities became available. Many of the modern and rapid methods with which we are concerned in this Symposium are different because they concern these other levels of genetic expression and I will now look briefly at the kinds of techniques which are available. The list is not complete, neither are the categories in any sense mutually exclusive.

At the first level genetic information is expressed in terms of the DNA molecule and the genome of the cell. If we heat the double-stranded DNA in the microbial cell, the two strands separate and will re-combine on cooling because of the mutual attraction of their complementary base pairs. If we heat DNA and then fix the single-stranded molecules in space so that they cannot re-combine on cooling, we can then determine the extent to which these single-stranded molecules will hybridize with single-stranded DNA from other organisms. The extent of hybridization will be a measure of the similarity of base sequences shared between DNAs of different micro-organisms and hence a direct measure of their relatedness. Similar techniques allow us to examine the ability of DNA and RNA to hybridize.

Micro-organisms possessing many base sequences in common will clearly have similar gross compositions for their DNAs. The determination of the guanine plus cytosine content of DNAs provides a figure which reflects the degree of relationship and the G + C percentage has come to feature prominently in taxonomic studies.

There are other ways of studying the genome of the cell. We can, for instance, ask questions about the ability of the genome of a particular cell to accept genetic material from another organism, perhaps in the form of a plasmid or carried by a bacteriophage.

At the second level genetic information is expressed first in terms of the amino acid sequence of proteins. This in turn determines the shape, the size and the surface electrical charge on the molecules. Amino acid sequencing is expensive and time-consuming and has not as yet found much application in microbial classification. Several methods, however, are available for characterizing protein molecules. Particularly important have been the methods based on determining the ability of protein molecules to migrate under the influence of an electrostatic field through a gel, the pores of which are of molecular dimensions—gel electrophoresis. The patterns of proteins resulting from gel electrophoretic analysis of microbial cell extracts are often highly complex and their interpretation requires the use of mechanical measurement and statistical analysis. Some workers prefer to concentrate on specific proteins, analysing particular enzymes by gel electrophoresis rather than the whole protein content of the cell. If we ask of a group of aerobic spore-forming bacteria whether they produce esterase and catalase, the replies will be 'yes' and 'yes' and the information will be virtually useless for taxonomic purposes. If, however, we ask what kind of molecule is produced in order to carry out esterase and catalase activity, then we shall achieve much more information and such methods have been applied extensively in microbial classification and identification.

Of course, we can also characterize proteins immunologically using the techniques of immunodiffusion and immunoelectrophoresis.

At the third level, the compositional level, a whole gamut of new methods have become available and much of our Symposium will be concerned with studies at this level. They may concern the analysis of whole cells or of parts of cells. The determination of amino acid pools, cell wall composition and cell lipids, have played specific and important roles as have the techniques like infrared absorption spectroscopy, pyrolysis gas-liquid chromatography and pyrolysis mass spectrometry which are concerned with the overall composition of microbial cells. Often when we study bacteriocins or bacteriophage we are in reality asking questions about the presence of specific chemical structures, receptor sites, on the walls of bacteria.

Particularly important for the identification of specific cell components has been the development of serological methods and these will undoubtedly recur throughout our meeting. Perhaps the ultimate objective of identification techniques is the provision of methods allowing the identification of individual microbial cells, since the necessity for cells to grow before cultures can conveniently be handled for identification provides the major stumbling block for the development of rapid identification methods. There appear to be only two approaches capable of providing specific identification at the individual cell level. One of these is new; the use of differential light scattering techniques which are as yet largely untested and unproven in microbiology but may well have considerable potential. The other is old; the Neufeldt capsular swelling reaction is based upon the fact that when a pneumococcal cell is mixed with an antiserum specific for its capsular material, the appearance and staining characteristics of the capsule change and the specific change can be observed under the microscope. This technique has been applied in a variety of ways to different organisms and in its more refined form, the use of fluorescent labelled antisera, can provide a highly sensitive and specific technique for the identification of individual cells.

Finally, at the morphological and behavioural level the electron microscope has enabled us to ask more and different questions about the structure and morphology of cells. Little is yet known about the correlation of morphology with the results of other taxonomic studies but with electron microscopes becoming routine instruments in many laboratories, the potential of such methods is obvious.

Also at the functional level we have seen the development of several micromethods and multi-test systems designed to standardize the examination of microbial behaviour and to extend the range of tests that are available. New methods have included the analysis of products of fermentation using gas chromatography and a growing awareness of the taxonomic significance of enzyme systems as opposed to the individual enzymes of the cell.

One of the most important, perhaps the most important, development to influence microbial taxonomy over the past two decades has, however, nothing to do with the technology of testing. It is the rapid development of the electronic computer and data handling methods and the elaboration of the statistical theory underlying classification.

Early attempts at classification and identification were based on the use of carefully chosen characters and they provided strategies which were extremely valuable. Selections of characters were made by highly intelligent and experienced microbiologists who wished to evolve systems of characterization which were relevant to the part played by micro-organisms in the natural world. What these taxonomists were doing was imposing on to a natural population of micro-

organisms classifications based on heavily biased, weighted selections of characters. To overcome such bias Professor Sneath in the mid 1950s introduced
numerical taxonomy into microbiology and this development has had tremendous impact on our understanding and use of microbial classification.

The statistical concepts introduced by Professor Sneath and others go far
beyond the analysis of 'traditional' data. They are, indeed, germane to much of
the work to be discussed in this Symposium. Many of the new methods of
characterizing organisms produce quantities and types of data which require
statistical analysis and computer handling. I would like to emphasize the importance of this aspect of the subject by quoting from our own work on the use of
pyrolysis techniques.

If an organic molecule is heated under carefully controlled conditions in an
inert gas atmosphere it will break down in a way which is predictable and highly
repeatable. If the products of such a pyrolysis are then separated and analysed
by gas-liquid chromatography (PGLC) or mass spectrometry (PMS), the analytical data produced will form a reproducible fingerprint of the molecule which can
be used for identification purposes. Similarly, if the microbial cells grown under
carefully controlled conditions are pyrolysed and the products analysed, the
resulting pyrogram will provide a fingerprint in the form of a number of peaks
representing the different products of pyrolysis. These can readily be recorded in
terms of peak numbers and heights. Most micro-organisms produce essentially
the same peaks but the quantities of material in the peaks, represented by the
peak heights, vary significantly and reproducibly from one organism to another.
Comparison of pyrograms between organisms thus provides a basis for studying
the relationships among a group of isolates. The data included in these pyrograms
must be compared quantitatively and are originally available in a continuous
form—that is, a peak may have any height, unlike the positive or negative scores
associated with unit characters in more traditional taxonomic studies. The
problem then is how to handle such data so as to provide a classification and a
strategy for identification.

If our micro-organism produced two peaks then the data could easily be
represented as one point on a two-dimensional graph of which one axis represented the height of one peak and the second axis that of the second peak. By
plotting the positions of a collection of organisms in the same two-dimensional
space, we could then study the relationship of one organism to another and
clusters would become immediately apparent. With three peaks we could build
models and work in three dimensions but with the pyrolysis data we are handling
there may be 40, 50 or indeed 100 units of data for each organism. We must
therefore resort to the mathematical concept of multi-dimensional space representing each isolate by a single point in space of dimensions equal to the number
of variables (peaks) we are studying. However, the concept is still the same; the
point locus representing the position of each organism is plotted in the same

multi-dimensional space and the relationships between these points can then be studied by a variety of statistical methods. Principle components analysis, for instance, will extract from the multi-dimensional array a two- or three-dimensional snapshot which represents the clearest natural arrangement of our points into clusters. The resulting visualization of clusters of points is often poor but is a valuable first step in the analysis of a group of organisms. It is particularly helpful for the detection of mishandled data or abnormalities of other kinds. The results of such analysis can be presented in the form of visual groupings or can be used as a basis for calculating similarity coefficients for presentation as shaded matrices or dendrograms or minimum spanning trees in the normal way. It is important to emphasize that we are presenting information to the computer and asking it to tell us whether or not there are any obvious relationships between the organisms represented which are reflected in the data.

We can, however, ask a quite different kind of question. We can superimpose on to our data a previously determined classification. Now we can concentrate on defined groups rather than on the loci of individual organisms. By using canonical variates analysis we can instruct the computer to derive scaling coefficients for the individual variables (peak heights) in such a way as to maximize the separation of predetermined groups and minimize the separation of individual points within groups. We can then obtain a visual, two-dimensional display in which the separation of the groups of points is maximized. The question we are now asking takes the form of a statement that a certain number of groups of organisms do exist and we are asking whether there is any sense in which our own pyrolysis data reflect the existence of these groups. If the resulting clusters are well spaced and the points within them are closely clustered around their centres, then we can have confidence that the groups are indeed real and that they are reflected in our data. Clearly we can use canonical variates analysis to test the validity of taxa derived from principal components analysis or of taxa superimposed on the new data that derive from a previous classification (Fig. 2).

An important opportunity is now open to us. The two-dimensional plot of a canonical variates analysis is derived by the calculation of scaling coefficients for individual peak heights. If we now analyse an unknown organism, it is a simple matter to apply the same scaling coefficients to its data and superimpose it on to a canonical variates display. Depending on its position, we can now identify the new isolate. Since PGLC may take perhaps 45 minutes, PMS perhaps five minutes and the computation is measured in micro-seconds and both methods of analysis can be applied to portions of a single colony from an agar plate, we have potentially at our disposal a very rapid method of identifying micro-organisms.

Naturally we shall wish to compare any classification we may develop by such an approach with other classifications based on other, perhaps more traditional, characters. This is best done by determining correlation coefficients between the similarity values exhibited by the two systems.

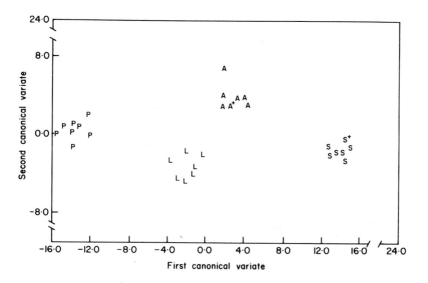

Fig. 2. Canonical variates plot of PGLC data from strains of *Bacillus subtilis* (S), *Bacillus pumilus* (P), *Bacillus licheniformis* (L) and *Bacillus amyloliquefaciens* (A). Unpublished data supplied by A. G. O'Donnell.

In the particular study illustrated in Fig. 2, pyrolysis gas-liquid chromatography was applied to a set of 32 strains representing the well-defined species *Bacillus subtilis, Bacl. pumilus* and *Bacl. licheniformis* and an organism of doubtful validity called *Bacl. amyloliquefaciens* which differ little, if at all, in standard tests from *Bacl. subtilis*. The analysis at level 3 of genetic expression was paralleled using the same strains at level 1 (DNA hybridization) and at level 4 (using the comprehensive range of biochemical tests available in the API systems for bacterial identification—API Laboratory Products Ltd., Farnborough, UK). In this particular case the classifications derived at the three levels coincided clearly indicating the existence of four well-defined taxa and confirming the separate existence of *Bacl. amyloliquefaciens*.

There have been few attempts to compare classifications derived from studies at the different levels of genetic expression and much of the interest in this Symposium will lie in such comparisons and in the possible definition of discriminatory ability of different taxonomic techniques.

I have chosen to talk about the specific case of pyrolysis data but the statistical concepts concerned can be applied generally to much of the information coming from the various techniques we shall be discussing during our Symposium.

As the Symposium unfolds, could I suggest that we bear in mind two questions and ask them each time we are considering a particular technique; they are 'At what level of expression of genetic information is this technique asking questions?'

and 'How are the data being handled?' In this way we shall gain a perspective on the present impact which newer methods are making on our understanding of the relationships between micro-organisms and perhaps an indication of the outstanding problems which remain to be tackled in this exciting and rapidly developing field.

DNA Reassociation and Base Composition

S. G. BRADLEY

Department of Microbiology, MCV Station, Virginia Commonwealth University, Richmond, Virginia, USA

Contents

1. Introduction

BACTERIAL CLASSIFICATION and identification are distinctly different processes, both in laboratory practice and in conceptual approach (Küster 1978). The goal of the applied bacteriologist is to determine whether a given isolate is the same as, or different from, another whose activities or attributes are of direct concern to man. A medical bacteriologist, for example, does not attempt to identify all of the members of the microbial flora in a clinical specimen.There is an immediate selection imposed by the applied bacteriologist by the procedures used to collect the sample, by the handling and transport of the sample to the laboratory, and by the nutritive medium and incubation conditions chosen for cultivation of the bacteria thought to be present in the sample, or whose absence needs to be established with some degree of confidence (Williams 1978). Currently the applied bacteriologist uses a sequence of selective or differential culture media and conditions, with appropriate observations and tests (Al-Diwany & Cross 1978). Their identification schemes are frequently presented as multicoloured charts with discriminatory results summarized succinctly. These practices are so familiar to applied bacteriologists that the underlying assumptions, and prejudices, are rarely considered. The applied bacteriologist may well consider that it is impossible to characterize and identify a large number of isolates from a sample from nature (Preobrazhenskaya *et al.* 1978). The time may come when this reservation is no longer valid. Automated scanners, colony pickers and inoculating devices may make it possible to characterize large numbers of isolates. Until this possibility is realized, the applied bacteriologist should recognize that current practice in culture identification is based upon a comparison of a highly selected isolate with a representative or type culture (Bradley 1975). Although the applied bacteriologist approaches the matter of identification by serial assessments of key characteristics, the laboratory worker observes many features not included in the conventional identification schemes, and uses these subconsciously as quality control references (Arai 1978). These auxiliary observations allow the experienced bacteriologist to accept some variation in the test results, while alerting him or her to spurious identifications arising from errors in scoring critical tests (Kutzner et al. 1978). The experienced bacteriologist, using a scheme that presumably relies upon less than a dozen characteristics, has probably considered nearly 100 observations (macroscopic morphology, microscopic morphology and secondary metabolites) before reaching a conclusion.

Given a set of reference cultures and a prescribed battery of tests, the experienced bacteriologist can identify readily an unknown culture or establish that it is not like any of the reference cultures (Bradley & Bond 1974). The utility of these identifications rests with the completeness of the available reference cultures and the discriminating tests. The process of identification is influenced by the intended utility of the information. The medical bacteriologist may need to determine whether a number of isolates from a population of patients in hospitals are probably of common origin. Alternatively, he may be concerned with establishing the possible infectious aetiology of a disease. It is at these decision points that nomenclature and classification become intertwined with identification (Lacey et al. 1978).

2. Genetic Approaches to Classification

The goal of classification is to arrange groups of organisms into useful, and hopefully, biologically relevant categories (Prauser 1978). Once clusters of related organisms have been formulated, names are assigned to these taxa according to the established rules for bacterial nomenclature (Lapage et al. 1975). A major impetus for applying new methods and new approaches to bacterial systematics has been the need to discover criteria that reveal biologically significant relatedness (Bradley 1965). Several of the newer methods are directed toward the characterization of the genetic composition of micro-organisms. In general terms, the genome of a bacterium may be analysed genetically or physicochemically (Bradley & Huitron 1973). The genetic approach to systematics has appeal because one widely used definition of a species in vertebrate animals and flowering plants is that it is an interbreeding population in nature (Ravin 1960). Until recently, gene transfer was considered so rare and limited among bacteria that it was felt to be of no consequence for bacterial populations in nature

(Bradley 1969). Accordingly, most taxonomists have concluded that the concept of genospecies, that is, a population sharing a common gene pool, is not applicable or useful in bacterial systematics (Sanderson 1976). During the past two decades, it has been recognized that gene transfer, particularly involving plasmids, can be a significant factor in bacterial variation, particularly where man has intervened, viz. the emergence and dissemination of multiple drug resistant bacteria (Falkow 1975). Plasmid transfer is so promiscuous that defining a population of bacteria which shares a pool of plasmids has essentially no value in delineating a genospecies.

Transfer of genes located on the fundamental genophore of a bacterium is limited by compatibility, competence and restriction factors, as well as by genetic homology (Bradley & Mordarski 1976). Genetic recombination therefore cannot be used routinely to delineate genospecies experimentally. An estimation of genetic homology can be obtained by measuring the extent and fidelity of pairing between single-stranded polydeoxyribonucleotides (Owen et al. 1978). A variety of methods are available for measuring DNA reassociation (Bradley & Enquist 1974). Each method has particular utilities and limitations (Enquist & Bradley 1970). It should be noted that homology measured by DNA pairing refers to similarity in nucleotide sequence whereas genetic homology refers to similarity in gene sequence.

3. DNA Base Composition

The first step in the physicochemical characterization of a DNA sample is to determine its mean nucleotide composition (Mordarski et al. 1978a). Base composition can be determined directly by hydrolysis of a DNA sample and separating the bases by chromatography or electrophoresis, or indirectly by calculating the mole per cent guanine plus cytosine (% G+C) from the hyperchromic shift accompanying thermal denaturation of DNA (the T_m or melting point of DNA), buoyant density in cesium chloride, and absorption differences at two wavelengths of the purines released by acid depurination of a DNA sample (Marmur et al. 1963). The % G+C of bacterial DNA samples can be determined readily and should constitute an essential item in the description of every taxon. The DNA from two biologically unrelated organisms may have the same % G+C, but the nucleotide composition of the DNA from two biologically related organisms cannot be markedly different.

DNA composition alone has provided definitive insights into particular taxonomic dilemmas (Mandel 1969). At one time, the so-called "slime-bacteria" were considered a well-defined group by virtue of their novel differentiation cycles (Dworkin 1966). The DNA of *Sporocytophaga* is 38 % G+C and that of *Myxococcus* is 68 % G+C, however, indicating that these two genera cannot be members of the same family (Cooney & Bradley 1965). At one time, *Staphylococcus* was considered closely related to *Micrococcus* but the % G+C of the former's DNA is *ca.* 35 % G+C (Pulverer *et al.* 1978) and that of *Micrococcus* is *ca.* 70 % G+C (Ogasawara-Fujita & Sakaguchi 1976). Until recently, *Listeria* was referred to as a coryneform bacterium and considered to be related to *Corynebacterium* (Seeliger 1961). The % G+C of the DNA of *Corynebacterium* is 50–60 % G+C whereas that of *Listeria* is *ca.* 38 % G+C (Stuart & Welshimer 1973).

Data on nucleotide composition are of value at the species as well as at the generic level. Gordon (1966) proposed that the members of the *'rhodochrous'* complex were a species in search of a genus. It was recognized that the *'rhodochrous'* complex consisted of two distinct populations, one having DNA with 60 to 63 % G+C and the other having DNA with 66 to 69 % G+C (Bradley 1966). Subsequent work has established that the *'rhodochrous'* complex contains about a dozen species (Mordarski *et al.* 1976; Goodfellow *et al.* 1978) and the *'rhodochrous'* complex has been elevated to the generic level as *Rhodococcus* (Goodfellow & Alderson 1977). Experience accumulated during the past two decades provides a firm basis for the guideline that the % G+C of DNA from the members of a species will vary by no more than 4 to 5 % G+C (Jones & Sneath 1970) and that of a genus by no more than 10 % G+C. The applied bacteriologist should be cautious, however, in making judgments dependent upon compilations of material in the literature that are based upon different methods and different reference standards (Henriksen 1976). The % G+C of *Escherichia coli* DNA is commonly used as a reference, but values ranging from 48 to 51 % G+C are used. This inconsistent practice constitutes one source of variation in published compilations of DNA compositions.

The nucleotide composition of plasmids is occasionally different from the nucleotide composition of the fundamental genome of the bacterium. Moreover, plasmid DNA may exist in multiple copies, equalling the amount found in the bacterial genophore in exceptional circumstances (Falkow 1975). It is conceivable, therefore, that the % G+C of the DNA from two bacteria, identical except for the presence or absence of a plasmid, would be quite different. To date, this problem has not arisen, possibly because plasmid DNA is lost preferentially in some extraction procedures and plasmid DNA is a minor component in the growing cells which are the source of most cellular material used for extraction of DNA.

Results from determinations of DNA composition can alert bacterial taxonomists to unsolved problems in classification. The genus *Bacillus* currently encompasses organisms whose DNA compositions range from 32 to 55 % G+C (Jones & Sneath 1970). Such a wide range indicates strongly that this genus needs to be divided into a minimum of two genera.

Base composition of DNA has limited utility in bacterial classification because this criterion can only be used to exclude an organism from a taxon, not as the sole criterion to include one. The size of the genome of bacteria can now be determined readily (Bak *et al.* 1970; Bradley *et al.* 1978). In general, the masses of bacterial genomes range between $1\text{-}5 \times 10^9$ daltons, representing $2 \times 10^6 - 1 \times 10^7$ nucleotide pairs (Bachman *et al.* 1976). Genome size has not proved to be a useful criterion in classifying bacteria because those of most bacteria have a mass of *ca.* 2.5×10^9 daltons. Moreover the masses of the genomes of closely related bacteria can vary appreciably (Brenner *et al.* 1972). From knowledge of the size of the genome and our present understanding of the manner in which genetic information is stored and retrieved, it is estimated that the genomes of most bacteria are able to code for 2000 to 5000 structural proteins and enzymes (Sanderson 1976). If the overlapping genes, and genes-within-genes that have been discovered in viruses are found elsewhere, then the number of structural proteins and enzymes coded for by the bacterial DNA may be substantially underestimated.

4. DNA Reassociation

In DNA reassociation, the nucleotide sequences of the entire genomes of two samples are compared. The nucleotide sequences include those coding for constitutive proteins, inducible proteins, regulatory genes and non-messenger RNA— that is, transfer RNA and ribosomal RNA (Fig. 1). DNA reassociation, therefore, compares the genetic potentials of the test organisms, not merely the overt

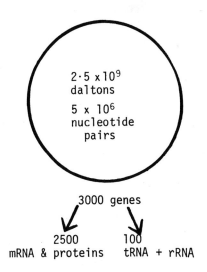

Fig. 1. Relationship between the bacterial genome and its products.

phenotypes of the organisms. Inferences based upon DNA reassociation, therefore, are based upon a consideration of several magnitudes more information than that considered by the numerical or the traditional taxonomist using cardinal or key characteristics. Moreover, DNA reassociation is not subject to the same kind of sampling error that confronts the numerical or traditional taxonomist who is considering only 5-10% of the attributes of an organism. DNA reassociation has a variety of utilities, but it is a technique that is applied most effectively to groups of organisms thought to be related at the generic or species level. Accordingly, DNA reassociation is a technique that is best used in concert with numerical taxonomy or traditional taxonomy (Goodfellow & Minnikin 1978).

DNA reassociation can be measured by a variety of techniques, each with its utility and its limitations (Moore 1974). When applied to bacterial classification, DNA reassociation is usually measured as radiolabelled DNA in solution bound to an unlabelled DNA immobilized on a menstruum such as nitro-cellulose, or as changes in absorbance as denatured DNA samples in solution reassociate. DNA reassociation in generally used to determine whether or not two organisms are related at the species level (Goodfellow & Minnikin 1977). To achieve this discrimination, the experimental conditions are contrived so that only stable DNA duplexes are formed, and the two DNA samples anneal extensively or not at all. The examples of this application of DNA reassociation to perplexing problems in bacterial classification are legion.

The all-or-none type of DNA reassociation assay has been used to establish that *Streptomyces coelicolor*, as defined by Kutzner & Waksman (1959), is different from *Streptomyces violaceoruber* (Monson *et al.* 1969). The trend in recent years has been to separate the taxon *Mycobacterium bovis* from *Mycobacterium tuberculosis*. Based upon DNA reassociation, the '*bovis*' strains must be considered a subspecies of *Mycobacterium tuberculosis* (Bradley 1975). The classification of the '*avium-intracellulare*' complex of mycobacteria has not yet been resolved (Baess 1978). Preliminary DNA reassociation data indicate that at least some members of *Mycobacterium avium* and some members of *Mycobacterium intracellulare* should be placed in the same species. The capacity of DNA samples, from the members of the '*rhodochrous*' complex having DNA composed of 60-62 % G+C, to reassociate has been examined (Bradley & Huitron 1973; Bradley *et al.* 1973). Cultures bearing five different epithets were brought into a single genospecies and it was further established that there are at least two other DNA homology groups in the subset of the '*rhodochrous*' complex having DNA composed of 60 to 62 % G+C. Goodfellow, Mordarski and their colleagues (Mordarski *et al.* 1977; Goodfellow *et al.* 1978) have applied extensive numerical phenetic surveys and DNA reassociation assays to members of the genus *Rhodococcus*. Although they were unable to relate directly homology and similarity values, there was good congruence between nucleotide sequence homology and phenetic data.

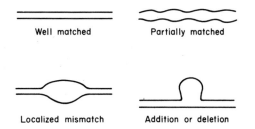

Fig. 2. Patterns of DNA/DNA reassociation.

DNA reassociation has also been applied successfully to classification of the genus *Bacteroides* (Johnson 1973). Workers using traditional approaches to the classification of this genus had lumped together a large number of isolates under the epithet *Bacteroides fragilis* (see Chapter 12). According to data from DNA reassociation studies, the *Bacteroides fragilis* complex consists of a number of DNA homology groups, which are now recognized as distinct species (Cato & Johnson 1976). This example demonstrates that taxonomists using traditional tests to characterize unfamiliar groups of bacteria, for example anaerobes, may consider organisms to be closely related because of inadequacies in test selection, design and scoring. DNA reassociation remains to be applied to a number of important taxonomic dilemmas at the species level. For example, in resolving the speciation of the *Nocardia asteroides* complex (Goodfellow & Minnikin 1978; Schaal & Reutersberg 1978). Moreover, DNA reassociation can provide insight into a number of additional biological problems that have an impact upon bacterial classification.

In DNA reassociation experiments, nucleotide pairs may be exactly matched, partially matched, or may be well matched in one but unmatched in another region [(Bradley & Enquist 1974) (Fig. 2)]. The degree of matching of nucleotide pairs can be assessed experimentally by measuring the stability of the reassociated DNA (Bradley & Enquist 1974). By prudently allowing DNA reassociation to occur under conditions that allow partially matched nucleotide sequences to pair, it is possible to detect biological relatedness beyond the species level. Farina & Bradley (1970) examined representatives of the sporangium-forming actinomycetes by DNA reassociation in order to ascertain the degree of relatedness among selected genera. This study demonstrated that some genera should be reduced to synonymy with established genera; for example, *Microellobosporia* should be reduced to synonymy with *Streptomyces*. Moreover, the flagella arrangements on the sporangiospores did not constitute a criterion that separated this group into a cluster of closely related genera with motile sporangiospores and a cluster of closely related genera with non-motile sporangiospores.

Nucleic acid reassociation studies focusing on specific genes or regions of the genophore are possible as well as those dealing with the entire genome. Ribosomal ribonucleic acid will associate with complementary sites on the appropriate DNA. Ribosomal RNA from any member of the family Enterobacteriaceae anneals well with *Escherichia coli* DNA, indicating that the nucleotide sequences of ribosomal RNA have been conserved during evolution (Moore 1974). It should be noted that transfer RNA cistrons are also conserved (Dubnau *et al.* 1965; Brenner *et al.* 1970). Because of the evolutionary conservation of certain nucleotide sequences, reassociation studies with ribosomal RNA are potentially useful for determining taxonomic relationships among distantly related bacteria.

5. Bases of Genomic Diversity

The stability and extent of nucleotide pairing can be used to differentiate between gene transfer and random mutation as a source of nucleotide sequence heterogeneity in particular bacterial populations (Enquist & Bradley 1970). It has been recognized for many years that the various attributes of a bacterial population, for example, morphology, biochemical characteristics, physiological responses, antigenic characteristics and virulence, are subject to change. The probability that any given bacterial gene will undergo a spontaneous, inherited alteration—a mutation—is generally considered to be once per 10^6 to 10^9 bacterial cell divisions. A single mutant cell will not be recognized unless its progeny becomes a substantial proportion of the population. This situation occurs most often when the bacteria are placed in an environment which favours mutants that happen to be present, or is detrimental to the wild type population, or both. The selective process may be severe and rapid, for example, selection of antibiotic resistant variants in milieu containing a drug, or may be virtually indiscernible. Strains derived from a single source, but propagated independently for many years, become dissimilar by virtue of subjective selection of variants during subculture, and as a result of genetic drift arising from variations in the constituents of the small samples used for inocula. DNA of variants generated by random mutation contains changes in nucleotide sequence scattered throughout the bacterial genome. This is in contrast to the DNA of variants generated by gene transfer in which new nucleotide sequences are acquired as added genetic information, for example, acquisition of a plasmid, or are substituted for pre-existing nucleotide sequences.

DNA reassociation studies involving a large number of strains of *Streptomyces* and *Nocardia* demonstrate that there is a progressive mis-pairing of nucleotides throughout DNA molecules as the test strains show more dissimilarity from reference strains (Bradley & Mordarski 1976). These results are consistent with the proposition that heterogeneity in the streptomycetes and nocardioform bacteria has arisen primarily by random mutation (Bradley 1975). It should be noted that there is a small proportion of DNA in the streptomycetes and nocardioform bacteria that appears to have been less susceptible to changes in nucleotide sequences. The conserved regions are probably those coding for ribosomal RNA (Moore 1974).

In the enteric bacteria, plasmids have been transferred from members of one genus to members of relatively unrelated genera (Jones & Sneath 1970). Moreover, genes on the fundamental genophore of *Esch. coli* have been transferred to strains of *Shigella, Salmonella* and *Proteus*. It is significant that the overall genome of *Shigella* is '85%' homologous with that of *Esch. coli,* that of *Salmonella typhimurium* is '45%' homologous with that of *Esch. coli,* and that of *Proteus mirabilis* is only '6%' homologous with that of *Esch. coli* (Sanderson 1976). Because the nucleotide composition of *Esch. coli* DNA is substantially different from that of *Prot. mirabilis* DNA (50 % G+C and 39 % G+C respectively), it is possible to separate *Esch. coli* genes from *Proteus* genes rather easily (Wohlhieter *et al.* 1975). In DNA samples from hybrids between *Esch. coli* and *Proteus,* those nucleotide sequences derived from the *Esch. coli* parent reassociate extensively and exactly with parental *Esch. coli* but not with the parental *Proteus* DNA. Accordingly, DNA reassociation techniques have made it possible to trace the transfer of genes, particularly those for antibiotic resistance, from one species to another, and sometimes across generic and family boundaries. A further discussion of the taxonomy of Enterobacteriaceae is given in Chapter 4.

6. Insertion Sequence Elements

Nucleotide sequence homologies among different DNA samples are usually determined indirectly. The usefulness of these indirect assays for measuring DNA reassociation is documented extensively in this essay; the ability to visualize DNA duplexes directly has greatly extended, however, the utility of DNA reassociation analyses (Westmoreland *et al.* 1969). By electron microscope heteroduplex techniques, the recognition sites at which the F factor molecules insert into the *Esch. coli* genophore during Hfr formation have been identified. These sites have been named insertion sequence elements. The *Esch. coli* genophore has been found to possess several different insertion sequence (IS) elements and multiple sets of each, for example, eight copies of IS-1 and five copies of IS-2 [(Starlinger & Saedler 1976) (Fig. 3)]. Insertion sequence elements do not exist as autonomous units but are normal constituents of both bacterial genophores and plasmids. Insertion sequence elements vary in size from 60 to 1700 nucleotide pairs in length and are probably always bracketed by inverted repetitious sequences. These 'inverted repeat' sequences, which may be

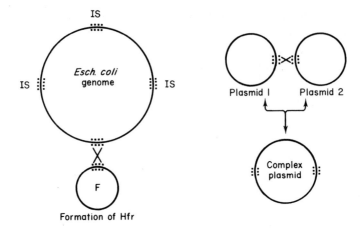

Fig. 3. Role of IS elements in the formation of Hfr strains and of complex plasmids.

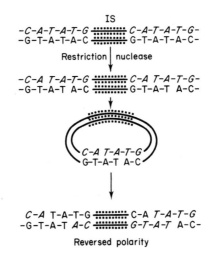

Fig. 4. Role of inverted repeat sequences in the transposition of an IS element. This example illustrates a change in polarity of the IS element at the same position of the genophore.

more than 20 nucleotides long, possess two-fold symmetry, the nucleotide sequence of one end of the DNA segment is repeated in the inverse order at the other end of the segment (Fig. 4). These 'inverted repeat' sequences allow the IS element to take a new position in the DNA in either of two orientations, permit joining of two plasmids that have the same IS elements, and facilitate dissociation of plasmids with more than one IS element (Figs. 3–5).

Insertion sequence segments are detected genetically when they transpose to a different location in the genophore. An IS element when transposed into a gene may provoke specific deletions immediately adjacent to the insertion site. In fact, IS elements may be responsible for most of the so-called spontaneous mutations in *Esch. coli* (Bukhari *et al.* 1977). In addition, phase variation in *Salmonella* may be regulated by an IS element. As noted before, IS elements may assume either of two orientations when they relocate in the genome (Fig. 5). Some IS elements contain an RNA polymerase binding region that can either promote or inhibit transcription depending on the IS segment. If an IS element is transposed to the beginning of the H_2 locus in one orientation, it could prevent the expression of the H_2 operon and allow the H_1 locus to be expressed. In the opposite orientation, the IS element could promote H_2 operon transcription and formation of an H_1 operon repressor. These newly described entities enhance profoundly the possibility of variation in a bacterial population.

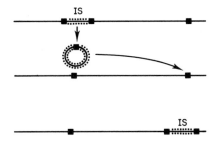

Fig. 5. Translocation of an IS element. The transposon can be inserted in either of two polarities.

7. Neutral Mutations

Although lack of DNA reassociation is interpreted as evidence for some degree of unrelatedness, it is conceivable that phenetically similar organisms might possess substantial amounts of genomic diversity as a consequence of mutations that generate synonymous codons or of mutations that encode for an amino acid substitution that does not alter appreciably the activity of an enzyme. For example, the amino acid, leucine, is coded for by the triplets, UUA, UUG, CUU, CUC, CUA and CUG (Watson 1976). A mutation affecting the 3' end of four of the six triplets for leucine will not change the amino acid specified. Moreover, a mutation affecting the 5' end of the codons for leucine will not change the amino acid specified 22% of the time. Overall, one-third of the mutations affecting leucine codons will generate synonymous codons. It should be noted that most other mutations, for example, UUA to the terminating codon UAG or UGA, will be lethal for the mutant if the change occurs in an essential gene.

```
                                protein
              -leu-glu-gly-pro-cys-asn-val-lys-
                                 mRNA
              -UUA-GAA-GGA-CCU-UGU-AAU-GUU-AAA-

                             DNA (29 % G C)
              -AAT-CTT-CCT-GGA-ACA-TTA-CAA-TTT-
              -TTA-GAA-GGA-CCT-TGT-AAT-GTT-AAA-

                                mRNA'
              -CUG-CAG-GGC-CCC-UGC-AAC-GUA-AAG-

                            DNA' (62.5% G C)
              -GAC-GTC-CCG-GGG-ACG-TTG-CAT-TTC-
              -CTG-CAG-GGC-CCC-TGC-AAC-GTA-AAG-
```

Fig. 6. Effect of synonymous mutations on the overall nucleotide composition of DNA.

Although it is conceivable that synonymous substitutions would be selected for or against because the concentrations of different transfer RNA molecules for a given amino acid differ, there is no experimental evidence to support the proposition. Although proteins having exactly the same primary structure could be coded for by nucleotide sequences differing in more than one third of their bases (Fig. 6), no example has been encountered. In fact, DNA reassociation has given regularly results consistent with other thorough taxonomic studies (Mordarski et al. 1978b). Accordingly, if neutral mutations occur frequently, back mutations must also occur frequently. Moreover, the availability of a particular deoxyribonucleotide might become selective for a given organism should the neutral mutations alter the DNA composition sufficiently that the demands on the deoxyribonucleotide pools would deplete one of the constituents. If mutations that generate synonymous codons are neutral, mutation rates in the sense of altered nucleotide sequences may be much more frequent than generally accepted. There are a number of examples of bacterial variation that occur with a greater frequency than expected for spontaneous mutation, for example, changes in growth rate in Salmonella (Atwood et al. 1951) and pigmentation variants in Serratia marcescens (Bunting 1946).

8. Concluding Remarks

DNA reassociation can be monitored by a variety of techniques, including binding of radiolabelled denatured DNA to an immobilized single stranded DNA, changes in absorbance of polydeoxyribonucleotides in solution during the course of duplex formation, and direct visualization of duplexes with the electron microscope. DNA reassociation studies compare the nucleotide sequences of DNA samples containing *ca.* 3000 genes. A diligent bacteriologist using a wide range of standard tests, aided by automated devices, can examine only 10% of the genetic capability of a bacterium. DNA reassociation therefore is a definitive technique for establishing biological relatedness at the species level. As more data from DNA reassociation studies are compiled, it may be possible to determine whether bacterial variation is continuous from one 'species' to another or whether bacterial 'species' reflect discrete populations or clusters (Jones & Sneath 1970). To date, results from DNA reassociation studies reveal few intermediate homology values, but few groups have been investigated intensively, and even these have encompassed far too few isolates to provide a meaningful answer. DNA reassociation studies can be manipulated to provide insights into biological relatedness at the generic level but has limited utility at the level of family, order and class. DNA reassociation is a powerful taxonomic tool when used in conjunction with other approaches to microbial classification. It is doubtful that DNA reassociation will find utility as a routine method for microbial identification but it will occasionally be useful for determining whether a particular pair of isolates or variants have a recent common ancestor (Mordarski *et al.* 1978*b*). Although it is not essential that a system for bacterial identification or classification reflect biological relatedness in order to be useful, a system of classification based upon DNA reassociation data can be both practical and reflect the evolutionary processes and affinities of the microbial populations being considered. It is now well established that a thorough characterization of the DNA of an organism, or a group of organisms, provides a reliable basis for formulating models about the organization of the genome, about the mechanisms that regulate the biological phenomena specified by the genome and the processes by which changes occur and become established in a population (Bradley *et al.* 1978).

This project was supported in part by a Senior International Fellowship award from the Fogarty International Center, National Institutes of Health, Bethesda, Maryland 20014, USA. This review was prepared while the author was a Visiting Worker, Department of Pharmacology, University of Cambridge, Cambridge CB2 2QD, UK.

9. References

AL-DIWANY, L. J. & CROSS, T. 1978 Ecological studies on nocardioforms and other actinomycetes in aquatic habitats. In *Nocardia and Streptomyces.* ed. Mordarski, M., Kuryłowicz, W. & Jeljaszewicz, J. pp. 153-160. Stuttgart: Gustav Fischer Verlag.
ARAI, T. 1978 Taxonomy and bioactivity of actinomycetes. In *Nocardia and Streptomyces.* ed. Mordarski, M., Kuryłowicz, W. & Jeljaszewicz, J. pp. 13-20. Stuttgart: Gustav Fischer Verlag.

ATWOOD, K. C., SCHNEIDER, L. K. & RYAN, F. J. 1951 Selective mechanisms in bacteria. *Cold Spring Harbor Symposium* 16, 345–354.

BACHMAN, B. J., LOW, K. B. & TAYLOR, A. L. 1976 Recalibrated linkage map of *Escherichia coli* K-12. *Bacteriological Reviews* 40, 116–167.

BAESS, I. 1978 DNA–DNA hybridization in mycobacteria. In *Genetics of the Actinomycetales.* ed. Freerksen, E., Tárnok, I. & Thumin, J. H. pp. 225–229. Stuttgart: Gustav Fischer Verlag.

BAK, A. L., CHRISTIANSEN, C. & STENDERUP, A. 1970 Bacterial genome sizes determined by DNA renaturation studies. *Journal of General Microbiology* 64, 377–380.

BRADLEY, S. G. 1965 Interspecific genetic homology in the actinomycetes. *International Bulletin of Bacteriological Nomenclature and Taxonomy* 15, 239–241.

BRADLEY, S. G. 1966 Genetics in applied microbiology. *Advances in Applied Microbiology* 8, 29–59.

BRADLEY, S. G. 1969 The genetic and physiologic bases for microbial variation. In *Proceedings of the Symposium on the Use of Drugs in Animals Feeds.* Publ. No. 1679 pp. 287–299. Washington, D.C.: National Academy of Science.

BRADLEY, S. G. 1975 Significance of nucleic acid hybridization to systematics of actinomycetes. *Advances in Applied Microbiology* 19, 59–70.

BRADLEY, S. G. & BOND, J. S. 1974 Taxonomic criteria for mycobacteria and nocardiae. *Advances in Applied Microbiology* 18, 131–190.

BRADLEY, S. G. & ENQUIST, L. W. 1974 Microbial nucleic acids. In *Molecular Microbiology.* ed. Kwapinski, J. B. G. pp. 47–100. New York: John Wiley & Sons.

BRADLEY, S. G. & HUITRON, M. E. 1973 Genetic homologies among nocardiae. *Developments in Industrial Microbiology* 14, 189–199.

BRADLEY, S. G. & MORDARSKI, M. 1976 Association of polydeoxyribonucleotides of deoxyribonucleic acids from nocardioform bacteria. In *Biology of the Nocardiae.* ed. Goodfellow, M., Brownell, G. H. & Serrano, J. A. pp. 310–336. London: Academic Press.

BRADLEY, S. G., BROWNELL, G. H. & CLARK, J. E. 1973 Genetic homologies among nocardiae and other actinomycetes. *Canadian Journal of Microbiology* 19, 1007–1014.

BRADLEY, S. G., ENQUIST, L. W. & SCRIBNER, H. E. III. 1978 Heterogeneity among deoxyribonucleotide sequences of actinomycetes. In *Genetics of the Actinomycetales.* ed. Freerksen, E., Tárnok, I. & Thumin, J. H. pp. 207–224. Stuttgart: Gustav Fischer Verlag.

BRENNER, D. J., FOURNIER, M. J. & DOCTOR, B. P. 1970 Isolation and partial characterization of the transfer ribonucleic acid cistrons from *Escherichia coli. Nature, London* 227, 448–451.

BRENNER, D. J., FANNING, G. R., SKERMAN, F. J. & FALKOW, S. 1972 Polynucleotide sequence divergence among strains of *Escherichia coli* and closely related organisms. *Journal of Bacteriology* 109, 953–965.

BUKHARI, A. I., SHAPIRO, J. & ADHYA, S. ed. 1977 *DNA Insertions.* Cold Spring Harbor, N.Y.: Cold Spring Harbor Laboratory.

BUNTING, M. I. 1946 The inheritance of color in bacteria, with special reference to *Serratia marcescens. Cold Spring Harbor Symposium* 11, 25–31.

CATO, E. P. & JOHNSON, J. L. 1976 Reinstatement of species rank for *Bacteroides fragilis, B. ovatus, B. distasonis, B. thetaiotaomicron,* and *B. vulgatus*: Designation of neotype strains for *Bacteroides fragilis* (Veillon and Zuber) Castellani and Chalmers and *Bacteroides thetaiotaomicron* (Diasto) Castellani and Chalmers. *International Journal of Systematic Bacteriology* 26, 230–237.

COONEY, W. J. & BRADLEY, S. G. 1965 Amino acid composition of the cell-proteins from streptomycetes and myxobacters. *Biochimica et Biophysica Acta* 100, 594–595.

DUBNAU, D., SMITH, I., MORELL, P. & MARMUR, J. 1965 Gene conservation in *Bacillus* species. I. Conserved genetic and nucleic acid base sequence homologies. *Proceedings of the National Academy of Science U.S.A.* 54, 491–498.

DWORKIN, M. 1966 Biology of the myxobacteria. *Annual Reviews of Microbiology* 20, 75–106.

ENQUIST, L. W. & BRADLEY, S. G. 1970 Nucleotide divergence in deoxyribonucleic acids of actinomycetes. *Advancing Frontiers of Plant Science* 25, 53–73.

FALKOW, S. 1975 *Infectious Multiple Drug Resistance*. London: Pion Ltd.

FARINA, G. & BRADLEY, S. G. 1970 Reassociation of deoxyribonucleic acids from *Actinoplanes* and other actinomycetes. *Journal of Bacteriology* 102, 30–35.

GOODFELLOW, M. & ALDERSON, G. 1977 The actinomycete-genus *Rhodococcus:* home of the "*rhodochrous*" complex. *Journal of General Microbiology* 100, 99–122.

GOODFELLOW, M. & MINNIKIN, D. E. 1977 Nocardioform bacteria. *Annual Reviews of Microbiology* 31, 159–180.

GOODFELLOW, M. & MINNIKIN, D. E. 1978 Numerical and chemical methods in the classification of *Nocardia* and related taxa. In *Nocardia and Streptomyces*. ed. Mordarski, M., Kuryłowicz, W. & Jeljaszewicz, J. pp. 43–51. Stuttgart: Gustav Fischer Verlag.

GOODFELLOW, M., MORDARSKI, M., SZYBA, K. & PULVERER, G. 1978 Relationships among rhodococci based upon deoxyribonucleic acid reassociation. In *Genetics of the Actinomycetales*. ed. Freerksen, E., Tárnok, I. & Thumin, J. H. pp. 231–234. Stuttgart: Gustav Fischer Verlag.

GORDON, R. E. 1966 Some strains in search of a genus–*Corynebacterium, Mycobacterium, Nocardia* or what? *Journal of General Microbiology* 43, 329–343.

HENRIKSEN, S. D. 1976 *Moraxella, Neisseria, Branhamella* and *Acinetobacter*. *Annual Reviews of Microbiology* 30, 63–83.

JOHNSON, J. L. 1973 Use of nucleic acid homologies in the taxonomy of anaerobic bacteria. *International Journal of Systematic Bacteriology* 23, 308–315.

JONES, D. & SNEATH, P. H. A. 1970 Genetic transfer and bacterial taxonomy. *Bacteriological Reviews* 34, 40–81.

KÜSTER, E. 1978 The concept of genus and species within the Actinomycetales. In *Nocardia and Streptomyces*. ed. Mordarski, M., Kuryłowicz, W. & Jeljaszewicz, J. pp. 21–24. Stuttgart: Gustav Fischer Verlag.

KUTZNER, H. J. & WAKSMAN, S. A. 1959 *Streptomyces coelicolor* Müller and *Streptomyces violaceoruber* Waksman and Curtis, two distinctly different organisms. *Journal of Bacteriology* 78, 528–538.

KUTZNER, H. J., BÖTTINGER, V. & HEITZER, R. D. 1978 The use of physiological criteria in the taxonomy of *Streptomyces* and *Streptoverticillium*. In *Nocardia and Streptomyces*. ed. Mordarski, M., Kuryłowicz, W. & Jeljaszewicz, J. pp. 25–29. Stuttgart: Gustav Fischer Verlag.

LACEY, J., GOODFELLOW, M. & ALDERSON, G. 1978 The genus *Actinomadura* Lechevalier and Lechevalier. In *Nocardia and Streptomyces*. ed. Mordarski, M., Kuryłowicz, W. & Jeljaszewicz, J. pp. 105–117. Stuttgart: Gustav Fischer Verlag.

LAPAGE, S. P., SNEATH, P. H. A., LESSEL, E. F., SKERMAN, V. B. D., SEELIGER, H. P. R. & CLARK, W. A. ed. 1975 *International Code of Nomenclature of Bacteria*. Washington, D.C.: American Society for Microbiology.

MANDEL, M. 1969 New approaches to bacterial taxonomy: Perspectives and prospects. *Annual Reviews of Microbiology* 23, 239–274.

MARMUR, J., FALKOW, S. & MANDEL, M. 1963 New approaches to bacterial taxonomy. *Annual Reviews of Microbiology* 17, 329–372.

MONSON, A. M., BRADLEY, S. G., ENQUIST, L. W. & CRUCES, G. 1969 Genetic homologies among *Streptomyces violaceoruber* strains. *Journal of Bacteriology* 99, 702–706.

MOORE, R. L. 1974 Nucleic acid reassociation as a guide to genetic relatedness among bacteria. *Current Topics in Microbiology and Immunology* 64, 105–128.

MORDARSKI, M., SZYBA, K., PULVERER, G. & GOODFELLOW, M. 1976 Deoxyribonucleic acid reassociation in the classification of the "*rhodochrous*" complex and allied taxa. *Journal of General Microbiology* 94, 235–245.

MORDARSKI, M., GOODFELLOW, M., SZYBA, K., PULVERER, G. & TKACZ, A. 1977 Classification of the "*rhodochrous*" complex and allied taxa based upon deoxyribonucleic acid reassociation. *International Journal of Systematic Bacteriology* 27, 31–37.

MORDARSKI, M., SCHAAL, K., TKACZ, A., PULVERER, G., SZYBA, K. & GOOD-FELLOW, M. 1978*a* Deoxyribonucleic acid base composition and homology studies on *Nocardia*. In *Nocardia and Streptomyces*. ed. Mordarski, M., Kuryłowicz, W. & Jeljaszewicz, J. pp. 91-97. Stuttgart: Gustav Fischer Verlag.

MORDARSKI, M., GOODFELLOW, M., SZYBA, K., PULVERER, G. & TKACZ, A. 1978*b* Deoxyribonucleic acid base composition and homology studies on *Rhodococcus* and allied taxa. In *Nocardia and Streptomyces*. ed. Mordarski, M., Kuryłowicz, W. & Jeljaszewicz, J. pp. 99-106. Stuttgart: Gustav Fischer Verlag.

OGASAWARA-FUJITA, N. & SAKAGUCHI, K. 1976 Classification of micrococci on the basis of deoxyribonucleic acid homology. *Journal of General Microbiology* **94**, 97-106.

OWEN, R. J., LEGROS, R. M. & LAPAGE, S. P. 1978 Base composition, size and sequence similarities of genome deoxyribonucleic acids from clinical isolates of *Pseudomonas putrefaciens*. *Journal of General Microbiology* **104**, 127-138.

PRAUSER, H. 1978 Considerations on taxonomic relations among Gram-positive, branching bacteria. In *Nocardia and Streptomyces*. ed. Mordarski, M., Kuryłowicz, W. & Jeljaszewicz, J. pp. 3-12. Stuttgart: Gustav Fischer Verlag.

PREOBRAZHENSKAYA, T. P., SVESHNIKOVA, M. A., TEREKHOVA, L. P. & CHOR-MONOVA, N. T. 1978 Selective isolation of soil actinomycetes. In *Nocardia and Streptomyces*. ed. Mordarski, M., Kuryłowicz, W. & Jeljaszewicz, J. pp. 119-123. Stuttgart: Gustav Fischer Verlag.

PULVERER, G., MORDARSKI, M., TKACZ, A., SZYBA, K., HECZKO, P. & GOOD-FELLOW, M. 1978 Relationships among some coagulase-negative staphylococci based upon deoxyribonucleic acid reassociation. *FEMS Microbiology Letters*, **3**, 51-56.

RAVIN, A. W. 1960 The origin of bacterial species. Genetic recombination and factors limiting it between populations. *Bacteriological Reviews* **24**, 201-220.

SANDERSON, K. E. 1976 Genetic relatedness in the family Enterobacteriaceae. *Annual Reviews of Microbiology* **30**, 327-349.

SCHAAL, K. P. & REUTERSBERG, H. 1978 Numerical taxonomy of *Nocardia asteroides*. In *Nocardia and Streptomyces*. ed. Mordarski, M., Kuryłowicz, W. & Jeljaszewicz, J. pp. 53-62. Stuttgart: Gustav Fischer Verlag.

SEELIGER, H. P. R. 1961 *Listeriosis*. Basel: S. Karger.

STARLINGER, P. & SAEDLER, H. 1976 IS-elements in microorganisms. *Current Topics in Microbiology and Immunology* **75**, 111-132.

STUART, S. E. & WELSHIMER, H. J. 1973 Intrageneric relatedness of *Listeria* Pirie. *International Journal of Systematic Bacteriology* **23**, 8-14.

WATSON, J. D. 1976 *The Molecular Biology of the Gene,* 3rd edn. New York: W. A. Benjamin.

WESTMORELAND, B., SZYBALSKI, W. & RIS, H. 1969 Mapping of deletions and substitutions in heteroduplex DNA molecules of bacteriophage lambda by electron microscopy. *Science*, Washington, D. C. **163** 1343-1348.

WILLIAMS, S. T. 1978 Streptomycetes in the soil ecosystem. In *Nocardia and Streptomyces*. ed. Mordarski, M., Kuryłowicz, W. & Jeljaszewicz, J. pp. 137-144. Stuttgart: Gustav Fischer Verlag.

WOHLHIETER, J. A., GEMSKI, P. & BARON, L. S. 1975 Extensive segments of the *Escherichia coli* K-12 chromosome in *Proteus mirabilis* diploids. *Molecular and General Genetics* **139**, 93-101.

Plasmids

C. R. HARWOOD

Department of Microbiology, University of Newcastle upon Tyne,
Newcastle upon Tyne, UK

Contents

1. Introduction

RICHMOND & WIEDEMAN (1974) have referred to plasmids as "the crucible of bacterial evolution". This quotation, reflecting the promiscuity and widespread occurrence of plasmids, as well as their ability to recombine with other molecules of deoxyribonucleic acid (DNA), echoes the views of others (Anderson 1966; Hedges 1972) that a substantial amount of bacterial evolution has occurred by the rearrangement of existing genes between individual bacteria, rather than simply by the development and selection of new protein structures. If gene rearrangements are shown to occur, then the evolutionary time-scale, devised purely in terms of the selection of new protein structures, will have to be foreshortened and clear evolutionary (cladistic) lines are unlikely to be recognized. Although it is likely to be some time before sufficient evidence is available to allow a critical assessment of the contribution of gene rearrangements to bacterial evolution, evidence available from work on members of the Enterobacteriaceae and the Pseudomonadaceae has already shown the potential of plasmids in such an evolutionary process.

The acceptance, either wholly or partly, of the view that gene rearrangements may contribute to bacterial evolution clearly has implications for the ways in which taxonomists classify bacteria. The purpose of this chapter, therefore is to discuss briefly the evidence for, and the mechanisms of, gene rearrangements and to show how plasmids are implicated. It also indicates the extent to which plasmids have been found in representatives of an increasingly wide range of bacterial taxa, often altering the phenotypic properties of strains and making their identification difficult. Methods used currently for the detection, isolation and characterization of plasmids are also discussed.

2. Basic Properties

Plasmids are molecules of DNA, with molecular weights ranging from *ca*. 1.5 to 400 million daltons, that exist in many, possibly the majority, of bacterial cells. Although plasmids are usually found to be stably inherited from one generation to the next, occasionally combinations of host and plasmid are found to be unstable. With a few notable exceptions, most plasmids exist autonomously of the host's main chromosome, although some plasmids may transiently associate with this chromosome (Moody & Runge 1972). Autonomous plasmids consist of supercoiled, covalently closed circular (ccc) molecules of double-stranded DNA, similar in many respects, but not in size, to the bacterial chromosome (Clowes 1972). Although often regarded as 'optional extras', under certain environmental conditions particular plasmids may be essential to the survival of their bacterial host.

A property of plasmids that makes them of fundamental interest and importance to both bacterial evolution and classification is their ability to transfer to new bacterial hosts. They are, of course, not unique in this respect since bacteriophages, with which they may share a common ancestry (Falkow 1975), are similarly transmissible. Their transfer properties have been used to classify plasmids into two very heterogeneous classes: those able to mediate conjugal transfer—*conjugative* plasmids, and those unable to do so—*non-conjugative* plasmids. All plasmids, however, are potentially transmissible between closely related bacterial cells, being transmitted by bacteriophage-mediated transduction and transformation (in experimental systems at least), as well as by conjugation. With non-conjugative plasmids, conjugal transfer may be mediated by a co-existing conjugative plasmid.

3. Chromosomal Mobilization Ability

The transfer of a conjugative plasmid by conjugation is often, but not always, accompanied by the transfer of host chromosomal genes. The transfer of host chromosomal genes usually occurs at frequencies that are very much lower than that of the plasmid, although some plasmids are exceptional in being able to mobilize host chromosomal genes with a frequency approaching 1, e.g. Hfr strains of *Escherichia coli*. The property of chromosomal mobilization ability (*cma*) is sometimes still referred to a sex factor activity. Chromosomal mobilization ability was one of the first plasmid borne traits to be recognized (Hayes 1953).

In contrast to plasmids, transferred host chromosomal genes are not normally capable of self-replication and their entry alone into the recipient is not sufficient to ensure their stable maintenance and subsequent expression. Stable inheritance of chromosomal genes occurs by integration into an existing replicon of the recipient. Integration normally occurs by homologous recombination between the incoming fragment of donor chromosome and a homologous region on the chromosome of the recipient. Homologous recombination is a reciprocal process and does not involve either a net loss or gain of DNA by the recipient; it has an absolute requirement for the product of the $recA^+$ gene. The efficiency of homologous recombination is dependant on the genetic relatedness of the bacteria involved; when closely related bacteria (e.g. *Esch. coli/Esch. coli* or *Esch. coli/ Shigella*) are involved, the efficiency with which a gene integrates may approach 50%. With less related bacteria (e.g. *Esch. coli/Salmonella*), recombination efficiency is low. $RecA^+$-dependant recombination is not detected in crosses between more distantly related genera (e.g. *Esch. coli/Proteus* and *Esch. coli/ Pseudomonas*).

The frequency at which different plasmids are able to mobilize host genes varies enormously. In crosses between strains of *Esch. coli*, many plasmids (including the fertility plasmid, F) bring about recombinant formation at a frequency that is *ca*. 10^{-5} times that at which the plasmid is transferred; genes situated at widely separate regions of the donor chromosome forming recombinants at more or less the same frequency. However, with high frequency recombination (Hfr) strains in which the plasmid is integrated at one of a limited number of sites on the host chromosome, the whole of the bacterial chromosome is mobilized so that a copy is transferred in a directed way to the recipient. The genes transferred most frequently are those adjacent to and to one side of the site of plasmid integration and increasing distance from the integration site results in a lower transfer frequency. Although rarely achieved, the transfer of the whole chromosome takes *ca*. 100 min at $37°C$, genes close to the site of integration forming recombinants with a frequency greater than 10^{-1} (Falkow 1975).

An intermediate situation is observed with F prime (F′) plasmids which are formed as a result of aberrant excision of an integrated F plasmid (Fig. 1). F′ plasmids consist of a complete or almost complete F plasmid together with DNA excised from the region of the host chromosome adjacent to the site of integration. The strain in which the F′ plasmid arises, the primary strain, has a corresponding deletion in its chromosome, however, if the plasmid is transferred to a secondary host strain this usually shows diploidy for the bacterial gene(s) carried by the plasmid. When F′ plasmids are transferred from secondary hosts they exhibit, not only the usual low frequency chromosomal mobilization ability of other autonomous plasmids (i.e. *ca.* 10^{-5}), but also directed transfer of genes adjacent to the site of integration in the primary strain, the frequency being only slightly lower than that found with Hfr strains. F′ directed transfer is $recA^+$-dependent. Some resistance (R) plasmids, including R1, have been shown to cause directed transfer of particular chromosomal genes (Hedén & Meynell 1976) even though they show no obvious large area of homology with the host chromosome. As with F′ plasmids, directed transfer by these R plasmids occurs at a relatively high frequency, but differs in being $recA^+$-independent and in being able to promote directed transfer in both *Esch. coli* and *Salm. typhimurium*. The DNA of these species share little recombinational or structural homology and the explanation of these observations appears to involve some form of non-homologous ($recA^+$-independent) recombination.

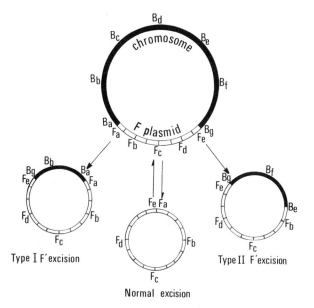

Fig. 1. Type I and Type II F′ excision and normal excision of the F plasmid from its integration site in the host chromosome.

4. Unit Structure of Plasmids

Plasmid molecules can be conveniently divided into sections which appear to be of functional significance.

A. *Maintenance and replication functions*

The stability of plasmids and their presence in relatively constant copy numbers indicates the functioning of an efficient regulatory process controlling not only their replication but also their segregation amongst the products of host cell division. Autonomous plasmids resemble bacterial chromosomes in forming independent units of replication, called replicons. Like the host chromosome, control of plasmid replication is probably achieved by regulating the frequency at which new rounds of replication are initiated, rather than controlling the rate at which their DNA is synthesized (Falkow 1975).

Surprisingly little is known about the underlying mechanisms that control plasmid replication. Plasmid molecules are present in defined numbers per host cell chromosome, although this may vary extensively for any particular plasmid according to the host and growth conditions. When multiple copies are present, individual plasmid molecules are selected randomly for replication (Bazaral & Helinski 1970; Gustafsson & Nordström 1975). Whether plasmid replication is restricted to a limited fraction of the cell cycle (Davis & Helmstetter 1973), or can occur throughout the whole cycle (Gustafsson *et al.* 1978) is not entirely clear. There is, however, increasing evidence (Gustafsson & Nordström 1978) for the involvement of an autorepressor in the control mechanism, as originally proposed by Pritchard *et al.* (1969).

Whatever the mechanism for controlling plasmid replication, it is clear that plasmids utilize host enzymes for their replication. Many plasmids appear to use the same DNA polymerases as their host, whilst others, like the colicinogenic (*col*) plasmid, *Col*E1, will not replicate in host cells lacking DNA polymerase I (*polA*). Plasmid *Col*E1 is able to replicate extensively in the absence of bacterial replication, for example in the presence of high levels of chloramphenicol and consequently can reach copy numbers of several thousand (Falkow 1975).

Plasmids can be classified according to the relatedness of their maintenance and replication (*inc*) genes. When two plasmids with related *inc* genes are introduced into the same bacterial cell they are incapable of co-existing and segregate amongst the bacterial progeny. Plasmids which are incapable of co-existing are described as *incompatible*. Plasmids with unrelated *inc* genes are, however, capable of co-existing in a single bacterium and are therefore referred to as *compatible*. Plasmids which are mutually incompatible form an *incompatibility group* (Datta 1975). Some of the major incompatibility groups are shown in Table 1

Although of some epidemiological significance, recent work on transposition [see (5)] has tended to reduce the value of incompatibility groups for classifying plasmids. The molecular mechanisms of incompatibility are not clear (De Vries *et al.* 1976).

B. Transfer ability and conjugation

Bacterial conjugation involves the transfer of DNA, usually of plasmid origin, from donor to recipient by cell-to-cell contact (Falkow 1975). Conjugation always seems to be plasmid-mediated in bacteria. The mechanisms of conjugation, which are still only poorly understood, have been studied most extensively for *inc*FI and *inc*FII plasmids in *Esch. coli.* Initial contact between donor and recipient appears to involve a plasmid-mediated sex pilus, which extends from the surface of the donor, and a specific receptor protein in the envelope

TABLE 1

Some incompatibility groups, pilus type and donor-specific 'phage sensitivity of Gram negative plasmids

Incompatibility group	Pilus type	Donor-specific 'phage sensitivity	Example
A	—	—	RA1
B	—	—	R16
C	—	—	R40a, R55
E	E	piE/R1	
FI	F	f1 & f2	F, ColV, R386
FII	F	f1 & f2	R1, R6, R100
FIII	F	f1 & f2	ColB-K98
FIV	F	f1 & f2	R124
FV	F	f1	F*olac*
H	—	—	R27, R726
I$_\alpha$	I	If1	R64, R144, ColIb-P9
L$_\gamma$	I	If1	R483, R621a
J	—	—	R391
K	—	—	R387
L	—	—	R471
M	—	—	R69, R446b
N	N?	Ike	N3, R15, R46
P1	P	PRR1	RP4
P2	—	—	OCT
Q	—	—	R310
S	—	—	R476a
T	—	—	Rtsl, R394, R401
V	—	—	R753
W	W	PR4	S-a. R7K, R388
X	—	—	R6K
Y	—	—	P1

Data mainly from Datta (1977) and Jacob *et al.* (1977).

of the recipient (Falkinham & Curtiss 1976). Transferred DNA is apparently always in the form of a single strand and is generated by a type of rolling circle replication (Curtiss 1969). A complementary strand is synthesized against the transferred DNA molecule in the recipient.

The ability of a conjugative plasmid to mediate conjugation is determined by its transfer (*tra*) genes. In the F plasmid these are located together on the plasmid. The genes required for the synthesis of F pili (*tra* A, L, E, K, B, C. F, H and 'in part' *tra*G), the surface exclusion phenomenon which makes F plasmid-carrying strains poor recipients (*tra*S and T), the formation of stable mating aggregates (*tra*G) and DNA transfer (*tra*D and I), comprise a single operon which is positively controlled by the product of the *tra*J gene (Achtman & Helmuth 1974). At least one *tra* gene, *tra*M, which is concerned with DNA transfer, is located outside this operon (Kennedy *et al*. 1977).

Whereas F exhibits high frequency transfer ability, many naturally occurring conjugative plasmids appear to mediate DNA transfer at a low frequency. Low frequency transfer is due to the production of a plasmid-mediated autorepressor complex which is absent in bacteria carrying F alone (Kennedy *et al*. 1977). An early observation was that the introduction of certain autorepressed conjugative plasmids into cells containing F resulted in the repression of F transfer. The resulting fertility inhibition was used to classify plasmids, in particular their transfer properties, into two groups, namely fin^+ (fi^+) and fin^- (fi^-). Although initially helpful, both fin^+ and fin^- groups have been subsequently shown to be heterologous (Falkow 1975; Gasson & Willetts 1976).

Plasmids of incompatibility groups *inc*Fl to FV (see Table 1) possess related transfer genes as judged by their ability to absorb a similar range of donor specific bacteriophages (i.e. MS2, M13) and the immune cross-reactivity of their sex pili. The relatedness of plasmids in these *inc* groups has been partially confirmed by DNA/DNA reassociation and heteroduplex studies (Sharp *et al*. 1972; Falkow *et al*. 1974).

Fin$^-$ plasmids form a very heterogeneous group both with respect to incompatibility and transfer. Only a few *fin*$^-$ plasmids have been studied in any detail with respect to transfer. Most notable are the 'I-like' plasmids associated with *inc* groups I_α and I_γ. 'I-like' plasmids determine sex pili which are related to each other on the basis of structure, immune cross-reactivity and adsorption of donor-specific bacteriophages, but which are clearly distinguishable from the sex pili determined by the 'F-like' plasmids of *inc*FI to FV. Plasmids of *inc*I_α and *inc*I_γ show a high degree of relatedness with each other on the basis of DNA/DNA reassociation studies, but a much lower level of relatedness with plasmids of other *inc* groups (Falkow *et al*. 1974). Sex pili have been recognized for a few other *inc* groups, including *inc*PI, *inc*W and *inc*E (Bradley 1974, 1976; To *et al*. 1975).

Although conjugation has been reported widely in representatives of Gram negative bacterial species, there are still relatively few reports of conjugation between Gram positive species. Conjugation, or at least a process requiring cell-to-cell contact, has been reported for *Streptococcus faecalis* (Jacob *et al.* 1975) and for the actinomycetes, *Streptomyces coelicolor* A3(2) (Hopwood *et al.* 1973), other *Streptomyces* spp. (Hopwood & Merrick 1977) and *Rhodococus erythropolis* (Brownell 1978). Conjugal transfer between Gram positive and Gram negative strains has not been reported although there is evidence from the origins of drug resistance determinants to indicate that some form of gene exchange between these broad groups may be possible (Davies *et al.* 1977). The importance of plasmids for bacterial classification and identification arises from their ability to transfer to and become expressed in new hosts, particularly in hosts which are not closely related to the donor. Unlike transferred chromosomal DNA, however, plasmids possess their own maintenance and replication genes and consequently do not need to integrate into a stable replicon in order to be maintained in the new host. Those plasmids that do integrate do so either by homologous (e.g. F′) or non-homologous (e.g. F) recombination.

When plasmids are transferred by transduction—as for example the staphylococcal plasmids frequently are—the effective host range of transfer is restricted to that of the bacteriophage vector. Although many of the classical bacteriophages have extremely restricted host ranges, many others have ranges that cross species boundaries. This is so for many bacteriophages of the genus *Bacillus* (Reanney & Teh 1976) and also for the coliphage, P1, which not only infects *Esch. coli*, but also *Salmonella, Shigella* and *Klebsiella* spp. (Reanney 1976). There is even a report of P1 transducing genes from *Esch. coli* to a myxobacterium strain (Kaiser & Dworkin 1975). Inter-species transfer clearly limits the taxonomic value of bacteriophages.

The host range of plasmid-mediated conjugation and gene transfer is generally wider than those of bacteriophages. It is not uncommon for plasmids isolated from *Esch. coli* to be freely transmitted to other members of the Enterobacteriaceae. The host range of plasmids varies extensively with different *inc* groups, certain plasmids of the *inc*P1 being particularly promiscuous. Originally isolated in species of *Pseudomonas*, *inc*P1 plasmids have been shown to be self-transmissible to a wide range of Gram negative genera including *Acinetobacter, Aeromonas, Agrobacterium, Azotobacter, Caulobacter, Erwinia, Klebsiella, Escherichia, Rhizobium, Rhodopseudomonas, Rhodospirillum, Salmonella, Shigella* and others (Jacoby & Shapiro 1977). The *inc*P1 plasmid R68.45—a derivative of R68, combines promiscuity with high chromosomal mobilization ability (Haas & Holloway 1976) and consequently has been successfully used to develop gene transfer systems in species that lack suitable ones of their own.

Enzymic restriction can affect the establishment of a transferred plasmid in its new host (Arber 1974). Restriction, which is mediated by restriction endonucleases, results in the cleavage of DNA into fragments that can be

subsequently degraded by non-specific exonucleases. Restriction endonucleases recognize specific short nucleotide sequences, either cleaving these sequences directly or at a random site nearby (Roberts 1977). One class of restriction
endonucleases generates polydeoxynucleotides with overlapping, cohesive ends.
The latter can anneal with other, similarly cleaved, polydeoxynucleotides of
any source, to be ligated into larger molecules of DNA. Restriction endonucleases of this class are often used for *in vitro* genetic engineering. Breakdown
and digestion of the host's own chromosomal DNA is prevented by the modification of the specific endonuclease recognition sites by the addition of methyl
groups. Methylation is due to the action of a modification methylase which
recognizes the same sequence as the corresponding endonuclease.

Restriction and modification systems are usually strain specific and until
recently their role was thought to be the provision of a mechanism that, whilst
allowing gene transfer between related strains (and which therefore have similarly
modified DNA), prevents the establishment in the cell of 'foreign' DNA. Restriction and modification would appear to form a major barrier to extensive interactions between heterologous nucleotide sequences. It is, therefore, paradoxical
that Reanney (1976) should propose a new role for restriction endonucleases, in
particular those that generate cohesive ends, namely that under suitable selection
pressures they may actually facilitate the emergence *in vivo* of novel gene combinations from heterologous DNA. In essence, Reanney proposes *natural* genetic
engineering, predicting that incoming unmodified DNA may be cleaved at its
restriction sites and then integrated between similarly cleaved sites either on
the host bacterial chromosome or on a resident plasmid. Evidence for restriction endonuclease-mediated gene rearrangements comes from the work of Chang &
Cohen (1977) who have shown the formation *in vivo* of hybrid plasmids, under
the direction of the restriction endonuclease, *Eco* Rl.

The stable establishment of a plasmid, or other transferred DNA, in a new
host is not in itself sufficient to ensure full functional expression of its phenotypic traits. The latter depends upon the information processing machinery of
the host and also the suitability of its biochemical environment.

C. Phenotypic traits

Plasmids are often initially detected by the phenotypic traits they confer upon
otherwise well-characterized bacterial strains. A number of traits either known
or suspected of being plasmid-mediated are listed in Table 2. In theory at least,
there are few limits to the types of genes that may become integrated within the
replicon of a plasmid, a view seemingly confirmed by F' production and more
recently by *in vitro* genetic engineering. In nature, however, only those genes or
combinations of genes with a positive selective advantage are likely to survive in

TABLE 2

Plasmid-mediated traits (known and suspected)

Resistance to antibiotics	Chromosomal donor ability

e.g. Kanomycin
   Penicillin
   Tetracycline
   and at least 20 others
	Utilization of
	   Lactose
	   Nopaline
	   Octopine
	   Sucrose
	   Raffinose

Resistance to cations
   Antimony
   Bismuth
   Cadmium
   Cobalt
   Lead
   Mercury
	Production of
	   Antibiotics
	   Bacteriocins
	   Coagulase
	   Enterotoxin
	   Exfoliative toxin
	   Fibrinolysin
	   Haemolysin
	   Hydrogen sulphide
	   Pigment

Resistance to anions
   Arsenate
   Arsenite	

Resistance to mutagenic agents
   Acridine
   Ethidium bromide
   u.v. light
	Others
	   Gas vacuoles
	   Nitrogen fixation
	   Plant virulence (host specificity)
	   Restriction and modification
	   Surface antigens K88, K99 (adhesion)
	   Tumour induction (crown gall)

Degradation of hydrocarbons
   Camphor
   Naphthaline
   Nicotine-Nicotinate
   Octane
   Salicylate
   Toluene
   Xylene	

sufficient numbers to be detectable. The implications for classification and identification of individual, plasmid-mediated, phenotypic traits are discussed later (pp. 39–45).

5. Transposition

In 1967 Saedler & Starlinger made an observation that has influenced profoundly the understanding of how the DNA of bacteria, plasmids and bacteriophages may interact. They noted that *Esch. coli* possessed specific sequences of naked DNA that could insert into a large number of sites in the galactose (*gal*) operon resulting in strong polar mutations. These sequences are now known as insertion sequence (IS) elements. Several copies of IS elements have been detected in the chromosome of *Esch. coli* K12 and in some of its plasmids (e.g. F and R1) and bacteriophages–e.g. λ–(Schwesinger 1977). More recently IS elements have been detected in other replicons, including the chromosomes of *Salmonella typhimurium* and *Citrobacter freundii* (Starlinger 1977).

At least 5 distinct IS elements have been noted to date. Designated IS1 etc., they do not encode any known phenotypic traits and appear unable to exist autonomously. IS elements vary in length from *ca.* 60 to 1700 nucleotide pairs. Each integrated IS element retains the potential to transpose from one site to another, either on the same or on a separate, co-existing molecule of DNA. Currently, little is known of the mechanism by which IS elements transpose except that their terminal sequences appear to play an important role in the process. The termini of IS1 consist of virtually identical sequences 18 nucleotide pairs long, but inverted with respect to each other (*inverted repeats*).

Although IS elements are able to integrate at a large number of sites on DNA, their integration specificity is not random. Their integration is intermediate between that of lysogenic bacteriophages like λ, which have just one main integration site, and bacteriophage Mu-1, which does appear to integrate at random. IS1, for example, has been shown to integrate between direct repeat sequences 9 nucleotide pairs long (see Sherratt 1978). If IS1 integrates within a structural gene it induces a strong polar mutation, regardless of its orientation. IS2 also causes strong polar mutations when integrated in one orientation, but acts as a transcriptional promoter site in the opposite orientation (Saedler *et al.* 1974). These polar mutations appear to result from rho-dependant termination of transcription, rather than termination of translation as observed with nonsense mutations (Reyes *et al.* 1976).

The insertion of IS elements involves a precise form of non-homologous recombination that is independent of the bacterial $recA^+$ system. Excision may be as precise as integration, restoring the normal function of the 'host' gene, or may be imprecise, causing secondary aberrations. These aberrations usually take the form of deletions originating at a fixed point corresponding to one end of the IS element, but extending for a variety of lengths into the adjacent DNA.

Transposition (Tn) elements, often referred to as 'transposons', have many properties in common with IS elements, but are more complex with lengths varying from 2500 to 25 000 nucleotide pairs. At least 14 Tn elements have been recognized to date, each possessing additional genes unrelated to insertion functions. As each Tn element is isolated from nature so it is assigned an independent isolation number e.g. Tn1, Tn2, Tn3 . . . etc., as a result some of the Tn elements already recognized have very similar properties. Tn1, Tn2 and Tn3, for example, are all approximately 4800 nucleotide pairs long and carry a TEM β-lactamase (penicillinase) gene.

The termini of the Tn elements studied to date consist of identical DNA sequences. These sequences are usually inverted repeats, but sometimes are direct repeats, a configuration that appears to be less stable (Schwesinger 1977). Tn1,

Tn2 and Tn3 have inverted repeats 140 nucleotide pairs long; Tn9 has direct repeats of IS1 while Tn10 has inverted repeats of IS10. Tn elements with inverted repeats show a characteristic 'pan-handle' structure under electron microscopy after denaturation and renaturation treatment.

Tn elements show similar transposition properties to IS elements but are detected more easily because of their accompanying phenotypic traits. Like IS elements, the integration of Tn elements is by precise, non-homologous recombination and is $recA^+$-independent. The number of integration sites are numerous, but limited, indicating a requirement for specific nucleotide sequences. For example, Tn5 like IS1 has been shown to integrate between direct repeat sequences which are 9 nucleotide pairs long.

6. Detection, Isolation and Characterization

When bacteria are physically screened for the presence of ccc plasmid DNA, the results are often positive. However, it is usually difficult to ascribe specific phenotypic traits to plasmids detected by physical screening methods and they are referred to as cryptic plasmids. Many, if not the majority, of cryptic plasmids, however, probably determine traits that confer a selective advantage on their host, at least under some growth conditions. More usually the presence of a plasmid is suspected because of the additional phenotypic traits they confer upon their host, such as the ability to degrade particular hydrocarbons or to resist the action of normally inhibiting antibiotics.

Although it is not always easy to confirm the plasmid-borne nature of a particular phenotypic trait, recent advances in DNA separation technology has made the screening of ccc plasmid DNA relatively convenient. Initially, however, it is usual to adopt a genetic approach, which usually involves obtaining answers to the following questions.

1. Is the trait additional to the characteristics of an otherwise well defined taxon?
2. Is the trait lost *irreversibly* by some of the clones after repeated sub-culture or after treatment with agents like acridine, ultra-violet light, mitomycin C, high temperature etc.?
3. Is the trait transmissible and, if so, is it transmitted in the absence of linkage with known chromosomal genes?
4. Is the trait known to be plasmid-borne in representatives of any other bacterial taxa?

A positive answer to any of these questions is indicative, but not proof, of the involvement of a plasmid.

A large number of physical methods are now available for detecting, isolating and analysing plasmid DNA. Until recently the standard methods for detecting and isolating plasmid DNA were time-consuming and relatively costly. These involve the lysis of the host and treatment of the resulting lysate to separate the smaller ccc plasmid molecules from the larger mass of linear chromosomal molecules, to produce a cleared lysate. Sedimentation of plasmid DNA is achieved either by rate zonal centrifugation through an alkaline or neutral sucrose gradient, or by isopycnic centrifugation in cesium chloride with the intercalating dye, ethidium bromide (Freifelder 1970; Helinski & Clewell 1972). Both methods work very satisfactorily as plasmid isolation methods, but have major disadvantages for routine screening. Isopycnic centrifugation usually takes several days to reach equilibrium, although it does have the advantage that the plasmid DNA can be visualized directly by ultra-violet light. Sucrose gradient centrifugation, although less time-consuming, usually requires the use of labelled DNA.

Two methods for detecting ccc plasmid DNA more rapidly have been developed, one involves gel electrophoresis the other relatively low speed centrifugation. In the electrophoretic method, the ethanol-precipitated DNA of a cleared lysate, which has had chromosomal DNA removed by sodium lauryl sulphate/NaCl lysis and centrifugation at 17 000g, is subjected to electrophoresis in a 0.7% agarose gel. Each species of plasmid DNA migrates as a distinct band, the distance from the origin being inversely related to the log of the molecule's molecular weight (Meyers et al. 1976). The bands are visualized by staining with ethidium bromide and observing under ultra-violet light. Some workers claim to have adapted this method so that plasmid DNA may be detected in a single colony isolated from an agar plate (Barnes 1977; Telford et al. 1977).

The centrifugation method involves analysing cleared lysates by rate zonal centrifugation through neutral sucrose gradients containing ethidium bromide (Hughes & Meynell 1977). When viewed by ultra-violet light, the ethidium bromide indicates the position of the plasmid band(s) in the centrifuge tube. The position of the plasmid band is dependent on its molecular weight, the density gradient of the sucrose and the duration of centrifugation.

Once isolated, plasmid DNA can be concentrated (Humphreys et al. 1975) and characterized in a number of ways, including heteroduplex analysis, visualization by electron microscopy after treatment with formamide (Sharp et al. 1972) and restriction endonuclease (Greene et al. 1974) and denaturation mapping (Hsu 1974). Combining the techniques of restriction endonuclease digestion and hybridization has proved a powerful tool for comparing plasmids of different origins (Southern 1975; Heinaru et al., 1978).

7. Significance of Plasmids for Bacterial Classification and Identification

Although plasmids have been detected in well over 50 species of bacteria and have been shown to determine more than 50 different phenotypic traits, it has not been possible to evaluate their significance for bacterial classification in

anything other than the simplest of terms. Although a plasmid may account for
as much as 10 to 15% of the total DNA of a bacterium, much of the plasmid's
coding capacity may go unaccounted for.

The most obvious way in which a plasmid can affect classification is by con-
ferring additional traits on its host, which are considered when constructing a
classification. In numerical phenetic studies, where a large number of characters
are employed (see pp. 78), the effect of a plasmid in a classification based on
many independent characters would normally be expected to be small. In diagnostic tests, however where relatively small numbers of characters are used and
often considerable weight placed on individual features (e.g. lactose fermentation), a plasmid borne trait may influence significantly the outcome of the test.
The potential of plasmids to interfere with physico-chemical taxonomic methods,
like %G+C content of DNA/DNA reassociation studies, appears to be limited at
present. In the case of %G+C content, for example, determinations are now usually made from melting temperature curves in which heterologous DNA, like
that of a plasmid, usually causes a discernible shoulder on the melting profile. In
DNA/DNA reassociation studies, the DNA extraction procedures are designed
where possible to separate plasmid DNA from the chromosomal DNA under
investigation. Significantly it is often the larger plasmids that are more difficult
to separate from chromosomal DNA.

Examples of the variety of plasmid encoded traits are discussed here by taxonomic groups in relation to their effect, if any, on the classification and identification of their host. The discussion is based on the classification in the current
eighth edition of *Bergey's Manual of Determinative Bacteriology* (Buchanan &
Gibbons 1974).

A. Bacilliaceae

One of the earliest reports of plasmids in representatives of this taxon involved
plasmids that interfere with sporulation of *Bacillus pumilus* (Lovett 1973).
Plasmids are frequently detected by physical screening methods. Between 10
to 40% of *Bacl. pumilus* and *Bacillus subtilis* strains studied in two laboratories
contained ccc plasmid DNA, but no functions could be attributed to these
plasmids (Lovett & Bramucci 1975). More recently, two plasmids have been
isolated from strains of *Bacillus cereus*, one had a mol. wt of 45×10^6 daltons
and determined the production of a bacteriocin, the other had a mol. wt of
2.8×10^6 daltons and determined resistance to tetracycline (Bernhard *et al*.
1978). The tetracycline resistance plasmid was successfully transferred by
transformation to *Bacl. subtilis* 168, where it was stably maintained and fully
expressed.

B. Enterobacteriaceae

A high proportion of plasmids detected in this taxon determine resistance to antibiotics. Resistance (R) plasmids have been found in the genera *Enterobacter, Escherichia, Klebsiella, Proteus, Salmonella, Serratia* and *Shigella* (Jacob *et al.* 1977). Many of these R plasmids are conjugative and are able to transfer widely amongst other members of the Enterobacteriaceae. The majority of these plasmids encode determinants for multiple drug resistance, sometimes as many as seven resistances being determined by a single plasmid. Although R plasmids are usually of greatest significance to diagnostic bacteriologists, there is some evidence that they may sometimes by responsible for the formation of phenetic sub-groups (Grimont & Dulong de Rosnay 1972).

Plasmids from members ot the Enterobacteriaceae have been shown to determine the enzymes necessary for the fermentation of various carbohydrates. For example, genes for lactose utilization have been found widely amongst representatives of species which are not normally or characteristically considered to possess the ability to ferment this sugar e.g. *Enterobacter hafniae, Klebsiella aerogenes, Proteus morganii, Proteus rettgeri, Salmonella oranianbourg, Salmonella typhimurium, Serratia liquefaciens* and *Yersinia enterocolitica* (Guiso & Ullman 1976; Cornelis *et al.* 1978). The possession of this trait by normally lactose-negative pathogens is of concern to diagnostic bacteriologists as it interferes with routine identification. It is therefore of particular interest that lactose genes have been found located on a transposition element (Cornelis *et al.* 1978).

The ability of some clinical isolates of *Salmonella* and *Esch. coli* to ferment sucrose has also been found to be plasmid-mediated. Wohlhieter *et al.* (1975) isolated sucrose fermenting strains of *Salmonella senftenburg* and *Salm. typhimurium* and have shown this trait to be plasmid-determined in each case. These workers have also isolated a strain of *Salmonella tennessee* possessing a single plasmid which determines the ability to ferment sucrose and lactose (Johnson *et al.* 1976). Palchaudhuri *et al.* (1977) have isolated a strain of *Esch. coli* which has sucrose-fermenting activity determined by a plasmid.

At least two other metabolic plasmids have been detected in isolates of *Esch. coli*, one determining the ability to utilize raffinose (Smith & Parsell 1975), the other the production of hydrogen sulphide (Ørskov & Ørskov 1973).

Several plasmids have been shown to contribute to the pathogenicity of members of the Enterobacteriaceae and have been particularly well studied in strains of *Esch. coli*. Pathogenic plasmids include those which determine the production of enterotoxin (Ent), haemolysin (Hly) and the surface antigens, K88 and K99, which have been implicated in cellular adhesion (Ørskov & Ørskov 1966; Smith & Linggood 1971; Gyles *et al.* 1974; Goebel *et al.* 1974; Falkow 1975). Two classes of enterotoxin have been recognized, one heat stable (ST) and the other heat labile (LT); the heat labile toxin shares a partial antigenic identity with the enterotoxin of *Vibrio cholerae* (Smith & Sack 1973; Gyles 1974). Although the Ent, Hly and surface adhesion properties contribute individually to pathogenicity and are often found together in the same isolate, they usually appear to be associated with separate plasmids (Falkow 1975). So (see Saunders 1978) has presented evidence that the gene determining the heat stable enterotoxin is located on a transposition element which is flanked by two IS1 elements.

C. Halobacteriaceae
Simon (1978) has reported the presence of 3 plasmids in *Halobacterium* strain 5. The molecular weights of the plasmids are 26, 44 and 86 $\times$ 10^6 daltons and pre-liminary evidence indicates that the 44 $\times$ 10^6 dalton plasmid determines or con-trols the production of gas vacuoles in this strain.

D. Lactobacillaceae

Plasmids have been detected in *Lactobacillus casei* and *Lact. casei* subsp. *rham-nosus* (Chassy *et al*. 1976) although no bacterial phenotypes have been ascribed to these plasmids. Hofer (1977), however, has shown that the ability of *Lact. casei* strain 1185 to utilize lactose is plasmid-determined. The plasmid borne nature of lactose utilization in this strain is of commercial as well as taxonomic importance since strain 1185 is commonly used by the dairy industry in the production of cheese.

E. Micrococcaceae

Resistance plasmids are frequently detected in clinical isolates of *Staphylococcus aureus*. In addition to antibiotic resistance, staphylococcal plasmids frequently determine resistance to various cations (e.g. Sb, Cd, Pb, Hg) and anions (e.g. arsenate, arsenite). Staphylococcal plasmids have also been shown to determine other traits, including the production of β-haemolysin (Dornbusch 1971), entero-toxin B (Shalita *et al*. 1977) and exfoliative toxin (Warren *et al*. 1975). It has also been suggested, but not proved conclusively, that pigment, coagulase and fibrinolysin production may also be plasmid-determined traits (Falkow 1975). As coagulase production is a character to which a great deal of import-ance is still attached in the identification of staphylococci, the possibility of this trait being plasmid-mediated is of great significance to diagnostic bacterio-logists. No staphylococcal plasmids have been shown to mediate conjugation and their usual method of transfer is by transduction (Falkow 1975).

F. Pseudomonadaceae

Despite having a genome no larger than that of *Esch. coli*, pseudomonads show enormous metabolic and nutritional versatility. When, for example a taxonomic study was carried out on more than 250 representatives of this genus, most were found capable of utilizing at least half of the 147 compounds offered as sole carbon and energy source (Stanier *et al*. 1966). At least part of this diversity is due to the presence of degradative plasmids amongst pseudomonads. Degradative plasmids have been detected in isolates of *Pseudomonas putida* which enable this host to utilize compounds like naphthaline (NAH), salicylate (SAL), camphor (CAM), *n*-alkanes such as hexane, octane and decane (OCT), *p* or *m*-xylene (XYL) and *p* or *m*-toluene (TOL) (Chakrabarty 1976). Recently a plasmid capable of mediating nicotine-nicotinate (NIC) degradation has been isolated from *Pseudomonas convexa* (Thacker *et al.* 1978). Some of these plasmids have been shown to be conjugative, whilst others are apparently incapable of self-transfer. Although easily transferred to other species of *Pseudomonas*, there has been some doubt about the ability of degradative plasmids to be readily transferred to other genera (Chakrabarty 1976), however, recent work with the TOL plasmid shows that it can be transferred to, and stably maintained in, *Esch. coli* although the toluate degradative genes are not expressed (Benson & Shapiro 1978). CAM and OCT plasmids are in the same incompatibility group (*inc*P2) and cannot therefore co-exist in a single host cell. Similarly, SAL and TOL plasmids are also incompatible, although their *inc* group is unknown. The remaining degradative plasmids are compatible and it is therefore possible for single host cells to possess a number of these plasmids. If, as now seems possible, degradative plasmids are ubiquitous amongst pseudomonads, they are likely to influence profoundly the classification of this taxonomically-complex group by altering properties of individual strains and causing spurious relationships between separate strains carrying the same plasmid. Perhaps more significantly, the degradative plasmids of pseudomonads may represent evolutionary intermediates in the inter-species transposition of genes from one genome to another (see pp. 36–39).

R plasmids are commonly detected in clinical isolates of *Pseudomonas aeruginosa*. These R plasmids normally determine resistance to antibiotics like kanamycin, carbenicillin and tetracycline, but resistance to mercuric ions and ability to mobilize chromosomal genes are frequently encountered (Jacoby & Shapiro 1977).

G. Rhizobiaceae

Rhizobium spp. are still largely classified on the basis of their host specificity. The suggestion that plasmids may be involved in the ability of *Rhizobium* to form root nodules in their plant host is therefore of significance for their classification, although the evidence is far from complete (Dunican & Cannon 1971). The suggestion has also been made that another taxonomically important feature of *Rhizobium* strains, namely the ability to fix nitrogen, might be plasmid determined (Dunican & Tierney 1974). Large plasmid molecules, with molecular weights between 70 and 400 x 10^6 daltons, have been isolated from a number of *Rhizobium* spp., including *Rhizobium dolichos, Rhizobium japonicum, Rhizobium leguminosarum* and *Rhizobium trifolii*, however, currently no phenotypic traits have been found to be associated with the presence of these plasmids (Nuti *et al.* 1977).

Plasmids have been more clearly implicated in the virulence of *Agrobacterium tumefaciens* (Zaenen *et al.* 1974), which causes crown-gall tumour disease in a variety of dicotyledon plants. Tumour-inducing (Ti) plasmids have been shown to be transmissible from tumouragenic to non-tumouragenic strains of *Agbc. tumefaciens* (Kerr, 1969; Kerr *et al.* 1977) and recently to *Esch. coli* as a hybrid with RP4 (Holsters *et al.* 1978). An *Esch. coli* harbouring the Ti:RP4 hybrid plasmid is unable to induce crown-gall tumours in plants. Utilization of octopine and nopaline is nearly always associated with Ti plasmids.

H. Streptococcaceae

A number of properties have been found to be plasmid-mediated in this taxon. Of greatest taxonomic significance are two plasmids detected in *Streptococcus faecalis* var. *zymogenes* which determine β-haemolysin and bacteriocin production. Unusually for Gram positive bacteria, these plasmids are capable of mediating conjugation (Jacob & Hobbs 1974; Jacob *et al.* 1975). Also since β-haemolysin activity is the only property that distinguishes the variety *zymogenes*, it has been suggested that this varietal status be dropped.

The ability of *Streptococcus lactis* (McKay *et al.* 1976) and *Streptococcus cremoris* (Anderson & McKay 1977) to utilize lactose has been shown to correlate with the presence of a plasmid. It has also been suggested that the production of the polypeptide antibiotic nisin by *Stco. lactis* might also be plasmid-mediated. Evidence is based largely on the occurrence of non-nisin producing derivatives, both spontaneously and after induction by acridine, ethidium bromide and high temperature, in nisin producing parental strains (Kozak *et al.* 1974). However, although some nisin-negative derivatives lack plasmid DNA which is present in their nisin-positive parental strain, others do not (Fuchs *et al.* 1975).

I. Streptomycetaceae

The extensively studied *Streptomyces coelicolor* A3(2) (Hopwood *et al.* 1973) has been shown genetically to carry two plasmids concerned with its fertility, SCP1 and SCP2. Only a single species of ccc plasmid DNA, mol. wt 18 to 20 x

10^6 daltons, has been physically detected in strain A3(2) and this has been shown to correlate with the presence of SCP2 (Schrempf *et al.* 1975; Bibb *et al.* 1977). SCP1 can acquire bacterial chromosome insertions (e.g. SCP1-*cys*B$^+$) and is transmissible as SCP1 or SCP1' to other *Streptomyces* spp., including *Streptomyces lividans* and *Streptomyces parvulus* (Hopwood & Merrick 1977).

Streptomyces coelicolor A3(2) produces two antibiotics, methylenomycin and actinorhodin; the former is determined by SCP1 (Wright & Hopwood 1976) and is produced in species to which SCP1 is transferred, the latter is determined by chromosomal genes. Claims have been made that the antibiotics produced by several other species of *Streptomyces* are determined by plasmids, usually based on the loss, either spontaneously or by induction, of antibiotic production by some of their clones. On this basis it is suggested that plasmids are involved in the ability of *Streptomyces aureofaciens* to produce chlortetracycline, *Streptomyces griseus* to produce streptomycin and cephamycin, *Streptomyces kasugaensis* to produce aureothricin (Okanishi & Umezawa 1978; Parag 1978), *Streptomyces hygroscopicus* to produce turimycin (Noack *et al.* 1978), *Streptomyces rimosus* to produce oxytetracycline (see Hopwood & Merrick 1977) and *Streptomyces venezuelae* to produce chloramphenicol (Akagawa *et al.* 1975).

Finally, melanin production by *Streptomyces glaucescens* has been shown to involve a plasmid. The plasmid appears to act in the regulation rather than synthesis of this pigment (Baumann & Kocher 1976).

8. Plasmids and Bacterial Evolution

Two processes that contribute to bacterial evolution are mutation, which results in changes in protein structure, and recombination, which leads to a rearrangement of existing genes. At present, however, the relative contributions of mutation and recombination to bacterial evolution is not clear. Changes in protein structure resulting from mutation and subsequent positive or neutral selection pressures have been well documented (see Reanney 1976). Mutational changes can be detected by a variety of methods, including amino acid and DNA sequence analysis, immune cross-reactivity and micro-complement fixation. By comparison little attention has been paid to the possible involvement of gene rearrangements in evolution. The reason for this failure has been, until recently, the difficulty in detecting such rearrangements. With plasmids at least, the existence of molecular relationships that result from transposition can be revealed by methods such as heteroduplex formation, restriction endonuclease mapping and *in situ* hybridization (Southern 1975; Heinaru *et al.* 1978).

One very obvious way to assess the contribution of gene rearrangements to bacterial evolution is to compare directly the DNA sequences of related and unrelated strains. Whilst this would be too time-consuming to be of any practical value, Sanderson (1976) has used the partially elucidated genetic linkage maps of several members of the Enterobacteriaceae for a preliminary study. He concludes that mutation rather than gene rearrangements account for most of the divergence found within this group and substantiates this view with three main types of evidence: 1. The linkage maps of even quite distantly related members of the Enterobacteriaceae show a great deal of similarity. Where major differences do occur, as between *Esch. coli* and *Salm. typhimurium*, they can be accounted for by specific events like the inversion of a whole region of the linkage map. Gene rearrangements occurring at a significant level would, he argues, have been reflected in the order of genes on the linkage map; 2. Members of the Enterobacteriaceae show differences in the amino acid sequences of their proteins that correlate well with relationships based on other criteria. 3. Finally, low, rather than high, polynucleotide sequence relatedness exists between members of this taxon.

Evidence for the involvement of gene rearrangements in bacterial evolution comes mainly from work on Tn elements. Genes coding for TEM-like β-lactamases, which confer resistance to penicillins, have been detected in R plasmids belonging to 14 different *inc* groups; however, the products of many of these genes have been shown by analytical isoelectric focusing studies to be closely related (Mathews & Hedges 1976). These genes are widely distributed in nature, having been isolated in four continents and detected on plasmids in representatives of at least 17 different taxospecies, including *Haemophilus influenzae* and *Neisseria gonorrhoeae* (Datta 1977). Under laboratory conditions at least, TEM β-lactamase genes are able to integrate into the chromosome of *Esch. coli* by $recA^{+}$-independent, non-homologous recombination. TEM β-lactamase genes also occur in Tn1, Tn2 and Tn3.

The isolation of lactose plasmids in representatives of normally lactose-negative Enterobacteriacae provides further evidence for the involvement of gene rearrangements in evolution. Each of the lactose plasmids determine a β-galactosidase and a β-galactoside permease, which, on the basis of biochemical, genetical and immunological evidence, are related to the enzymes determined by *Esch. coli*, but unrelated to those determined by the normally lactose-positive species, *Klebsiella oxytoca* and *Enterobacter aerogenes* (Guiso & Ullman 1976). Originally, these observations were interpreted as evidence that the plasmid-determined lactose enzymes were derived from *Esch. coli* however, recently Cornelis *et al.* (1978) have shown that the lactose genes of a plasmid isolated from *Yers. entercolitica* are located on a Tn element (Tn951), 16 000 nucleotide pairs long. Only 5600 nucleotide pairs, corresponding to the lactose genes, share homology with the chromosome of *Esch. coli* and lack of homology between the rest of the Tn element and genes adjacent to the lactose operon of *Esch. coli*, argues against the view that these genes originated in this bacterium. Instead *Esch. coli* may have acquired its lactose genes by transposition.

Recently the toluene degrading (*tol*) genes of a *Pseudomonas* TOL plasmid have been found to be transposable (Jacoby *et al.* 1978). The transposable nature of the *tol* genes accounts for the heterogeneity found amongst independently isolated TOL plasmids. The *tol* genes have been transposed to the promiscuous plasmid RP4 and the resulting RP4:*tol* plasmid shown to be capable of transfer to, and expression in *Esch. coli*. Provided a suitable selective advantage exists, there appears to be no reason why *tol* genes should not be transposed from this plasmid to the chromosome of *Esch. coli*.

Recent work on transposable genes taken together with the wide host-range of many plasmids, demonstrates the possible involvement of gene rearrangements in bacterial evolution. The extent to which transpositions occur is probably highly variable and is likely to be dependant on selection pressures in the environment. For example, transpositions may occur more frequently between bacteria in heterogeneous habitats like the soil, rather than in relatively homogeneous habitats like the mammalian gut. This is best illustrated by considering the ways by which a soil bacterium might cope with changes in its chemical environment. At least three mechanisms are available. Firstly, the bacterium may shelve the problem by forming a spore or a resting body (e.g. members of the genera *Bacillus, Streptomyces* and *Azotobacter*). Secondly, it may already possess the capacity to utilize a large variety of potential carbon and nitrogen sources, in which case it will have a heavy genetic load (i.e. unexpressed DNA). Thirdly, it may acquire appropriate genes from the environment by gene transfer and possibly also transposition. The latter would appear in part to be the solution adopted by members of the genus *Pseudomonas*. In contrast, gut inhabiting bacteria are usually only subject to relatively subtle changes in their chemical environment. However, where extreme changes do occur, as with the therapeutic use of antibiotics, gene transfer and transposition are clearly involved in disseminating antibiotic resistance genes.

9. Final Remarks

There is now good evidence to show how plasmids can interfere with the identification of bacterial taxa, particularly in diagnostic schemes where few characters are used. For example, the occurrence of lactose plasmids in representatives of normally lactose-negative pathogens like *Salm. typhimurium*, haemolysin plasmids in *Esch. coli*, and possibly, coagulase plasmids in *Staphylococcus aureus*. Consequently, there are good reasons for either avoiding the use of potentially plasmid-determined characters in identification, or at least giving them less weight.

In contrast, it is still not possible to assess the effects of plasmids on bacterial classification. If plasmids result in significant levels of gene rearrangements between bacteria, clear cladistic lines may not exist. Bacteria isolated from the same or similar environments may show relationships that reflect their ability, *via* gene transfer and transposition to acquire genes from a common gene pool, rather than simply their sharing a common ancestor (cf. plants, see Reanney 1976).

The question arises as to whether known or suspected plasmid-determined characters should be included in the reclassification of taxa using methods such as numerical taxonomy. The strong case for inclusion can be based on the following arguments:

1. In constructing a general purpose classification, all characters should be given equal weight.
2. Present methods do not allow all plasmid-determined characters to be easily recognized.
3. The possession of a particular plasmid-determined character may have a taxonomic relevance.
4. If plasmid-determined characters are not considered presumably the question would soon arise as to whether known transposon-determined characters should be included in such studies!

10. References

ACHTMAN, M. & HELMUTH, R. 1974 The F factor carries an operon of more than 15 × 10^6 daltons coding for deoxyribonucleic acid transfer and surface exclusion. *Microbiology 1976*, 95–103.

AKAGAWA, H., OKANISHI, M. & UMEZAWA, H. 1975 A plasmid involved in chloramphenicol production in *Streptomyces venezuelae*: evidence from genetic mapping. *Journal of General Microbiology* 90, 336–346.

ANDERSON, D. G. & McKAY, L. L. 1977 Plasmids, loss of lactose metabolism and appearance of partial or full lactose-fermenting revertants of *Streptococcus cremoris*. *Journal of Bacteriology* 129, 367–377.

ANDERSON, E. S. 1966 Possible importance of transfer factors in bacterial evolution. *Nature, London* 209, 637–638.

ARBER, W. 1974 DNA modification and restriction. *Progress in Nucleic Acid Research and Molecular Biology* 14, 1–37.

BARNES, W. M. 1977 Plasmid detection and sizing in single colony lysates. *Science, New York* 195, 393–394.

BAUMANN, R. & KOCHER, H. P. 1976 Genetics of *Streptomyces glaucescens* and regulation of melanin production. In *Second International Symposium on the Genetics of Industrial Micro-organisms* ed. Macdonald, K. D. pp. 535–552, London and New York: Academic Press.

BAZARAL, M. & HELINSKI, D. R. 1970 Replication of a bacterial plasmid and an episome in *Escherichia coli. Biochemistry* 9, 399–406.

BENSON, S. & SHAPIRO, J. 1978 TOL is a broad-host-range plasmid. *Journal of Bacteriology* 135, 278–280.

BERNHARD, K., SCHREMPF, H. & GOEBAL, W. 1978 Bacteriocin and antibiotic resistance plasmids in *Bacillus cereus* and *Bacillus subtilis*. *Journal of Bacteriology* 133, 897–903.

BIBB, M. J., FREEMAN, R. F. & HOPWOOD, D. A. 1977 Physical and genetical characterisation of a second sex factor, SCP2 for *Streptomyces coelicolor. Molecular and General Genetics* 154, 155–166.

BRADLEY, D. E. 1974 Adsorption of bacteriophages specific for *Pseudomonas aeruginosa* R-factors. *Biochemical and Biophysical Research Communications* 57, 893-900.
BRADLEY, D. E. 1976 Adsorption of R-specific bacteriophage PR4 to pili determined by a drug resistance plasmid of the W compatibility group. *Journal of General Microbiology* 95, 181-185.
BROWNELL, G. H. 1978 Genetic interactions in the genus *Nocardia*, In *Genetics of the Actinomycetales* ed. Freerksen, E., Tárnok, I. & Thumin, J. H. pp. 103-148 Stuttgart: Gustav Fischer, Verlag.
BUCHANAN, R. E. & GIBBONS, N. E. (eds). 1974 *Bergey's Manual of Determinative Bacteriology, 8th edn.* Baltimore: Williams & Wilkins.
CHAKRABARTY, A. M. 1976 Plasmids of *Pseudomonas*. *Annual Reviews of Genetics* 10, 7-30.
CHANG, S. & COHEN, S. N. 1977 *In vivo* site specific, genetic recombination promoted by the *Eco*R1 restriction endonuclease. *Proceedings of the National Academy of Science, U.S.A.* 74, 4811-4815.
CHASSY, B. M., GIBSON, E. & GIUFFRIDS, A. 1976 Evidence for extrachromosomal elements in *Lactobacillus*. *Journal of Bacteriology* 127, 1576-1578.
CLOWES, R. C. 1972 Molecular structure of bacterial plasmids. *Bacteriological Reviews* 36, 361-405.
CORNELIS, G., GHOSAL, D. & SAEDLER, H. 1978 Tn951: A new transposon carrying a lactose operon. *Molecular and General Genetics* 160, 215-224.
CURTISS, R. 1969 Bacterial conjugation. *Annual Reviews of Microbiology* 23, 69-136.
DATTA, N. 1975 Classification of plasmids into compatibility groups: Distribution of genes among the groups. In *Microbial Drug Resistance* ed. Mitsuhashi, S. & Hashimoto, H. pp. 83-91. Tokyo: University of Tokyo Press.
DATTA, N. 1977 Classification of plasmids as an aid to understanding their epidemiology and evolution. *Journal of Antimicrobial Chemotherapy (Suppl. C)* 3, 19-23.
DAVIES, J., COURVALIN, P. & BERG, D. 1977 Thoughts on the origins of resistance plasmids. *Journal of Antimicrobial Chemotherapy (Suppl. C)* 3, 7-17.
DAVIS, D. B. & HELMSTETTER, C. E. 1973 Control of F'*lac* replication in *Escherichia coli* B/r. *Journal of Bacteriology* 114, 294-299.
DE VRIES, J. K., PFISTER, A., HAENNI, C., PALCHAUDHURI, S. & MAAS, W. 1976 F incompatibility. *Microbiology–1976*, 166-170.
DORNBUSCH, K. 1971 Genetic aspects of methicillin resistance and toxin production in a strain of *Staphylococcus aureus*. *Annals of the New York Academy of Science* 182, 91.
DUNICAN, L. K. & CANNON, F. C. 1971 The genetic control of symbiotic properties in *Rhizobium*: evidence for plasmid control. *Plant and Soil* (special volume) 73-79.
DUNICAN, L. K. & TIERNEY, A. B. 1974 Genetic transfer of nitrogen fixation from *Rhizobium trifolii* to *Klebsiella aerogenes*. *Biochemical and Biophysical Research Communications* 57, 62-72.
FALKINHAM, J. O. & CURTISS, R. 1976 Isolation and characterisation of conjugation-deficient mutants of *Escherichia coli* K12. *Journal of Bacteriology* 126, 1194-1206.
FALKOW, S. 1975 *Infectious Multiple Drug Resistance*. London: Pion Ltd.
FALKOW, S., GUERRY, P., HEDGES, R. W. & DATTA, N. 1974 Polynucleotide sequence relationships amongst plasmids of the I compatibility complex. *Journal of General Microbiology* 83, 65-76.
FREIFELDER, D. 1970 Isolation of extrachromosomal DNA from bacteria. In *Methods in Enzymology*, vol. 21. Nucleic acids. ed. Grossman, L. & Moldave, K. pp. 153-160. New York: Academic Press.
FUCHS, P. G., ZAJDEL, J. & DOBRZANSKI, W. T. 1975 Possible plasmid nature of the determinant for production of the antibiotic nisin in some strains of *Streptococcus lactis*. *Journal of General Microbiology* 88, 189-192.
GASSON, M. & WILLETTS, N. 1976 Transfer gene expression during fertility inhibition of the *Escherichia coli* K12 sex factor F by the I-like plasmid R62. *Molecular and General Genetics* 149, 329-333.

GOEBEL, W., ROYER-POKORA, B., LINDEUMAKER, W. & BUJARD, H. 1974 Plasmids controlling synthesis of haemolysin in *Escherichia coli*: Molecular properties. *Journal of Bacteriology* 118, 964–973.

GREENE, P. J., BETLACH, M. C., BOYER, H. W. & GOODMAN, H. M. 1974 The *Eco*R1 restriction endonuclease. In *Methods in Molecular Biology*, vol. 7. ed. Wickner, R. B. pp. 87–111. New York: Marcel Dekker.

GRIMONT, P. A. D. & DULONG de ROSNAY, H. L. C. 1972 Numerical study of 60 strains of *Serratia*. *Journal of General Microbiology* 72, 259–268.

GUISO, N. & ULLMAN, A. 1976 Expression and regulation of lactose genes carried by plasmids. *Journal of Bacteriology* 127, 691–697.

GUSTAFSSON, P. & NORDSTRÖM, K. 1975 Random replication of the stringent plasmid R1 in *Escherichia coli*. *Journal of Bacteriology* 123, 443–448.

GUSTAFSSON, P. & NORDSTRÖM, K. 1978 Temperature-dependant amber copy mutants of plasmid R1*drd*19 in *Escherichia coli*. *Plasmid* 1, 134–144.

GUSTAFSSON, P., NORDSTROM, K. & PERRAM, J. W. 1978 Selection and timing of replication of plasmids R1*drd*19 and F'*lac* in *Escherichia coli*. *Plasmid* 1, 187–203.

GYLES, C. L. 1974 Relationships among heat-labile enterotoxins of *Escherichia coli* and *Vibrio cholerae*. *Journal of Infectious Diseases* 129, 277–288.

GYLES, C. L., SO, M. & FALKOW, S. 1974 The enterotoxin plasmids of *Escherichia coli*. *Journal of Infectious Diseases* 130, 40–49.

HAAS, D. & HOLLOWAY, B. W. 1976 R factor variants with enhanced sex factor activity in *Pseudomonas aeruginosa*. *Molecular and General Genetics* 144, 243–251.

HAYES, W. 1953 Observations of a transmissible agent determining sexual differentiation in *Bacterium coli*. *Journal of General Microbiology* 8, 72–88.

HEDÉN, L-O. & MEYNELL, E. 1976 Comparative study of R1-specific chromosomal transfer in *Escherichia coli* K12 and *Salmonella typhimurium* LT2. *Journal of Bacteriology* 127, 51–58.

HEDGES, R. W. 1972 The pattern of evolutionary change in bacteria. *Heredity* 28, 39–48.

HEINARU, A. L., DUGGLEBY, C. J. & BRODA, P. 1978 Molecular relationships of degradative plasmids determined by *in situ* hybridization of their endonuclease-generated fragments. *Molecular and General Genetics* 160, 347–351.

HELINSKI, D. R. & CLEWELL, D. B. 1972 Circular DNA. *Annual Reviews of Biochemistry* 40, 899–942.

HOFER, F. 1977 Involvement of plasmids in lactose metabolism in *Lactobacillus casei* suggested by genetic experiments. *FEMS Microbiological Letters* 1, 167–170.

HOLSTERS, M., SILVA, B., VAN VLIET, F., HERNALSTEENS, J. P., GENETELLO, C., VAN MONTAGU, M. & SCHELL, J. 1978 *In vivo* transfer of the Ti-plasmid of *Agrobacterium tumefaciens* to *Escherichia coli*. *Molecular and General Genetics* 162, 335–338.

HOPWOOD, D. A. & MERRICK, M. J. 1977 Genetics of antibiotic production. *Bacteriological Reviews* 41, 595–635.

HOPWOOD, D. A., CHATER, K. F., DOWDING, J. E. & VIVIAN, A. 1973 Advances in *Streptomyces coelicolor* genetics. *Bacteriological Reviews* 37, 371–405.

HSU, M-T. 1974 Electron microscopic analysis of partial denaturation of F factor deoxyribonucleic acid. *Journal of Bacteriology* 118, 425–433.

HUGHES, C. & MEYNELL, G. G. 1977 Rapid screening for plasmid DNA. *Molecular and General Genetics* 151, 175–179.

HUMPHREYS, G. O., WILLSHAW, G. A. & ANDERSON, E. S. 1975 A simple method for the preparation of large quantities of pure plasmid DNA. *Biochimica et Biophysica Acta* 383, 457–463.

JACOB, A. E. & HOBBS, S. J. 1974 Conjugal transfer of plasmid-borne multiple drug resistance in *Streptococcus faecalis* var. *zymogenes*. *Journal of Bacteriology* 117, 360–372.

JACOB, A. E., DOUGLAS, G. J. & HOBBS, S. J. 1975 Self-transferable plasmid determining the hemolysin and bacteriocin of *Streptococcus faecalis* var. *zymogenes*. *Journal of Bacteriology* 121, 863–872.

JACOB, A. E., SHAPIRO, J. A., YAMAMOTO, L., SMITH, D. I., COHEN, S. N. & BERG, D. 1977 Appendix B: Bacterial plasmids. b. Plasmids studied in *Escherichia coli* and other enteric bacteria. In *DNA Insertion Elements, Plasmids and Episomes.* ed. Bukhari, A. I., Shapiro, J. A. & Adhya, S. pp. 607–638. New York: Cold Spring Harbor Laboratory, Cold Spring Harbor.

JACOBY, G. A. & SHAPIRO, J. A. 1977 Appendix B: Bacterial plasmids. c. Plasmids studied in *Pseudomonas aeruginosa* and other pseudomonads. In *DNA Insertion Elements, Plasmids and Episomes.* ed. Bukhari, A. I., Shapiro, J. A. & Adhya, S. pp. 639–656. New York: Cold Spring Harbor Laboratory, Cold Spring Harbor.

JACOBY, G. A., ROGERS, J. E., JACOB, A. E. & HEDGES, R. W. 1978 Transposition of *Pseudomonas* toluene-degrading genes and expression in *Escherichia coli. Nature, London* 274, 179–180.

JOHNSON, E. M., WOHLHIETER, J. A., PLACEK, B. P., SLEET, R. B. & BARON, L. S. 1976 Plasmid-determined ability of a *Salmonella tennessee* strain to ferment lactose and sucrose. *Journal of Bacteriology* 125, 385–386.

KAISER, D. & DWORKIN, M. 1975 Gene transfer to a myxobacterium by *Escherichia coli* phage P1. *Science, New York* 187, 653–654.

KENNEDY, N., BEUTIN, L., ACHTMAN, M., SKURRAY, R., RAHMSDORF, U. & HERRLICH, P. 1977 Conjugation proteins encoded by the F sex factor. *Nature, London* 270, 580–585.

KERR, A. 1969 Transfer of virulence between strains of *Agrobacterium. Nature, London* 223, 1175–1176.

KERR, A., MANIGAULT, P. & TEMPE, J. 1977 Transfer of virulence *in vivo* and *in vitro* in *Agrobacterium. Nature, London* 265, 560–561.

KOZAK, W., RAJCHERT-TRZPIL, M. & DROBRZANSKI, W. T. 1974 The effect of proflavin, ethidium bromide and an elevated temperature on the appearance of nisin-negative clones in nisin-producing strains of *Streptococcus lactis. Journal of General Microbiology* 83, 295–302.

LOVETT, P. S. 1973 Plasmids in *Bacillus pumilus* and enhanced sporulation of plasmid-negative variants. *Journal of Bacteriology* 115, 291–298.

LOVETT, P. S. & BRAMUCCI, M. G. 1975 Plasmid deoxyribonucleic acid in *Bacillus subtilis and Bacillus pumilus, Journal of Bacteriology* 124, 484–490.

McKAY, L. L., BALDWIN, K. A. & EFSTATHIOU, J. D. 1976 Transductional evidence for plasmid linkage of lactose metabolism in *Streptococcus lactis* C2. *Applied and Environmental Microbiology* 32, 45–52.

MATHEWS, M. & HEDGES, R. W. 1976 Analytical isoelectric focussing of R-factor-determined lactamase: correlation with plasmid compatibility. *Journal of Bacteriology* 125, 713–718.

MEYERS, J. A., SANCHEZ, D., ELWELL, L. P. & FALKOW, S. 1976 Simple agarose gel electrophoretic method for the identification and characterisation of plasmid deoxyribonucleic acid. *Journal of Bacteriology* 127, 1529–1537.

MOODY, E. E. M. & RUNGE, R. 1972 The integration of autonomous transmissible plasmids into the chromosome of *Escherichia coli* K12. *Genetical Research* 19, 181–186.

NOACK, D., ROTH, M. & ZIPPEL, M. 1978 Extrachromosomal control of growth and antibiotic production in *Streptomyces hygroscopicus* JA6599. In *Genetics of the Actinomycetales* ed. Freerksen, E., Tárnok, I. & Thumin, J. H. pp. 15–17, Stuttgart: Gustav Fischer, Verlag.

NUTI, M. P., LEDEBOER, A. M., LEPIDI, A. A. & SCHILPEROORT, A. A. 1977 Large plasmids in different *Rhizobium* species. *Journal of General Microbiology* 100, 241–248.

OKANISHI, M. & UMEZAWA, H. 1978 Plasmid involved in antibiotic production in streptomycetes. In *Genetics of the Actinomycetales,* ed. Freerksen, E., Tárnok, I. & Thumin, J. H. pp. 19–38, Stuttgart: Gustav Fischer Verlag.

ØRSKOV, I. & ØRSKOV, F. 1966 Episome carried surface antigen K88 of *Escherichia coli*. I. Transmission of the determinant of K88 antigen and influence on the transfer of chromosomal markers. *Journal of Bacteriology* 91, 69–75.

ØRSKOV, I. & ØRSKOV, F. 1973 Plasmid-determined H_2S character in *Escherichia coli* and its relation to plasmid-carried fermentation and tetracycline resistance characters. *Journal of General Microbiology* 77, 487–499.

PALCHAUDHURI, S., RAHN, S., SANTOS, D. S. & MAAS, W. K. 1977 Characterization of plasmids in a sucrose-fermenting strain of *Escherichia coli*. *Journal of Bacteriology* **130**, 1402–1403.

PARAG, Y. 1978 Genetic recombination and possible plasmid controlled antibiotic production in *Streptomyces griseus* NRRL 3851, a cephamycin producer. In *Genetics of the Actinomycetales*. ed. Freerksen, E., Tárnok, I. & Thumin, J. H. pp. 47–50. Stuttgart: Gustav Fischer Verlag.

PRITCHARD, R. H., BARTH, P. T. & COLLINS, J. 1969 Control of DNA synthesis in bacteria In *Microbial Growth*. ed. Meadow, P. & Pirt, S. J. Society for General Microbiology Symposium vol. 19 pp. 263–297. Cambridge University Press.

REANNEY, D. 1976 Extrachromosomal elements as possible agents of adaption and development. *Bacteriological Reviews* **40**, 552–590.

REANNEY, D. & TEH, C. K. 1976 Mapping pathways of possible phage-mediated genetics interchange among soil bacilli. *Soil Biology and Biochemistry* **8**, 305–311.

REYES, O., GOTTESMAN, M. & ADHYA, S. 1976 Suppression of polarity of insertion mutations in the *gal* operon and N mutations in bacteriophage Lambda. *Journal of Bacteriology* **126**, 1108–1112.

RICHMOND, M. H. & WIEDEMAN, B. 1974 Plasmids and bacterial evolution. In *Evolution in the Microbial World*. ed. Carlile, M. J. & Skehel, J. J. Society for General Microbiology Symposium Vol. 24. pp. 59–85. Cambridge: University Press.

ROBERTS, R. J. 1977 Appendix D: Restriction endonucleases. 1. Restriction and modification enzymes and their recognition sequences. In *DNA Insertion Elements, Plasmids and Episomes*. ed. Bukhari, A. I., Shapiro, J. A. & Adhya, S. pp. 757–768. New York: Cold Spring Harbor Laboratory, Cold Spring Harbor.

SAEDLER, H. & STARLINGER, P. 1967 O^c-mutations in the galactose operon of *Escherichia coli*. I. Genetic characterization. *Molecular and General Genetics* **100**, 178–189.

SAEDLER, H., REIF, H. J., HU, S. & DAVIDSON, N. 1974 IS2, a genetic element for turn-off and turn-on of gene activity in *Escherichia coli*. *Molecular and General Genetics* **132**, 265–289.

SANDERSON, K. E. 1976 Genetic relatedness in the family Enterobacteriaceae. *Annual Reviews of Microbiology* **30**, 327–349.

SAUNDERS, J. R. 1978 New transposons. *Nature, London* **274**, 211.

SCHREMPF, H., BUJARD, H., HOPWOOD, D. A. & GOEBEL, W. 1975 Isolation of covalently closed circular deoxyribonucleic acid from *Streptomyces coelicolor* A3(2). *Journal of Bacteriology* **121**, 416–421.

SCHWESINGER, M. D. 1977 Additive recombination in bacteria. *Bacteriological Reviews* **41**, 872–902.

SHALITA, Z., HERTMAN, I. & SARID, S. 1977 Isolation and characterization of a plasmid involved with enterotoxin B production in *Staphylococcus aureus*. *Journal of Bacteriology* **129**, 317–325.

SHARP, P. A., HSU, M-T., OHTSUBO, E. & DAVIDSON, N. 1972 Electron microscopic heteroduplex studies of sequence relations among plasmids of *Escherichia coli*. I. Structure of F-prime factors. *Journal of Molecular Biology* **71**, 471–497.

SHERRATT, D. J. 1978 Illegitimate recombination legitimised. *Nature, London* **274**, 213–214.

SIMON, R. D. 1978 *Halobacterium* strain 5 contains a plasmid which is correlated with the presence of gas vacuoles. *Nature, London* **273**, 314–317.

SMITH, H. W. & LINGGOOD, M. A. 1971 Observation on the pathogenic properties of the K88, Hly and Ent plasmids of *Escherichia coli* with particular reference to porcine diarrhoea. *Journal of Medical Microbiology* **4**, 469–485.

SMITH, H. W. & PARSELL, Z. 1975 Transmissible substrate-utilizing ability in enterobacteria. *Journal of General Microbiology* **87**, 129–140.

SMITH, N. & SACK, R. B. 1973 Immunologic cross-reaction of enterotoxins from *Escherichia coli* and *Vibrio cholerae*. *Journal of Infectious Diseases* **127**, 164–170.

SOUTHERN, E. M. 1975 Detection of specific sequences among DNA fragments separated by gel electrophoresis. *Journal of Molecular Biology* **98**, 503–517.

STANIER, R. Y., PALLERONI, N. J. & DOUDOROFF, M. 1966 The aerobic pseudomonads: a taxonomic study. *Journal of General Microbiology* **43**, 159–273.

STARLINGER, P. 1977 DNA rearrangement in procaryotes. *Annual Review of Genetics* **11**, 103–126.

TELFORD, J., BOSELEY, P., SCHAFFNER. W. & BIRNSTIEL, M. 1977 Novel screening procedure for recombinant plasmids. *Science, New York* **195**, 391–393.

THACKER, R., RØRVIG, O., KAHOLN, P. & GUNSALUS, I. C. 1978 NIC, a conjugative nicotine-nicotinate degradative plasmid in *Pseudomonas convexa*. *Journal of Bacteriology* **135**, 289–290.

TO, C-M., TO, A. & BRINTON, C. C. 1975 A new epiviral pilus, the E pilus and a new RNA pilus phage piE/R1. *Abstracts of the American Society for Microbiology Annual meeting—1975* p. 259.

WARREN, R., ROYOLSKY, M., WILEY, R. R. & GLASGOW, L. A. 1975 Isolation of extrachromosomal deoxyribonucleic acid for exfoliative toxin production from phage group II *Staphylococcus aureus*. *Journal of Bacteriology* **122**, 99–105.

WOHLHIETER, J. A., LAZERE, J. R., SNELLINGS, N. J., JOHNSON, E. M., SYNENKI, R. M. & BARON, L. S. 1975 Characterization of transmissible genetic elements from sucrose-fermenting *Salmonella* strains. *Journal of Bacteriology* **122**, 401–406.

WRIGHT, L. F. & HOPWOOD, D. A. 1976 Identification of the antibiotic determined by the SCP1 plasmid of *Streptomyces coelicolor* A3(2). *Journal of General Microbiology* **95**, 96–106.

ZAENEN, I., VAN LAREBEKE, N., TEUCHY, H., VAN MONTAGUE, M. & SCHELL, J. 1974 Super-coiled circular DNA in crown gall inducing *Agrobacterium* strains. *Journal of Molecular Biology* **86**, 109–127.

Progress in Classification and Identification of Neisseriaceae Based on Genetic Affinity

K. BØVRE

Kaptein W. Wilhelmsen og Frues, Bakteriologiske Institutt, Rikshospitalet, Oslo, Norway

Contents

1. Introduction

THE LITERATURE on the taxonomy, habitats and possible ecological roles of the less distinct pathogens or nonpathogens of the family Neisseriaceae is extensive, and at the same time rather confusing. This is particularly so with both the oxidase-positive and oxidase-negative rod-shaped taxa, but also applies to the coccal bacteria classified in this family. For excellent reviews of the literature the reader is referred to Berger (1963) and Henriksen (1973, 1976).

As a consequence of the poor classification, wide distribution and possible medical and veterinary importance of species other than the gonococci and meningococci, members of the Neisseriaceae have been extensively studied by taxonomists, particularly during the last two decades. The progress made is clearly illustrated in the two latest editions of *Bergey's Manual of Determinative Bacteriology* (Breed *et al.* 1957; Buchanan & Gibbons 1974) and in many research reports and proposals for taxonomic change. During this period the composition of the family Neisseriaceae has been altered; the family contained only the genera *Neisseria* and *Veillonella* in 1957 but now comprises the genera *Moraxella, Acinetobacter* and *Kingella* (Henriksen & Bøvre 1976), in addition to *Neisseria* and *Branhamella*. The taxon *Branhamella* was derived from part of the old *Neisseria* concept, and is classified in the genus *Moraxella* by some taxonomists (Henriksen & Bøvre 1968b; Bøvre 1979). *Veillonella* and a *Neisseria* spp. currently classified as *Gemella haemolysans* have been removed from the family. In contrast, several new species have become recognized, particularly in the genus *Moraxella* (Bøvre 1979). A large number of names have been reduced to synonyms of existing taxa. One such synonym was *Mima*, which contained organisms corresponding both to *Acinetobacter* and *Moraxella* (Henriksen 1973).

The rapid progress made in the reclassification of the Neisseriaceae is to a large extent due to the successful application of some of the new taxonomic methods that provide data on which natural classification can be constructed, i.e., techniques for measuring genetic affinity or nucleic acid similarity (see Chapter 1). This chapter is concerned mainly with the application of genetic transformation, nucleic acid hybridization and determination of average DNA base composition to the classification of the Neisseriaceae. The results of fatty acid analyses of whole cells have been evaluated in light of the genetic data and have been found to be of some value in the classification of these organisms. Both genetic transformation and fatty acid analysis have also been applied successfully to the identification of these bacteria, the two methods should be used in combination. Indirectly, these techniques have helped in the search for biochemical, cultural and chemical tests which can be applied in the routine laboratory.

2. Classification Using Genetic Methods

A. Genetic transformation

As early as 1952, my predecessor and for many years collaborator at the University of Oslo, Professor S. D. Henriksen, considered that the rod-shaped bacteria classified in the genus *Moraxella* on the basis of morphological appearance and some physiological similarities, might have an affinity to the genus *Neisseria*. He suggested the inclusion of *Moraxella* in the family Neisseriaceae (Henriksen 1952), a proposal that was later formally made on the basis of genetic data (Henriksen & Bøvre 1968b).

In 1961, Catlin & Cunningham detected two genetic groups within *Neisseria*. One species, *Neisseria catarrhalis*, was distinguished from the others in the genus (*Neis. caviae* and *Neis. ovis* were not examined) as it did not show any compatibility with them in genetic transformation (with streptomycin resistance = Str^r as marker), and was also clearly distinct on the grounds of overall base composition of DNA. Bøvre (1963) was the first to report Str^r transformation results which showed that *Moraxella* was related to *Neis. catarrhalis*, a finding subsequently independently confirmed by Catlin (1964).

When *Moraxella* was investigated systematically, a large group of strains labelled *Moraxella nonliquefaciens* was examined by quantitative Str^r transformation (Bøvre 1964a,b). All but two of these strains shared high affinities with the reference strains. The two deviating strains were incompatible both with the main

TABLE 1

Str transformation of* Moraxella nonliquefaciens *type/reference recipients**

Donor strains	Ratios of interstrain to homologous transformation
Moraxella	
nonliquefaciens	0.3–1
osloensis	10^{-5}†
phenylpyruvica	10^{-6}†
atlantae	10^{-6}†
'Moraxella'	
urethralis	Neg

*After Bøvre (1964b, 1965d, 1967a), Bøvre & Henriksen (1967b), Bøvre et al. (1976), and Lautrop et al. (1970): Bøvre (unpublished experiments). †, transformants detected; Neg, transformants not detected.

group and with each other, and formed the nucleus of the distinct species *Mrxl. osloensis* (Bøvre 1965d; Bøvre & Henriksen 1967a) and *Mrxl. phenylpyruvica* (Bøvre & Henriksen 1967b); the main group represents the revised taxon *Mrxl. nonliquefaciens* (Bøvre & Henriksen 1967a). The differences between the homo-logous affinities of *Mrxl. nonliquefaciens* strains and the heterologous affinities are shown in Table 1. The affinities of *Mrxl. nonliquefaciens* to strains classified in the recently described taxa *Mrxl. atlantae* (Bøvre et al. 1976) and *Mrxl. urethralis* (Lautrop et al. 1970) are also shown. It is evident that on the basis of the genetic data, *Mrxl. nonliquefaciens* is distinct from these species but it does share a close genetic similarity to strains classified as *Mrxl. lacunata* and *"Mrxl." bovis* (Bøvre 1965a,c; Table 2). Strains formerly known as *Mrxl. liquefaciens* were transferred to the taxon *Mrxl. lacunata* mainly on the basis of genetic data (Bøvre 1965c; Henriksen & Bøvre 1968b). *Moraxella. lacunata, Mrxl. bovis* and *Mrxl. nonliquefaciens* may be referred to as 'classical moraxellae' or as the '*Mrxl. lacunata* group'.

TABLE 2

Str transformation of 'classical moraxellae'* ('Moraxella lacunata *group*')*

Donor organisms	Ratios of interstrain to intrastrain transformation, with recipient organisms		
	lacunata	*bovis*	*nonliquefaciens*
Moraxella			
lacunata	0.1–1	1×10^{-2}	2×10^{-3}
bovis	1×10^{-3}	0.5–1	3×10^{-3}
nonliquefaciens	5×10^{-3}	2×10^{-3}	0.3–1

*After Bøvre (1965a,c); † average values.

TABLE 3
Str^r transformation in the genus Moraxella

Donor organisms	†Ratios* of interstrain to intrastrain transformation, with recipient organisms				
	Moraxella (Branhamella)				
	catarrhalis	ovis	cuniculi	new coccus† pigs	new coccus†, sheep
Moraxella (Branhamella)					
catarrhalis		4×10^{-5}	2×10^{-3}	8×10^{-3}	5×10^{-4}
ovis	1×10^{-5}		2×10^{-4}	1×10^{-3}	5×10^{-3}
cuniculi†	1×10^{-3}	2×10^{-4}		2×10^{-2}	2×10^{-4}
sp. (new coccus, pigs)	6×10^{-4}	9×10^{-4}	5×10^{-3}		5×10^{-5}
sp. (new coccus, sheep)	8×10^{-4}	6×10^{-3}	2×10^{-4}	1×10^{-3}	
caviae	2×10^{-5}	5×10^{-4}	5×10^{-5}	5×10^{-4}	3×10^{-4}
Moraxella					
sp. (new rod, cattle)	5×10^{-4}	2×10^{-3}	4×10^{-4}	5×10^{-4}	1×10^{-3}
lacunata	5×10^{-5}	2×10^{-4}	1×10^{-4}	2×10^{-4}	1×10^{-3}
bovis	4×10^{-5}	8×10^{-4}	3×10^{-5}	1×10^{-4}	1×10^{-3}
nonliquefaciens	5×10^{-5}	9×10^{-5}	7×10^{-5}	2×10^{-5}	3×10^{-4}
sp. (new rod, pigs)	5×10^{-5}	8×10^{-6}	2×10^{-5}	6×10^{-6}	5×10^{-4}
osloensis	2×10^{-5}	3×10^{-5}	1×10^{-5}	1×10^{-5}	1×10^{-5}
phenylpyruvica	8×10^{-6}	1×10^{-5}	2×10^{-5}	3×10^{-5}	6×10^{-5}
atlantae	8×10^{-6}	1×10^{-5}	2×10^{-5}	3×10^{-5}	4×10^{-5}

*Approximate and partly preliminary figures, based either on the average of several observations or only on single experiments (Bøvre 1965b, 1967a, Bøvre *et al.* 1976; Bøvre unpublished studies. †, Not formally proposed as belonging to genus *Moraxella*.

It can be seen from Table 3 that the rod-shaped organisms of the genus *Moraxella*, particularly those in the *Mrxl. lacunata* group, share a close genetic affinity to the coccal species previously known as the 'false neisseriae'. The latter include organisms labelled *Neis. catarrhalis, Neis. caviae, Neis. ovis* and *Neis. cuniculi* (Bøvre 1965b; Bøvre unpublished data). Some of these affinities appear to be as high as, or higher than, those between the coccal species, and some of them are distinctly higher than those found between the rod-shaped *Moraxella* spp. (Bøvre 1967a; Bøvre *et al.* 1976; see also Table 1). It has been proposed, therefore, that these cocci be included in the genus *Moraxella* (Henriksen & Bøvre 1968b). Catlin's (1970) proposal to create the genus *Branhamella* for *Neis. catarrhalis* does not do justice to these affinities to the rod-shaped organisms, nor does this genus include the other 'false neisseriae'. As a compromise, it has been proposed recently that the genus *Moraxella* be subdivided into the sub-genus *Moraxella* containing the rod-shaped organisms and the subgenus *Branhamella* comprising coccal organisms (Bøvre 1979). This classification will be followed in this chapter.

During the last few years my collaborators and I have studied the distribution and genetic relations in Str[r] transformation of oxidase-positive rods and cocci found as parasites in animals, and have paid particular attention to isolates sharing affinities with the genus *Moraxella*. Several new homogeneous taxa have been detected and preliminary transformation data on four of them are included in Table 3. It can be seen that one of the groups containing rod-shaped isolates (from the eyes of cattle) shares a higher general affinity to the subgenus *Branhamella* than strains in the *Mrxl. lacunata* group. The affinity of this new taxon to the *Mrxl. lacunata* group has been found to be distinctly lower, a result which adds to the pattern of closeness and overlapping affinities found between the cocci and rods. Excluded from Table 3 is the coccal strain NCTC 4103, which represents a separate genetic subunit within *Branhamella* (Catlin & Cunningham 1964; Bøvre 1965*b*).

The oxidase-negative acinetobacters have a distinct but very low transformation activity on *Branhamella* recipients. The acinetobacters are usually somewhat less active on a *Moraxella (Branhamella) catarrhalis* recipient than are *Mrxl. osloensis* donors, and are at least 10 times less active on a *Mrxl. (Bran.)ovis* recipient (Bøvre 1967*b*). These results suggest that the genus *Acinetobacter* should be classified in the same family as *Moraxella*. Transformation studies by Juni (1972), who used nutritional DNA markers, indicate that the genus *Acinetobacter* is rather homogeneous from a genetical point of view. Our own preliminary results with quantitative Str[r] transformation of the same recipient strain suggest that there are genetic sub-units within the genus (Bøvre *et al.* 1976; Bøvre, unpublished data). It remains to be seen, however, whether such studies will lead to a subdivision of *Acinetobacter calcoaceticus*, the only species presently recognized within *Acinetobacter* (Lautrop 1974).

Str[r] transformation has been performed recently with donors from a group of psychrophilic oxidase-positive rods isolated from poultry, fish and clinical material from humans. The genetic affinity of the psychrophilic group to the subgenus *Branhamella* appears to be similar to that shown by *Acinetobacter* strains. This new group has little or no affinity to *Acinetobacter* but shows a higher affinity to *Mrxl. osloensis* than acinetobacters. These genetical studies have still to be published; the psychrophilic isolates have been described, however, on the basis of conventional criteria (Bøvre *et al.* 1974) and in terms of their wax content and composition (Bryn *et al.* 1977).

The oxidase-positive "*Mrxl.*" *urethralis* (Lautrop *et al.* 1970), a taxon still to be validly named, has no distinct affinity on the basis of genetic transformation to presently known species classified in the family Neisseriaceae, and may belong to a separate genus. Juni (1977) found that this organism had some activity on *Mrxl. (Bran.) catarrhalis, Mrxl. osloensis* and *Acinetobacter* recipients in Str[r] transformation. The author is able to confirm a marginal activity on a *Mrxl. (Bran.) catarrhalis* recipient, but not on recipients of the two other organisms (Bøvre, unpublished data).

The oxidase-positive saccharolytic *Kingella kingae*, previously *Mrxl. kingae*, was recognized as a new species belonging to a new genus of Neisseriaceae partly because the results with Str^r transformation indicated no genetic affinity to other genera classified in the family (Henriksen & Bøvre 1968a, 1976). However, recent results with the same technique suggest a very low affinity to *Neisseria* (Bøvre et al. 1977a). The other two species classified in the genus *Kingella* (Snell & Lapage 1976), *King. indologenes* and *King. denitrificans*, were not proposed on the basis of genetic transformation data and their genetic affinities have still to be established.

Like *Kingella*, the genus *Neisseria* ('true neisseriae') shows no genetic affinity either to the subgenera of the genus *Moraxella* or to *Acinetobacter* in Str^r transformation (Catlin & Cunningham 1961; Bøvre 1965b, 1967a,b; Bøvre & Holten 1970; Bøvre et al. 1977a), except for a few possible transformant colonies with *Mrxl. (Bran.) catarrhalis* as donor (Siddiqui & Goldberg 1975). The early work of Catlin & Cunningham (1961) showed, however, that most species of *Neisseria* were closely interrelated once *Branhamella* strains had been removed from the genus. *Neis. perflava* and *Neis. flava* have been reduced to synonyms of *Neis. subflava* (Henriksen & Bøvre 1968b; Reyn 1974) partly on the basis of genetic data and recent results with Str^r and nutritional marker transformation indicate that *Neis. meningitidis, Neis. gonorrhoeae* and *Neis. lactamica* are very closely related (Siddiqui & Goldberg 1975).

The close and overlapping genetic affinities found between rod-shaped and coccal taxa classified in the genus *Moraxella* indicated that cellular shape and mode of division are not the fundamental taxonomic criteria they were considered to be originally. Additional evidence that a genus may contain both rod-shaped and coccal species was provided by the discovery of *Neis. elongata*. Although strains of this organism are rod-shaped, they show very high genetic affinities to all of the coccal taxa of the 'true' genus *Neisseria* that have been tested in Str^r transformation (Bøvre & Holten 1970; Bøvre et al. 1977a). It can

TABLE 4

Some affinities of the rod-shaped Neisseria elongata *in Str^r transformation**

Donor organisms	Ratios of interstrain to intrastrain transformation, with recipient organisms		
	elongata	*meningitidis*	*subflava†*
Neisseria			
elongata		3×10^{-2}	1×10^{-1}
meningitidis	1×10^{-2}		3×10^{-1}
subflava†	2×10^{-2}	5×10^{-2}	

*After Bøvre & Holten (1970), Bøvre et al. (1977a), and Catlin & Cunningham (1961).
† Designated *Neisseria flava* in previous publications.

be seen from Table 4 that these affinities may be in the same order of magnitude as those between the coccal species of the genus. It has been observed that *Neis. elongata* is genetically heterogeneous (Bøvre *et al.* 1972; Bøvre, unpublished data), and the affinities to coccal *Neisseria* may be close to, or even higher than, those found between the genetic subunits of *Neis. elongata* itself. On the other hand, no affinity in Str[r] transformation has been detected between *Neis. elongata* and other genera of the Neisseriaceae, apart from a possible very low affinity to *King. kingae* (Bøvre *et al.* 1977*a*).

B. Nucleic acid hybridization

The partly overlapping genetic affinities found between the subgenera *Branhamella* and *Moraxella* of the genus *Moraxella* have been supported by data from pulse-RNA (mRNA)/DNA hybridization experiments (Bøvre 1970*a*; Tables 5,6). This technique is very precise with respect to the degree of relatedness needed for the formation of RNase-resistant hybrids, more so than ribosomal (rRNA) to DNA pairing which expresses similarities between relatively conserved regions of the genome (Doi & Igarashi 1965; Dubnau *et al.* 1965). The method is also more exacting than some types of DNA/DNA pairing methods (Bøvre 1970*a*). Particularly high nucleotide affinities were obtained between taxa classified in the *Mrxl. lacunata* group, up to 34% of an intrastrain reaction, where as in the subgenus *Branhamella*, species were found to show less similarity to one another (Table 5).

TABLE 5

Pulse-RNA to DNA hybridization in Neisseriaceae*

Origin of DNA	RNA-DNA hybrid in % of intrastrain hybrid with RNA donor organisms		
	nonliquefaciens	*catarrhalis*	*ovis*
Moraxella			
nonliquefaciens		8.4	4.9
lacunata subsp. *lacunata*	22.2	4.5	7.5
lacunata subsp. *liquefaciens*	34.2	4.6	8.7
bovis	13.1	2.4	9.7
(Bran.) catarrhalis	6.5		2.4
(Bran.) ovis	7.1	2.9	
(Bran.) caviae	4.5	2.5	5.5
osloensis	2.0		
phenylpyruvica	1.2		
Acinetobacter	0.6		
Kingella kingae	0.4		
Neisseria	0.3–0.8	0–0.7	0.2–0.7

*After Bøvre (1970*a*).

TABLE 6

Pulse-RNA to DNA hybridization with the rod-shaped Neisseria elongata*

Origin of DNA	RNA-DNA hybrid in % of homologous hybrid with RNA donor organisms	
	Neisseria elongata	*Moraxella nonliquefaciens*
Neisseria elongata		0.5
Neisseria subflava†	11.0	0.8
Moraxella nonliquefaciens	0.4	

*After Bøvre (1970a). †Designated *Neis. flava* in original publication.

Both *Mrxl. (Bran.) ovis* and *Mrxl. (Bran.) catarrhalis* have a higher affinity to most of the *Mrxl. lacunata* group than to the other *Branhamella* taxa. The affinities found between *Mrxl. nonliquefaciens* and *Mrxl. osloensis/Mrxl. phenyl-pyruvica* are very low (1.2-2%). The genera *Acinetobacter, Kingella* and *Neisseria* show very low activities with the genus *Moraxella* (Table 5). However, the reactions of the genus *Neisseria* with the genus *Moraxella*, including *Branhamella*, may indicate a distant genetic relationship as *Escherichia coli* DNA included in parallel studies was inactive. It can be seen from Table 6 that the nucleic acid hybridization data confirm the genetic affinities found between *Neis. elongata* and the 'true' coccal neisseriae; there is a relatively pronounced hybrid formation with *Neis. subflava* DNA, but only a very low homology with *Mrxl. nonliquefaciens*. Parallel studies with *Esch. coli* DNA were negative (Bøvre 1970a).

Kingsbury (1967) found a distant relationship between the subgenus *Branhamella* of the genus *Moraxella* and the genus *Neisseria* in DNA/DNA pairing experiments performed under nonexacting conditions. These experiments were carried out using a more exacting temperature in a subsequent study (Kingsbury *et al.* 1969). In the first of the two studies, species of the genus *Neisseria* formed hybrids with each other at the level of 30-80% of the homologous reaction; some of these homology values were reduced, however, to 8-15% under the exacting conditions. Very stable duplexes were formed between *Neis. meningitidis* and *Neis. gonorrhoeae*. These results are generally consistent with those obtained in genetic transformation.

A third important study was concerned mainly with the examination of *Acinetobacter* strains by competitive DNA/DNA pairing (Johnson *et al.* 1970). On the basis of the DNA pairing data it was concluded that the acinetobacters be classified in a single genus, a finding supported by subsequent transformation studies (Juni 1972). Johnson and his colleagues were able to divide the genus into several subgroups showing intergroup homology values of around 20% or less; the significance of these subgroups has yet to be established. In the same investigation low DNA homologies were found between some strains of *Acinetobacter* and *Moraxella* spp. (up to 12% for *Mrxl. osloensis*), but no homology was recorded between *"Mrxl." urethralis* ATCC 17960 (Lautrop *et al.* 1970) and *Acinetobacter* or *Mrxl. osloensis.* In competitive rRNA to DNA pairing experiments Johnson *et al.* (1970) found homology values of around 68% between *Acinetobacter* on the one hand and *Mrxl. osloensis/Mrxl. (Bran.) catarrhalis* on the other. These results suggest a relationship between the oxidase-negative and -positive organisms corresponding to different genera of the same family, and are in good agreement with the transformation data with the Str^r marker discussed earlier.

In general, the nucleic acid hybridization techniques have not been used extensively in the classification of Neisseriaceae but the results reported appear to confirm the genetic transformation data.

C. Determination of average base composition of DNA

The moles per cent of guanine + cytosine bases (% G+C) of the total bases of DNA for taxa classified in the family Neisseriaceae are presented in Table 7. The results are compiled from determinations based upon chromatographic, buoyant density and thermal denaturation techniques (Catlin & Cunningham 1964; Hill 1966; De Ley 1968; Bøvre *et al.* 1969, 1976, 1977*a*; Bøvre & Holten 1970; Lautrop *et al.* 1970; Johnson *et al.* 1970; Reyn 1974; Snell & Lapage 1976). It can be seen that all four genera have wide G+C ranges which overlap.

It is remarkable that strains classified in the subgenera *Moraxella* and *Branhamella* of the genus *Moraxella* have identical G+C ranges, a result which underlines their closeness and overlapping compatibities in genetic transformation and nucleic acid hybridization experiments. The high value of 50.5% reported for *Branhamella* was obtained with a strain of *Mrxl. (Bran.) caviae* (LaMacchia & Pelczar 1966) examined using a chromatographic technique. Another strain of

TABLE 7
Moles per cent guanine + cytosine of DNA in Neisseriaceae *

Genus *Moraxella*	
Subgenus *Moraxella (lacunata, bovis, nonliquefaciens, osloensis, phenylpyruvica, atlantae)*	40–47.5
Subgenus *Branhamella (catarrhalis, caviae, ovis, cuniculi†)*	40–47.5 (50.5)‡
'*Mrxl.' urethralis*	46–47
Genus *Acinetobacter*	38–47
Genus *Kingella (kingae, indologenes, denitrificans)*	44.5–55
Genus *Neisseria (gonorrhoeae, meningitidis, sicca, subflava, flavescens, cinerea, mucosa, elongata)*	46.5–53.5

*Results with different methods compiled from literature (Bøvre *et al.* 1969, 1976, 1977*a*; Bøvre & Holten 1970; Catlin & Cunningham 1964; De Ley 1968; Hill 1966; Johnson *et al.* 1970; Lautrop *et al.* 1970; Reyn 1974; Snell & Lapage 1976). † Not yet formally proposed as belonging to genus *Moraxella*. ‡ Highest value obtained by LaMacchia & Pelczar (1966) for *Mrxl. (Bran.) caviae.*

this species was found, however, to contain 44.5% G+C using the buoyant density method (Bøvre *et al.* 1969) and 47.5% using the thermal denaturation method (Snell & Lapage 1976; J. J. S. Snell, personal communication).

It should be noted that almost all upper and lower limits of the group ranges shown in Table 7 are represented by organisms that have been shown to be distinctly related genetically to strains with different G+C ratios within the respective ranges. The main exception is the upper value for *Kingella*. This is due to *King. denitrificans* which has yet to be compared with *King. kingae* and *King. indologenes* (Snell & Lapage 1976) in genetic transformation and hybridization studies.

As the taxa classified in the family Neisseriaceae have wide and overlapping G+C ratios, base composition studies can only be interpreted in the light of other taxonomic criteria such as genetic transformation and nucleic acid pairing studies. Used in this way, the base ratio data have been of considerable value in separating *Branhamella* from the genus *Neisseria* and allocating it to the genus *Moraxella*. Again, the high G+C content of *Neis. elongata* strains, 53-53.5%, (Bøvre & Holten 1970; Bøvre *et al.* 1977a) strongly supports the genetic transformation and hybridization data which indicate that, despite the fact it contains rod-shaped bacteria, this species be classified in the genus *Neisseria*.

The results of base ratio determinations have to be interpreted with care as individual strains examined by different methods can differ in value by 3% or more (see Chapter 1). Such discrepancies have been observed with several *Acinetobacter* strains (Bøvre *et al.* 1969; Johnson *et al.* 1970; Jantzen *et al.* 1975), and the type strain of *King. kingae* was found to have a G+C value of 44.5 by the buoyant density method (Bøvre *et al.* 1969) but up to 49% using the thermal denaturation method (Snell *et al.* 1972). When a strain is examined by the same method in the same laboratory, however the variation is generally within 1% and is often considerably less. Under these standard conditions variation within an otherwise genetically homogeneous species or subspecies is usually within the range of 1-2% (Bøvre *et al.* 1969; Johnson *et al.* 1970). These observations are particularly important in cases where fine distinctions in terms of G+C content are part of an incipient genetic classification, as is the case with *Acinetobacter* (De Ley 1968; Pinter & De Ley 1969).

3. Congruence Between Genetic Data and Other Criteria

A. Cellular lipid composition

In 1968, Lewis *et al.* noted that *Mrxl. (Bran.) catarrhalis* had a unique cellular fatty acid pattern compared to species belonging to the genus *Neisseria*. Since 1970 my colleagues and I have been determining the value of gas-liquid chromatography of whole cell fatty acids in bacterial classification and identification

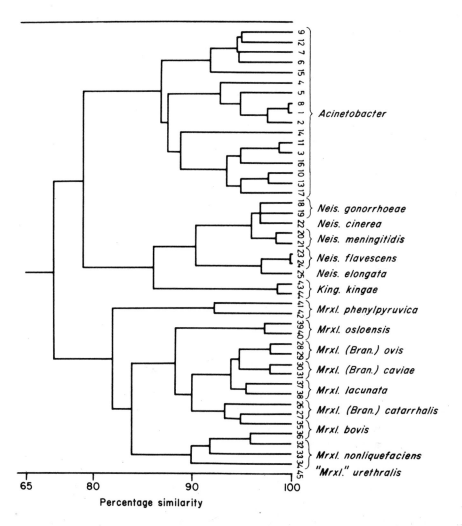

Fig. 1. Dendrogram showing similarities of fatty acid patterns in Neisseriaceae (after Jantzer *et al.* 1974*a,b*, 1975).

with particular reference to the family Neisseriaceae (Jantzen *et al.* 1974*a,b*, 1975, 1978). The qualitative and quantitative distribution of fatty acids has been found to reflect the spectrum of genetic affinities mentioned above (Fig. 1). The relative concentrations of some 20 fatty acids, as percentages of the total whole cell fatty acids, were first transformed using the formula $y = ln (x+1)$, where x is the actual concentration in per cent and y the figure used for comparing the strains (numbered 1–45) using a correlation coefficient.

This formula was used to give relatively more weight to fatty acids occurring in small quantities. Affinity coefficients were calculated and the phenogram constructed on the basis of 'unweighted pair group analysis' (Jantzen *et al.* 1974*b*, 1975).

Some of the strains examined genetically and by gas-liquid-chromatography are shown in Fig. 1. It can be seen that *Acinetobacter, Neisseria* and *Moraxella* form three distinct clusters on the basis of the fatty acid data and that these clusters correspond to the genetic classification. It is particularly interesting that the rod-shaped *Neis. elongata* falls into the *Neisseria* cluster and that the subgenera *Moraxella* and *Branhamella* of the genus *Moraxella* are recovered in a single cluster. *Kingella kingae* is associated with the *Neisseria* cluster, indicating that further comparative studies are required to determine the relationship between these two genera. *"Mrxl." urethralis* is clearly separated from the other Neisseriaceae, a result that is consistent with its genetic incompatibility.

Additional comparative fatty acid analyses have been applied to some large collections of Neisseriaceae identified using genetical criteria. A remarkable reproducibility of the fatty acid patterns has been found within the species examined to date (Bøvre *et al.* 1976, 1977*b*). These findings indicate that fatty acid analyses are valuable aids both in the classification and identification of Neisseriaceae strains. Gas-liquid chromatography of whole cell fatty acids is particularly valuable for distinguishing strains which are comparatively different in terms of fatty acid profiles, e.g. *Mrxl. osloensis* and *"Mrxl." urethralis.* Pitfalls exist, however, since an isolate belonging to an undescribed species may have a similar fatty acid pattern to strains of an established taxon. This was the case with *Mrxl. atlantae* which is clearly distinct from *Mrxl. phenylpyruvica* genetically and in several other ways but not on the basis of fatty acid analysis. The presence of a wax alcohol in the profile of *Mrxl. atlantae* strains provides, however, a way of distinguishing this taxon from *Mrxl. phenylpyruvica* (Bøvre *et al.* 1976). Good agreement has been found between the distribution of waxes in taxa classified in the family Neisseriaceae and the genetic data, including the separation of the genera *Neisseria* (which lacks waxes) and *Moraxella*, and the similarity found between the subgenera *Moraxella* and *Branhamella* (Bryn *et al.* 1977).

When isolates with unknown taxonomic affinities are studied fatty acid analysis may be used as a first line of attack with appropriate transformation tests performed secondarily. Fatty acid profiles are particularly important in the exclusion of candidates from the family Neisseriaceae. Thus, several groups of oxidase-positive bacteria have been shown to possess prominent amounts of branched fatty acids, which may be consistent with a relationship to *Flavobacterium* (Oshima & Miyagawa 1974; Bøvre, unpublished studies).

B. *Physiological, immunological and conventional cultural-biochemical criteria*

Comprehensive studies by Baumann *et al.* (1968*a,b*) on the nutritional and other physiological properties of *Moraxella* (including subgenus *Branhamella*) and

Acinetobacter strains gave results largely consistent with the genetic data. The enzymological studies of Holten (1973, 1974*a*) on glucokinase, glucose-6-phosphate dehydrogenase and glutamate dehydrogenases in *Neisseria* and *Branhamella* strains have shown fundamental differences between them, and his immunological comparison of glutamate and malate dehydrogenase, among the same bacteria further confirmed the genetic distinction between the two groups (Holten 1974*b*).

In most cases, the new species circumscribed in the family Neisseriaceae by genetic means have been confirmed by others using conventional tests. These species include the components of the previous *Mrxl. nonliquefaciens* concept: the revised species *Mrxl. nonliquefaciens*, *Mrxl. osloensis* and *Mrxl. phenylpyruvica*, as well as "*Mrxl.*" *urethralis* (Snell *et al.* 1972; Lautrop 1974; Riley *et al.* 1974). It is possible, however, that the key reactions in the conventional identification of these and other species are less stably associated with the genetic circumscription than hitherto believed. This would not be surprising if those strains examined genetically were pre-selected with respect to some key reactions. Recent results obtained with nonselective identification of all oxidase-positive rods from a defined habitat indicate that key reactions like phenylalanine and tryptophan deaminase may be absent in *Mrxl. phenylpyruvica* (Bøvre *et al.* 1977*b*). In such a case, the search for other criteria, apparently more stably associated with genotype, has been successful, such as the ability of *Mrxl. phenylpyruvica* to grow at high salt concentrations and at 4°C (Bøvre *et al.* 1976). Similar adjustments of diagnostic criteria may prove necessary for other species, with the genetic identity as a yardstick.

At the genetic level the separation of *King. kingae* from *Moraxella* is well substantiated using conventional tests, the former being saccharolytic and catalase-negative. Likewise, independent investigations by Berger & Piotrowski (1974) on the biochemical properties of *Neis. elongata* have confirmed that it belongs to the 'true' *Neisseria* and can be identified by conventional methods. In many cases, therefore, the genetic studies have just guided the application of less specialized methods of classification and identification, and no conflict has arisen.

4. The Use of Genetic Transformation in Identification

Several of the known species of Neisseriaceae, including "*Mrxl.*" *urethralis* and *Acnb. calcoaceticus*, can be identified by genetic transformation techniques (Bawdon *et al.* 1977; Bøvre 1964*b*, 1965*a,b,c,d*, 1970*b*; Bøvre *et al.* 1972, 1976, 1977*a,b*; Brooks & Sodeman 1974; Henriksen & Bøvre 1968*a*; Janik *et al.* 1976; Juni 1972, 1974, 1977; Riley *et al.* 1974). Successful identification depends on the existence of transformable or competent strains that have the ability to take up DNA efficiently. As competence can be lost on subcultivation, it is important for some types of diagnostic transformation that competence in some species is associated with distinct colony types, e.g. in *Mrxl. nonliquefaciens, Mrxl. bovis,*

King. kingae (Bøvre & Frøholm 1971, 1972), *Neis. gonorrhoeae* (Sparling 1966) and *Neis. elongata* (Bøvre *et al.* 1977*a*).

In one technical modification of such an identification, reference Str^r–transforming DNAs are tested on a competent isolate as recipient. This can be done semi-quantitatively, the DNA with the highest activity representing the probable identity of the isolate. If transforming DNA is available from a mutant of the isolate, the test can be performed quantitatively, the reference donor DNA giving close to the intrastrain transformation frequency representing the definite identity. A combination of these methods has been used in several studies on the occurrence of the genetic groups in human habitats (Bøvre 1970*b*; Bøvre *et al.* 1972, 1977*b*) and in animals (Bøvre, unpublished data).

Another and more extensively used modification of genetic identification by transformation is independent of the competence of the isolate. This employs reference recipients maintained in the laboratory. Transforming DNA is extracted from the isolate (in nutritional marker transformation) or a mutant thereof (in Str^r transformation). This approach, as utilized by Juni and collaborators, is as follows: reference auxotrophic recipients are exposed to DNA of the unidentified strain, which is simply a crude lysate produced within 30 min from as little material as one bacterial colony. Prototrophic transformant yield gives the species of the isolate within one or two days. This method has been used successfully in organisms with the ability to grow on defined media, i.e. *Acinetobacter* (Juni 1972; Brooks & Sodeman 1974), *Mrxl. osloensis* (Juni 1974; Riley *et al.* 1974), "*Mrxl.*" *urethralis* (Juni 1977) and *Neis. gonorrhoeae* (Janik *et al.* 1976; Bawdon *et al.* 1977). The use of such crude lysate DNA has also been adapted for identification by Str^r transformation (Bøvre, unpublished data). Usually, however, DNA is extracted and partly purified from a mutant of the isolate (Bøvre 1964*a*) before application in Str^r transformation of reference recipients.

As transformation of 'ribosomal' markers like Str^r can occur with decreased frequency between related species (Dubnau *et al.* 1965), one is often dependent on a quantitative or semi-quantitative assay of transformants in relation to an homologous parallel. This is ideal in transformation with nutritional markers as well, as heterologous transfer of such markers is also known (Siddiqui & Goldberg 1975; Janik *et al.* 1976). Although the tendency is less than for the Str^r marker, particularly when certain nutritional markers are used, it should not be underestimated when identification is performed in a group of species with relatively high mutual affinities, such as the genus *Neisseria*. A Str^r transformation system may be easier to quantify and has a wider applicability with respect to fastidious bacterial species.

5. Present Status and Future Prospects

It can be concluded that the genetic approach, together with newer chemotaxonomic methods, has an important potential role to play in improving both the classification and identification of the Neisseriaceae. The application of these methods has already helped to clarify the classification of the family, although it can be argued that more problems sometimes seem to be created than solved by the detection of taxa which have gone unrecognized using conventional methods. Such problems are probably only temporary; the main fact seems to be that the previous methods have been inadequate and that large collections of unidentified and unclassified strains have accumulated. The aim must be a complete classification of the Neisseriaceae, with taxa identifiable by genetic and chemotaxonomic procedures. Improved classification and identification will lead to progress in other fields. It will, for instance, facilitate the study of the distribution of members of the Neisseriaceae in natural habitats and lead to an understanding of speciation in relation to host and other habitats, and help to elucidate the roles of these bacteria as free living organisms and as parasites of animals.

6. References

BAUMANN, P., DOUDOROFF, M & STANIER, R. Y. 1968a Study of the *Moraxella* group I Genus *Moraxella* and the *Neisseria catarrhalis* group. *Journal of Bacteriology* **95**, 58–73.

BAUMANN, P., DOUDOROFF, M., & STANIER, R. Y. 1968b A study of the *Moraxella* group II Oxidase-negative species (Genus *Acinetobacter*). *Journal of Bacteriology* **95**, 1520–1541.

BAWDON, R. E., JUNI, E. & BRITT, E. M. 1977 Identification of *Neisseria gonorrhoeae* by genetic transformation: a clinical laboratory evaluation. *Journal of Clinical Microbiology* **5**, 108–109.

BERGER, U. 1963 Die anspruchslosen Neisserien. *Ergebnisse der Mikrobiologie, Immunitats-, forschung und experimentelle Therapie* **36**, 97–167.

BERGER, U. & PIOTROWSKI, H. D. 1974 Die biochemische Diagnose von *Neisseria elongata* (Bøvre und Holten 1970). *Medical Microbiology and Immunology* **159**, 309–316.

BØVRE, K. 1963 Affinities between *Moraxella* spp, and a strain of *Neisseria* catarrhalis as expressed by transformation. *Acta Pathologica et Microbiologica Scandinavica* **58**, 528.

BØVRE, K. 1964a Studies on transformation in *Moraxella* and organisms assumed to be related to *Moraxella*. 1. A method for quantitative transformation in *Moraxella* and *Neisseria*, with streptomycin resistance as the genetic marker. *Acta Pathologica et Microbiologica Scandinavica* **61**, 457–473.

BØVRE, K. 1964b Studies on transformation in *Moraxella* and organisms assumed to be related to *Moraxella*. 2. Quantitative transformation reactions between *Moraxella nonliquefaciens* strains, with streptomycin resistance marked DNA. *Acta Pathologica et Microbiologica Scandinavica* **62**, 239–248.

BØVRE, K. 1965a Studies on transformation in *Moraxella* and organisms assumed to be related to *Moraxella*. 3. Quantitative streptomycin resistance transformation between *Moraxella bovis* and *Moraxella nonliquefaciens* strains. *Acta Pathologica et Microbiologica Scandinavica* **63**, 42–50.

BØVRE, K. 1965b Studies on transformation in *Moraxella* and organisms assumed to be related to *Moraxella*. 4. Streptomycin resistance transformation between asaccharolytic *Neisseria* strains. *Acta Pathologica et Microbiologica Scandinavica* **64**, 229–242.

BØVRE, K. 1965c Studies on transformation in *Moraxella* and organisms assumed to be related to *Moraxella*. 5. Streptomycin resistance transformation between serum-liquefying nonhaemolytic moraxellae, *Moraxella bovis* and *Moraxella nonliquefaciens*. *Acta Pathologica et Microbiologica Scandinavica* **65**, 435–449.

BØVRE, K. 1965d Studies on transformation in *Moraxella* and organisms assumed to be related to *Moraxella*. 6. A distinct group of *Moraxella nonliquefaciens*-like organisms (the 19116/51 group). *Acta Pathologica et Microbiologica Scandinavica* **65**, 641–652.

BØVRE, K. 1967a Studies on transformation in *Moraxella* and organisms assumed to be related to *Moraxella*. 7. Affinities between oxidase positive rods and neisseriae, as compared with group interactions on both sides. *Acta Pathologica et Microbiologica Scandinavica* 69, 92–108.

BØVRE, K. 1967b Studies on transformation in *Moraxella* and organisms assumed to be related to *Moraxella*. 8. The relative position of some oxidase negative, immotile diplobacilli (*Achromobacter*) in the transformation system. *Acta Pathologica et Microbiologica Scandinavica* 69, 109–122.

BØVRE, K. 1970a Pulse-RNA-DNA hybridization between rod-shaped and coccal species of the *Moraxella-Neisseria* groups. *Acta Pathologica et Microbiologica Scandinavica* 78B, 565–574.

BØVRE, K. 1970b Oxidase positive bacteria in the human nose. Incidence and species distribution, as diagnosed by genetic transformation. *Acta Pathologica et Microbiologica Scandinavica* 78B, 780–784.

BØVRE, K. 1979 Proposal to divide the genus *Moraxella* Lwoff 1939 emend. Henriksen and Bøvre 1968 into two subgenera—subgenus *Moraxella* (Lwoff 1939) Bøvre 1979 and subgenus *Branhamella* (Catlin 1970) Bøvre 1979. *International Journal of Systematic Bacteriology* 29, 403–406.

BØVRE, K & FRØHOLM, L. O. 1971 Competence of genetic transformation correlated with the occurrence of fimbriae in three bacterial species. *Nature, New Biology* 234, 151–152.

BØVRE, K. & FRØHOLM, L. O. 1972 Competence in genetic transformation related to colony type and fimbriation in three species of *Moraxella*. *Acta Pathologica et Microbiologica Scandinavica* 80B, 649–659.

BØVRE, K. & HENRIKSEN, S. D. 1967a A new *Moraxella* species, *Moraxella osloensis,* and a revised description of *Moraxella nonliquefaciens*. *International Journal of Systematic Bacteriology* 17, 127–135.

BØVRE, K. & HENRIKSEN, S. D. 1967b A revised description of *Moraxella polymorpha* Flamm 1957 with a proposal of a new name *Moraxella phenylpyruvica* for this species. *International Journal of Systematic Bacteriology* 17, 343–360.

BØVRE, K. & HOLTEN, E. 1970 *Neisseria elongata* sp. nov., a rod-shaped member of the genus *Neisseria*. Re-evaluation of cell shape as a criterion in classification. *Journal of General Microbiology* 60, 67–75.

BØVRE, K., FIANDT, M. & SZYBALSKI, W. 1969 DNA base composition of *Neisseria, Moraxella,* and *Acinetobacter,* as determined by measurement of buoyant density in CsCl gradients. *Canadian Journal of Microbiology* 15, 335–338.

BØVRE, K., FUGLESANG, J. E. & HENRIKSEN, S. D. 1972 *Neisseria elongata*. Presentation of new isolates. *Acta Pathologica et Microbiologica Scandinavica* 80B, 919–922.

BØVRE, K., FUGLESANG, J. E., HENRIKSEN, S. D., LAPAGE, S. P., LAUTROP, H. & SNELL, J. J. S. 1974 Studies on a collection of Gram-negative bacterial strains showing resemblance to moraxellae: examination by conventional bacteriological methods. *International Journal of Systematic Bacteriology* 24, 438–446.

BØVRE, K., FUGLESANG, J. E., HAGEN, N., JANTZEN, E. & FRØHOLM, L. O. 1976 *Moraxella atlantae* sp. nov. and its distinction from *Moraxella phenylpyruvica*. *International Journal of Systematic Bacteriology* 26, 511–521.

BØVRE, K., FRØHOLM, L. O., HENRIKSEN, S. D. & HOLTEN, E. 1977a Relationship of *Neisseria elongata* subsp. *glycolytica* to other members of the family Neisseriaceae. *Acta Pathologica et Microbiologica Scandinavica* 85B, 18–26.

BØVRE, K., HAGEN, N., BERDAL, B. P. & JANTZEN, E. 1977b Oxidase positive rods from cases of suspected gonorrhoea. A comparison of conventional, gas chromatographic and genetic methods of identification. *Acta Pathologica et Microbiologica Scandinavica* 85B, 27–37.

BREED, R. S., MURRAY, E. G. D. & SMITH, N. R. 1957 *Bergey's Manual of Determinative Bacteriology* 7th edn. Baltimore: Williams & Wilkins.

BROOKS, K. & SODEMAN, T. 1974 Clinical studies on a transformation test for identification of *Acinetobacter (Mima and Herellea)*. *Applied Microbiology* 27, 1023–1026.

BRYN, K., JANTZEN, E. & BØVRE, K. 1977 Occurrence and patterns of waxes in Neisseriaceae. *Journal of General Microbiology* 102, 33–43.

BUCHANAN, R. E. & GIBBONS, N. E. 1974 *Bergey's Manual of Determinative Bacteriology* 8th edn. Baltimore: Williams & Wilkins.

CATLIN, B. W. 1964 Reciprocal genetic transformation between *Neisseria catarrhalis* and *Moraxella nonliquefaciens. Journal of General Microbiology* 37, 369–379.

CATLIN, B. W. 1970 Transfer of the organism named *Neisseria catarrhalis* to *Branhamella* gen. nov. *International Journal of Systematic Bacteriology* 20, 155–159.

CATLIN, B. W. & CUNNINGHAM, L. S. 1961 Transforming activities and base contents of deoxyribonucleate preparations from various neisseriae. *Journal of General Microbiology* 26, 303–312.

CATLIN, B. W. & CUNNINGHAM, L. S. 1964 Genetic transformation of *Neisseria catarrhalis* by deoxyribonucleate preparations having different average base compositions. *Journal of General Microbiology* 37, 341–352.

DE LEY, J. 1968 DNA base composition and taxonomy of some *Acinetobacter* strains. *Antonie van Leeuwenhoek* 34, 109–114.

DOI, R. H. & IGARASHI, R. T. 1965 Conservation of ribosomal and messenger ribonucleic acid cistrons in *Bacillus* species, *Journal of Bacteriology* 90, 384–390.

DUBNAU, D., SMITH, I., MORELL, P. & MARMUR, J. 1965 Gene conservation in *Bacillus* species I. Conserved genetic and nucleic acid base sequence homologies. *Genetics* 54, 491–498.

HENRIKSEN, S. D. 1952 *Moraxella*: Classification and taxonomy. *Journal of General Microbiology* 6, 318–328.

HENRIKSEN, S. D. 1973 *Moraxella, Acinetobacter,* and the *Mimeae. Bacteriological Reviews* 37, 522–561.

HENRIKSEN, S. D. 1976 *Moraxella, Neisseria, Branhamella,* and *Acinetobacter. Annual Reviews of Microbiology* 30, 63–83.

HENRIKSEN, S. D. & BØVRE, K. 1968*a Moraxella kingii* sp. nov., a haemolytic saccharolytic species of the genus *Moraxella. Journal of General Microbiology* 51, 377–385.

HENRIKSEN, S. D. & BØVRE, K. 1968*b* The taxonomy of the genera *Moraxella* and *Neisseria. Journal of General Microbiology* 51, 387–392.

HENRIKSEN, S. D. & BØVRE, K. 1976 Transfer of *Moraxella kingae* Henriksen & Bøvre to the genus *Kingella* gen. nov. in the family Neisseriaceae. *International Journal of Systematic Bacteriology* 26, 447–450.

HILL, L. R. 1966 An index to deoxyribonucleic acid base compositions of bacterial species. *Journal of General Microbiology* 44, 419–437.

HOLTEN, E. 1973 Glutamate dehydrogenases in genus *Neisseria. Acta Pathologica et Microbiologica Scandinavica* 81B, 49–58.

HOLTEN, E. 1974*a* Glucokinase and glucose 6-phosphate dehydrogenase in *Neisseria. Acta Pathologica et Microbiologica Scandinavica* 82B, 201–206.

HOLTEN, E. 1974*b* Immunological comparison of NADP-dependent glutamate dehydrogenase and malate dehydrogenase in genus *Neisseria. Acta Pathologica et Microbiologica Scandinavica* 82B, 849–859.

JANIK, A., JUNI, E. & HEYM, G. A. 1976 Genetic transformation as a tool for detection of *Neisseria gonorrhoeae. Journal of Clinical Microbiology* 4, 71–81.

JANTZEN, E., BRYN, K. & BØVRE, K. 1974*a* Gas chromatògraphy of bacterial whole cell methanolysates. IV. A procedure for fractionation and identification of fatty acids and monosaccharides of cellular structures. *Acta Pathologica et Microbiologica Scandinavica* 82B, 753–766.

JANTZEN, E., BRYN, K., BERGAN, T. & BØVRE, K. 1974*b* Gas chromatography of bacterial whole cell methanolysates. V. Fatty acid composition of neisseriae and moraxellae. *Acta Pathologica et Microbiologica Scandinavica* 82B, 767–779.

JANTZEN, E., BRYN, K., BERGAN, T. & BØVRE, K. 1975 Gas chromatography of bacterial whole cell methanolysates. VII. Fatty acid composition of *Acinetobacter* in relation to the taxonomy of Neisseriaceae. *Acta Pathologica et Microbiologica Scandinavica* 83B, 569–580.

JANTZEN, E., BRYN, K., HAGEN, N., BERGAN, T. & BØVRE, K. 1978 Fatty acids and monosaccharides of Neisseriaceae in relation to established taxonomy. *National Institute of Public Health Annals (Norway) (no. 2)*, 59–71.

JOHNSON, J. L., ANDERSON, R. S. & ORDAL, E. J. 1970 Nucleic acid homologies among oxidase-negative *Moraxella* species. *Journal of Bacteriology* **101**, 568–573.

JUNI, E. 1972 Interspecies transformation of *Acinetobacter*: genetic evidence for a ubiquitous genus. *Journal of Bacteriology* **112**, 917–931.

JUNI, E. 1974 Simple genetic transformation assay for rapid diagnosis of *Moraxella osloensis*. *Applied Microbiology* **27**, 16–24.

JUNI, E. 1977 Genetic transformation assays for identification of strains of *Moraxella urethralis*. *Journal of Clinical Microbiology* **5**, 227–235.

KINGSBURY, D. T. 1967 Deoxyribonucleic acid homologies among species of the genus *Neisseria*. *Journal of Bacteriology* **94**, 870–874.

KINGSBURY, D. T., FANNING, G. R., JOHNSON, K. E. & BRENNER, D. J. 1969 Thermal stability of interspecies *Neisseria* DNA duplexes. *Journal of General Microbiology* **55**, 201–208.

LAMACCHIA, E. H. & PELCZAR, M. J. 1966 Analyses of deoxyribonucleic acid of *Neisseria caviae* and other *Neisseria*. *Journal of Bacteriology* **91**, 514–516.

LAUTROP, H. 1974 Genus III. *Moraxella* Lwoff 1939 173. Genus IV. *Acinetobacter* Brisou and Prèvot 1954, 727. In *Bergey's Manual of Determinative Bacteriology* 8th edn. ed. Buchanan, R. E. & Gibbons, N. E. Baltimore: Williams & Wilkins.

LAUTROP, H., BØVRE, K. & FREDERIKSEN, W. 1970 A *Moraxella*-like microorganism isolated from the genito-urinary tract of man. *Acta Pathologica et Microbiologica Scandinavica* **78B**, 255–256.

LEWIS, V. J., WEAVER, R. E. & HOLLIS, D. G. 1968 Fatty acid composition of *Neisseria* species as determined by gas chromatography. *Journal of Bacteriology* **96**, 1–5.

OSHIMA, M. & MIYAGAWA, A. 1974 Comparative studies on the fatty acid composition of moderately and extremely thermophilic bacteria. *Lipids* **9**, 476–480.

PINTER, M. & DE LEY, J. 1969 Overall similarity and DNA base composition of some *Acinetobacter* strains. *Antonie van Leeuwenhoek* **35**, 209–214.

REYN, A. 1974 Genus I. *Neisseria* Trevisan 1885 105. In *Bergey's Manual of Determinaitve Bacteriology*, 8th edn. ed. Buchanan, R. E. & Gibbons, N. E. Baltimore: Williams & Wilkins.

RILEY, P. S., HOLLIS, D. G. & WEAVER, R. E. 1974 Characterization and differentiation of 59 strains of *Moraxella urethralis* from clinical specimens. *Applied Microbiology* **28**, 355–358.

SIDDIQUI, A. & GOLDBERG, I. D. 1975 Intrageneric transformation of *Neisseria gonorrhoeae* and *Neisseria perflava* to streptomycin resistance and nutritional independence. *Journal of Bacteriology* **124**, 1359–1356.

SNELL, J. J. S. & LAPAGE, S. P. 1976 Transfer of some saccharolytic *Moraxella* species to *Kingella* Henriksen and Bøvre 1976, with descriptions of *Kingella indologenes* sp. nov. and *Kingella denitrifcans* sp. nov. *International Journal of Systematic Bacteriology* **26**, 451–458.

SNELL, J. J. S., HILL, L. R. & LAPAGE, S. P. 1972 Identification and characterization of *Moraxella phenylpyruvica*. *Journal of Clinical Pathology* **25**, 959–965.

SPARLING, P. F. 1966 Genetic transformation of *Neisseria gonorrhoeae* to streptomycin resistance. *Journal of Bacteriology* **111**, 705–716.

Numerical Methods in the Classification and Identification of Bacteria with Especial Reference to the Enterobacteriaceae

DOROTHY JONES AND M. J. SACKIN

Department of Microbiology, University of Leicester, Leicester, UK

Contents

1 Introduction

THE APPLICATION of high speed electronic computers to taxonomic data has been one of the major advances in the fields of bacterial classification and identification over the past 20 years. The method is referred to as computer assisted taxonomy, taxometrics, Adansonian taxonomy or more usually as numerical taxonomy. A detailed discussion of the principles and procedures of the method is given by Sneath & Sokal (1973). Modifications and refinements are constantly being published and the whole field is still in active development.

Numerical taxonomy may be defined as the computer assisted numerical evaluation of similarities between groups of organisms and the ordering or arranging of these groups into clusters or taxa based on these similarities. On the basis of this clustering a classification can be constructed. The groups or taxa can be quantitatively defined and their relationships to each other measured by their intra- and inter-group similarity or dissimilarity values. As numerical taxonomy is based on phenetic evidence, i.e. on the similarities between organisms based on observable and recordable characters without regard to ancestry, the relationship between the organisms and any hierarchies constructed from these relationships are phenetic not phylogenetic.

The computer may also be programmed to print out lists of character values for the individual groups or taxa (as defined by the investigator) and from these the characters most useful for identification can be selected. This information can be used to construct different kinds of identification systems ranging from simple diagnostic keys to those which rely on computer storage of the data (see Sneath & Sokal 1973; Sneath 1978b).

Although numerical taxonomic methods have been used for studies on most of the major groups of organisms they have been found particularly useful amongst the bacteria. The paucity of morphological detail in bacteria coupled with the almost complete lack of phylogenetic data have led to the development of numerous tests based on physiological, biochemical, chemical and serological data for the classification of these organisms. The sheer volume of this material often makes its evaluation, other than by computer, very difficult if not impossible.

Numerical taxonomic methods were first applied to bacteria more than 20 years ago by P.H.A. Sneath (1957a,b). Since that time ca. 500 publications have appeared on numerical taxonomic studies of bacteria (see Colwell 1970, 1973; Sneath & Sokal 1973; Johnson et al. 1975; Sneath 1976a; Grimont et al. 1977; Austin et al. 1978; Jones 1978a for some references to the literature). These investigations have been of three main kinds:

I.   Studies of large groups of bacteria in an attempt to define taxa within the groups and to clarify the relationships of these taxa to each other, e.g. the enterobacteria (Johnson et al. 1975); the coryneform bacteria (Jones 1975); the nocardioform bacteria (Goodfellow 1971); the Gram negative, non-motile, coccoid rods (Thornley 1967).

II.   More restricted studies (not necessarily in numbers of organisms) which have been undertaken to investigate the taxonomic validity of certain accepted genera or species, e.g. streptococci (Colman 1968); propionibacteria (Malik et al. 1968); Corynebacterium renale (Yanagawa 1975).

III.   What may be termed ecological studies. These include the examination of large numbers of bacteria isolated from particular environments, either to assess the numbers of different bacterial forms present and identify them with already well-described taxa, or to establish their status as new taxa, e.g. soil bacteria (Goodfellow 1969); phylloplane bacteria (Austin et al. 1978) and seawater bacteria (Pfister & Burkholder 1965).

It would clearly be impossible to mention, much less evaluate, the impact of numerical taxonomic methods on the studies of all the bacterial groups in all these areas. The present paper is, therefore, confined to certain of the studies which have been carried out with the enterobacteria. Before these studies are considered, however, the steps involved and some of the computer methods available for numerical taxonomic studies of bacteria will be briefly described. Although the contributions to this book are more concerned with the impact of the various techniques on bacterial classification and identification than with the techniques themselves, the outcome of a numerical taxonomic study depends very much on the way in which the study is planned from the initial selection of the data, through the computer processing to the final evaluation.

2. Planning Numerical Taxonomic Studies

All numerical taxonomic studies, for whatever purposes they are designed, should be planned very carefully. The strength of the technique relies on the ability of computers to handle large amounts of data on large numbers of strains. Thus the taxonomic groups produced should be, and generally are, stable because they are based on large amounts of information and no single character is essential for membership nor sufficient to exclude a strain from a group. All phenetic characters reflect a portion, however small, of the genetic (chromosomal and plasmid) constitution of a bacterium. Therefore, theoretically, the more phenetic characters that are examined the better will be the measure of the phenetic and therefore probably the genetic relatedness between the bacteria.

Unfortunately, the capacity of computers to handle large quantities of data has led, in some instances, to the thoughtless processing of any sort of data from any collection of bacteria by whatever computer program is available, in the belief that the groupings so produced will form the basis of good and useful classifications. This approach indicates a lack of comprehension of the role of the computer. The computer works out the quantitative comparisons between organisms in respect of a large number of characters simultaneously in a manner which depends on the way it is programmed. In other words it does the arithmetic for us, but at high speed and without bias. Computers have by no means made the skill and experience of bacteriologists redundant.

The steps in a numerical taxonomic study are almost identical to the planning, execution and processes of deduction which the bacteriologist uses for conventional classification studies and should be monitored as carefully if not more carefully. There is always the danger that the definition of a group of bacteria by a numerical value (per cent similarity, %S) confers on that group a substance it does not necessarily possess. Factors that should be considered are: (a) the effects of the choice of clustering method (see below), and (b) the choice of S-value for defining the group. It is often advisable to choose an S-value that cuts a long branch of the phenogram, so that the group so defined is a relatively tight subgroup without strains near the edge.

The results of any taxonomic study, including numerical ones, must be evaluated against other available evidence on the organisms—genetic, serological, chemical—before the groups can be accepted. Various factors are known or suspected to affect the classification of bacteria in numerical taxonomic studies.

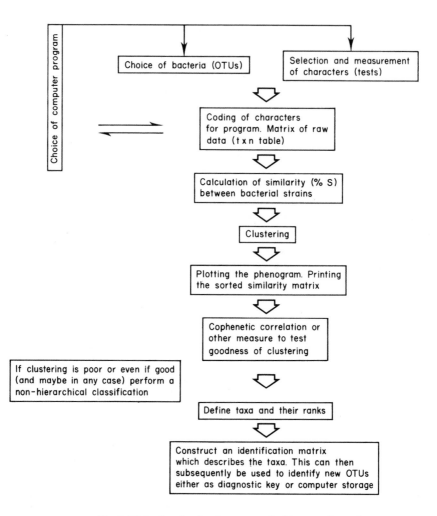

Fig. 1. Main steps in planning a numerical taxonomic study.

These include choice and number of strains and tests, test reproducibility and choice of computer program (see Sneath 1972, 1974, 1978a,b; Sneath & Johnson 1972; Sneath & Sokal 1973; Jones 1978a).

At the earliest stage in planning a numerical taxonomic study there should be close liaison between the bacteriologist and the computer scientist for reasons which will become apparent. The logical steps in planning a numerical taxonomic study are set out in Fig. 1.

A. *Choice of strains*

No common term is available for all the entities which can be classified by computer assisted techniques—strains, species, local populations, man-made objects, business enterprises—they are therefore called Operational Taxonomic Units (OTUs). The type of bacteriological study to be undertaken will obviously influence the kind and number of strains (OTUs) to be included. The capacity of the available computer will limit the amount of computer processing that can be performed and in particular also limit the number of OTUs. This should be checked at an early stage.

Whether the study is concerned with taxonomic revision of known forms or with the study of apparently new forms, the worker will need to obtain strains from other collections for comparative purposes. It is essential that reference strains (often referred to as 'marker' or 'predictor' strains) suitable to the study should be included. These should be type strains, i.e. reference strains for named species or subspecies. If type strains are not included, and it is surprising how frequently they are not, evaluation of the results against other numerical taxonomies or taxonomic studies based on other criteria is very difficult. Where type strains are not available, authentic representatives of species or subspecies should be included. Such strains should also be included, where possible, in addition to the type strains because it is known that in some cases the designated type strain is not typical of the taxon (see Krieg & Lockhart 1966). For the same reason it is also important to include new isolates in studies which are concerned with the taxonomic revision of known forms. Additional strains are also very useful because if several strains of a taxon cluster tightly it gives greater confidence to the conclusions.

It is desirable to choose, for comparison with the OTUs to be studied, not only representatives of those species thought to be present but also strains of taxa which are known or suspected to be related to those in the study. These strains help to orientate the study and also show if some of the collection belong to these other taxa. The choice of such strains is not always an easy exercise when the study is concerned with the revision of known forms and it can be very difficult in ecological type investigations. In ecological studies clustering of the new isolates with 'marker' or 'predictor' strains frequently does not occur. In these cases it is then necessary to reprocess the data with different 'marker' strains in the light of the information gained from the first study.

The cultures should be pure and this should be checked on a non-selective medium at periods throughout the study. The strains should preferably be maintained as frozen or freeze-dried cultures in addition to working stocks. It is known that continued subculture not only increases the risk of contamination, but also the properties of the bacteria can change through loss or gain of plasmids, and in the case of the streptomycetes, if heterokaryons occur in nature (see Hopwood 1967) these might segregate during the course of a study. To guard against these chances and to provide an internal check on reproducibility, a proportion, e.g. 10 strains in a study of 200, should be divided into two cultures and treated as two separate OTUs.

B. Choice of tests

The data required for classifying the bacteria to be studied are the properties of the organisms. These properties (or characters) are assessed by tests which are either observational or depend on the results of various techniques. The tests should be selected so as to reflect as much phenetic information as possible. The greater the number of tests performed the greater the amount of information which is obtained. In order to obtain enough information to be discriminatory at least 50 and preferably several hundred characters are desirable but with higher numbers the gain in information falls off disproportionately to the effort involved in obtaining the data (Sneath 1978a). However, abundance of characters is not sufficient: they should be of high quality. Bacteriological work is particularly subject to a variety of errors in addition to those which arise from contaminated or mishandled cultures.

The method of testing for a particular property should be chosen carefully. In some cases there are more than 20 different methods of performing the same test in the literature (see Skerman 1969). The test chosen should be repeatable and easily standardized. The composition of the basal medium, temperature of incubation, growth conditions (aerobic or anaerobic), time of reading and especially in rapid test methods the size of inoculum, are known to affect the results [see Sneath (1972) for a detailed discussion]. Many rapid testing methods have now been developed to yield a large quantity of data relatively quickly [see Sneath (1972) p. 46].

It is not possible to make generalizations about the sort of characters [with the possible exception of those based on morphological observations; see Wilkinson & Jones (1977)] which are the most influenced by lack of test reproducibility. Many tests are reliable with some groups of bacteria but not with others. It is now usual, but not universal, practice to repeat the tests on at least 25% of the strains in the study to assess which tests are particularly unreliable [see Lapage et al. (1970); Sneath & Johnson (1972) and Wilkinson & Jones (1977)].

C. Evaluation of raw data

When the data have been obtained they should be evaluated carefully before proceeding further. As noted earlier, sophisticated computing techniques can in no way correct for poor quality data. The results obtained from the repeat tests should be carefully scrutinized, as should be those from the split cultures. Highly unreliable tests should be dropped but before doing this the bacteriologist should balance carefully the effect of the lack of test reproducibility on the clustering of the strains [see Sneath & Johnson (1972) for discussion] against the information lost by the omission of these results for all strains [see Wilkinson & Jones (1977)]. A statistical technique, Analysis of Variance [see Sneath (1972) p. 47] can be valuable in detecting unsuspected sources of error [see Sneath & Collins (1974)].

Although inconstant characters are best avoided for obvious reasons, there are some characters which are inconstant because they are plasmid borne. It is obviously not possible to be certain of these characters in a particular study. It is also debatable whether or not they should be dropped even if it is known they are plasmid borne. Currently there is not enough information on the effect of such characters on classifications (see Harwood, Chapter 2) but where information is available (McKell 1977) the effect does not seem to be great. As discussed by Jones (1978b), however, the selection of such characters for identification purposes could have adverse effects if a simple diagnostic key is employed rather than a computer stored identification matrix.

3. Numerical Methods of Classification

A. Coding of data

For all but the smallest sizes of data it is quite impracticable to apply numerical taxonomic and related methods by hand, and a computer program is necessary. Many different programs have been written, but their usual starting point is a $t \times n$ matrix—a table of the number (t) of OTUs (Operational Taxonomic Units—in Microbiology, usually strains) each described in terms of the number (n) of characters in the study. [Our terminology in general follows that of Sneath & Sokal (1973)]. It is, however, very important to decide early on which variety of program is to be used. One major area of variation among programs is in the types of character that are permitted. Thus, some programs require all characters to be 2-state, i.e. 'all or none' characters, e.g. result negative, result positive—often scored respectively as 0, 1. Some programs allow multistate characters, e.g. negative, weak positive, strong positive, typically scored respectively as 0, 1, 2. Indeed, the program may allow characters with more than 3 states. Further, many programs allow missing-data entries, termed NC ('no comparison'). This is a very useful feature. NCs may arise for a variety of logical or practical reasons [Sneath & Sokal (1973) p. 178]. Although many missing entries present difficulties in the interpretation of the classification results, a few missing entries are often unavoidable, and it would be unfortunate to have to exclude a whole character or OTU just because it contained one or two missing entries.

The final $t \times n$ matrix will normally be punched for example onto cards for the program to read. It is very important to ascertain exact details of the layout of the data before writing it out for punching. Most professional punch operators require data to be written out on supplied coding forms. The user may have a computing colleague—perhaps even the writer of the program himself—who can help with these matters. Indeed, for the microbiologist unfamiliar with computing it is very important to consult with a biologically-oriented computer expert or a computer-minded biologist at all stages from the selection of OTUs and characters, *via* the choice of program and coding of the data, through to the running of the data and the interpretation of the results.

B. Hierarchical classification methods

Most computer classifications in microbiology are *hierarchical,* i.e. nested, and they are also non-overlapping, i.e. taxa at any one rank are mutually exclusive. Not all data sets, however, lend themselves to such an arrangement, and it is strongly recommended that any hierarchical method used should include within itself a test for the hierarchicalness of the data (see *Interpretation of clustering results* below). If the data fit a hierarchical structure reasonably well then a hierarchical classification has much to recommend it, simply on grounds of working ease of the resulting taxonomy. Whether or not a hierarchical method is used it is probably best that the method is non-overlapping. Overlapping methods [e.g. in Jardine & Sibson (1968)], despite some theoretical elegance, yield results too complex for the purposes of most practising taxonomists.

In general, hierarchical classifications are performed in two stages:
(I) starting with the $t \times n$ matrix, calculating a measure of resemblance between every pair of OTUs, resulting in a triangular *resemblance matrix,* and
(II) using the resemblance matrix to cluster the OTUs into a hierarchical structure.

(i) Computation of resemblance

Many different coefficients of resemblance have been devised [Sneath & Sokal (1973) pp. 116 *et seq.*]. However, only a few have found general favour in microbiology, and they are presented here.

Simple matching coefficient, S_{SM}. The simple matching coefficient is a very commonly used coefficient for use on 2-state data only. It is simply the proportion of characters that have the same state (both negative or both positive) in the pair of OTUs being compared. Any characters for which either or both OTUs have a missing entry (NC) are ignored.

Jaccard's coefficient, S_J.   Jaccard's coefficient is the same as S_{SM} except that matching negative results are ignored, i.e. treated the same as a comparison involving an NC entry. The choice between S_{SM} and S_J is somewhat arbitrary. S_J may be preferred when classifying slow-growing organisms, i.e. when there are many negative results in the data. Under these circumstances a pair of matching negatives may seem much less of a similarity than a pair of matching positives. At the same time there is no reason to treat a pair of matching negatives as a *dis*similarity; hence the strategy of ignoring them altogether in the computation of S_J. On the other hand valuable information may be lost when using S_J rather than S_{SM}.

Gower's coefficient, S_G.   This is a generalization of both S_{SM} and S_J to allow for multistate characters. Whereas in S_{SM} and S_J each character contributes either a similarity or no similarity, in S_G a character may contribute a partial similarity. For example, if a given character can have states (a) negative, (b) weak positive, (c) strong positive, coded, say, as 0, 1, 2 respectively, then a pair of identical states will yield a whole similarity: a 0 compared with a 2 will give no similarity, and a 0 compared with a 1, or a 1 compared with 2, will give a half similarity. Two-state characters may be treated either like S_{SM} or like S_J. As with S_{SM} and S_J all the resulting similarity contributions are averaged over all the characters. S_G does not allow matching zero states to be ignored in multistate characters, and in practice, in studies where there is a mixture of two-state and multistate characters it is common to count matching zeros (i.e. as in S_{SM}, not S_J) in the two-state characters as well. S_G also allows non-ordered multistate characters—e.g. colour, with states red, brown, green—in which a match (e.g. same colour) counts as a whole similarity and a mismatch counts as no similarity. Thus there are no partial similarities. These characters are rarely of microbiological application, and it is usually more satisfactory to break them down into several 2-state characters [Sneath & Sokal (1973) pp. 149–152]. A formal definition of S_G is given in Gower (1971).

Pattern difference, D_P.   There are two pattern difference coefficients, analogous respectively to S_{SM} and S_J. They attempt to measure the difference between a pair of OTUs but leaving out the component of the difference that is due merely to differences in vigour. Both these coefficients are defined for 2-state characters only, and the vigour of an OTU is simply the number of its characters that show positive results. The vigour difference, D_V, between two OTUs is the difference between the two vigours, averaged over all the characters (and for the S_J-based measure ignoring matching negatives), and the pattern difference coefficient D_P is defined according to the formula:

$$D_P{}^2 = D_T{}^2 - D_V{}^2$$

where D_T (total difference) is $1 - S_{SM}$ or $1 - S_J$ depending on which of the two measures is being used. As with the other coefficients, any characters that are NC for one or both of the OTUs are ignored [See Sneath (1968) and Sneath & Sokal (1973) pp. 176, for a discussion on the S_{SM}-based measure].

It will be noticed that S_{SM}, S_J and S_G are similarity coefficients but D_P is a difference or dissimilarity coefficient. Each of the three similarity coefficients can range from 1 (complete similarity) down to 0 (complete dissimilarity). For D_P the range is reversed—0 for identical OTUs, 1 for a maximally dissimilar pair. This presents no problems for clustering, which will always group together OTUs with a mutually high similarity or resemblance.

It will be noted that all the coefficients here described give equal weighting to the characters in accordance with Adansonian principles (after the eighteenth century botanist Michel Adanson who followed similar principles). The logical basis for this approach is the notion that each character contains the same amount of information and in any case that any *a priori* character weighting would be uncomfortably subjective.

(ii) *Clustering*

Most clustering methods begin by finding the pair of closest OTUs. (Some variants of the method, though preferred in some respects, are sensitive to the choice of pair if more than one pair ties for highest similarity). This pair then forms a group or cluster. The similarities between this group and each of the remaining OTUs are computed and the process cycles back, treating the newly-formed group as an OTU, to find the new highest similarity. The new highest similarity may thus be either between two single OTUs or between an OTU and the group just formed during the first clustering cycle. The process finishes after $t - 1$ cycles when all the OTUs will have joined to form one group.

Variants of the method arise over the definition of the similarity between an OTU and a group and in general between two groups. The commonly used *single linkage* defines the similarity between two groups as the similarity of the two most similar OTUs, one in each group. *Average linkage* takes the average of all the similarities across the two groups. The average linkage method itself has variants. For most purposes *unweighted average linkage* (UPGMA) is recommended. Here the simple arithmetic average of the similarities across the two groups is taken, each similarity having equal weight. In *weighted average* each of the two subgroups forming the group has equal weight [for details see Sneath & Sokal (1973) pp. 218, 234]. Other variants of average linkage are *centroid sorting* and *median sorting* (pp. 218). A minimum variance method (Delabre 1973) is discussed below, under *Numerical Taxonomic Studies of Some Enterobacteria*.

The results of clustering are normally presented in the form of a *dendrogram* or tree diagram. Where, as here, the dendrogram represents the results of a phenetic classification, it is called a *phenogram,* to distinguish it from the various kinds of phylogenetic trees or *cladograms* (Sneath 1974). In a phenogram the tips of the branches represent the OTUs, and the axis at right angles to the tips is the similarity axis, and it shows the similarity values at which the groups form. Many clustering programs plot the phenogram automatically. It is also useful to print the whole similarity matrix with the OTUs in the same order as along the tips of the corresponding phenogram.

An alternative way of representing the results of a single linkage clustering is by way of a *minimum spanning tree.* In fact a minimum spanning tree tells more than a single linkage phenogram; for each pair of groups that join it tells which pair of OTUs, one in each group, has the highest similarity, i.e. which pair of OTUs under the single linkage criterion was responsible for the two groups joining.

(iii) *Interpretation of clustering results*

As mentioned above, it is important to examine to what extent the data are indeed suitable for hierarchical clustering. Intuitively the data are suitably hierarchical if a phenogram can be constructed in which most or all of the resemblance coefficients between the OTUs have values close to the corresponding values (the *cophenetic values*) derived from the phenogram alone. The cophenetic value between two OTUs is the resemblance level at which their two constituent groups join, and it is thus the lowest similarity value that is passed when travelling along the tree between the two corresponding tips. The most common measure of hierarchicalness is the *cophenetic correlation coefficient,* which is the correlation coefficient between these two sets of values. It can be proved that UPGMA clustering always gives the highest cophenetic correlation of all possible clustering methods of the kind described in the first paragraph above (Farris 1969). This in itself is a strong recommendation of this method.

Hierarchical clustering is unsatisfactory, for example, when intermediate groups occur amongst the strains being clustered, or more generally when straggly clusters occur. The clustering sequence with which a straggly cluster forms in UPGMA may suggest sub-clusters to which less credence should be given than a cursory examination of the phenogram would suggest. In single linkage the whole cluster would join up over a narrow similarity range. Although it can be recommended that UPGMA be used because of its property of maximizing the cophenetic correlation, the interpretation of the results should be backed up by the corresponding single linkage clustering and by a detailed examination of the UPGMA sorted similarity matrix. If the cophenetic correlation from the UPGMA method is 1—i.e. perfect correlation—then it will also be 1 for the single linkage method and the two results will be identical. The lower the UPGMA cophenetic correlation the more different will be the two trees and the more unsatisfactory will be a hierarchical scheme to describe the data, but detailed examination of the two trees will indicate which areas of the trees fare worst, and this may lead to ways of improvement. For example, a poor result on S_{SM} coefficients may lead to a re-examination of growth rates and a trial with D_P. Typical cophenetic correlation coefficient values are in the range 0.6 to 0.95. Above about 0.8 is usually reasonably good; below about 0.7 is sufficiently poor for no more than limited credence to be given to the hierarchical structure.

For completeness it should be added that higher cophenetic correlation can sometimes be obtained if 'inversions' are allowed, i.e. if not necessarily the two closest groups join in a given clustering cycle. However, inversions are not very desirable for most taxonomic purposes.

(iv) Nomenclature and ranking

The results of a hierarchical clustering may be used to give names to bacterial groups and to help decide on their taxonomic rank. The question of which groups one chooses to name and how the ranks relate to the similarity levels between the groups, are related. For full discussion see Sneath & Sokal (1973). A numerical taxonomy may also allow one to refine previous classifications.

C. Non-hierarchical methods

Non-hierarchical methods are useful when the data are not suitable for hierarchical analysis and also in their own right. Within the Enterobacteriaceae, Grimont & Dulong de Rosnay (1972) have used both hierarchical and non-hierarchical methods for classifying *Serratia*. In the non-hierarchical method they represented the groups in the form of *taxonomic maps*. In this method (Carmichael & Sneath 1969) the groups are represented as circles whose diameter is proportional to the two most distant OTUs in the group or—in a variant of the method which they also use (Liston *et al.* 1963)—proportional to the complement of the mean intra-group similarity. Taxometric maps are informative but they must be drawn by hand, leaving to the investigator arbitrary decisions on the positioning of the groups to maximize the extent of the inter-group relationships that are depicted. Clearly not all the relationships can in general be represented in two dimensions.

D. Ordination

Like taxometric maps this method—or group of methods—is designed to represent the data in a reduced number of dimensions, say 2 or 3. Unlike taxometric maps, however, most ordination methods are fully automatic, but they do not classify the OTUs into groups. They simply form an optimal representation of the relationships between the OTUs in a specified number of dimensions. The most widely used ordination method is Principal Components Analysis, together with its close relative Principal Coordinates Analysis.

E. Identification

A central application of any satisfactory classification is Identification which is the process of determining to which group or taxon a new OTU belongs. The new OTU will normally have been examined only after the groups have been formed by a classificatory process. The most reliable methods of microbial identification are computer-based. Such non-computer methods as keys are more portable but are subject to big identification errors if there is one mistake in coding up the test results on the unknown. Actually there are key-generating computer programs, used mainly in botany, which are designed to minimize the likelihood of such errors, but most favoured identification methods in microbiology work on Adansonian principles and thereby avoid the identification hinging on one particular test result.

A number of computer identification programs exist. Typically, information on the groups is stored away as an *identification matrix* in a form to which the program has access. There will also be facilities to update an identification matrix (for a case study on coryneform bacteria see Hill *et al.* 1978). For good identification the groups should be reasonably tight and well separated from one another. Sneath (1977) describes measures of overlap between groups in relation to their tightness.

F. Comparisons of classifications

Various computer techniques are available to compare pairs of classifications on the same set of OTUs or even classifications where some of the OTUs appear in only one of the two sets [see especially Bobisud & Bobisud (1972), Boorman & Olivier (1973), Farris (1973) and Rohlf (1974)]. Some of these cited methods compare phenograms; others compare non-hierarchical groups (Sackin & Jones, unpublished). Some of the methods for comparing phenograms compare the topologies only; others [notably those in Boorman & Olivier (1973)] also use the actual similarity values at which the clusters form. These methods, though not much used in microbiology so far, are very useful in comparing different classification methods, and they can even be used to compare a numerical classification with a classical, non-numerical one.

4. Numerical Taxonomic Studies of Some Enterobacteria

Some studies with the enterobacteria will be considered in an attempt to illustrate the points discussed in the preceding section and to evaluate the importance of computer assisted methods in the classification and identification of these bacteria.

The family Enterobacteriaceae comprises a large, diverse group of Gram negative, rod-shaped bacteria which when motile have peritrichous flagella. Their metabolism is both respiratory and fermentative and acid is produced from the fermentation of glucose and other carbohydrates. Not all the members of the family are intestinal parasites and certainly not all are pathogenic to their hosts. However, the group has been, and is, of wide interest to medical, veterinary and agricultural bacteriologists because it includes important pathogens for man, animals and plants. The taxonomy of the family has been influenced strongly by clinical workers who require rapid, convenient identification methods and in many cases the clinically important strains can be characterized in such detail that the source of an epidemic or epizootic can be determined accurately.

There is no consensus of opinion on the taxa which should be included in the family Enterobacteriaceae. Edwards & Ewing (1972) and Cowan (1974) recognized five major groupings (Tribes) within the family but the authors differ both on the names of the Tribes and their generic composition (see Table 1). The

TABLE 1

Composition of the family Enterobacteriaceae

Edwards & Ewing (1972) Tribe Genus		Cowan (1974) Tribe Genus	
(1)	Escherichieae	(1)	Escherichieae
	Escherichia		*Escherichia*
	Shigella		*Edwardsiella*
(2)	Edwardsielleae		*Citrobacter*
	Edwardsiella		*Salmonella*
(3)	Salmonelleae		*Shigella*
	Salmonella	(2)	Klebsielleae
	Arizona		*Klebsiella*
	Citrobacter		*Enterobacter*
(4)	Klebsielleae		*Hafnia*
	Klebsiella		*Serratia*
	Enterobacter	(3)	Proteeae
	Pectobacterium		*Proteus*
	Serratia	(4)	Yersinieae
(5)	Proteeae		*Yersinia*
	Proteus	(5)	Erwinieae
	Providencia		*Erwinia*

genera *Aeromonas, Vibrio* and *Photobacterium* are usually excluded from the family Enterobacteriaceae because of their polar flagella and other characters [see Cowan (1974)].

A. Numerical taxonomic studies of whole group

Owing to the large size of the family grouping it is hardly surprising that there have been very few numerical taxonomic studies of the whole group. Krieg & Lockhart (1966) studied a relatively small number (53) of strains representing 12 genera of the family Enterobacteriaceae (*Arizona, Citrobacter, Enterobacter, Erwinia, Escherichia, Klebsiella, Paracolobactrum, Proteus, Providencia, Salmonella, Serratia, Shigella*) together with four members of the genus *Aeromonas*. At least two species from most of the listed genera were included. For 18 of the species, the type, neotype or proposed neotype strains were used. A total of 105 characters was determined for each organism (OTU). The coded data were analysed by simple matching coefficient and single-linkage clustering. However, a variant of the clustering was used in that all the similarities at which the clusters were formed were rounded down to the nearest multiple of 5%. This rounding process economizes on computer storage, and it was used in many earlier numerical taxonomy programs and is still used in some programs designed to handle very large sized data. However, this practice can blur much detail in the results (Bascomb *et al.* 1971), especially when the rounding step is as large as 5%. It also compounds any test errors which in this study are unassessed. (They are unassessed in all the studies discussed in this section.) The results indicated a single large cluster formed at the 90% similarity level in the single linkage phenogram and comprising representatives of the genera *Enterobacter, Escherichia, Salmonella, Arizona, Citrobacter* and *Shigella*. Within this cluster the results showed some internal structure. All the *Salmonella* strains, with the exception of *Salm. typhi*, formed a distinct subcluster at the 95% level. Representatives of the genera *Aeromonas, Klebsiella, Paracolobactrum, Proteus, Providencia, Serratia* and *Erwinia* (except the strain of the type species of the last genus), fell outside the single large grouping and appeared as distinct taxa, as did the one strain of *Shgl. dysenteriae*.

On the basis of this study Krieg & Lockhart (1966) concluded that most of the family Enterobacteriaceae (i.e. the genera *Enterobacter, Escherichia, Arizona, Citrobacter, Shigella* and *Salmonella*) appear to consist of a spectrum of related organisms and these authors were of the opinion that their results provided no justification for the separation of this portion of the family into numerous species organized into genera and tribes.

Although designed primarily as a study of the classification of the tribe Klebsielleae, so many representatives of other enterobacteria were included that the study of Bascomb *et al.* (1971) deserves mention. The data comprised 177 strains on which 50 tests were performed. All the tests were coded as ordered multistate characters, and Gower's similarity coefficients were computed. Clustering was by median sorting and single linkage. After comparing the two clusterings the authors depicted taxa mainly on the basis of the median sorting which was performed at rounding steps of 2½% similarity. With the exception of four strains (including two simulated *Kluyvera* strains) the remainder clustered at or above the 80% level. Within this main grouping the tribe Klebsielleae appeared as a distinct cluster (85%S) containing subclusters corresponding to the traditional taxa *Enterobacter, Klebsiella, Chromobacterium typhiflavum* and *Serratia*. Another distinct cluster was formed by strains of *Hafnia alvei* (87.5%S), and a third cluster comprised strains of *Escherichia coli, Edwardsiella tarda, Salm. typhi, Shigella* spp. and Alkalescens-Dispar group (82.5%S). The %S values are the similarity values at which the respective clusters formed.

The next main numerical taxonomic survey of members of the family Enterobacteriaceae was that of Colwell *et al.* (1974). These authors examined 105 strains of the genera *Citrobacter, Enterobacter, Edwardsiella, Erwinia, Escherichia, Hafnia, Klebsiella, Proteus, Salmonella, Serratia* and *Shigella* (representing the family Enterobacteriaceae) together with strains of the genera *Vibrio* and *Aeromonas*. Only six culture collection strains were included as reference cultures. Jaccard coefficients were calculated for 128 characters and single linkage clustering was performed. Thus, in respect of the clustering, the method of computation was similar to that used by Krieg & Lockhart (1966) albeit with 1% clustering steps rather than 5%. Unlike 5% steps, 1% steps lose very little detail. The results of the study sorted the strains into three main areas corresponding to (1) *Edwardsiella tarda* (90%S); (2) six clusters viz. *Escherichia coli*–Alkalescens–Dispar (79%S), *Citrobacter freundii* (87%S), the tribe Klebsielleae (82%S), *Serratia marcescens* including *Enterobacter liquefaciens* (84%S), *Hafnia alvei* (77%S) and *Shigella* spp. (79%S); (3) *Vibrio alginolyticus* (87%S). The three main clusters joined at 64%S.

A later study by Johnson *et al.* (1975) involved a large number (384) of strains representing the majority of the genera in the family Enterobacteriaceae (Edwards & Ewing 1972) and members of the genus *Yersinia*. The data were processed by the same computer techniques as those used by Colwell *et al.* (1974). The majority of the strains clustered into three main groups which together contained 32 clusters. Group A corresponded to the tribe Klebsielleae (Edwards & Ewing 1972) and consisted of the genera *Enterobacter, Klebsiella* and *Serratia* (77%S). Group B comprised the tribes Edwardsielleae, Salmonelleae and Escherichieae (Edwards & Ewing 1972) together with members of the genus *Yersinia* (80%S). Group C corresponded to the tribe Proteeae (Edwards & Ewing 1972) and contained clusters corresponding to members of the genera *Proteus and Providencia* (82%S); Another cluster composed of strains of *Hafnia alvei* joined Groups B and C at 77%S.

For identification purposes, the 32 clusters contained in the Groups A, B and C together with the cluster of *Hfna alvei* strains were differentiated by 56 characters [although only 50 of these are listed in Table 2 of Johnson *et al.* (1975)]. As the characters had been chosen subjectively, the authors were aware that it could be argued that a completely different arrangement might be obtained if the strains were clustered on the basis of these characters alone. However, rather than reprocessing all the strains the authors derived hypothetical median strains (HMSs) from each of the 33 clusters. A data matrix comprised of these 33 HMSs scored for 56 characters was then analysed by the simple matching coefficient rather than Jaccard and clustered by single linkage. In the event there was little difference in the relationships between the clusters as depicted in the phenogram arising from this process and in the previous phenogram based on all the strains. However, there were some differences [see Johnson *et al.* (1975), Figs. 1 & 6].

This study involved a great deal of work but, in addition to the lack of internal monitoring for test error and test reproducibility, it is open to criticism on three main counts: (1) Very few reference strains were used: (2) The data contained a high number of 'non-comparable' entries due to missing characters for many of the strains: (3) If the simple matching coefficient was preferred in the computation of the data based on the 33 HMSs why was it not used on the bulk data?

A later study from the same laboratory (Sakazaki *et al.* 1976) included, in addition to representatives of the genera examined by Johnson *et al.* (1975), representatives of more recently described species of the genera *Levinea*, *Citrobacter* and *Enterobacter*. The data were analysed by the simple matching coefficient, and clustering was by single linkage. In general the results obtained were similar to those of Johnson *et al.* (1975), but to test the stability of the clusters formed by those strains not included in the previous study (Johnson *et al.* 1975), all of the strains were combined with a further 196 strains from the previous study. Computer analysis of the new total of 337 strains again indicated a similar arrangement of clusters to that described by Johnson *et al.* (1975).

The numerical taxonomic study of representative strains of the family Enterobacteriaceae and some strains of the family Vibrionaceae by Véron & Le Minor (1975) is of interest because the characters used to examine the strains were solely the ability to use 146 organic substrates as the sole source of carbon and energy. The data were analysed by a dissimilarity (or distance) coefficient equal to 1-S_J and the minimum variance clustering method of Delabre *et al.* (1973). For background to the clustering method see Benzécri (1973). In this method the formula for the resemblance between groups is designed so as to minimize the gain in intra-group variance at each clustering cycle. In fact, the distance D_{JK} between two groups J and K is given by:

$$D_{JK}^2 = \frac{t_J\,t_K}{t_J + t_K} \sum_{j=1}^{t_J} \sum_{k=1}^{t_K} d_{jk}^2$$

where d_{jk} is the distance (i.e. the value of $1\text{-}S_J$) between OTU j of group J and OTU k in group K, and t_J, t_K are the numbers of OTUs in the two groups. The groups formed by this method tend to be at least as homogeneous and well separated as in unweighted average linkage (Delabre *et al.* 1973), but the resemblances (here the distances D_{JK}) bear a less close relationship to the resemblances of the constituent OTUs across the groups than in average or, in practice, single linkage. In particular, the distance between large groups (t_J and t_K large in the formula) will tend to be much larger than the distances between most pairs of OTUs, one in each group. Hence the method is more reluctant to join up large groups than isolated OTUs, so that the resulting phenogram tends to be more symmetric—i.e. it contains groups of roughly equal size at a given resemblance level, and with few isolated OTUs—than in average and especially single linkage. Although symmetric phenograms are usually more convenient to work with than skew (i.e. non-symmetric) ones (Sackin 1972), the tendency of this method to absorb isolated OTUs into clusters at an early stage of the clustering is liable to be misleading.

Véron & Le Minor (1975) scaled the distances at which clusters formed to yield dissimilarity values ranging from 100% dissimilarity at the level at which the final join took place to 0% dissimilarity at the tips. The 100% dissimilarity level partitioned the strains into two distinct areas. One comprised the genera *Aeromonas* and *Vibrio* (exhibiting 45% dissimilarity from each other) which joined the genera *Proteus* and *Providencia* at the 50% dissimilarity level. All four of these genera joined the genus *Shigella* at approximately 75% dissimilarity. The other area contained a cluster comprising *Klebsiella, Enterobacter, Serratia, Hafnia* and *Levinea* (*ca.* 25% dissimilarity) which joined a cluster comprising *Klebsiella–Enterobacter* and *Yersinia* strains (internal dissimilarity 23%) at slightly greater than 54% dissimilarity. Both these clusters joined the remaining main clusters (comprising *Escherichia* and *Salmonella* strains internal dissimilarity 25%), at approx. 60% dissimilarity. The results of Véron & Le Minor (1975) are in good accord with the other studies, particularly in the fine structure of the groups. The main differences can probably be explained by the differences in the clustering method, in particular the relative reluctance of the minimum variance method to join up large groups and its relative readiness to join isolated OTUs.

The study of Delabre *et al.* (1973) was an investigation of aquatic bacteria which included only a few enterobacteria. It is interesting to note that Delabre *et al.* (1973) used $1\text{-}S_{SM}$, not $1\text{-}S_J$, as distance measure, for similar reasons to ours for preferring usually to count negative matches.

B. Numerical taxonomic studies of certain groups within the Enterobacteriaceae

There have been a number of numerical taxonomic studies of 'groups' of bacteria in the family Enterobacteriaceae, e.g. the tribe Klebsielleae (Bascomb *et al.* 1971), the genera *Erwinia* (Lockhart & Koenig 1965), *Obesumbacterium* (Priest *et al.* 1973), *Proteus* (McKell & Jones 1976), *Salmonella* (Lockhart & Holt 1964), *Serratia* (Colwell & Mandel 1965; Grimont & Dulong de Rosnay 1972; Grimont *et al.* 1977), and *Yersinia* (Stevens & Mair 1973). Studies of only two genera will be described here because space does not allow all the studies to be treated in detail. The studies, those of Grimont and his colleagues (Grimont & Dulong de Rosnay 1972; Grimont *et al.* 1977) with the genus *Serratia,* and that of McKell & Jones (1976) with the genus *Proteus,* have been chosen because in several ways they illustrate points of importance in numerical taxonomic studies.

(i) *The genus* Serratia

As many as 42 species names have been associated with the genus *Serratia* and 27 species were listed in the 4th edition of *Bergey's Manual of Determinative Bacteriology* (1934). However, Ewing *et al.* (1959) recommended that only one species, *Sera. marcescens,* be recognized. This view was supported by a number of workers including Colwell & Mandel (1965), who based their views on the results of a numerical taxonomic study of 33 strains of the genus *Serratia* despite the fact that four of their strains differed quite markedly from the homogeneous *Sera. marcescens* cluster (Colwell & Mandel 1965). The classification of Ewing *et al.* (1959) is retained in the 8th edition of *Bergey's Manual of Determinative Bacteriology* (Buchanan & Gibbons 1974) although other studies had indicated that there were more than one species in the genus (Brisou & Cadeillan 1959; Fulton *et al.* 1959; Grimont 1969; Bascomb *et al.* 1971).

Mainly because they could not identify some red-pigmented isolates with *Sera. marcescens,* Grimont & Dulong de Rosnay (1972) used numerical taxonomic methods to investigate the relationships between 60 strains of red-pigmented and non-pigmented bacteria which resembled *Sera. marcescens.* The strains were coded for 55 characters, and Jaccard similarity coefficients were computed. Clustering was by weighted average linkage and single linkage. From the clustering results taxometric maps were constructed (see *Non-hierarchical methods,* pp. 84). All but one of the strains clustered into three distinct groups, A, B and C, in both the phenograms. The two taxometric maps also indicated the same overall relationship between the three groups. Strains in group A resembled *Sera. marcescens* of Ewing *et al.* (1959, 1962), the single homogeneous cluster of Colwell & Mandel (1965), biotype 1 of Bascomb *et al.* (1971) and, except that cellobiose was not readily fermented, strains of pattern 1 of Fulton *et al.* (1959). Strains which clustered in group B, although differing in a number of respects, most closely resembled those strains designated biotype II by Bascomb *et al.* (1971). Group C strains resembled those of pattern 2 of Fulton *et al.* (1959). However, as noted by Grimont & Dulong de Rosnay (1972), if pigment formation was discounted the strains in group C most closely resembled *Enterobacter liquefaciens* as described by Edwards & Ewing (1962).

An omission in the design of the study of Grimont & Dulong de Rosnay (1972) was the lack of authentic reference strains. In an attempt to rectify this omission, the authors compared the properties of three reference strains with each of the three groups. *Serratia marcescens* (NCTC 1377) corresponded to group A; *Serratia* sp. (NCTC 9493) corresponded to group C; the third strain (which they received from P.H.A. Sneath as D119, *Sera. marinorubra,* type strain) corresponded to group B.

Grimont & Dulong de Rosnay (1972) also commented on the number of subgroups in group A in particular. They noted that evidence from other studies in their laboratory indicated that the subdivision of group A (*Sera. marcescens*) into subgroups might well be due to the presence of plasmids in some strains— e.g. the majority of subgroup A_1 strains carried R factors.

A further numerical taxonomic study was undertaken (Grimont *et al.* 1977) in an attempt to confirm the validity of the groupings (species) noted in the genus *Serratia* (Grimont & Dulong de Rosnay 1972), to define these species and also to circumscribe the genus *Serratia.* Grimont *et al.* (1977) examined 156 strains labelled *Serratia* spp. and related bacteria including *Enterobacter liquefaciens, Enbc. cloacae, Enbc. aerogenes, Erwinia caratovora, Erwn. chrysanthemi, Erwn. herbicola* and *Erwn. nimipressuralis,* for 233 characters. Authentic reference cultures were included. The coded data were analysed by the simple matching coefficient and by both unweighted average and single linkage clustering. The same clustering methods were performed on the pattern difference coefficient D_P (see pp. 81). The results indicated four groups of species rank within the genus *Serratia.* These corresponded to *Sera. marcescens, Sera. marinorubra* and *Sera. plymuthica* which all produce prodigiosin, and one colourless species *Sera. liquefaciens.* The % G+C values for each group are listed. A summary of properties allowing differentiation between the species in the genus *Serratia* and between *Serratia* and other groups (*Erwinia, Enterobacter, Klebsiella* and pectobacteria) are given.

The unweighted average and single linkage methods gave approximately the same results. The pattern difference coefficient did not greatly alter the resemblances between strains. There were, however, a few minor changes within the main groups. The main point of interest highlighted by use of the D_P coefficient is that the strain *Sera. marinorubra* (NCTC 10912) did not cluster with strains of group B when the simple matching coefficient was used, but when D_P was used it clustered as a typical strain of this group. This strain has low vigour compared with group B strains, but when the vigour difference is discounted its 'pattern' of test results is typical of this group.

The earlier paper (Grimont & Dulong de Rosnay 1972) used weighted average linkage, whereas the later one (Grimont et al. 1977) used unweighted average linkage. This change makes the results difficult to compare in detail. Results on the two methods tend in practice to be very similar [see examples in Sneath & Sokal (1973)], but the weighted method gives more weight to the most recent arrival in a cluster. Sneath (1976b) has demonstrated the effects on the unweighted method of reduplicating one OTU varying numbers of times. Such duplications will have no effect on the weighted method, as is immediately seen from the definition of the method in which stems rather than OTUs have equal weight.

The study of Grimont et al. (1977) is a good example of a well-designed numerical taxonomic study. The study was devised after all the evidence from previous pertinent numerical (Bascomb et al. 1971; Grimont & Dulong de Rosnay 1972) and non-numerical studies (Brisou & Cadeillan 1959; Fulton et al. 1959; Hamon et al. 1970; Ewing et al. 1972, 1973) had been taken into account. Suitable reference strains and fresh isolates from a number of sources were used. The characters were chosen on as broad a basis as possible. Clustering was performed after the similarity between strains had been assessed with and without allowance for differences in vigour between the strains. The study (i) resulted in the recognition of four groups worthy of species rank in the genus Serratia; (ii) yielded tables of characters suitable for the differentiation of these taxa and therefore for the identification of new isolates; and (iii) indicated the possible effect of plasmids on the classification of these bacteria. In addition, the study raised some interesting points of nomenclature.

The results of the numerical taxonomic study were in general agreement with the DNA relatedness studies amongst the genera Serratia and Enterobacter (Steigerwalt et al. 1976) although representatives of not all the species recognized by Grimont et al. (1977) were included in the relatedness study.

A subsequent study (Goullet 1978) of the esterases produced by the four Serratia spp. recognized by Grimont et al. (1977) indicated that the esterases of the four species were characterized by distinct electrophoretic patterns. On this basis the strains of Sera. marcescens, Sera. liquefaciens and Sera. plymuthica appeared to be more closely related to one another than to Sera. marinorubra, an observation which is in accord with the relationships of these species as depicted in the phenogram in the numerical study of Grimont et al. (1977).

(ii) Proteus-Providence group

The numerical taxonomic study of McKell & Jones (1976) was undertaken in an attempt to clarify the confused taxonomic situation of bacteria of the Proteus-Providence group. A detailed discussion of the controversial taxonomy of these bacteria may be found in the publications of Ewing [Ewing (1958, 1962); Edwards & Ewing (1972)] Coetzee (1972); Kauffmann & Edwards (1952); and Rauss (1936). Prior to the study of McKell & Jones (1976) the only numerical taxonomic studies of the area had been carried out as part of studies of the enterobacteria. Representatives of the Proteus-Providence group were therefore limited in number and inconclusive results were obtained (Krieg & Lockhart 1966; Johnson et al. 1975).

One hundred and six strains from the Proteus-Providence group and 27 other strains representing other taxa in the family Enterobacteriaceae were tested for 178 characters covering a broad spectrum of tests. Suitable reference strains were used and internal checks on error and test reproducibility were included. The coded data were analysed by Gower's coefficient (S_G) and clustered by both unweighted average linkage and single linkage. The results indicated the division of Proteus-Providence strains into two main areas, A and B, at a similarity level of 72%. The cophenetic correlation between the similarity matrix and the relationships depicted in the phenogram was 0.81. This value indicates that the main clusters are a fairly good summary of the information in the similarity matrix (Sneath & Sokal 1973). This figure is not given for any other study dealt with in this chapter, but it is very important that it should be quoted.

Area A contained two clusters corresponding to the species *Prts. vulgaris* and *Prts. mirabilis*. Area B contained four clusters corresponding to *Prts. morganii; Prts. rettgeri;* a cluster corresponding to Ewing's subgroup A [*Providencia alcalifaciens,* Ewing (1962)] and the four *Shgl. dysenteriae* strains included in the study; a cluster corresponding to Ewing's subgroup B [*Providencia stuartii,* Ewing (1962)]. The clustering of the strains of *Shgl. dysenteriae* with the *Prov. alcalifaciens* group was very surprising. A further study was therefore undertaken to elucidate the relationship of *Shigella* spp. to Providence bacteria. The results did not indicate any close relationship between *Shgl. dysenteriae* strains and *Prov. alcalifaciens* (Dodd & Jones 1978). Although the unweighted average linkage phenogram suggests that the *Shgl. dysenteriae* are unexpectedly closely related to *Prov. alcalifaciens* strains, examination of the corresponding sorted similarity matrix [in McKell (1977)] reveals that the relationship is only marginally stronger than that between any other pair of the following groups: *Prts. morganii, Prts. rettgeri, Prov. alcalifaciens, Shgl. dysenteriae, Prov. stuartii.* In the single linkage clustering all these groups join up over a very narrow similarity range. Results of this kind emphasize the need for constant monitoring and evaluation of the groupings derived from numerical taxonomic studies (See *Interpretation of Clustering Results,* pp. 83).

The results of the study of McKell & Jones (1976) are in accord with the views of Edwards & Ewing (1972) regarding the number of taxa amongst the Proteus-Providence bacteria. The close association of the taxospecies *Prts. morganii, Prts. rettgeri, Prov. alcalifaciens* and *Prov. stuartii* is in accord with the results of studies based on other criteria [see McKell & Jones (1976) and McKell (1977)]. The main objection to a close taxonomic relationship between *Prts. morganii* and the taxospecies *Prts. rettgeri*, *Prov. alcalifaciens* and *Prov. stuartii* is on the basis of the %G+C content of the DNA of these taxa: *Proteus morganii*, 50%; *Prts. rettgeri*, 39%; *Providencia*, 41% [see Hill (1966)]. It is now well established that homogeneous phenetic groups are also homogeneous in the % G+C content of their DNA. A difference of % G+C of 5% usually implies at least a species difference (De Ley 1969; Sneath 1972). De Ley (1969) estimated that a difference of 20 to 30% in G+C ratio means there are practically no nucleotide sequences in common between two bacteria. However, as yet, there is little firm evidence on which to equate generic rank with % G+C differences. The study of McKell & Jones (1976), while indicating the presence of four distinct taxa worthy of species in the genus *Proteus* and two in the genus *Providencia*, casts little light on the higher taxonomic groupings of these clusters.

The study of McKell & Jones (1976) is of interest, however, for a number of other reasons. It is a good example of the necessity of using sorted similarity matrices and phenograms derived from unweighted average and single linkage clustering in the interpretation of the results. Further, a program called IGROUPS was used to print character-value statistics on specified groups of bacteria (in this case the groupings in the genera *Proteus* and *Providencia*). This program aids the selection of tests important in the characterization of taxonomic groups and produces them in a form suitable for computer based identification systems such as that of Bascomb *et al.* (1973). The study is also one of the few with the enterobacteria where sampling and test error were monitored.

It is of interest that the results of McKell & Jones (1976) were in very good agreement with the groupings of strains of the genera *Proteus* and *Providencia* obtained by Véron & Le Minor (1976) in their numerical taxonomic study of enterobacteria referred to earlier, although quite different characters were used by these authors. Electrophoretic studies of the esterases of representatives of all the *Proteus* and *Providencia* spp. (Goullet 1975) also indicated the presence of at least four species in the genus *Proteus* and two in the genus *Providencia*. However, while the esterase patterns of strains of *Prts. mirabilis*, *Prts. morganii* and especially *Prov. stuartii* suggested that these groups were homogeneous, the patterns of representatives of the species *Prts. vulgaris*, *Prts. rettgeri* and *Prov. alcalifaciens* indicated that these taxa were heterogeneous. This difference in the homogeneity of the groups is not apparent from the phenogram depicting the relationship between these taxa in the study of McKell & Jones (1976).

This heterogeneity in the species *Prts. vulgaris*, *Prts. rettgeri* and *Prov. alcalifaciens* was also noted by Brenner *et al.* (1978) on the basis of DNA relatedness studies. These studies also indicated that protei are only distantly related to other members of the family Enterobacteriaceae. Brenner *et al.* (1978) discuss the case for creating another family for the protei [except *Prts. morganii* for which they suggest a new genus *Morganella*, Fulton (1943)]. However, they decided against such a change because protei share the morphological and biochemical properties of the family Enterobacteriaceae, and because conserved DNA sequences, such as those that specify for ribosomal ribonucleic acid, are highly conserved between protei and other enterobacteria (Brenner *et al.* 1978) but substantially less related between the enterobacteria and members of other families (Kato & Bolton 1964; Pace & Campbell 1971).

C. Ecological studies

There do not appear to be any numerical taxonomic studies specifically designed to investigate the numbers and kinds of different enterobacteria in a given environment, but there are a number of studies which have resulted in the isolation and identification of enterobacteria, along with other bacteria, from different environments. Some of these studies are reviewed by Colwell (1973).

Numerical techniques with computer assisted analysis of the data are invaluable in ecological studies because they facilitate the handling of the large amount of data which these studies generate. However, as noted by Colwell (1973) and Jones (1978*a*), identification of the clusters obtained from such studies with 'marker' or 'predictor' strains frequently does not occur. This can be due to the selection of the 'wrong' reference strains but it also often occurs because the selected reference or type strains from culture collections are themselves nontypical of the strains found in nature because they have rarely been designated on the basis of a large sample of strains for a wide variety of characters. Furthermore, the bias of medical microbiology influences ecological studies since nearly all the tests used for identification were originally designed for the characterization of strains pathogenic for man [see Colwell (1973)].

The study of Ercolani (1978) on the bacterial flora of olive leaves is of especial interest because of the way in which the large amount of data (1789 isolates from four different sampling sets in each of three consecutive years, each isolate screened for 210 characters) was processed. The total data were first analysed by the Jaccard coefficient and single linkage clustering. This step determined the number and size of the phenetically related groups of isolates in each sampling set. The quantitative variations of different groups in time were then assessed by deriving a hypothetical median organism for each phenetically related group by comparing the phenotypic data of all isolates within the cluster and coding the HMO 0, 1 and NC for those characters that were shared by 0 to 33, 67 to 100 and 34 to 66% of the isolates in the cluster, respectively. The coded HMOs were then entered into a $t \times n$ matrix which was itself analysed by the Jaccard coefficient and single linkage clustering. Clusters of HMOs formed above the 70% level were defined and the mean similarity values within and between each of these clusters were calculated as before. For each of these HMO clusters a bar graph was prepared with the identification symbol of the HMO as the abscissa and the number of isolates in the parent cluster (i.e. the cluster from which the HMO was derived) as the ordinate.

The next step, designed to identify the isolates representative of the defined phenetic groups, involved the rearrangement of the isolates in the clusters obtained by the processing of the whole data into aggregate clusters. These aggregate clusters included only those isolates which had generated HMOs belonging to the same cluster. Within each aggregate cluster, the isolate with the highest average %S value with the HMOs generated from that aggregate was selected as a type isolate. These type isolates were identified at least to genus level by the use of such sources as the 8th edition of *Bergey's Manual of Determinative Bacteriology* (Buchanan & Gibbons 1974).

The results of the study indicated the presence on olive tree leaves of bacteria of 14 taxa and also indicated that there were seasonal fluctuations in the populations of most of the bacteria. Taxa of the family Enterobacteriaceae included *Erwinia herbicola, Klebsiella pneumoniae* and *Sera. marcescens.*

The study provides an interesting way of dealing with sizes of data that are too large to be analysed in a single computer run. Ercolani (1978) preferred S_J to S_{SM} because S_J ignores matching absences whether they are coded as 'negative' or as 'no comparison'. Sometimes it is difficult to judge the status of matching negative characters, especially in a study which contains some bacteria which exhibit fermentative and some which possess a strictly oxidative metabolism. In general, however, we prefer S_{SM} [see Sneath (1972) p. 58 for further discussion]. Test error and reproducibilty were monitored by statistical analysis of 12 replicate sets of test data on six of the reference strains. On this sample a *p*-value (Sneath & Johnson 1972) of only 1.435% was achieved. Of interest is the successful identification of all the clusters in view of the limited number (15) of reference strains used. This contrasts with the experience of most other workers e.g. Austin *et al.* (1978).

5. Discussion

Numerical taxonomic studies have indicated that members of the family Enterobacteriaceae are all relatively closely related on phenetic grounds as reflected by their relationships expressed as similarity values (%S). The studies also indicate that within the group of bacteria designated as the family Enterobacteriaceae there are clusters of more closely related forms which correspond to the accepted genera *Escherichia, Shigella, Edwardsiella, Salmonella, Arizona, Citrobacter, Klebsiella, Enterobacter, Serratia, Proteus, Providencia, Morganella, Yersinia, Erwinia* and *Pectobacterium.* Within these clusters, subclusters corresponding to species have been recovered.

However, not all numerical taxonomic studies of this group of bacteria have given identical groupings, e.g. compare the study of Krieg & Lockhart (1966) with that of Johnson *et al.* (1975) or the study of Grimont *et al.* (1977) on the genus *Serratia*, with the study of Colwell & Mandel (1965) on the same genus. These differences are not surprising because, as pointed out earlier, only rarely have the same reference strains (or indeed any reference strains) been included in the different studies and there are also differences in laboratory tests (including monitoring for test error) and computer methodology. In view of these differences in design of the studies it is encouraging that the results of the majority of more recent numerical taxonomic studies of the enterobacteria have given broadly comparable results and that these results are also in general agreement with the groupings derived from intuitive interpretations of the same sort of data based on long laboratory experience [e.g. the classification of Edwards & Ewing (1972)].

In order to evaluate the soundness of the classification groupings indicated by computer analysis of phenetic data it is necessary to evaluate the results in the context of relationships based on other criteria. There is now sufficient evidence based on comparisons of a number of numerical taxonomic surveys of many bacteria with other classes of evidence for the same bacteria, to indicate that four of these are of especial interest, i.e. serology, chemotaxonomy, DNA and genetics.

Serological analysis, based on antisera raised against somatic, surface and flagella antigens, has of course long been used to classify and identify bacteria of the family Enterobacteriaceae and indeed in some taxa (e.g. *Salmonella*) the procedure is so refined that individual strains can be identified as the source of an epidemic. However, while serological methods have proved of great value for subdividing such groups or taxa, there are so many cross-reactions between the groups in the family Enterobacteriaceae that a proper evaluation of serological data of this kind in the context of classification, rather than identification, must await further analysis. Computational methods which can handle complex serological data are still in their infancy. The present method of data scoring in particular needs attention before numerical taxonomic analyses of such data can be undertaken with confidence.

The results of serological studies with purified enzymes or enzyme subunits from representatives of some genera of the family Enterobacteriaceae show good accord with the relationships between the taxa demonstrated by numerical taxonomic analysis of phenetic data (Murphy & Mills 1969; Rocha *et al.* 1972; Reyes & Rocha 1977). Thus, of the taxa studied using antisera raised against *Esch. coli* enzymes, the closest relationships were demonstrated between *Esch. coli* and *Shgl. dysenteriae*, the most distant between *Esch. coli* and *Sera. marcescens* and *Erwn. caratovora* (taken as a group), whereas *Salm. typhimurium, Enterobacter aerogenes* and *Prts. vulgaris* (as a group) showed an intermediate relationship to *Esch. coli; Sera. marcescens* and *Erwn. caratovora* being more closely related than *Prts. vulgaris*.

Relationships between members of the Enterobacteriaceae as demonstrated by the electrophoretic mobilities of certain enzymes are also in broad accord with the results of numerical taxonomic studies (Bowman et al. 1967; Baptist et al. 1969). However, as with numerical taxonomic techniques, care must be taken in interpreting relationships between bacteria by both these techniques and more data are required before they can be properly evaluated [see Baptist et al. (1969) and Rocha et al. (1972) for discussion].

In the past twenty years studies of bacterial deoxyribonucleic acid have been used increasingly to establish taxonomic relationships. Deoxyribonucleic acid base ratio values are now available for the great majority of enterobacteria (see Colwell & Mandel 1965; Hill 1966; Rosypal & Rosypalova 1966; Starr & Mandel 1969; Jones & Sneath 1970; Priest et al. 1973; Cowan 1974). Values for most of the family are in the range 49 to 60% G+C. However, % G+C values for strains of the genera *Proteus* (with the exception of *Prts. morganii*, 50% G+C) and *Providencia* are considerably lower (39 to 42% G+C). Similarity in the % G+C values between bacteria does not of course necessarily indicate a close genetic relationship but differences in % G+C values indicate that the bacteria are not closely related (Bradley, Chapter 1). The lack of homogeneity in % G+C values of DNA in the genus *Proteus* does not correlate with the relationship between the taxa in this genus as indicated by numerical taxonomic studies. For example, in the study of Johnson et al. (1975) all the strains of enterobacteria including those of the genera *Proteus* and *Providencia* showed at least a 75% similarity to each other and all *Proteus* and *Providencia* strains showed an intra-group similarity of 78.5%. Similarly in the study of McKell & Jones (1976), although the clusters representing taxa in the genera *Proteus* and *Providencia* showed about 64% similarity to other enterobacteria, their similarity to each other was 75%.

More recently, a considerable amount of work has been done on establishing relationships amongst the enterobacteria by DNA:DNA base pairing studies. Studies of this kind with the enterobacteria have been reviewed by Brenner & Falkow (1971). All the studies indicate that while strains of *Esch. coli* and of the *Shigella*–Alkalescens–Dispar group exhibit an 80 to 90% relatedness to *Esch. coli* DNA all the other enterobacteria studied, including strains of *Salmonella,* exhibited less than 50% relatedness. There thus appears to be an extensive divergence in the total DNA of the enterobacteria (Brenner et al. 1977). In contrast, the ribosomal ribonucleic acid cistrons (r DNA) appear to be highly conserved amongst the enterobacteria (Moore & McCarthy 1967; Kohne 1968; Brenner & Falkow 1971) as do the genes for transfer ribonucleic acid (t DNA) and 5 S ribonucleic acid (5 S DNA), although t DNA and 5 S DNA appear to be conserved to a lesser degree that r DNA genes (Brenner et al. 1977). The latter using labelled *Esch. coli* B t RNA cistrons demonstrated 82% binding with t RNA cistrons of *Shgl. flexneri,* 65% binding with *Salm. typhimurium,* 42%

binding with *Proteus mirabilis* and 43% binding with *Yersinia pestis*. Brenner
et al. (1977) suggest that r DNA cistrons can be used to group bacteria at the
family or supra-family level, t DNA or 5 *S* DNA cistrons can be used to separate
bacteria at the family or tribe level, but species separation is best obtained by
determining total DNA relatedness [e.g. see Crosa *et al.* (1973)]. However, it
must be noted that DNA pairing study results are also subject to laboratory
variations in technique [see Bradley (1975)]. On the whole, the results of whole
DNA pairing are highly congruent with indications of phenetic similarity obtained
by numerical taxonomic analysis (Jones & Sneath 1970). The relationship is not
linear, however, as there is very little pairing at phenetic similarities below 50%,
but it can be made linear by suitable transformation of the data as indicated and
discussed by Sneath (1972).

It is generally agreed that the degree of genetic exchange between organisms
is a good indication of the degree of relatedness between them. Indeed, in higher
organisms it forms the basis of the species concept. Until relatively recently little
was known about genetic exchange between bacteria. However, in the last
twenty years much information has become available and there have been
numerous reports of genetic exchange between enterobacteria [see Jones &
Sneath (1970)]. As pointed out by these authors most of the members of the
family Enterobacteriaceae conform to the bacterial genospecies concept of
Ravin (1963). This view does not contradict the high degree of similarity exhibited between all members of the family in the majority of the numerical taxonomic studies. However all numerical taxonomic studies of this group of bacteria
have indicated the presence of much internal structure and resulted in the recognition of many well defined taxa which can be equated with taxospecies as
defined by Ravin (1963) and taxogenera, defined as groups of closely related
taxospecies which do not overlap with other such groups. On the whole these
taxa correspond well with the long recognized genera amongst the enterobacteria and their component species [see Edwards & Ewing (1972)]. However,
some of these taxospecies and taxogenera had not been recognized, or their
existence was disputed, before data on a large number of strains of the Enterobacteria were processed by numerical techniques. Examples are the different
taxospecies in the genus *Serratia* (Grimont *et al.* 1977) and the separation of the
taxa *Proteus* and *Providencia* (McKell & Jones 1976). If, as seems likely from
genetic evidence, it is accepted that most of the family forms a genospecies because genetic exchange occurs between the majority of the members, the bacteriologist will still need to distinguish for practical purposes between the
shigellae and the salmonellae, the protei and the klebsiellae, etc.

A major contribution of numerical taxonomic techniques to the classification
and identification of the enterobacteria has been in the development of practical
classifications of these bacteria and of an identification system based upon a
number of characters either in the form of tables [see McKell & Jones (1976);

Johnson *et al.* (1975); Grimont *et al.* (1977)] or as a computer based identification system such as that developed by Bascomb *et al.* (1973).

Numerical taxonomic studies can accommodate large amounts of data in the form of high numbers of strains and large numbers of characters. In the past enterobacteria of interest to the clinician, veterinarian or plant pathologist have been studied in isolation by a few tests developed to give the best identifications for the particular field. The use of computers to handle the data has resulted in the comparison of strains from a variety of sources by as many tests as possible. The groupings thus obtained are polythetic groupings based on a large number of phenetic characters. The intra- and inter-relationships of these groups can be quantified by a %S value. Liston *et al.* (1963) and Skerman (1967) suggested correlating %S values with taxonomic rank. Care is needed here because the %S depends on a number of factors, particularly the choice of tests and the choice of similarity coefficient.

Computer assisted identification has the great advantage of allowing much larger sized information to be stored on the groups. It also allows the information to be used in a polythetic way. These two factors greatly reduce the chances of gross misidentification as a result of laboratory error or plasmid borne characters which may be lost or gained (see Harwood, Chapter 2).

Studies such as that of Ercolani (1978) have already shown the potential of numerical taxonomic techniques in ecological studies on bacteria. As pointed out by Colwell (1973), numerical taxonomic techniques will doubtless play an even greater role in the analysis of ecological data in the future.

There is no doubt that numerical taxonomic methods have resulted in a greater insight into the classification and identification of the enterobacteria than any other modern method. While genetic and DNA pairing studies have raised some interesting issues of evolutionary significance [see Jones & Sneath (1970); Brenner *et al.* (1977)] , and techniques such as protein electrophoresis suggest that there may be promise in this and allied approaches especially if combined with serology, the data available in these areas are as yet sparse. The most comprehensive appreciation of stable groups and subgroups within the enterobacteria has to date undoubtedly resulted from the use of numerical techniques because they have provided the requisite information for classification and identification in a quantity not approached by serological and DNA studies.

6. References

AUSTIN, B., GOODFELLOW, M. & DICKINSON, C. H. 1978 Numerical taxonomy of phylloplane bacteria isolated from *Lolium perenne. Journal of General Microbiology* **104**, 139–155.

BAPTIST, J. N., SHAW, C. R., MANDEL, M. 1969 Zone electrophoresis of enzymes in bacterial taxonomy. *Journal of Bacteriology* **99**, 180–188.

BASCOMB, S., LAPAGE, S. P., WILLCOX, W. R. & CURTIS, M. A. 1971 Numerical classification of the tribe Klebsielleae. *Journal of General Microbiology* **66**, 279–295.

BASCOMB, S., LAPAGE, S. P., CURTIS, M. A. & WILLCOX, W. R. 1973 Identification of bacteria by computer: Identification of reference strains. *Journal of General Microbiology* **77**, 291–315.

BENZÉCRI, J-P. 1973 *L'Analyse des Données. I La Taxinomie.* Paris: Dunod.

BOBISUD, H. M. & BOBISUD, L. E. 1972 A metric for classifications. *Taxon* **21**, 607–613.

BOORMAN, S. A. & OLIVIER, D. C. 1973 Metrics on spaces of finite trees. *Journal of Mathematical Psychology* **10**, 26–59.

BOWMAN, J. E., BRUBAKER, R. R., FRISCHER, H. & CARSON, P. E. 1967 Characterization of enterobacteria by starch-gel electrophoresis of glucose-6-phosphate dehydrogenase and phosphogluconate dehydrogenase. *Journal of Bacteriology* **94**, 544–551.

BRADLEY, S. G. 1975 Significance of nucleic acid hybridization to systematics of actinomycetes. *Advances in Applied Microbiology* **19**, 59–70.

BREED, R. S., HAMMAR, B. W., HUNTOON, F. M., MURRAY, E. G. D. & HARRISON, F. C. (ed.) 1934. *Bergey's Manual of Determinative Bacteriology* 4th edn. Baltimore: Williams & Wilkins.

BRENNER, D. J. & FALKOW, S. 1971 Molecular relationships among members of the Enterobacteriaceae. *Advances in Genetics* **16**, 81–118.

BRENNER, D. J., FANNING, G. R., STEIGERWALT, A. G., SODD, M. A. & DOCTOR, B. P. 1977 Conservation of transfer ribonucleic acid and 5 S ribonucleic acid cistrons in the Enterobacteriaceae. *Journal of Bacteriology* **129**, 1435–1439.

BRENNER, D. J., FARMER, J. J., FANNING, G. R., STEIGERWALT, A. G., KLYKKEN, P., WATHEN, H. G., HICKMAN, F. W. & EWING, W. H. 1978 Deoxyribonucleic acid relatedness of *Proteus* and *Providencia* species. *International Journal of Systematic Bacteriology* **28**, 269–282.

BRISOU, J. & CADEILLAN, J. 1959 Étude sur les *Serratia*. A propos de quartre souches isolées en medecine vétérinaire. *Bulletin de l'Association des Diplomés de Microbiologie de la Faculté de Pharmacie de Nancy* **75**, 34–39.

BUCHANAN, R. E. & GIBBONS, N. E. (ed.) 1974 *Bergey's Manual of Determinative Bacteriology*, 8th edn. Baltimore: Williams & Wilkins.

CARMICHAEL, J. W. C. & SNEATH, P. H. A. 1969 Taxometric maps. *Systematic Zoology* **18**, 402–415.

COETZEE, J. N. 1972 Genetics of the *Proteus* group. *Annual Review of Microbiology* **26**, 23–54.

COLMAN, G. 1968 The application of computers to the classification of streptococci. *Journal of General Microbiology* **50**, 149–158.

COLWELL, R. R. 1970 Numerical analysis in microbial identification and classification. *Developments in Industrial Microbiology* **11**, 154–160.

COLWELL, R. R. 1973 Genetic and phenetic classification of bacteria. *Advances in Applied Microbiology* **16**, 137–175.

COLWELL, R. R. & MANDEL, M. 1965 Adansonian analysis and deoxyribonucleic acid base composition of *Serratia marcescens. Journal of Bacteriology* **89**, 454–461.

COLWELL, R. R., JOHNSON, R., WAN, L., LOVELACE, T. E. & BRENNER, D. J. 1974 Numerical taxonomy and deoxyribonucleic acid reassociation in the taxonomy of some Gram-negative fermentative bacteria. *International Journal of Systematic Bacteriology* **24**, 422–433.

COWAN, S. T. 1974 The family Enterobacteriaceae Rahn. In *Bergey's Manual of Determinative Bacteriology*, 8th edn. ed. Buchanan, R. E. & Gibbons, N. W. Baltimore: Williams & Wilkins.

CROSA, J. H., BRENNER, D. J., EWING, W. H. & FALKOW, S. 1973 Molecular relationships among the salmonellae. *Journal of Bacteriology* **115**, 307–315.

DELABRE, M., BIANCHI, A. & VÉRON, M. 1973 Étude critique de methodes de taxonomie numerique. Application á une classification de bactéries aquicoles. *Annales de Microbiologie (Institut Pasteur)* **124A**, 489–506.

DE LEY, J. 1969 Compositional nucleotide distribution and the theoretical prediction of homology in bacterial DNA. *Journal of Theoretical Biology* **22**, 89–116.

DODD, C. E. R. & JONES, D. 1978 A numerical taxonomic study of the genus *Shigella*. *Journal of Applied Bacteriology* 45, xxvii.

EDWARDS, P. R. & EWING, W. H. 1962 *Identification of Enterobacteriaceae*. 2nd edn. Minneapolis: Burgess Publishing Co.

EDWARDS, P. R. & EWING, W. H. 1972 *Identification of Enterobacteriaceae*. 3rd edn. Minneapolis: Burgess Publishing Co.

ERCOLANI, G. L. 1978 *Pseudomonas savastanoi* and other bacteria colonizing the surface of olive leaves in the field. *Journal of General Microbiology* 109, 245–257.

EWING, W. H. 1958 The nomenclature and taxonomy of the Proteus and Providence groups. *International Bulletin of Bacteriological Nomenclature and Taxonomy* 8, 17–22.

EWING, W. H. 1962 The tribe Proteae, its nomenclature and taxonomy. *International Bulletin of Bacteriological Nomenclature and Taxonomy* 12, 93–102.

EWING, W. H., DAVIS, B. R. & REAVIS, R. W. 1959 *Studies on the* Serratia *Group*. Atlanta: Center for Disease Control.

EWING, W. H., DAVIS, B. R. & JOHNSON, J. G. 1962 The genus *Serratia*: its taxonomy and nomenclature. *International Bulletin of Bacteriological Nomenclature and Taxonomy* 12, 47–52.

EWING, W. H., DAVIS, B. R. & FIFE, M. A. 1972 *Biochemical Characterization of* Serratia liquefaciens *and* Serratia rubidaea. Atlanta: Center for Disease Control.

EWING, W. H., DAVIS, B. R., FIFE, M. A. & LESSEL, E. F. 1973 Biochemical characterization of *Serratia liquefaciens* (Grimes & Hennerty) Bascomb *et al.* (formerly *Enterobacter liquefaciens*) and *Serratia rubidaea* (Stapp) comb. nov. and designation of the type and neotype strains. *International Journal of Systematic Bacteriology* 23, 217–225.

FARRIS, J. S. 1969 On the cophenetic correlation coefficient. *Systematic Zoology* 18, 279–285.

FARRIS, J. S. 1973 On comparing the shapes of taxonomic trees. *Systematic Zoology* 22, 50–54.

FULTON, M. 1943 The identity of *Bacterium columbensis* Castellani. *Journal of Bacteriology* 46, 79–81.

FULTON, M., FORNEY, C. E. & LEIFSON, E. 1959 Identification of *Serratia* occurring in man and animals. *Canadian Journal of Microbiology* 5, 269–275.

GOODFELLOW, M. 1969 Numerical taxonomy of some heterotrophic bacteria isolated from a pine forest soil. In *The Soil Ecosystem* ed. Sheals, J. G. Systematics Association Publication No. 8, pp. 83–105 London: The Systematics Association.

GOODFELLOW, M. 1971 Numerical taxonomy of some nocardioform bacteria. *Journal of General Microbiology* 69, 33–80.

GOULLET, Ph. 1975 Esterase zymograms of *Proteus* and *Providencia*. *Journal of General Microbiology* 87, 97–106.

GOULLET, Ph. 1978 Characterization of *Serratia marcescens*, *S. liquefaciens*, *S. plymuthica* and *S. marinorubra* by the electrophoretic patterns of their esterases. *Journal of General Microbiology* 108, 275–281.

GOWER, J. C. 1971 A general coefficient of similarity and some of its properties. *Biometrics* 27, 857–871.

GRIMONT, P. A. D. 1969 *Les Serratia: étude taxometrique*. Thèse de Médecine, Bordeaux.

GRIMONT, P. A. D. & DULONG DE ROSNAY, H. L. C. 1972 Numerical study of 60 strains of *Serratia*. *Journal of General Microbiology* 72, 259–268.

GRIMONT, P. A. D., GRIMONT, F., DULONG DE ROSNAY, H. L. C. & SNEATH, P. H. A. 1977 Taxonomy of the genus *Serratia*. *Journal of General Microbiology* 98, 39–66.

HAMON, Y., Le MINOR, L. & PERON, Y. 1970 Les bactériocines d'*Enterobacter liquefaciens*. *Comptes Rendus Hebdomadaire des Séances de l'Academie des Sciences, Paris*. 270, 886–889.

HILL, L. R. 1966 An index to deoxyribonucleic acid base compositions of bacterial species. *Journal of General Microbiology* 44, 419–437.

HILL, L. R., LAPAGE, S. P. & BOWIE, I. S. 1978 Computer assisted identification of coryneform bacteria. In *Coryneform Bacteria* ed. Bousfield, I. G. & Callely, A. G. London: Academic Press.

HOPWOOD, D. A. 1967 Genetic analysis and genome structure in *Streptomyces coelicolor. Bacteriological Reviews* 31, 373-403.

JARDINE, N. & SIBSON, R. 1968 The construction of hierarchic and non-hierarchic classifications. *Computer Journal* 11, 177-184.

JOHNSON, R., COLWELL, R. R., SAKAZAKI, R. & TAMURA, K. 1975 Numerical taxonomy study of the Enterobacteriaceae. *International Journal of Systematic Bacteriology* 25, 12-37.

JONES, D. 1975 A numerical taxonomic study of coryneform and related bacteria. *Journal of General Microbiology* 87, 52-96.

JONES, D. 1978a An evaluation of the contributions of numerical taxonomic studies to the classification of coryneform bacteria. In *Coryneform Bacteria* ed. Bousfield, I. J. & Callely, A. G. London: Academic Press.

JONES, D. 1978b Composition and differentiation of the genus *Streptococcus*. In *Streptococci* ed. Skinner, F. A. & Quesnel, L. B. London: Academic Press.

JONES, D. & SNEATH, P. H. A. 1970 Genetic transfer and bacterial taxonomy. *Bacteriological Reviews* 34, 40-81.

KATO, Y. & BOLTON, E. T. 1964 Relationships among some enterobacteria. *Carnegie Institute, Washington, Yearbook* 63, 372-373.

KAUFFMANN, F. & EDWARDS, P. R. 1952 Classification and nomenclature of Enterobacteriaceae. *International Bulletin of Bacteriological Nomenclature and Taxonomy* 2, 2-8.

KOHNE, D. E. 1968 Isolation and characterisation of bacterial ribosomal RNA cistrons. *Biophysical Journal* 8, 1104-1118.

KRIEG, R. E. & LOCKHART, W. R. 1966 Classification of enterobacteria based on overall similarity. *Journal of Bacteriology* 92, 1275-1280.

LAPAGE, S. P., BASCOMB, S., WILLCOX, W. R. & CURTIS, M. A. 1970 Computer identification of bacteria. In *Automation, Mechanization and Data Handling in Microbiology* ed. Baillie, A. & Gilbert, R. J. London: Academic Press.

LISTON, J., WIEBE, W. & COLWELL, R. R. 1963 Quantitative approach to the study of bacterial species. *Journal of Bacteriology* 85, 1061-1070.

LOCKHART, W. R. & HOLT, J. G. 1964 Numerical classification of *Salmonella* serotypes. *Journal of General Microbiology* 35, 115-124.

LOCKHART, W. R. & KOENIG, K. 1965 Use of secondary data in numerical taxonomy of the genus *Erwinia. Journal of Bacteriology* 90, 1638-1644.

MALIK, A. C., REINBOLD, G. W. & VEDAMUTHU, E. R. 1968 An evaluation of the taxonomy of *Propionibacterium. Canadian Journal of Microbiology* 14, 1185-1191.

McKELL, J. 1977 *A taxonomic study of the Proteus-Providence group with especial reference to the role of plasmids.* Ph.D. Thesis, University of Leicester, UK.

McKELL, J. & JONES, D. 1976 A numerical taxonomic study of the Proteus-Providence bacteria. *Journal of Applied Bacteriology* 41, 143-161.

MOORE, R. L. & McCARTHY, B. J. 1967 Comparative study of ribosomal ribonucleic acid cistrons in enterobacteria and myxobacteria. *Journal of Bacteriology* 94, 1066-1074.

MURPHY, T. M. & MILLS, S. E. 1969 Immunochemical and enzymatic comparisons of the tryptophan synthase subunits from five species of Enterobacteriaceae. *Journal of Bacteriology* 97, 1310-1320.

PACE, P. & CAMPBELL, L. L. 1971 Homology of ribosomal ribonucleic acid of diverse bacterial species with *Escherichia coli* and *Bacillus stearothermophilus. Journal of Bacteriology* 107, 543-547.

PFISTER, R. M. & BURKHOLDER, P. R. 1965 Numerical taxonomy of some bacteria isolated from Antarctic and tropical seawaters. *Journal of Bacteriology* 90, 863-872.

PRIEST, F. G., SOMERVILLE, H. J., COLE, J. A. & HOUGH, J. S. 1973 The taxonomic position of *Obesumbacterium proteus*, a common brewery contaminant. *Journal of General Microbiology* 75, 295-307.

RAUSS, K. 1936 The systematic position of Morgan's bacillus. *Journal of Pathology and Bacteriology* **42**, 183–192.

RAVIN, A. W. 1963 Experimental approaches to the study of bacterial phylogeny. *American Naturalist* **97**, 307–318.

REYES, G. R. & ROCHA, V. 1977 Immunochemical comparison of phosphoribosylanthranilate isomerase–indoglycerol phosphate synthetase among the Enterobacteriaceae. *Journal of Bacteriology* **129**, 1448–1456.

ROCHA, V., CRAWFORD, I. P. & MILLS, S. E. 1972 Comparative immunological and enzymatic study of the tryptophan synthetase β_2 subunit in the Enterobacteriaceae. *Journal of Bacteriology* **111**, 163–168.

ROHLF, F. J. 1974 Methods of comparing classifications. *Annual Review of Ecology and Systematics* **9**, 101–113.

ROSYPAL, S. & ROSYPALOVA, A. 1966 Genetic, phylogenetic and taxonomic relationships among bacteria as determined by their deoxyribonucleic acid base composition. *Folia, Faculty of Science, National University, Purkyne, Brno.* **7**, 1–91.

SAKAZAKI, R., TAMURA, K., JOHNSON, R. & COLWELL, R. R. 1976 Taxonomy of some recently described species in the family Enterobacteriaceae. *International Journal of Systematic Bacteriology* **26**, 158–179.

SACKIN, M. J. 1972 "Good" and "bad" phenograms. *Systematic Zoology* **21**, 225–226.

SKERMAN, V. B. D. 1967 *A Guide to the Identification of the Genera of Bacteria.* 2nd edn. Baltimore: Williams & Wilkins.

SKERMAN, V. B. D. 1969 *Abstracts of Microbiological Methods.* London: Wiley–Interscience.

SNEATH, P. H. A. 1957a Some thoughts on bacterial classification. *Journal of General Microbiology* **17**, 184–200.

SNEATH, P. H. A. 1957b The application of computers to taxonomy. *Journal of General Microbiology* **17**, 201–226.

SNEATH, P. H. A. 1968 Vigour and pattern in taxonomy. *Journal of General Microbiology* **54**, 1–11.

SNEATH, P. H. A. 1972 Computer taxonomy. In *Methods in Microbiology* eds. Norris, J. R. & Ribbons, D. W. London: Academic Press.

SNEATH, P. H. A. 1974 Phylogeny of microorganisms. In *Evolution in the Microbial World.* Symposium of the Society of General Microbiology **24**, 1–39.

SNEATH, P. H. A. 1976a An evaluation of numerical taxonomic techniques in the taxonomy of *Nocardia* and allied taxa. In *The Biology of the Nocardiae* ed. Goodfellow, M., Brownell, G. H. & Serrano, J. A. London: Academic Press.

SNEATH, P. H. A. 1976b Phenetic taxonomy at the species level and above. *Taxon* **25**, 437–450.

SNEATH, P. H. A. 1977 A method for testing the distinctness of clusters: a test of the disjunction of two clusters in euclidean space as measured by their overlap. *Mathematical Geology* **9**, 123–143.

SNEATH, P. H. A. 1978a Classification of microorganisms. In *Essays in Microbiology* ed. Norris, J. R. & Richmond, M. H. London: John Wiley.

SNEATH, P. H. A. 1978b Identification of microorganisms. In *Essays in Microbiology* ed. Norris, J. R. & Richmond, M. H. London: John Wiley.

SNEATH, P. H. A. & COLLINS, V. G. 1974 A study in test reproducibility between laboratories: report of a *Pseudomonas* Working Party. *Antonie van Leeuwenhoek* **40**, 481–527.

SNEATH, P. H. A. & JOHNSON, R. 1972 The influence on numerical taxonomic similarities of errors in microbiological tests. *Journal of General Microbiology* **72**, 377–392.

SNEATH, P. H. A. & SOKAL, R. R. 1973 *Numerical Taxonomy: The Principles and Practice of Numerical Classification.* San Francisco: W. H. Freeman.

STARR, M. P. & MANDEL, M. 1969 DNA base composition and taxonomy of phytopathogenic and other enterobacteria. *Journal of General Microbiology* **56**, 113–123.

STEIGERWALT, A. G., FANNING, R. G., FIFE-ASBURY, M. A. & BRENNER, D. J. 1976

DNA relatedness among species of *Enterobacter* and *Serratia*. *Canadian Journal of Microbiology* **22**, 121–137.

STEVENS, M. & MAIR, N. S. 1973 A numerical taxonomic study of *Yersinia enterocolitica* strains. In *Contributions to Microbiology and Immunology*. Vol. 2, ed. Winblad, S. Basel: S. Karger.

THORNLEY, M. J. 1967 A taxonomic study of *Acinetobacter* and related genera. *Journal of General Microbiology* **49**, 211–257.

VÉRON, M. & Le MINOR, L. 1975 Nutrition et taxonomie des Enterobacteriaceae et bactéries voisines. II Résultats d'ensemble et classification. *Annales de Microbiologie (Institut Pasteur)* **126B**, 111–124.

WILKINSON, B. J. & JONES, D. 1977 A numerical taxonomic survey of *Listeria* and related bacteria. *Journal of General Microbiology* **98**, 399–421.

YANAGAWA, R. 1975 A numerical taxonomic study of the strains of three types of *Corynebacterium renale*. *Canadian Journal of Microbiology* **21**, 824–827.

Citrate Synthase and Succinate Thiokinase in Classification and Identification

P. D. J. WEITZMAN

Department of Biochemistry, University of Bath, Bath, UK

Contents

1. Introduction

I FEEL RATHER like Daniel in the lions' den amongst so many professional bacterial taxonomists, classifiers and identifiers. Most of the participants in this meeting have been bred and brought up on bacterial diversity and are anxious to explore the means and methods of identifying bacteria and of probing natural relationships between them. On the other hand, I seem to have come to this area from quite the opposite direction from enzymological studies, stumbling onto enzymic diversity and pattern and realizing that this led to the classification of diverse organisms into distinct groupings.

Unlike higher organisms, gross morphology has limited value in bacterial classification, but it may be possible to use molecular morphology as an additional taxonomic tool. The methods of protein sequence studies, electrophoretic characterization (see Chapter 11), immunological examination and DNA pairing (see Chapter 1) all probe similarities and differences in the structural make-up of cellular macromolecules, while many biochemical tests are designed to reveal the presence or absence of particular enzymes and metabolic reactions.

The studies which will be described show that both gross structural features, such as the subunit composition of polymeric enzymes, as well as catalytic and regulatory behaviour, are properties linked to the metabolic 'life-style' and taxonomic groups of the source organism. Although biochemists have tended to stress the unity of metabolism, there is undoubtedly a very considerable diversity in the fine details of the structural and functional properties of the tools of cellular metabolism—the enzymes. It is some aspects of this diversity which will be discussed in this chapter. I hope that they will prove of relevance and interest to this book so that, like Daniel's, my survival may be ensured!

2. The Citric Acid Cycle

The near universal occurrence of the citric acid (Krebs) cycle in living cells makes it a very suitable metabolic pathway for comparative studies between diverse organisms. The cycle is at the very core of cellular metabolism and fulfils, simultaneously, a multifunctional role in the provision of energy, reducing power and several starting materials for various biosynthetic pathways. Thus while the chemical natures of the intermediates of the cycle and their interconversions are fixed, the different uses of the cycle which diverse organisms may make and the different emphasis which they may place on its several metabolic functions might be expected to be reflected in differences in the fine detail of the enzymic apparatus of the cycle. Structural, catalytic and regulatory features of cycle enzymes might be similar within one group of organisms and yet distinctively different in another taxon. The demonstration of enzyme patterns could be helpful in establishing natural relationships between different organisms.

The cycle is shown conventionally in Fig. 1 and in greatly simplified form in Fig. 2. The present paper is concerned with two enzymes of the cycle—citrate

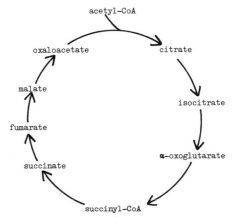

Fig. 1. The citric acid cycle.

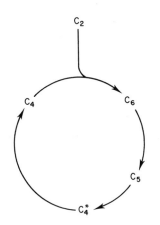

Fig. 2. Simplified scheme of the citric acid cycle.

synthase and succinate thiokinase—and the cycle has been drawn in a simplified form to focus attention on the function of these two enzymes.

The scheme in Fig. 2 indicates that the cycle is initiated by condensation of a 2-carbon fragment with a 4-carbon compound to form a 6-carbon compound. This is the reaction catalysed by citrate synthase, and it may be represented as:

$$\text{acetyl-CoA} + \text{oxaloacetate} + H_2O = \text{citrate} + \text{CoA-SH}$$

It is this reaction which serves to introduce carbon atoms, in the form of acetyl units, into the cycle. Two successive decarboxylation reactions serve to convert the C_6 compound (actually the isomer of citrate, isocitrate) first into α-oxoglutarate (C_5) and then into succinyl-CoA (designated C_4^*), both of which are important for biosynthesis. In a sense, the cycle might stop at this stage. The two carbon atoms introduced initially have been oxidized away to CO_2, reduced pyridine nucleotides have been produced—both NADPH for biosynthesis and NADH for reoxidation coupled to ATP production—as well as the biosynthetic starting materials α-oxoglutarate and succinyl-CoA. However, by rearrangement of the 4-carbon compound C_4^* to the 4-carbon starting compound C_4, a complete cycle is put into operation and a new round of chemical conversions can be undertaken. By this device it is possible for the pathway to oxidize, and thereby produce energy from, a supply of C_2 'fuel' without running out of the requisite 'carrier' compound C_4. The reaction which initiates this 'rearrangement' phase of the cycle is the conversion of succinyl-CoA to succinate and is catalysed by the enzyme succinate thiokinase:

$$\text{succinyl-CoA} + \text{ADP/GDP} + \text{phosphate} = \text{succinate} + \text{ATP/GTP} + \text{CoA-SH}$$

The enzyme may utilize ADP or GDP (see section 4B below) and catalyses their phosphorylation to the nucleoside triphosphates.

In addition to their special 'initiating' roles in the cycle, citrate synthase and succinate thiokinase have in common the fact that they both utilize an acyl-CoA compound as substrate. Citrate synthase uses the energy available on the hydrolysis of acetyl-CoA to form a covalent carbon-carbon bond of the citric acid molecule, whereas succinate thiokinase couples the energy of hydrolysis of succinyl-CoA to the phosphorylation of a nucleoside diphosphate. In the discussion which follows it will be shown that these two enzymes display a remarkable diversity among bacteria and that this diversity has considerable value in classification and identification.

3. Citrate Synthase

The oxidative function of the cycle, effected by dehydrogenase enzymes, results in the production of the reduced nucleotide NADH, whose subsequent reoxidation to NAD^+ is coupled to ATP production. NADH may therefore be considered the 'end-product' of the cycle and, by analogy with non-cyclic metabolic pathways, it might be anticipated that this end-product would inhibit the activity of the initial enzyme of the cycle-citrate synthase. Initial experiments on citrate synthase from *Escherichia coli* revealed that NADH is indeed a powerful and specific inhibitor of this bacterial enzyme (Weitzman 1966). As neither yeast nor pig heart citrate synthases showed NADH inhibition it seemed possible that the inhibition by NADH is a property exclusive to the bacterial enzyme. This possibility was explored by studying citrate synthases from a very wide range of organisms, a task facilitated by the availability of a very convenient assay method applicable to crude cell extracts.

The CoA-SH produced in the course of the reaction catalysed by citrate synthase contains a free thiol (SH) group, and this provides a means of assaying the enzymic activity. Thiol groups react rapidly, non-enzymically and stoicheiometrically with the chromogenic reagent 5, 5'-dithiobis-(2-nitrobenzoate), DTNB, resulting in a thiol-disulphide exchange reaction (Ellman 1959):

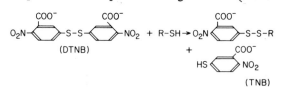

At pH values above 7, the thionitrobenzoate (TNB) formed is yellow and absorbs strongly at 412 nm (ϵ_M = 13 600 1mol cm). Thus, by including DTNB in the citrate synthase reaction mixture, the formation of CoA-SH may be monitored colorimetrically and continuously by following the increase in light absorbance at 412 nm. This simple assay readily permits the determination of citrate synthase activity and of the effects on this activity of any metabolite inhibitors or activators, and hence makes it possible to examine large numbers of bacterial strains for the behavioural characteristics of their citrate synthases.

TABLE 1

Regulatory patterns among bacterial citrate synthases

AMP reactivation	No AMP reactivation
NADH inhibition	
Acinetobacter { *anitratus* / *calcoaceticus* / *lwoffi*	*Aeromonas formicans*
Azotobacter vinelandii	*Arizona arizonae*
Brevibacterium leucinophagum	*Erwinia uredovora*
Cellulomonas rossica	*Escherichia coli*
Chromobacterium violaceum	*Hafnia alvei*
Corynebacterium nephridii	*Klebsiella* { *(Aerobacter) aerogenes* / *pneumoniae*
Flavobacterium devorans	*Pasteurella pseudotuberculosis*
Moraxella { *calcoacetica* / sp. *(Vibrio 0-1)*	*Proteus* { *rettgeri* / *vulgaris*
Pseudomonas { *aeruginosa* / *fluorescens* / *ovalis* / *syringae*	*Salmonella* { *anatum* / *cholere-suis* / *typhimurium*
Rhodopseudomonas { *capsulata* / *spheroides*	*Serratia marcescens*
Rhodospirillum rubrum	*Thiobacillus* A2
Vibrio tyrosinaticus	
Xanthomonas hyacinthi	
No NADH inhibition	
Achromobacter liquefaciens	*Haemophilus vaginalis*
Arthrobacter { *atrocyaneus* / *globiformis* / *nicotianae*	*Kurthia zopfii*
	Microbacterium thermosphactum
Bacillus { *cereus* / *megaterium* / *stearothermophilus** / *subtilis*	*Micrococcus* { *luteus* / sp.
Brevibacterium { *flavum*† / *linens*	*Mycobacterium* { *phlei* / *rhodochrous* / *smegmatis*
Cellulomonas cellasea	*Nocardia* { *corallina* / *farcinica*
Clostridium acidi-urici	*Pseudomonas iodinum*
Corynebacterium { *equi* / *fascians* / *michiganense*	*Staphylococcus aureus*
	Streptomyces { *somaliensis* / *viridochromogens*

Weitzman & Danson (1976); *Higa & Cazzulo (1976); †Shiio *et al.* (1977).

A. Regulatory patterns—energy control

In the early examination of various bacterial citrate synthases it became clear that inhibition of citrate synthase by NADH was by no means a general feature of the enzyme from bacteria; the enzyme from a number of species appeared to be insensitive to inhibition. As the work progressed, however, a remarkable and quite unexpected pattern emerged (Weitzman & Jones 1968). Only citrate synthases from Gram negative bacteria were inhibited by NADH; Gram positive bacteria possessed citrate synthases which were totally unaffected. But there was even more correlation pattern to come. One of the NADH-sensitive citrate synthases was fortuitously tested with AMP and was found to show a de-inhibition or re-activation response to AMP. In other words, the inhibition by NADH could be specifically overcome by AMP, though AMP, on its own in the absence of NADH, was not an activator of the enzyme. When all the NADH-sensitive citrate synthases were examined for this property it was found that those enzymes which were reactivated by AMP came from strictly aerobic Gram negative bacteria, while those citrate synthases with which NADH inhibition could not be overcome by AMP were products of facultatively-anaerobic Gram negative bacteria. Table 1 presents these results together with additional data from subsequent studies which have supported the pattern of correlation. It is worth commenting that the Gram positive bacterial citrate synthases, in lacking sensitivity to NADH, resemble the citrate synthases of all eukaryotic organisms examined to date. A simplified scheme (Table 2) can, therefore, be presented to show the correlation between NADH and AMP effects on citrate synthase and the taxonomic status and 'metabolic life-style' of the source organism.

TABLE 2
Regulatory patterns in citrate synthases

NADH inhibition		No NADH inhibition
Gram negative bacteria		Gram positive bacteria and all eukaryotic organisms
AMP reactivation	No AMP reactivation	
Strict aerobes	Facultative anaerobes	

Can any rationale be proposed for this striking pattern of enzyme regulatory properties? As far as the distinction between Gram negative and Gram positive bacteria is concerned the answer to this question is currently negative. One can only be intrigued by the remarkable conservation of behaviour within the two groups of bacteria. The inhibition of citrate synthase by NADH is certainly an attractive and plausible feedback mechanism whereby the activity of the citric acid cycle may be self-regulated, but why this control device should occur in

Gram negative bacteria and not in other organisms remains a puzzle. To speculate perhaps the answer lies in differences in overall metabolic organization or compartmentation between the different types of cells.

However, for the correlation of the AMP effect with the subdivision between aerobic and anaerobic Gram negative bacteria a rationale based on the pathways of energy metabolism adopted by these two classes of organisms has been suggested (Weitzman & Jones 1968). The strict aerobes are absolutely dependent on the citric acid cycle for energy production and it is therefore logical that the first enzyme of the cycle should be sensitive to positive effector action by AMP, a 'low-energy' metabolic signal. By acting as a de-inhibitor of citrate synthase, AMP can promote the ultimate formation of ATP and hence redress the energy imbalance of which it is itself an indicator. The facultative anaerobes, on the other hand, can generate energy by fermentation alone and possess key glycolytic enzymes sensitive to activation by low-energy signals; they have no obligatory dependence on the citric acid cycle for energy and therefore have no need for an AMP response at the level of citrate synthase.

B. Regulatory patterns—biosynthetic control

Thus far the discussion has concentrated on the energy-yielding role of the citric acid cycle and the associated regulation of citrate synthase. However, attention was earlier drawn to the multifunctional role of the cycle and to its significance in biosynthesis. Two major biosynthetic starting materials generated by the cycle are α-oxoglutarate and succinyl-CoA, required for amino acid and porphyrin biosynthesis. Referring to Fig. 1, α-oxoglutarate and succinyl-CoA are the compounds designated C_5 and C_4^*. When the complete citric acid cycle is operative, these compounds are intermediates rather than end-products, but situations do exist in which the cycle is incomplete owing to the absence of α-oxoglutarate dehydrogenase, the enzyme complex catalysing the oxidation of α-oxoglutarate to succinyl-CoA. This incompleteness alters the status of both α-oxoglutarate and succinyl-CoA to that of end-products and under such circumstances it might be anticipated that they would act as additional regulators of the initial enzyme citrate synthase.

One instance of this situation is provided by *Esch. coli*. During anaerobic growth this organism generates energy by fermentation; α-oxoglutarate dehydrogenase is absent and the citric acid cycle is modified to a branched, noncyclic pathway (Fig. 3) which can meet the biosynthetic demands for both α-oxoglutarate and succinyl-CoA (Amarasingham & Davis 1965). Only α-oxoglutarate formation, however, requires the participation of citrate synthase, succinyl-CoA being produced from oxaloacetate by a reversed set of reactions. α-Oxoglutarate is now the end-product of a short sequence of reactions initiated by citrate synthase acting in an exclusively biosynthetic role.

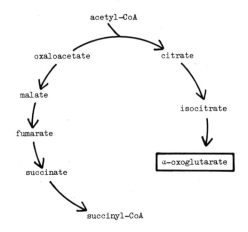

Fig. 3. Modified citric acid cycle operative in *Escherichia coli* growing anaerobically.

TABLE 3

Inhibition of citrate synthases by α-oxoglutarate

Inhibition by α-oxoglutarate	No inhibition by α-oxoglutarate
Arizona arizonae	Gram negative bacteria
Escherichia coli	*Acinetobacter lwoffi*
Klebsiella (Aerobacter) aerogenes	*Azotobacter vinelandii*
Proteus vulgaris	*Moraxella calcoacetica*
Salmonella anatum	*Pseudomonas* { *aeruginosa* / *fluorescens* / *ovalis*
	Gram positive bacteria
	Bacillus megaterium
	Kurthia zopfii
	Streptomyces somaliensis
	Eukaryotes
	Baker's yeast
	Wheat germ
	Pig heart

Weitzman & Dunmore (1969*a*).

Inhibition of citrate synthase by α-oxoglutarate was first reported by Wright *et al.* (1967) and subsequently confirmed and extended by Weitzman & Dunmore (1969*a*). The latter workers found that the effect was restricted to citrate synthases from Gram negative facultatively-anaerobic bacteria; no inhibition was shown by the enzymes from aerobic Gram negative bacteria, Gram positive bacteria or eukaryotic organisms (Table 3). It was therefore concluded that the effect of α-oxoglutarate on citrate synthase represents a typical case of end-product inhibition of the initial enzyme of a pathway and is consistent with the operation of the split pathway illustrated in Fig. 3. Taylor (1970) has observed α-oxoglutarate inhibition of citrate synthase from strictly autotrophic Gram negative thiobacilli, even though they are strict aerobes. These organisms also lack α-oxoglutarate dehydrogenase, so that α-oxoglutarate is here, too, a bio-synthetic end-product of citrate synthase action. It thus appears that α-oxogluta-rate inhibition of citrate synthase is typical of Gram negative bacteria which can function with the incomplete citric acid cycle.

A somewhat different situation occurs in the blue-green bacteria (cyanobac-teria). These organisms lack the enzyme complex α-oxoglutarate dehydrogenase and do not use the citric acid cycle for energy production. It would therefore be expected that blue-green bacterial citrate synthases would not be regulated by NADH, and this has indeed been found to be the case (Taylor 1973; Lucas & Weitzman 1975). α-Oxoglutarate is an end-product of citrate synthase action and, as in the cases described above, is an inhibitor of the enzymic activity (Taylor 1973; Lucas & Weitzman 1975). The formation of succinyl-CoA in blue-green bacteria, however, is believed to occur not by the split pathway of Fig. 3 but *via* the operation of the glyoxylate cycle (Pearce *et al.* 1969; Lucas 1974) as illu-strated in Fig. 4. In this scheme, citrate synthase functions as the initial enzyme of the branched pathway leading not only to α-oxoglutarate but also to succinyl-CoA, and it was therefore gratifying to observe that succinyl-CoA is an additional specific and powerful inhibitor of blue-green bacterial citrate synthases (Lucas & Weitzman 1977).

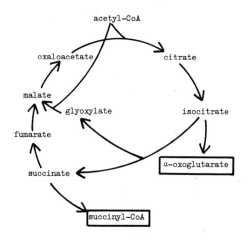

Fig. 4. Modified citric acid cycle operative in blue-green bacteria.

C. Allosteric nature of controls

Several experimental approaches have indicated that the various effectors of Gram negative bacterial citrate synthases—NADH, AMP, α-oxoglutarate and succinyl-CoA—all exert their influences on enzymic activity by an allosteric mechanism, i.e. there appear to be receptor sites on the different enzymes with which these effectors may specifically interact and, probably as a result of conformational changes in the enzyme structure, produce alterations in the catalytic process. What emerges from all these studies is that citrate synthase is finely constructed so as to display diverse regulatory properties attuned to its diverse metabolic functions in different organisms. Thus organisms which are metabolically and taxonomically related may possess distinctive and characteristic properties of this enzyme which, as will be shown, may serve as aids to classification and identification.

D. Molecular size patterns

The Gram negative bacteria, among which may be included the blue-green bacteria, have citrate synthases which are clearly more complex in functional behaviour than are their counterparts from Gram positive bacteria and eukaryotes. This suggests that the former enzymes may well be more complex in structure and prompted work to explore this possibility at the gross level of molecular size (Weitzman & Dunmore 1969b). By carrying out gel filtration on a column of Sephadex G-200 with suitable marker proteins (catalase and lactate dehydrogenase) it was possible to estimate the molecular weights of various citrate synthases.

All the citrate synthases examined fall into two groups—'large' and 'small'. The 'large' enzymes have a molecular weight in the region of 250 000 and are eluted from the Sephadex G-200 column ahead of catalase, whereas the 'small' enzymes have a mol. wt of around 100 000 and are eluted after lactate dehydrogenase. Most significantly, the 'large' citrate synthases are all derived from Gram negative bacteria, while the enzymes from Gram positive bacteria and eukaryotes are, without exception, of the 'small' type. Subsequent studies have furnished data on an increasing number of different citrate synthases and support this scheme correlating molecular size of the enzyme and the Gram staining status of the organism. Table 4 divides bacteria according to the size of their citrate synthase and clearly demonstrates this correlation. Currently available data suggest that the 'small' and 'large' citrate synthases are respectively dimeric and tetrameric in structure and that the subunits within each type of molecule are probably identical.

It is thus clear that the more complex regulatory properties observed in the

TABLE 4
Molecular sizes of bacterial citrate synthases

'Large' citrate synthases

Acetobacter xylinum	Cyanobacteria
Acinetobacter spp.	*Anabaena flow-aquae*
Aeromonas formicans	*Anabaenopsis circularis*
Azotobacter vinelandii	*Anacystis nidulans*
Escherichia coli	*Aphanocapsa* 6714
Klebsiella (Aerobacter) aerogenes	*Chlorogloea fritschii*
Moraxella calcoacetica	*Coccochloris elabens*
Pseudomonas spp.	*Gloeocapsa alpicola*
Rhodopseudomonas spheroides	*Mastigocladus laminosus*
Salmonella anatum	*Nostoc* sp. E
Serratia marcescens	*Plectonema boryanum*

'Small' citrate synthases

Arthrobacter globiformis	Haemophilus vaginalis
Bacillus { stearothermophilus* / subtilis	Kurthia zopfii
	Mycobacterium rhodochrous
Brevibacterium { flavum † / linens	Streptomyces somaliensis

Weitzman & Danson (1976); Lucas & Weitzman (1977);
*Higa & Cazzulo (1976); †Shiio *et al.* (1977).

Gram negative bacterial citrate synthases are indeed reflected in a more complex quaternary structure. A molecular structural characteristic is therefore available as an aid to bacterial recognition, additional to the functional characteristics discussed above.

E. Rapid characterization techniques

The potential usefulness of the regulatory and molecular patterns of citrate synthase for bacterial typing prompted studies to examine the speeding-up of the tests for these enzymic properties so that they might be suitable as an additional taxonomic tool in bacteriological laboratories. It was found possible to observe regulatory properties of citrate synthase *in situ* in bacteria rendered permeable to enzyme substrates by treatment with toluene (Harford *et al.* 1976). This procedure allows rapid examination of the enzyme in bacteria grown on an agar plate and obviates the necessity to grow larger quantities of bacteria in liquid media and to prepare a sonic extract for investigation. The method is simple and readily permits rapid examination of a number of different bacteria. At the same time it was shown to be possible to reduce very considerably the dimensions of the gel filtration column used to determine enzyme molecular size, so that the assignment of a particular citrate synthase to the 'large' or 'small' group may be made extremely rapidly. Total agreement was demonstrated between the results of these micro-methods and those obtained by the more conventional large-scale procedures, so that one may confidently employ the fast micro-methods in bacteriological testing.

F. Odd ones out

The perfectionist Sherlock Holmes could maintain that an exception disproves the rule. Most microbiologists, however, are only too familiar with biological anomalies and unpredictabilities, are less puristic about their rules and positively anticipate exceptions to them. It is therefore in order to comment on some apparent exceptions to the citrate synthase 'rules'.

Swissa & Benziman (1976) reported the insensitivity of *Acetobacter xylinum* citrate synthase to NADH, though this organism is unambiguously Gram negative. Experiments in the author's laboratory have confirmed this and have suggested that other *Acetobacter* spp. may similarly have anomalous citrate synthase. The enzyme is, however, of the 'large' type in keeping with the Gram staining status of *Acetobacter*. On the other hand, the citrate synthases of *Halobacterium* spp. and of *Thermus aquaticus* and related extreme thermophiles are not inhibited by NADH and are of the 'small' type, despite the fact that these bacteria are all Gram negative (Cazzulo 1973; Weitzman & Danson 1976; Weitzman 1978). It has been suggested that this irregularity may be related to the extreme halophilicity and thermophilicity of these organisms, factors which may not be conducive to the stable existence of the 'large' type of citrate synthase required for sensitivity to NADH (Weitzman & Danson 1976), and Cazzulo (1973) has speculated that, as inhibition of citrate synthase by NADH is in all cases overcome by high salt concentration, the evolutionary adaptation of halobacteria to halophilic life resulted in the loss of enzymic features which became physiologically ineffective. Another exception to have been reported is the insensitivity of *Thiobacillus novellus* citrate synthase to NADH (Taylor 1970) despite the Gram negative nature of this organism and its possession of a complete citric acid cycle. Future studies may throw more light on these apparent anomalies and may even allow investigators to trace some logic in the existence of these few exceptions to the general rules.

4. Succinate Thiokinase

A. Molecular size patterns

Both *Esch. coli* and pig heart succinate thiokinases have been investigated in some detail (Bridger 1974) and one striking difference to emerge is in their molecular weights. The *Esch. coli* enzyme has a mol. wt of 140 000–150 000 and is a tetramer. The pig heart enzyme, on the other hand, has a mol. wt of

70 000–75 000 and is dimeric in structure. This molecular size difference between the *Esch. coli* and eukaryotic succinate thiokinases is strongly reminiscent of the differences between citrate synthases and prompted an investigation into the molecular sizes of a range of bacterial and eukaryotic succinate thiokinases (Weitzman & Kinghorn 1978). This was done by gel filtration of cell-free extracts through Sephadex G-200 using lactate dehydrogenase as a marker protein. Succinate thiokinase may conveniently be assayed by a polarographic procedure (Weitzman 1969, 1976) in which the formation of coenzyme A is monitored continuously with a dropping mercury electrode. Consistent with the molecular weight values quoted above, the *Esch. coli* succinate thiokinase was eluted from the Sephadex column slightly ahead of lactate dehydrogenase (mol. wt 140 000) whereas the pig heart enzyme was eluted considerably later than lactate dehydrogenase. All other succinate thiokinases examined conform to one of these two types, i.e. 'large' or 'small'. The remarkable finding, however, was that the 'large' succinate thiokinase is encountered only in Gram negative (including blue-green) bacteria whereas the 'small' succinate thiokinase is associated with Gram positive bacteria and eukaryotic organisms. Table 5 lists the currently available data on the molecular size of various succinate thiokinases. There is thus a striking correlation between the incidence of 'large' and 'small'

TABLE 5

Molecular sizes of succinate thiokinases

'Large' enzyme	'Small' enzyme
Acinetobacter spp.	*Bacillus* { *megaterium* / *stearothermophilus* }
*Aerobacter aerogenes**	
Aeromonas formicans	*Brevibacterium linens*
Brevibacterium leucinophagum	*Corynebacterium rubrum*
Escherichia coli	
Halobacterium salinarium	Baker's yeast
*Herella vaginicola**	Cauliflower mitochondria
Pseudomonas { *aeruginosa* / *citronellolis** }	Wheat germ
Rhodopseudomonas spheroides	Pig { heart / liver }
Thermus aquaticus	
Xanthomonas hyacinthi	
Cyanobacteria	
Anabaenopsis circularis	
Anacystis nidulans	
Aphanocapsa 6714	
Chlorogloea fritschii	
Gloeocapsa alpicola	
Nostoc sp. E	
Plectonema boryanum	

Weitzman & Kinghorn (1978, 1979); *Kelly & Cha (1977).

succinate thiokinases and that of 'large' and 'small' citrate synthases (see Section 3D) and there is speculation that there may be some evolutionary link between the two enzymes (Weitzman & Kinghorn 1978).

It is noteworthy that the succinate thiokinases of *Halobacterium salinarium* and *Thermus aquaticus* are 'large' and typically ot the Gram negative type (Weitzman unpublished). This is in contradistinction to the anomalous 'small' citrate synthases found in these organisms (see Section 3F) and suggests that the 'large' succinate thiokinase may be an even more rigidly conserved feature of Gram negative bacteria than is the 'large' citrate synthase. Characterization of the molecular size of succinate thiokinase is thus of potential value in bacterial classification.

B. Nucleotide specificity patterns

It has been claimed that mammalian succinate thiokinase can use either GDP or IDP as substrate whereas plant and bacterial succinate thiokinases are specific for ADP (Bridger *et al*. 1969). In the course of studying the kinetics of succinate thiokinase from *Acinetobacter lwoffi* the author observed an extremely high Km

TABLE 6

Nucleotide specificity patterns of bacterial succinate thiokinases

Group	Organism	Gram reaction	K_mADP	K_mGDP
I	*Acinetobacter* spp. *Bordetella bronchiseptica* *Chromobacterium violaceum* *Mima polymorpha* *Xanthomonas hyacinthi*	—	very high	low
II	*Pseudomonas* { *aeruginosa* *fluorescens* *stutzeri* *Rhodopseudomonas spheroides* *Thermus aquaticus*	—	low	low
III	*Alcaligenes faecalis* *Arizona arizonae* *Escherichia coli* *Klebsiella (Aerobacter) aerogenes* *Serratia marcescens*	—	low	high
IV	*Arthrobacter simplex* *Bacillus* { *megaterium* *stearothermophilus* *Brevibacterium linens* *Kurthia zopfii*	+	low	no activity

Jaskowska-Hodges & Weitzman (unpublished).

value for ADP but a very much lower Km value for GDP. The *Acinetobacter* enzyme thus 'prefers' GDP to ADP and indicates that the claim that bacterial succinate thiokinase functions specifically with ADP requires some modification.

Examination of a range of bacterial succinate thiokinases (Jaskowska-Hodges 1975; Jaskowska-Hodges & Weitzman 1979) for their nucleotide substrate specificity has revealed the existence of several classes of enzyme. One group of bacteria resembled *Acinetobacter lwoffi* in having succinate thiokinases with a very high Km for ADP ($\sim$ 1 mM) but a much lower Km for GDP ($\sim$ 0.02 mM). Another group exhibited similar and low Km values ($<$0.05 mM) for both ADP and GDP, while a third group showed low Km values for ADP ($\sim$ 0.01 mM) but significantly higher Km values for GDP ($\sim$ 0.1 mM). A fourth group had succinate thiokinases which functioned with ADP (Km values $<$0.1 mM) but which displayed no activity with GDP as substrate. These groups, designated I-IV, are listed in Table 6. Kelly & Cha (1977) have also reported nucleotide substrate preferences for a few bacterial succinate thiokinases, and their results are essentially in agreement with those cited. It is worth noting that the same patterns of substrate utilization are observed if the succinate thiokinases are assayed in the reverse direction in which the hydrolysis of ATP or GTP is coupled to the formation of succinyl-CoA; whichever nucleoside diphosphate-ADP or GDP- is preferred in one direction, the equivalent triphosphate-ATP or GTP- is preferred in the reverse direction.

The results presented in Table 6 are derived from a fairly small sample of diverse bacteria so that it may be premature to read too much into the identity of organisms within the four groups. Nevertheless there are unmistakeable signs of some grouping together of bacteria with established taxonomic relatedness. Groups I, II and III are all Gram negative bacteria; the Gram positive bacteria appear to be confined to Group IV. The enterobacteria occur together in Group III and the pseudomonads in Group II. Several *Acinetobacter* spp. and a few other strict aerobes make up Group I and it is significant that all these Group I organisms share several additional highly distinctive features of two other citric acid cycle enzymes—isocitrate dehydrogenase and α-oxoglutarate dehydrogenase (Weitzman unpublished)—which are not exhibited by organisms in Groups II, III and IV. Many more bacterial succinate thiokinases will need to be studied in order to substantiate (or refute) these preliminary conclusions, but the results so far point to interesting patterns of enzyme diversity and to possible correlations with bacterial classification.

5. Applications of Enzyme Characterization to Classification and Identification

The structural and functional patterns displayed by citrate synthase and succinate thiokinase have been discussed in some detail in the hope that this information may be useful to those interested in bacterial classification and identifica-

TABLE 7

Characteristics of citrate synthase and succinate thiokinase from bacteria showing an equivocal Gram stain reaction

Organism	Inhibition by NADH	Reactivation by AMP	Citrate synthase molecular size	Succinate thiokinase molecular size	Gram staining status indicated by enzyme characteristics	Acutal Gram stain reaction of bacteria
Achromobacter liquefaciens	–	–	small	small	+	±
Cellulomonas rossica	+	+	large	large	–	∓
Corynebacterium nephridii	+	+	large	large	–	∓
Haemophilus vaginalis	–	–	small	small	+	∓
Pseudomonas iodinum	–	–	small	small	+	±

Weitzman & Jones (1975); Weitzman *et al.* (1979).

tion, and that the potential value of these patterns may be explored profitably in the future. Past efforts have been directed primarily to the collection of experimental data on the enzymes from known and well-characterized organisms and to the disclosure of their patterns of diversity. In passing, however, a few organisms of relatively unsettled identity have been examined and the enzyme data obtained used as evidence to support, or question, their classification. On several occasions the author has been sent bacterial cultures by other investigators and, on the basis of their citrate synthase and succinate thiokinase behaviour, has been able to comment on the Gram staining status of the organism, its aerobic or facultative type of metabolism and its possible relatedness to other bacteria. That the investigators seemed content with the information provided and were able to harmonize it satisfactorily with other tests and characterizations they performed would indicate that the enzyme tests do indeed have a role to play in classification and identification.

A formal study has also been made of several bacteria which either give an equivocal Gram reaction or whose taxonomic status is uncertain for other reasons (Weitzman & Jones 1975; Weitzman unpublished). Table 7 lists some of the organisms examined and Weitzman & Jones (1975) nave detailed the nature of the uncertainties surrounding their classification. The results on citrate synthase and succinate thiokinase from these bacteria unambiguously indicated a particular Gram staining status and, in each case, this is in agreement with the latest views on the identities of these organisms.

Examination of citrate synthase or succinate thiokinase may also offer a first clue to misclassification. An example of this is provided by studies on the enzymes from *Brevibacterium leucinophagum* (Jones & Weitzman 1974). The citrate synthase of this supposedly Gram positive bacterium was found to be of the Gram negative type, suggesting an error in classification, and a thorough examination, including electron microscopy studies, confirmed that this bacterium is indeed Gram negative. As a result of detailed enzyme characterization it was suggested that this organism be classified in the genus *Acinetobacter*.

6. Concluding Remarks

It was the author's interest in the metabolic regulation of citrate synthase which led to the discovery of different modes of such regulation and to the gradual awareness of patterns of diversity and correlations with bacterial taxonomy. Examination of other citric acid cycle enzymes has revealed further diversity patterns and it would seem likely that enzymes of other metabolic pathways might also display fine differences which can serve to distinguish different groups of micro-organisms. Indeed there are several well-known examples of enzyme regulatory diversity among different groups of organisms, the area of amino acid biosynthesis being a particularly well-studied one. While it is hoped that the results presented in this chapter are useful in themselves, perhaps they may be of further use in emphasizing the potential value of this type of study so that the characterization of the fine details of enzyme catalysts and their regulation may prove to be a growing tool for use in bacterial taxonomy and identification.

I am indebted to Dr Dorothy Jones, University of Leicester, for her patient guidance and advice and for introducing me to the labyrinths of the microbial world. I also acknowledge research support from the Science Research Council.

7. References

AMARASINGHAM, C. R. & DAVIS, B. D. 1965 Regulation of α-ketoglutarate dehydrogenase formation in *Escherichia coli*. *Journal of Biological Chemistry* **240**, 3664-3667.

BRIDGER, W. A. 1974 Succinyl-CoA synthetase. In *The Enzymes* 3rd edn, Vol. X, ed. Boyer, P. D. pp. 581-606. New York: Academic Press.

BRIDGER, W. A., RAMALEY, R. F. & BOYER, P. D. 1969 Succinyl coenzyme A synthetase from *Escherichia coli*. *Methods in Enzymology* **13**, 70-75.

CAZZULO, J. J. 1973 On the regulatory properties of a halophilic citrate synthase. *FEBS Letters* **30**, 339-342.

ELLMAN, G. L. 1959 Tissue sulfhydryl groups. *Archives of Biochemistry and Biophysics* **82**, 70-77.

HARFORD, S., JONES, D. & WEITZMAN, P. D. J. 1976 Rapid techniques for the examination of bacterial citrate synthases. *Journal of Applied Bacteriology* **41**, 465-471.

HIGA, A. I. & CAZZULO, J. J. 1976 The citrate synthase from *Bacillus stearothermophilus*. *Experientia* **32**, 1373-1374.

JASKOWSKA-HODGES, H. 1975 *The control of citric acid cycle enzymes in* Acinetobacter. Ph.D. Thesis, University of Leicester, UK.

JONES, D. & WEITZMAN, P. D. J. 1974 Reclassification of *Brevibacterium leucinophagum* Kinney and Werkman as a Gram-negative organism, probably in the genus *Acinetobacter*. *International Journal of Systematic Bacteriology* **24**, 113-117.

KELLY, C. J. & CHA, S. 1977 Nucleotide specificity of succinate thiokinases from bacteria. *Archives of Biochemistry and Biophysics* **178**, 208-217.

LUCAS, C. 1974 *The intermediary metabolism of facultatively heterotrophic blue-green algae*. Ph.D. Thesis, University College of Wales, Aberystwyth, UK.

LUCAS, C. & WEITZMAN, P. D. J. 1975 Citrate synthase from blue-green bacteria. *Biochemical Society Transactions* **3**, 379-381.

LUCAS, C. & WEITZMAN, P. D. J. 1977 Regulation of citrate synthase from blue-green bacteria by succinyl coenzyme A. *Archives of Microbiology* **114**, 55-60.

PEARCE, J., LEACH, C. K. & CARR, N. G. 1969 The incomplete tricarboxylic acid cycle in the blue-green alga *Anabaena variabilis*. *Journal of General Microbiology* **55**, 371-378.

SHIIO, I., OZAKI, H. & UJIGAWA, K. 1977 Regulation of citrate synthase in *Brevibacterium flavum*, a glutamate-producing bacterium. *Journal of Biochemistry* **82**, 395-405.

SWISSA, M. & BENZIMAN, M. 1976 Factors affecting the activity of citrate synthase of *Acetobacter xylinum* and its possible regulatory role. *Biochemical Journal* **153**, 173-179.

TAYLOR, B. F. 1970 Regulation of citrate synthase activity in strict and facultatively autotrophic thiobacilli. *Biochemical and Biophysical Research Communications* **40**, 957–963.

TAYLOR, B. F. 1973 Fine control of citrate synthase activity in blue-green algae. *Archives of Microbiology* **92**, 245–249.

WEITZMAN, P. D. J. 1966 Regulation of citrate synthase activity in *Escherichia coli*. *Biochimica et Biophysica Acta* **128**, 213–215.

WEITZMAN. P. D. J. 1969 Polarographic assay for malate synthase and citrate synthase. *Methods in Enzymology* **13**, 365–368.

WEITZMAN. P. D. J. 1976 Assay of enzyme activity by polarography. *Biochemical Society Transactions* **4**, 724–726.

WEITZMAN, P. D. J. 1978 Anomalous citrate synthase from *Thermus aquaticus*. *Journal of General Microbiology* **106**, 383–386.

WEITZMAN, P. D. J. & DANSON, M. J. 1976 Citrate synthase. In *Current Topics in Cellular Regulation* Vol. 10, ed. Horecker, B. L. & Stadtman, E. R. pp. 161–204. New York: Academic Press.

WEITZMAN, P. D. J. & DUNMORE, P. 1969*a* Regulation of citrate synthase activity by α-ketoglutarate. Metabolic and taxonomic significance. *FEBS Letters* **3**, 265–267.

WEITZMAN, P. D. J. & DUNMORE, P. 1969*b* Citrate synthases: allosteric regulation and molecular size. *Biochimica et Biophysica Acta* **171**, 198–200.

WEITZMAN, P. D. J. & JONES, D. 1968 Regulation of citrate synthase and microbial taxonomy. *Nature, London* **219**, 270–272.

WEITZMAN, P. D. J. & JONES, D. 1975 The mode of regulation of bacterial citrate synthase as a taxonomic tool. *Journal of General Microbiology* **89**, 187–190.

WEITZMAN, P. D. J. & KINGHORN, H. A. 1978 Occurrence of 'large' or 'small' forms of succinate thiokinase in diverse organisms. *FEBS Letters* **88**, 255–258.

WRIGHT, J. A., MAEBA, P. & SANWAL, B. D. 1967 Allosteric regulation of the activity of citrate synthetase of *Escherichia coli* by α-ketoglutarate. *Biochemical and Biophysical Research Communications* **29**, 34–38.

Cytochrome Patterns in Classification and Identification Including their Relevance to the Oxidase Test

C. W. JONES

Department of Biochemistry, University of Leicester, Leicester, UK

Contents

1. Historical Perspective

THE CYTOCHROMES were discovered in 1886 as a result of spectroscopic studies of mammalian muscle carried out by a Wolverhampton general practitioner, Charles MacMunn (1886). The work was, however, severely and erroneously criticized by several eminent organic chemists, including Hoppe-Seyler, and was subsequently forgotten until the rediscovery of the cytochromes by Keilin (1925) working in Cambridge almost forty years later. Although much of Keilin's earliest work on the cytochromes was restricted to examining their occurrence in animal, yeast and plant cells, he also reported that cytochromes were present in the obligately aerobic bacterium, *Bacillus subtilis,* and that they were absent from the obligate anaerobe, *Clostridium sporogenes.* This work was followed by further extensive spectroscopic investigations of whole bacteria by Warburg and his colleagues in Germany and by Tamiya in Japan (see for example, Yaoi & Tamiya 1928; Warburg *et al.* 1933), until by the mid 1930s the simple cytochrome spectra of approximately sixty species of bacteria had been examined.

These early studies of cytochromes were based almost entirely on the use of simple hand spectroscopes or, in Keilin's laboratory, of a microscope in which the eyepiece had been replaced by either a microspectroscope ocular or a Hartridge reversion spectroscope. In general these techniques produced relatively well-defined, absolute absorption spectra. The changes which were observed in these spectra as a result of altering the environment in which the cells were suspended pointed clearly to an oxidation-reduction (redox) role for the cytochromes in cellular respiration, a role which was later confirmed by Warburg's measurements of photochemical action spectra (i.e. spectra based on the relative abilities of different wavelengths of light to relieve the inhibition of respiration by carbon monoxide, a potent isostere of molecular oxygen).

These approaches were refined subsequently by the introduction initially of manual spectrophotometers and later of recording spectrophotometers, with split-beam, dual-wavelength or derivative facilities, which allowed accurate measurement of difference spectra (most importantly, reduced *minus* oxidized and reduced + CO *minus* reduced spectra) and photochemical action spectra. Difference spectra yield maximum information when determined at $66°K$ ('by immersing the cuvettes in liquid nitrogen to sharpen and intensify the spectral bands'), or when repetitive data are subjected to numerical analysis (Chance 1954; Castor & Chance 1959; Shipp 1972). Over two hundred species of bacteria have now been investigated with respect to their cytochrome spectra but, unfortunately, not all of these investigations have been as comprehensive as might be desired. The result is that the use of cytochrome patterns in bacterial classification and identification is still in its infancy.

A massive amount of evidence has accumulated over the last twenty years to show that the cytochromes are specialized forms of haemoproteins (see, for example, Lemberg & Barrett 1973). They can be assigned to four classes according to the structures of their haem prosthetic groups: cytochromes *a* (haem *a*); *b* (haem *b* or protohaem); *c* (haem *c* or mesohaem), and *d* (haem *d* or chlorin). Haem consists of four pyrrole rings joined by —CH= bridges to form a planar structure, porphyrin, at the centre of which is held a single iron atom. The outside edge of the porphyrin is subject to substitution by a variety of reactive groups, and it is these substitutions that are largely responsible for the different absorption characteristics and standard redox potentials (E_o') of the four types of cytochromes. The central iron atom is usually in the low-spin configuration which leads to the formation of octahedral co-ordination complexes with a maximum of six ligands. Four of these are invariably the pyrrole nitrogens, whereas the fifth and sixth ligands are often nitrogen or sulphur atoms from appropriately located amino acids (usually histidine and methionine respectively) in the apoprotein. In a few specialized cytochromes, the cytochrome oxidases, the sixth position is available to water, molecular oxygen or related inhibitory molecules such as carbon monoxide. The relative position of the haem within the apoprotein is further stabilized in the *c*-type of cytochromes by covalent linkages (thioether bridges) between specific vinyl groups on the porphyrin ring and cysteine residues in the protein backbone. In all of the cytochromes, the iron undergoes one-electron oxidation-reduction, oscillating between the reduced (Fe^{2+}) and oxidized (Fe^{3+}) states with their characteristic absorption spectra.

2. The Properties of Bacterial Cytochromes

Bacterial cytochromes are involved in a wide variety of redox processes including photosynthetic electron transfer, chemolithotrophic respiration (from a variety of inorganic electron donors to molecular oxygen), heterotrophic respiration (from a variety of organic electron donors to molecular oxygen) and anaerobic respiration (to a variety of organic and inorganic electron acceptors). In each case electron transfer occurs from a low redox potential donor to a higher redox potential acceptor with the concomitant release of free energy which can be conserved in the formation of a trans- or intra-membrane protonmotive force, and subsequently in the synthesis of ATP *via* photosynthetic or oxidative phosphorylation (Mitchell 1966). The cytochromes operate over a very wide redox potential range (E_0' –400 to +450mV) and can be separated functionally into those which: (i) transfer electrons to adjacent redox carriers, be the latter other cytochromes or even different types of redox carriers; (ii) transfer electrons to exogenous electron acceptors such as molecular oxygen (cytochrome oxidase) or nitrite (nitrite reductase), and (iii) insert one or more atoms of oxygen into specific substrates (various oxygenases and hydroxylases). Disruption and differential centrifugation studies indicate that most cytochromes, as expected from their involvement in electron transfer and energy conservation, are associated with the cytoplasmic membrane. It should be noted, however, that some c-type cytochromes, nitrite reductase and various oxygenases are located within either the cytoplasm or the periplasmic space.

Some properties of the most common bacterial cytochromes are outlined in Table 1. Five carbon monoxide-binding cytochromes with potential oxidase functions are known: cytochromes aa_3, a_1, o (an autoxidizable b-type cytochrome), c_{CO} and d. Measurements of the photochemical action spectra and oxidation-reduction kinetics of these cytochromes in respiratory membranes from a variety of bacteria have so far confirmed only cytochromes aa_3, o and d as oxidases; the oxidase status of cytochromes a_1 and c_{co} is uncertain at present. Interestingly, unlike the mitochondrial respiratory chains of higher organisms, bacterial respiratory systems contain a varied complement of terminal oxidases which occur either singly (e.g. aa_3 or o) or in various combinations (e.g. aa_3o, doa_1, $doaa_3$). There is increasing evidence that some of these oxidases have significantly different affinities for molecular oxygen or for classical inhibitors of oxidase activity such as cyanide. It is likely, therefore, that the flexibility inherent in multiple oxidase systems enables certain organisms to avoid the potentially deleterious effects of some growth environments and to take maximum advantage of others (see Jurtshuk *et al.* 1975; Jones 1977; Haddock & Jones 1977). Most of the b- and c-type cytochromes have no oxidase function

TABLE 1

Bacterial cytochromes

Haem type	Cytochrome	Function	Notes
a	aa_3	Oxidase	Contains Cu; a non-autoxidizable
	a_1	Possible oxidase	
b	b	Intermediate electron carriers	
	o	Oxidase	Autoxidizable b-type
	P-450	Oxygenase	
	P-420	Oxygenase	
c	$c\ c_2\ c'$	Intermediate electron carriers	High and low E_O' species known; single haem c
	$c_3'c_7'$	Intermediate electron carriers	Multihaem c
	c_{CO}	Possible oxidase	
d	d	Oxidase	Originally called a_2
mixed	cd_1	Nitrite reductase	Some oxidase activity; dihaem

and serve only as intermediate electron carriers between other respiratory chain components. It should be noted that, in this respect, some species of c-type cytochromes contain more than one molecule of haem c (e.g. cytochromes c_3 and 'c_7' from some anaerobic, sulphate-reducing bacteria contain four and three haems respectively), and nitrite reductase contains two different haems within a single cytochrome (i.e. haems c and d_1, the latter being a specialized form of haem d).

It should be clear from the above introduction that two basic methods are available which use cytochromes as a basis for classification and identification, the 'pattern' and the 'structural' approach. The former compares the cytochrome patterns of different bacterial species (e.g. $bcaa_3o$ cf. baa_3o, $bdoa_1$ etc.) as determined by conventional difference spectrophotometry (see Smith 1968; Meyer & Jones 1973); the latter compares the primary structures, and, where possible, the tertiary structures, of easily purified cytochrome c as determined by amino acid sequence and X-ray diffraction analyses (see Ambler 1976; Dickerson *et al.* 1976). Both approaches have been used to some advantage although they have significant limitations. Thus, the widely-used 'pattern' approach is limited by the quality of the available data as noted above; the structural approach suffers from the relatively small number of c-type cytochromes that have so far been analysed, particularly at the tertiary level.

3. Cytochrome Patterns in Classification and Identification

Qualitative analyses of the cytochrome compositions of over two hundred species of bacteria have now been done. The results indicate some general trends in cytochrome distribution (Fig. 1). Most strikingly, the Gram positive heterotrophs (Bergey's Manual, Eighth edition Parts 14–17) comprise a rather homogeneous grouping with cytochromes $bcaa_3o$ forming the predominant pattern. One of the two cytochrome oxidases is occasionally absent from some species, and cytochrome c is often absent, particularly from the facultative anaerobes and pathogens. *Bacillus popilliae* is a notable exception to this general pattern. It contains cytochromes doa_1 in place of aa_3o and further taxonomic investigation would appear to be merited. The cytochrome content of obligate anaerobes is very variable. Thus, the genus *Clostridium* exhibits no cytochromes and the Lactobacillaceae contain only cytochrome b when cultured on a haem-containing medium. In contrast, the Propionibacteriaceae exhibit a cytochrome bda_1 pattern and the sulphate/sulphite-reducer, *Desulfotomaculum nigrificans*, contains cytochromes bcd. In the last two cases it is likely that the cytochrome oxidase d functions principally to scavenge traces of molecular oxygen from the environment, a function to which its high affinity for oxygen is perfectly suited (Jones 1978; Rice & Hempfling 1978). It is interesting to note in a slightly different context that the cytochrome $bcaa_3o$ pattern of logarithmic growth-phase cells of an aerobe, *Arthrobacter globiformis*, changes to $bcaa_3od$ when the cells become oxygen-limited and lose their ability to retain the crystal violet-iodine complex in the Gram stain. It is noteworthy that the ability to synthesize cytochrome oxidase d is characteristic of many Gram negative organisms (see below).

In contrast to the previous group, the Gram negative heterotrophs (Bergey's Manual, Eighth Edition Parts 7–11), appear to form a much less homogeneous group with respect to their cytochrome composition (Fig. 1). The majority of these organisms contain cytochrome oxidases doa_1, in place of cytochrome oxidases aa_3o, and have the pattern $bcdoa_1$ from which c may often be absent. This absence appears to be particularly characteristic of the Enterobacteriaceae in which the only significant amount of c-type cytochrome is a soluble, low redox potential form found during anaerobic growth. It appears to be associated with formate-nitrite reductase or formate-hydrogen lyase activity (see Haddock & Jones 1977). It should be noted, however, that a c-type cytochrome has been detected in membranes derived from *Escherichia coli* which had been grown under conditions of sulphate-limitation (Poole & Haddock 1975). This basic $bcdoa_1$ pattern, with omissions, is also characteristic of the families Azotobacteriaceae, Vibrionaceae, Neisseriaceae, and of a number of genera *incertae sedis*, *Haemophilus*, *Alcaligenes* (with the exception of *Alca. eutrophus*) and *Acetobacter* (d absent). The family Pseudomonadaceae also contains several species with this cytochrome pattern but only a few of these are in the genus *Pseudomonas* (e.g. *Psmn. malto-*

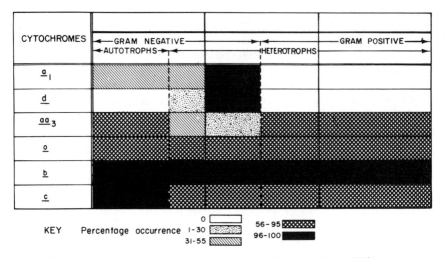

Fig. 1. The distribution of cytochromes in bacteria (after Meyer & Jones, 1973).

philia, Psmn. riboflavina and possibly *Psmn. aeruginosa* and *Psmn. putida*). In contrast, the genera *Halobacterium*, *Rhizobium* and *Paracoccus*, and many species within the genus *Pseudomonas*, exhibit a pattern $bcaa_3o$ with c occasionally absent that is more typical of the Gram positive heterotrophs. Some species of *Pseudomonas* (e.g. *fluorescens* and *syringae*) apparently fail to synthesize any a or d-type cytochromes and are characterized by the simple patterns bco or bo.

   The influence of the growth environment on bacterial cytochrome patterns cannot be over-emphasized, particularly within the nutritionally versatile groups of Gram negative organisms. Thus, for example, *Paracoccus denitrificans (Micrococcus denitrificans)* contains soluble cytochrome c_{co} when cultured autotrophically on methanol (Van Verseveld & Stouthamer 1978) and soluble cytochrome cd_1, nitrite reductase, when grown anaerobically with nitrate as the terminal electron acceptor. Interestingly, cytochrome c_{co} appears to be characteristic of methylotrophs (Higgins *et al.* 1976) and is thus found in the family Methylomonadaceae and in some species of *Pseudomonas* (e.g. *Psmn. extorquens* and AM1). It is also present, however, in the non-methylotrophic genera *Beneckea* and *Chromobacterium* (Weston & Knowles 1973; Niven *et al.* 1975). Cytochrome cd_1 is found in all bacteria growing under denitrifying conditions, including several species of *Pseudomonas* (e.g. *aeruginosa* and *denitrificans*), where it catalyses the reduction of NO_2^- to N_2 (Payne 1973).

   Two major subgroups of Gram negative bacteria remain, those which obtain energy from (a) light (photoanaerobes; Bergey's Manual, Eighth Edition Part 1), and (b) the oxidation of inorganic electron donors (chemolithotrophs, Bergey's

Manual, Eighth Edition Part 12). The phototrophs are either facultative photo-anaerobes (Rhodospirillaceae) or obligate photoanaerobes (Chromatiaceae and Chlorobiaceae). When grown photosynthetically, all three families contain cyto-chromes b and c for cyclic electron transfer, but cytochrome oxidases are synthesized only by the facultatively photoanaerobic Rhodospirillaceae (aa_3 and/or o), maximum concentrations being observed when these organisms are grown aerobically in the dark. The obligately aerobic chemolithotrophs (Nitro-bacteriaceae, Thiobacillaceae and Siderocapsaceae, and organisms such as $Alca.$ $eutrophus$, $Paco.$ $denitrificans$ and $Psmn.$ $saccharophila$ which are able to oxidize molecular hydrogen) exhibit the cytochrome pattern $bcaa_3oa_1$ with one or more terminal oxidases occasionally absent. There is convincing evidence that in $Nitro-$ $bacter$ cytochrome a_1 catalyses the initial oxidation of NO_2^- to NO_3 with the subsequent reduction of cytochrome c and hence of cytochrome oxidase aa_3, i.e. a_1 is probably not a terminal oxidase in this organism (see Aleem 1977). Neither the phototrophs nor the chemolithotrophs show any evidence of being able to synthesize cytochrome d (Fig. 1).

4. The Oxidase Test

The oxidase test (Kovacs 1956) measures the ability of an organism to catalyse the oxidative synthesis of indophenol blue from a mixture of a methylated p-phenylenediamine and α-naphthol (Fig. 2). The speed with which the blue colour appears has been shown to correlate with the terminal respiratory activity of the organism. A positive reaction within approx. 30 s generally reflects the presence of a membrane-bound, high-potential cytochrome c linked to an active cytochrome oxidase (Jurtshuk et $al.$ 1975). Although the effective absence of either of these two redox components leads to a negative reaction, in practice the latter is more usually associated with the inability of the test organism to synthesize cytochrome c. Thus, for example, within the genus $Pseudomonas$ only $Psmn.$ $syringae$ and $Psmn.$ $maltophilia$ fail to exhibit a positive oxidase reac-tion; both organisms contain active cytochrome oxidases but are deficient in cytochrome c.

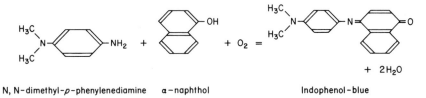

N, N-dimethyl-p-phenylenediamine α-naphthol   Indophenol-blue

Fig. 2. The oxidase reaction. Note that N,N,N',N'-tetramethyl-p-phenylenediamine is often used in place of the dimethyl derivative shown above.

It has been shown that the presence in bacterial respiratory systems of a high-potential cytochrome c linked to cytochrome oxidases aa_3 or o is an obligatory prerequisite to energy coupling at site 3 (Jones *et al.* 1975, 1977). In the absence of cytochrome c, or when cytochrome d is the terminal oxidase, site 3 is absent. Interestingly, therefore, site 3 appears to be present in the oxidative phosphorylation systems of predominantly aerobic bacteria (e.g. *Alcaligenes, Arthrobacter, Azotobacter* and most species of *Pseudomonas*) which normally grow in such nutritionally poor environments as soil or water, and which must conserve their energy resources. In contrast it often appears to be absent from many facultative anaerobes (e.g. most Enterobacteriaceae, many pathogens and some species of *Bacillus*) whose natural habitats encompass such relatively rich environments as the animal gut or decomposing organic matter where there is an abundant supply of both energy and reducing power. It is likely therefore that the oxidase test reflects not only terminal respiratory properties, but also—although not infallibly—the presence or absence of a third energy coupling site.

5. Cytochrome Structures in Classification and Identification

Current knowledge of bacterial cytochrome structures is very much more limited than that of bacterial cytochrome patterns. This limitation is due partly to the fact that studies are restricted to cytochromes that can be purified easily (i.e. the readily-solubilized c-type cytochromes) and partly to the much more sophisticated methods which are necessary for determining protein structures, particu-

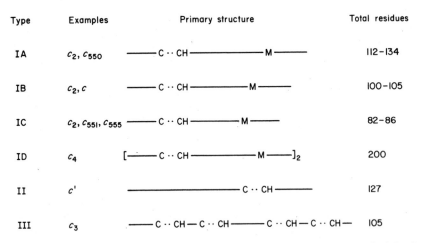

Fig. 3. Primary structures of c-type cytochromes. The amino terminus is on the left, the carboxy terminus on the right. C, cysteine; H, histidine; M, methionine; P, proline, and dots, unspecified amino acids (after Dickerson *et al.* 1976).

larly at the tertiary level. At the moment, the primary structures of 35 bacterial cytochromes c are known (including 14 species of cytochrome c_2 from the family Rhodospirillaceae, plus *Chlorobium thiosulphatophilum* c_{555}, *Pseudomonas* c_{551}, c_4 and c_5, *Paco. denitrificans* c_{550} and *Azotobacter vinelandii* c_{551}), but only 4 tertiary structures have been determined [(*Rhodospirillum rubrum* c_2, *Chlb. thiosulphatophilum* c_{555}, *Psmn. aeruginosa* c_{551} and *Paco. denitrificans* c_{550}) (Salemme *et al.* 1973, Dickerson *et al.* 1976; Timkovitch & Dickerson 1976; Korszun & Salemme 1977)]. The results of these sequence studies indicate that bacterial c-type cytochromes can be classified (Fig. 3) into 3 types: Type I (a single haem attached to the peptide close to the N-terminal end, with the sixth iron ligand to the haem provided by a methionine positioned about three-quarters of the way along the molecule); Type II (a single haem attached to the peptide close to the C-terminal end, e.g. cytochrome c'), and Type III (multiple haems, e.g. cytochrome c_3). Type I cytochromes can be divided further into four subgroups according to their size: (1) Large (112-134 amino acid residues; includes *Rhsp. rubrum* c_2, *Rhodopseudomonas capsulata* c_2 and *Paco. denitrificans* c_{550}); (2) Medium (100-105 amino acid residues; includes *Rhsp. molischianum* c_2, *Rhps. viridis* c_2 and mitrochondrial c); (3) Small (82-86 amino acid residues; includes *Rhsp. tenue* c_2. *Rhps. gelatinosa* c_2, *Chlb. thiosulphatophilum* c_{555} and *Psmn. aeruginosa* c_{551}), and (4) Duplicated (up to 200 amino acid residues, probably the result of gene duplication; includes *Psmn. aeruginosa* c_4).

Intriguingly, tertiary structure analyses of Type I large, medium and small c-type cytochromes indicate that they have very similar folding patterns in spite of the large variations in their size. Thus *Rhsp. rubrum* c_2 (112 amino acid residues) and *Paco. denitrificans* c_{551} (134 amino acid residues) differ only by the presence of an enlarged loop in the 20s region of the c_{551} peptide; furthermore, they differ from mitochondrial cytochrome c only by the latter having smaller loops in the 50s and 80s regions. *Chlorobium thiosulphatophilum* c_{555} (86 amino acid residues) and *Psmn. aeruginosa* c_{551} (82 amino acid residues) show very similar folding patterns to each other and differ from the large cytochrome c, which is present in mitochondria, mainly by the absence of the loop in the 50s region. The similarities in the primary and tertiary structures of cytochrome c from different bacteria and from eukaryote mitochondria have serious implications for the origin and evolution of electron transfer systems. Indeed, Dickerson and his colleagues (1976) have presented compelling evidence to support the hypothesis that present-day bacterial respiratory systems have evolved from the photosynthetic/respiratory systems of the purple bacteria by the loss of photosynthesis. There is also good evidence that the mitochondrial respiratory chain has evolved from the respiratory system of an ancestral bacterium closely related to *Paco. denitrificans* (John & Whatley 1975, 1977).

The use of cytochrome structures in bacterial taxonomy has so far been restricted largely to the facultatively phototrophic genera *Rhodospirillum* and *Rhodopseudomonas,* each of which contains species that have either the large, medium or small cytochrome c_2 as part of their photosynthetic apparatus (Ambler 1976). It would appear, therefore, that some *Rhodospirillum* and *Rhodopseudomonas* spp. are related more closely to each other than to other species in the genera to which they are assigned mainly on morphological criteria. As there is no evidence that such a blurring of taxonomic relationships has occurred by the lateral exchange of cytochrome c_2 genes between these two genera (a separate analysis has indicated that it does not even occur between different species of the genus *Pseudomonas*), it must be concluded that the detailed taxonomy of these photographs merits further investigation.

6. Conclusions

The use of bacterial cytochrome patterns as an aid to gross taxonomy has now reached a relatively advanced stage. The basic cytochrome patterns of many of the major families of bacteria are well established, although little attention has as yet been paid to the rather more obscure families such as those which comprise the gliding, sheathed and budding bacteria or the spirochetes, the rickettsias and the mycoplasmas. Several instances have been reported of bacteria containing one or more cytochromes which are not present in other, supposedly closely-related organisms, thus raising doubts as to the correctness of their current taxonomic positions. In contrast, little use has been made, or is likely to be made, of cytochrome patterns for identification, mainly because bacteria contain relatively few types of spectrally-distinct cytochromes. Thus, although it may be possible to identify certain genera by single-colony scanning techniques the unambiguous identification of individual species is clearly not a feasible proposition.

The much greater time required currently for the determination of cytochrome structures compared with the analysis of cytochrome patterns clearly rules out the immediate use of the former as an aid to gross taxonomy. Instead, cytochrome structures lend themselves much more to the fine taxonomic analysis of small groups of organisms, e.g. of single families or closely-related genera which may exhibit largely indistinguishable cytochrome patterns (as for example has been done with the family Rhodospirillaceae), or to the fine positioning of species which are currently classed as *incertae sedis*. Indeed, it seems increasingly likely that analysis of molecular structure, particularly of easily obtainable proteins such as *c*-type cytochromes, will become the precision taxonomic tool *par excellence* of the next decade.

I am deeply indebted to Dr Dorothy Jones (Department of Microbiology, University of Leicester) for her patient tutoring in the intricacies of bacterial taxonomy and for her constant encouragement with this work. I also wish to thank my ex-colleague, Dr David Meyer, for his efforts in the early stages of the cytochrome pattern work, and the Science Research Council for financial assistance.

7. References

ALEEM, M. I. H. 1977 Energy coupling in chemolithotrophic bacteria. In *Microbial Energetics* ed. Haddock B. A. & Hamilton W. A. *Symposium 27 of the Society for General Microbiology*, pp. 351–381. Cambridge: Cambridge University Press.

AMBLER, R. P. 1976 In *Handbook of Biochemistry and Molecular Biology* ed. Tasman, G. D. 3rd edn, Vol. III, pp. 292–307. Cleveland: Chemical Rubber Company Press.

CASTOR, L. N. & CHANCE, B. 1959 Photochemical determinations of the oxidases of bacteria. *Journal of Biological Chemistry* 234, 1587–1592.

CHANCE, B. 1954 Spectrophotometry of intracellular respiratory pigments. *Science, New York* 120, 767–775.

DICKERSON, R. E., TIMKOVICH, R. & ALMASSEY, R. J. 1976 The cytochrome fold and the evolution of bacterial energy metabolism. *Journal of Molecular Biology* 100, 473–491.

HADDOCK, B. A. & JONES, C. W. 1977 Bacterial respiration. *Bacteriological Reviews* 41, 47–99.

HIGGINS, I. J., KNOWLES, C. J. & TONGE, G. M. 1976 Enzymic mechanisms of methane and methanol oxidation in relation to electron transport systems in methylotrophs; purification and properties of methane oxygenase. In *Microbial Production and Utilisation of Gases* ed. Schlegel, H. G., Pfennig, N. & Gottschalk, G. pp. 389–402. Gottingen: E. Goltze Verlag.

JOHN, P. & WHATLEY, F. R. 1975 *Paracoccus denitrificans* and the evolutionary origin of the mitochondrion. *Nature, London* 254, 494–498.

JOHN, P. & WHATLEY, F. R. 1977 The bioenergetics of *Paracoccus denitrificans*. *Biochimica et Biophysica Acta* 463, 129–153.

JONES, C. W. 1977 Aerobic respiratory systems in bacteria. In *Microbial Energetics* ed. Haddock, B. A. & Hamilton, W. A. *Symposium 27 of the Society for General Microbiology*, pp. 23–59. Cambridge: Cambridge University Press.

JONES, C. W. 1978 Microbial oxidative phosphorylation. *Biochemical Society Transactions* 6, 361–363.

JONES, C. W., BRICE, J. M., DOWNS, A. J. & DROZD, J. W. 1975 Bacterial respiration-linked proton translocation and its relationship to respiratory chain composition. *European Journal of Biochemistry* 52, 265–271.

JONES, C. W., BRICE, J. M. & EDWARDS, C. 1977 The effect of respiratory chain composition on the growth efficiencies of aerobic bacteria. *Archives of Microbiology* 115, 85–93.

JURTSHUK, P., MUELLER, T. J. & ACORD, W. C. 1975 Bacterial terminal oxidases. *Cleveland Rubber Company, Critical Reviews in Microbiology* 3, 399–468.

KEILIN, D. 1925 On cytochrome, a respiratory pigment, common to animals, yeast and higher plants. *Proceedings of the Royal Society Series B* 98, 312–319.

KORSZUN, Z. R. & SALEMME, F. R. 1977 Structure of cytochrome c_{555} of *Chlorobium thiosulphatophilum*: primitive low-potential cytochrome *c*. *Proceedings of the National Academy of Sciences, U.S.A.* 74, 5244–5247.

KOVACS, N. 1956 Identification of *Pseudomonas pyocyanea* by the oxidase reaction. *Nature, London* 178, 703.

LEMBERG, R. & BARRETT, J. 1973 *Cytochromes*. London: Academic Press.

MacMUNN, C. A. 1886 Researches on myohaematin and the histohaematins. *Philosophical Transactions* 177, 167–298.

MEYER, D. J. & JONES, C. W. 1973 Distribution of cytochromes in bacteria: relationship to general physiology. *International Journal of Systematic Bacteriology* 23, 459–467.

MITCHELL, P. 1966 Chemiosmotic coupling in oxidative and photosynthetic phosphorylation. *Biological Reviews of the Cambridge Philosophical Society* 41, 445–502.

NIVEN, D. F., COLLINS, P. A. & KNOWLES, C. J. 1975 The respiratory system of *Chromobacterium violaceum* grown under conditions of high and low cyanide evolution. *Journal of General Microbiology* 90, 271–285.

PAYNE, W. J. 1973 Reduction of nitrogenous oxides by micro-organisms. *Bacteriological Reviews* 37, 409–452.

POOLE, R. K. & HADDOCK, B. A. 1975 Effects of sulphate-limited growth in continuous culture on the electron transport chain and energy conservation in *Escherichia coli* K12. *Biochemical Journal* 152, 537–546.

RICE, C. W. & HEMPFLING, W. P. 1978 Oxygen-limited continuous culture and respiratory energy conservation in *Escherichia coli*. *Journal of Bacteriology* 134, 115–124.

SALEMME, F. R., KRAUT, J. & KAMEN, M. D. 1973 Structural bases for function in cytochromes *c*. An interpretation of comparative x-ray and biochemical data. *Journal of Biological Chemistry* 248, 7701–7716.

SHIPP, W. S. 1972 Absorption bands of multiple *b* and *c* cytochromes in bacteria detected by numerical analysis of absorption spectra. *Archives of Biochemistry and Biophysics* 150, 482–488.

SMITH, L. 1968 The respiratory chain system of bacteria. In *Biological Oxidations* ed. Singer T. P. pp. 55–122. New York: Interscience Publishers Inc.

TIMKOVICH, R. & DICKERSON, R. E. 1976 The structure of *Paracoccus denitrificans* cytochrome C_{550}. *Journal of Biological Chemistry* 251, 4033–4046.

VAN VERSEVELD, H. W. & STOUTHAMER, A. H. 1978 Electron transport chain and coupled oxidative phosphorylation during autotrophic growth of *Paracoccus denitrificans*. *Archives of Microbiology* 118, 13–20.

WARBURG, O., NEGELEIN, E. & HAAS, E. 1933 Spektroskopischer Nachweiss des sauerstoffubertragenden Ferments neben Cytochrom. *Biochemische Zeitschrift* 266, 1–8.

WESTON, J. A. & KNOWLES, C. J. 1973 A soluble CO-binding *c*-type cytochrome from the marine bacterium *Beneckea natriegens*. *Biochimica et Biophysica Acta* 305, 11–18.

YAOI, H. & TAMIYA, H. 1928 On the respiratory pigment, cytochrome, in bacteria. *Proceedings of the Imperial Academy of Japan* 4, 436–439.

Micromorphology and Fine Structure of Actinomycetes

S. T. WILLIAMS AND E. M. H. WELLINGTON

Department of Botany, University of Liverpool, Liverpool, UK

Contents

I. Introduction

THE ADVENT OF electron microscopy played a major role in the discovery of the fundamental differences between eukaryotic and prokaryotic cells, resulting in the erection of the Kingdom Procaryotae (Murray 1968, 1974). Since this major development, however, morphology and fine structure have probably contributed less to advances in bacterial classification and identification than other methods, such as numerical and chemical taxonomy (Chapters 1, 4, 8, 9). Therefore consideration will be given not only to information provided by modern observational techniques but also the present status of morphology and ultrastructure in the light of other advances in bacterial classification and identification.

For convenience, the 'higher bacteria' with relatively diverse morphology can be distinguished from the 'lower bacteria' where morphological distinctions are fewer (Cowan & Liston 1974). Morphology clearly plays a more significant role in the classification and identification of the former group, of which the actinomycetes are a good example. It is for this reason and our own interest in these organisms that most emphasis will be placed on actinomycetes. It is hoped, however, that the actinomycetes will serve to illustrate the principles and

139

problems of using morphological characters in bacterial classification and identification in general.

The distinction between classification and identification (Cowan 1965) has practical consequences. Examination of ultra-thin sections by the electron microscope can enable the taxonomist to evaluate and modify classification, but such a procedure is unlikely to be used for routine identification. The electron microscope is not yet an everyday tool of the routine bacteriologist (Cowan & Liston 1974) and it is debatable whether it ever will be. In contrast, examination of living and stained cells by light microscopy is a routine procedure in identification and "the desirability of careful morphological examination of strains as a first step in identification cannot be over-emphasised" (Cowan & Liston 1974). Thus 12 of the 19 major bacterial taxa listed in *Bergey's Manual* (Buchanan & Gibbons 1974) include a morphological feature in their description. The reader is also advised that if an "outstanding morphological feature" is observed, further efforts at identification should be directed to the taxa possessing that feature. It is often a relief to find that an unknown culture consists of Gram positive rods with endospores!

2. Range of Form of Actinomycetes

It is necessary to consider briefly the range of form shown by actinomycetes. The considerable diversity of these bacteria has been conveniently categorized by Prauser (1976, 1978). Prauser's speculative proposals for the stages in the evolution of actinomycete morphology are as follows (see also Fig. 1):

(i)   blockage of cell division in single-celled coryneform bacteria leading to hyphal development;

(ii) changes in cell envelope leading to branching and mycelial development. The mycelium propagates by total fragmentation into rods and cocci; aerial mycelium when produced does likewise; bacteria which have these features are termed 'nocardioform' actinomycetes (e.g. *Actinomyces, Mycobacterium, Rhodococcus, Nocardia*), and

(iii) the mycelium becomes stable with spore formation in limited regions. Organisms in this category are termed 'sporoactinomycetes' and show a range of spore forms (Cross 1970), including aleuriospores (e.g. *Micromonospora*), arthrospores (e.g. *Streptomyces*), spores in vesicles (e.g. *Actinoplanes*) and endospores (e.g. *Thermoactinomyces*).

Three evolutionary levels, coryneforms, nocardioforms and sporoactinomycetes were therefore recognized. Prauser (1976) suggested that three evolutionary lines of sporoactinomycetes through nocardioforms from coryneforms may have existed. This proposal is supported to some extent by Schleifer & Kandler (1972) who found that the peptidoglycan type of *Streptomyces* was identical to

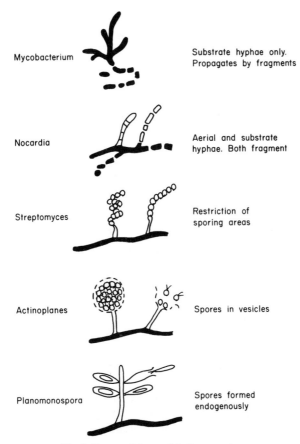

Mycobacterium — Substrate hyphae only. Propagates by fragments

Nocardia — Aerial and substrate hyphae. Both fragment

Streptomyces — Restriction of sporing areas

Actinoplanes — Spores in vesicles

Planomonospora — Spores formed endogenously

Fig. 1. Range of form of Actinomycetes.

that of certain coryneforms. Similarity between the peptidoglycans of many anaerobic actinomycetes and those of coryneforms was also reported by these authors.

3. Morphology and Wall Composition in the Definition of Actinomycete Genera

Morphology has always been a major criterion for the delimitation of actinomycete genera. Its role has been somewhat reduced by the development of chemotaxonomy (e.g. Lechevalier *et al*. 1971; Lechevalier *et al*. 1977; Minnikin & Goodfellow 1976). Most schemes for the identification of genera are now based on morphology and wall composition (Lechevalier & Lechevalier 1970). It is interesting to consider the results obtained using these two criteria (Table 1). The morphological groupings given for some genera are approximate but never-

TABLE 1

Morphology and wall composition of aerobic actinomycete genera

Morphological group	Wall chemotype*			
	I	II	III	IV
Nocardioforms	Nocardioides Sporichthya		Dermatophilus Geodermatophilus Nocardiopsis	Nocardia Mycobacterium Rhodococcus
Aleuriospores		Micromonospora	Thermomonospora	Saccharomono- spora
Arthrospores	Chainia Streptomyces Streptovert- icillium		Actinomadura Microbispora Microtetraspora	Actinopolyspora Micropolyspora Pseudonocardia Saccharopolyspora
Spores in vesicles	Microellobo- spora	Actinoplanes Ampullariella	Spirillospora Streptosporangium	
Spores formed endogenously		Dactylosporangium	Planomonospora Planobispora Thermoactinomyces	

*After Lechevalier & Lechevalier (1970).

theless there seems to be little evidence of correlation. The lack of nocardioform actinomycetes with a wall chemotype II or vesicle and endogenous spore producers with a wall chemotype IV are the most notable points.

Closer examination of genus definitions shows that often morphology and wall composition are not used jointly but that one or the other is given emphasis. Nocardioform genera are distinguished primarily by their chemical composition, as their morphology is variable (see Chapter 9) and not always diagnostic (Table 2). With such genera, analysis of mycolic acids has proved to be a useful adjunct to determination of the wall chemotype (e.g. Lechevalier *et al.* 1973a; Minnikin *et al.* 1975; Minnikin & Goodfellow 1976). Sporoactinomycetes, on the other hand, have more distinctive and stable morphological features, with less variation in the primary structure of their peptidoglycans (Prauser 1978). Therefore morphology is used to distinguish genera with identical wall components (Table 3). Thus the significance of morphology in genus delimitation varies considerably within the actinomycetes.

Reliance on one particular character can create problems. A good example is the genus *Micropolyspora* (*Mips.*), which was originally defined solely by the formation of short chains of spores on the substrate and aerial nyphae (Lechevalier *et al.* 1961). Chemical analysis has shown that although all species have a wall chemotype IV, *Mips. brevicatena* (the type species) has mycolic acids (Lechevalier 1977) while *Mips. faeni* and *Mips. rectivirgula* do not (Mordarska *et al.* 1972; Collins *et al.* 1977).

TABLE 2

Morphology and wall composition of some nocardioform genera

Genus	Morphology		Cell walls	
	Substrate hyphae	Aerial hyphae	Chemotype*	Mycolic acids†
Mycobacterium	Fragmenting to rods and cocci	Usually absent	IV	Long chained (60–90 carbons)
Nocardia	Fragmenting to rods and cocci	Fragmenting or absent	IV	Shorter chained (46–60 carbons)
Nocardioides	Fragmenting to rods and cocci	Fragmenting	I	None
Rhodococcus	Fragmenting to rods and cocci	Usually absent	IV	Shorter chained (34–66 carbons)

*After Lechevalier & Lechevalier (1970). † Data from Minnikin *et al.* (1978).

TABLE 3

Morphology of some sporoactinomycetes with chemotype III

Genus	Morphology
Thermomonospora	Single aleuriospores on substrate and aerial hyphae
Actinomadura	Chains of arthrospores (2–30) on aerial hyphae
Microbispora	Paired arthrospores on aerial hyphae
Microtetraspora	Chains of 4 arthrospores on aerial hyphae
Streptosporangium	Spores in vesicle ('sporangia') on aerial hyphae
Planomonospora	Single motile spores endogenously produced in elongated sporangium
Thermoactinomyces	Single heat-resistant endospores on substrate and aerial hyphae

4. Stability and Reproducibility of Morphological Characters

A. Plasticity of nocardioform actinomycetes

Nocardioform bacteria develop hyphae which eventually fragment into smaller, various shaped units. The rate and timing of the fragmentation processes are influenced by many environmental factors and the consequent variability of morphological features reduces their taxonomic value. Conditions controlling fragmentation have been studied extensively (see Williams *et al.* 1976). Factors claimed to retard fragmentation include the presence of polysaccharides or hydrocarbons as carbon sources, and nitrate nitrogen as nitrogen source. Those promoting fragmentation include glucose as a carbon source and ammonium nitrogen.

Heinzen & Ensign (1975) studied the morphology of a culture labelled *Nocardia corallina* on a defined medium with glucose supplemented with various amino acids, organic acids and sugars. Growth rates were low (doubling time, 44 h) in the basal medium and all cells were coccoid. Addition of substrates which increased growth rates (doubling time, 3-7 h) resulted in a coccus-rod-coccus cycle.

In our studies of *Rhodococcus rhodochrous*, the growth cycle on agar media followed the familiar hypha-rod-coccus sequence. Hyphae predominated during the active growth phase (Table 4). With the same organism grown in a chemostat, low dilution rates induced cocci and high rates induced hyphae (Table 4). A

TABLE 4

Morphology of Rhodococcus rhodochrous *in batch culture and in a chemostat*

Cell forms	24	48	72	96	120 k
	(%)	(%)	(%)	(%)	(%)
Hyphae	80	14	0	0	0
Rods	20	28	17	14	9
Cocci	0	58	83	86	91

Growth in a chemostat

Dilution rate (h^{-1})	Predominant cell form
0.049	Cocci
0.054–0.118	Rods and cocci
0.165–0.235	Hyphae

similar effect was obtained with *Arthrobacter* strains, the transition from rods to cocci being controlled by growth rate (Luscombe & Gray 1974). The transition point was characteristic for each species and was related to their maximum specific growth rates. Growth rate, therefore, has a major influence on cell form. Some other factors which control fragmentation of hyphae may well act through their effects on growth rates. As rates of growth are rarely measured or controlled in taxonomic studies, reproducibility of morphological features is poor.

Reproducibility of morphological characters in Streptomyces *spp.*

Streptomyces is the most studied genus of the sporoactinomycetes. The distinctive morphology of the sporing structures has been widely used in species delimitation. Spore chain (or sporophore) morphology, determined by light microscopy, and spore surface ornamentation, determined by electron microscopy, have figured prominently in many classifications (e.g. Ettlinger *et al.* 1958; Waksman 1961; Hütter 1967; Pridham & Tresner 1974). One of the most recent and extensive taxonomic studies of this genus was carried out in the International Streptomyces Project (I.S.P.). Over 400 type strains were examined independently by standardized procedures in three different laboratories; morphological criteria figured in the resulting species descriptions (Shirling & Gottlieb 1968*a*; *b*; 1969; 1972).

Spore chain morphology and spore surface ornamentation were also included in a current numerical taxonomic study of streptomycetes. This investigation includes over 500 representative strains, 394 of which were also examined in the I.S.P. As our observations were made using the methods of the I.S.P., direct comparisons of determinations are possible and allow a 'between laboratory' assessment of the reproducibility of morphological characters. In addition, 'within laboratory' reproducibility was assessed by comparing records for selected species made at two different times.

(i) *Spore chain morphology*

The spore chains borne on the aerial mycelium can be categorized according to their shape. Pridham *et al.* (1958) proposed four sections: straight to flexuous (Rectus-Flexibilis), hooks, loops or open spirals (Retinaculum-Apertum), tight spirals or coils (Spira) and verticillate (Biverticillus) (Fig. 2). Verticillate strains are placed in the genus *Streptoverticillium* by some workers. Several international co-operative studies found these categories to be taxonomically useful (Küster 1961; Gottlieb 1961; Szabo & Marton 1964) and they were adopted for the I.S.P. The stability of these features was high, no changes of category occurring by either spontaneous or induced mutation (Krassilnikov *et al.* 1961). Natural or induced variability may occur, however, with increased antibiotic activity (Kalakoutskii & Nikitina 1977), but the changes are usually in the degree of development of aerial mycelium rather than in spore chain shape. Impaired differentiation of a few species was observed in the I.S.P. (Shirling & Gottlieb 1977). Spore chain type also appears to be consistent on different media (Gottlieb 1961; Shirling & Gottlieb 1977).

In the present study, spore chain shape was determined on I.S.P. media by light microscopy. Results were checked using scanning electron microscopy (SEM). Assessment of reproducibility showed spore chain morphology to be among the less reproducible of the characters used (Table 5). The percentage agreement between duplicate determinations on selected species was the same as that for chitin degradation and spore colour. Chitin degradation zones are sometimes difficult to read and spore colour is difficult to determine objectively (Gottlieb 1961; Pridham 1965).

Comparison of data with those of the I.S.P. showed a similar proportion of

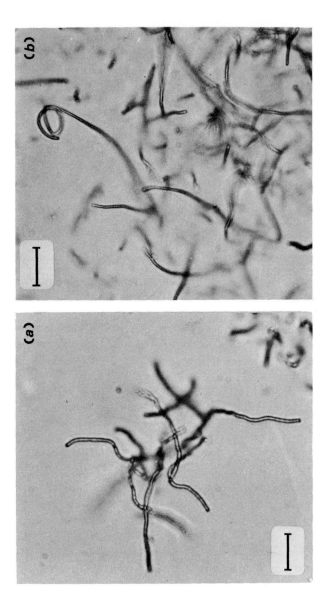

Fig. 2. Streptomycete spore chains examined by light microscopy (bar marker, 10 μm): (a) Rectus-flexibilis (*Streptomyces purpeofuscus*, I.S.P. 5283); (b) Retinaculum-Apertum (*Streptomyces vinaceus*, I.S.P. 5515); (c) Spira (*Streptomyces spheroides*, I.S.P. 5292); and (d) Biverticillus (*Streptomyces mora-okaensis*, I.S.P. 5503).

TABLE 5

Reproducibility of various taxonomic tests applied to 44 Streptomyces *spp. at two different times*

Test	Percentage agreement
Production of melanin pigment	100.00
Spore surface ornamentation	95.5
Presence of distinctive pigmentation of substrate mycelium	93.2
Antibiosis against 5 bacteria, 2 yeasts and 1 mould	92.0
Hippurate degradation	91.0
Nitrate reduction	91.0
Pectin degradation	91.0
Chitin degradation	86.4
Spore chain morphology	84.1
Colour of spore mass	84.1

TABLE 6

Results of two independent determinations of the spore chain morphology and spore surface ornamentation of 394 Streptomyces *spp.*

	Percentage occurrence	
	International Streptomyces Project	Present authors
Spore chain categories		
Rectus-Flexibilis	31.5	31.7
Retinaculum-Apertum	3.6	12.4
Spira	35.0	42.1
Biverticillus	5.8	5.8
Combined categories	20.3	4.1
Non-sporing	3.8	3.8
Spore surface categories		
Smooth	70.6	76.4
Warty	1.8	2.5
Spiny	15.2	12.9
Hairy	2.3	4.6
Combined categories	6.1	0.0
Non-sporing	4.1	3.6

strains placed in the various categories (Table 6). Most species placed in combined categories by the I.S.P. were placed into Retinaculum-Apertum or Spira in the present study. Levels of agreement for the categories Biverticillus, Rectus-Flexibilis and Spira were high (Table 7), showing that these groups were clearly defined. In contrast, Retinaculum-Apertum was not easily distinguished during the I.S.P. (Shirling & Gottlieb 1977) and the present interpretation of this category obviously differs from that of the I.S.P. The morphological range within Retinaculum-Apertum is intermediate between Rectus-Flexibilis and

TABLE 7

Agreement between morphological determinations of Streptomyces *spp. made by present authors and the International Streptomyces Project.*

	Percentage agreement
Spore chain categories	
Biverticillus	100.0
Rectus-Flexibilis	98.1
Spira	95.3
Retinaculum-Apertum	30.4
Spore surface categories	
Smooth	96.8
Spiny	93.5
Hairy	63.6
Warty	50.0

Spira and its delimitation from these groups is not always clear. The relatively poor reproducibility of spore chain shape (Table 5) was almost entirely confined to strains placed in Retinaculum-Apertum on one of the two determinations. Similar conclusions were reached by Gottlieb (1961) and Shirling & Gottlieb (1977). It has been suggested that the categories Retinaculum-Apertum and Spira should be combined to increase reproducibility of determinations of spore chain shape (Szabo & Marton 1976). Spore chain morphology has nevertheless proved to be a consistent and reliable character for the majority of strains included in both the present study and the I.S.P.

(ii) *Spore surface ornamentation*

The spores in a chain are enveloped by a thin fibrillar sheath which is distinct from the spore wall and which may bear ornaments. Four types of spore surface were recognized initially, smooth, spiny, hairy and warty [(Küster 1953; Ettlinger *et al.* 1958; Tresner *et al.* 1961) (Fig. 3)] and these have been widely used in species descriptions. The features are stable under various growth conditions (Lechevalier & Tikhonenko 1960; Mordarski & Kudrina 1961).

In the present investigation, ornamentation was recorded following the scheme of Tresner *et al.* (1961) and determinations were made using SEM of undisturbed growth on glass coverslips (Williams & Davies 1967). Most previous studies, including the I.S.P., used transmission electron microscopy (TEM) of the silhouettes of detached spore chains. This causes more disruption of the sheath and misinterpretations of the information provided by the silhouette can occur. The value of SEM in examining streptomycete spore surfaces has been noted (Dietz & Mathews 1969, 1971) but it has not been used previously to examine such a wide range of species. Reproducibility of spore surface ornamentation determinations was high, and it was one of the most reliable of tests used (Table 5). Similar conclusions were reached by Küster (1961) and Shirling & Gottlieb (1977). Comparison of the present data with those of the I.S.P. shows that the

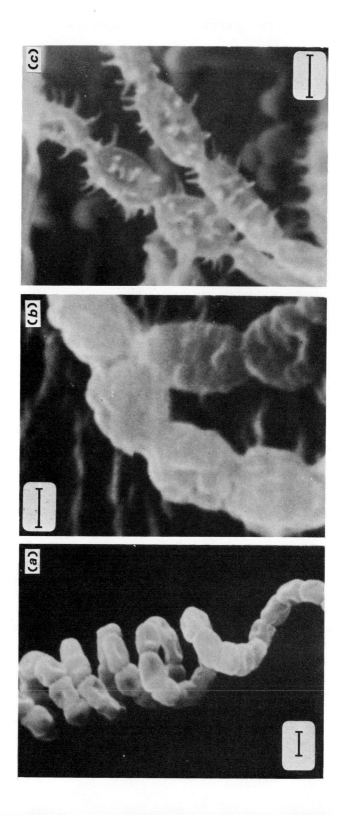

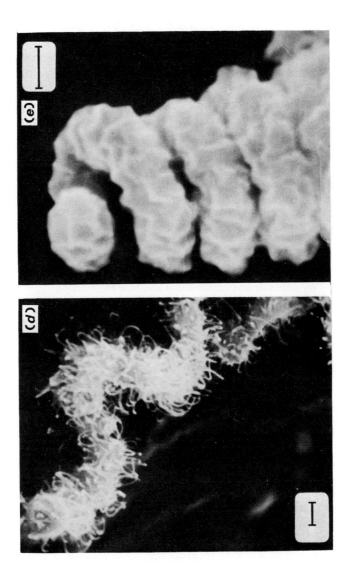

Fig. 3. Streptomycete spore surface examined by scanning electron microscopy (bar marker, 0.5 μm): (a) Smooth (*Streptomyces niveus*, I.S.P. 5088); (b) Warty (*Streptomyces pulcher*, I.S.P. 5566); (c) Spiny (*Streptomyces filipinensis*, I.S.P. 5112); (d) Hairy (*Streptomyces finlayi*, I.S.P. 5218); and (e) Ridged (*Streptomyces violaceus-niger*, I.S.P. 5563).

distributions of spore ornamentation categories were similar (Table 6). Agreement was good for the smooth and spiny spore categories but poor for hairy and warty types (Table 7).

Some variations within the categories occur, for example in length of spines and hairs, and further subdivisions based on these differences have been proposed (Dietz & Mathews 1972, 1977). Lyons & Pridham (1971) proposed a 'knobby' category for intermediates between spines and warts. The size of the ornaments is influenced, however, by growth conditions (Mordarski & Kudrina 1961).

The discrepancies between the present findings and those of the I.S.P. could be largely due to the differences in observational techniques. A number of strains described as 'warty' from silhouettes examined in the I.S.P. appeared smooth under the scanning electron microscope. Un-ornamented but wrinkled sheaths were sometimes misinterpreted when viewed as silhouettes. Misinterpretations due to loss of sheath were also less likely when large numbers of relatively undisturbed chains were examined by SEM.

The most notable discrepancies between results from scanning and transmission electron microscopy were found with species forming 'hygroscopic' spore masses. SEM showed the spore sheath to be ridged (Fig. 3), while I.S.P.

TABLE 8

Spore surface ornamentation of 'hygroscopic' Streptomyces spp. determined by scanning and transmission electron microscopy

	Spore surface determined by	
Taxon	Scanning electron microscopy*	Transmission electron microscopy†
Streptomyces		
bluensis	Ridged	Spiny
endus	Ridged	Warty
hygroscopicus	Ridged	Warty
melanosporofaciens	Ridged	Smooth-warty
sparsogenes	Ridged	Spiny
violaceus-niger	Ridged	Smooth

*Present authors. †International Streptomyces Project.

determinations placed them in various categories (Table 8). The present observations support those of Dietz & Mathews (1962, 1968, 1971) who proposed a new category, 'rugose', for such spores.

Present evidence indicates that spore surface ornamentation is a very stable and useful character. Most of the problems in its determination are overcome when SEM is used.

C. Genetical control of morphology in Streptomyces spp.

Cross & Attwell (1975) suggested that use of morphology in classification and identification of actinomycetes was generally justifed as morphological features were complex, relatively stable and determined by several genes. Although such opinions are not often stated by taxonomists, they are usually implicit in the classifications which they devise. At present there is little direct evidence for the genetical control of morphology of actinomycetes. The studies on *Streptomyces coelicolor* are a notable exception (Hopwood *et al.* 1970; Chater 1972; Chater & Hopwood 1973; Hopwood *et al.* 1973; McVittie 1974; Chater 1975).

Genetic mapping of mutants lacking aerial mycelium (*bld*) indicated that its formation was controlled by chromosomal genes rather than plasmids (Chater & Hopwood 1973). Other mutants (*whi*) were produced which formed aerial mycelium but were defective in spore production. Eight *whi* genes were detected and the morphology of the various mutants resembled that of intermediate stages in normal spore formation. It is possible, therefore, that the genes control various steps in spore formation. A tentative scheme was proposed by Hopwood *et al.* (1973):

  (i)   growth of aerial hyphae (*bld A*);
 (ii)   initiation of coiling of aerial hyphae (*whi G*);
(iii)   completion of coiling (*whi H*);
 (iv)   formation of sporulation septa (*whi A, B, I*);
  (v)   wall thickening and rounding-off at spore junctions (*whi D*);
 (vi)   rounding-off of spores (*whi F*); and
(vii)   production of grey pigment, maturation and release of spores (*whi E*).

Thus several genes were involved in the formation of *Streptomyces* spore chains and McVittie (1974) commented that more genes were involved than at first supposed.

Kalakoutskii & Nikitina (1977) reviewed evidence for natural variability in development patterns of streptomycetes and emphasized its taxonomic relevance. It is difficult at present, however, to make an overall assessment of this variability. Taxonomists may encounter it but do not seek it, while those interested in it select strains which show it. The present survey of the morphological features of a wide range of streptomycetes (see pp. 148) indicates that they are not particularly variable if standardized conditions are used.

Some workers have studied the properties of so-called 'fructose' varieties of streptomycetes (summarized by Kalakoutskii & Nikitina 1977). About 10% of streptomycetes showed variants when grown on D-fructose as the carbon source. At first normal colonies developed but after 5–12 days, 'bald' nocardioform areas arose within the original growth. Some of these when subcultured retained their nocardioform features when grown in the absence of fructose. They were found to differ from the parent cultures in ultrastructure, frequency of septation and the production of extracellular tubular material. The basic cause of these variants remains to be established.

5. Correlation of Morphology with Other Taxonomic Information

The degree of correlation between morphology and wall composition has been discussed previously. Correlations with information provided by some other modern taxonomic methods will now be considered.

A. Numerical taxonomy

Phenetic studies of actinomycetes have so far been concentrated on coryneform and nocardioform taxa. It is therefore difficult to make an overall assessment of the role of morphology in phenon characterization. Most numerical studies have included features such as cell shape and colony form in the battery of tests applied (e.g. Goodfellow 1971; Bousfield 1972; Holmberg & Hallander 1973; Holmberg & Nord 1975; Jones 1975). Holmberg & Nord (1975), who studied *Actinomyces* and *Arachnia*, noted that morphological features were among the least reproducible of tests used. Morphological characters nevertheless sometimes play a part in phenon definition. Formation of aerial hyphae helps to distinguish many *Nocardia* strains from other nocardioform genera such as *Rhodococcus* and *Oerskovia* (Goodfellow 1971) but detection of aerial growth is influenced by the observational method used (Williams *et al*. 1976). The ability of nocardioform genera to form a mycelium may distinguish them from coryneform bacteria (Jones 1975). However, this distinction is not always sharp (Prauser 1978). It will be interesting to discover the role of morphology in phenetic classification of sporoactinomycetes.

B. DNA reassociation

A strong theoretical case has been made for the importance of DNA homologies in bacterial taxonomy (Brenner 1973), although methods are complex and prone to errors (see Chapter 1). Generally good agreement is shown between DNA homologies and phenetic similarities, particularly in the range of 75-100% homology and similarity (Staley & Colwell 1973; Goodfellow *et al*. 1978). Brenner (1973) suggested that strains within a good taxospecies should show at least 70% DNA homology and that lower levels reflect genetic divergence.

There have been few attempts to define DNA homologies of morphologically similar sporoactinomycetes. DNA homologies among genera of the Actino-

TABLE 9

DNA homologies among morphologically similar Streptomyces *spp.*

Morphological features		Homology index (%)
	Streptomyces	
Straight to flexuous spore chains	*griseus*	100
	chrysomallus	86
Spore surface	*vinaceus*	¯64
smooth	*felleus*	51
	rutgerensis	38
	Streptoverticillium	
	kentuckense	100
	netropsis	70
Verticillate	*ehimense*	39
	abikoense	37
	eurocidicum	33

Data selected from Okanishi *et al.* (1972) and Toyama *et al.* (1974).

planaceae were studied by Farina & Bradley (1970). They concluded that there was better correlation with wall composition than morphology. *Streptomyces* and *Streptoverticillium* spp. sharing several morphological and other features (Table 9) were studied by Okanishi *et al.* (1972) and Toyama *et al.* (1974). Within both morphological groups, homologies covered a wide range and were often below 70%. Although the species within each group were not regarded as identical, they did share several cultural and physiological features in addition to their morphological attributes. DNA reassociation has regularly given results consistent with other thorough taxonomic studies (see Chapter 1). Therefore one possible explanation for these results is that the morphological and other features used to group together these streptomycetes did not produce a sound phenetic classification. Alternatively phenetically similar bacteria might have differences in their nucleotide sequences as a result of neutral mutations which do not affect their genotype or phenotype (see Chapter 1). Definition of taxa on the basis of DNA reassociation can be difficult. Results depend on the methods used, pairing falls to a low level for organisms only moderately different phenetically (Jones & Sneath 1970; Staley & Colwell 1973), and it is very sensitive to small differences within higher level phena (Sneath 1976).

Despite these difficulties of interpretation, the limited data available indicate that the value of morphological criteria in recognition of actinomycete genera and species may have been over-estimated.

C. Serology

Information on correlation between morphology and serology is very limited. A serological study of morphologically similar streptomycetes was done by Guthrie *et al.* (1971). Species with blue, spiny spores in spiral chains, which were melanin positive and had similar sugar utilization patterns, differed in only one or two of their wall antigens. It was concluded that morphology and pigmentation correlated better with serology than did sugar utilization patterns.

6. Contributions of Modern Observational Methods

A. Study of spore development by electron microscopy

The form, arrangement and location of spores, as determined by light microscopy, play a major role in genus recognition. It has been suggested that this is no longer sufficient without studies of spore development and fine structure (Cross & Attwell 1975). Two examples will serve to illustrate this point.

The genera *Thermoactinomyces* and *Thermomonospora* have a similar gross morphology and wall composition; they differ considerably in the mode of development and heat resistance of their spores (Table 10). This led to their separation into the families Thermoactinomycetaceae and Thermomonosporaceae (Cross & Goodfellow 1973).

TABLE 10

Differentiation of the genera Thermoactinomyces *and* Thermomonospora*

	Wall chemotype†	Gross morphology (light microscopy)	Spore development (electron microscopy)
Thermoactinomyces	III	Single spores on substrate and aerial hyphae	Heat-resistant spores formed endogenously within sporangium
Thermomonospora	III	Single spores on substrate and aerial hyphae	Aleuriospores formed by outgrowth of parent hypha

*, Data from Cross *et al.* (1968), Cross (1970), Cross & Attwell (1975). †, After Lechevalier & Lechevalier (1970).

The family Actinoplanaceae contains strains which appear, from light microscopy, to form sporangia (Table 11). Studies with transmission and scanning electron microscopy have shown, however, that two distinct types of 'sporangia' occur (Lechevalier *et al.* 1966; Sharples *et al.* 1974). In type A, spores are produced essentially in the same manner as in nonsporangial strains classified in genera such as *Streptomyces*; these sporangia are more accurately termed vesicles (Cross 1970). Type B involves truly endogenous spore formation with the parent hypha acting as a sporangium. Consequently the sporangia are comparatively

TABLE 11

Characteristics of genera in the family Actinoplanaceae

Genus	Wall chemotype*	Gross morphology (light microscopy)	Spore developmental pattern† (electron microscopy)
Actinoplanes	II	Globose-irregular sporangia with many spores	A
Ampullariella	II	Cylindrical sporangia with many spores	A
Dactylosporangium	II	Elongated sporangia with 3–4 spores	B
Planobispora	III	Elongated sporangia with 2 spores	B
Planomonospora	III	Elongated sporangia with 1 spore	B
Spirillospora	III	Variable shaped sporangia with many spores	A
Streptosporangium	III	Globose sporangia with many spores	A

*, After Lechevalier & Lechevalier (1970). †, Developmental patterns (from Lechevalier *et al.* 1966; Sharples *et al.* 1974): A, arthrospores formed by fragmentation of hypha within its expanded sheath (*'vesicle'*); and B, spores formed endogenously within parent hypha which acts as a sporangium.

regular in shape, release the spores at a defined apical point and retain their integrity after spore release. Thus the morphological definition of the family Actinoplanaceae is questionable.

B. Definitive identification of morphological structures

The nature of a number of structures initially detected by light microscopy has been subsequently elucidated by electron microscopy.

The genus *Chainia* (Thirumalachar 1955) was characterized by the formation of sclerotia. For some years, there was debate about the validity of applying the term 'sclerotium' to what may have been disorganized hyphal aggregations. Electron microscopic studies showed that these structures bore a close resemblance to fungal sclerotia in their development and structure (Lechevalier *et al.* 1973b; Sharples & Williams 1976). This information supported the retention of *Chainia* as a distinct genus.

The genus *Intrasporangium* was characterized by the production of intercalary sporangia (Kalakoutskii *et al.* 1967). Examination of the ultrastructure of the 'sporangia' indicated that they were vesicles which did not contain spores and their taxonomic significance was questioned (Lechevalier & Lechevalier 1969).

The genus *Dactylosporangium* forms sporangia and also produces globose bodies on its substrate hyphae (Thiemann *et al*. 1967). The latter were proposed as an additional morphological feature of the genus (Thiemann 1970). Study of their fine structure indicated that they were products of abnormal development, possibly induced by phage infection (Sharples & Williams 1974).

C. Fine structure of spore surfaces

The importance of using SEM rather than TEM accurately to determine the surface features of streptomycete spores has already been discussed (pp. 152). A similar misinterpretation resulted when spores of *Thermoactinomyces vulgaris* were examined as silhouettes with the transmission electron microscope. The spores were described as being 'spiny' but examination by SEM revealed that they were ridged (Williams & Davies 1967; Williams 1970). Subsequent examinations of freeze-etched preparations showed that the ridges formed in pentagonal and hexagonal patterns (Cross *et al*. 1971; McVittie *et al*. 1972).

Many actinomycetes form their spores within a fine fibrillar sheath. When the fine structure of this sheath is examined by carbon replication, negative staining or freeze-etching, a distinctive fibrillar pattern is often visible. The patterns vary in different *Streptomyces* spp. (e.g. Wildermuth *et al*. 1971; Williams *et al*. 1972). Clear patterns have also been observed in *Streptoverticillium* (Cross *et al*. 1973), *Actinomadura* (Williams *et al*. 1974b), *Nocardia* (Takeo & Uesaka 1975), *Micropolyspora* (Takeo 1976) and several other genera. The taxonomic value of these patterns remains to be assessed, as so far few strains have been examined.

D. Colony morphology

The form of bacterial colonies on agar media, determined by eye or low power light microscopy, is a long established aid in identification. Within the actinomycetes, such features are most commonly used for coryneform and nocardioform bacteria. Distinctions are usually based on colony shape and texture. Such characters can vary with growth conditions and also with the method of determination. Considerable variations occur in the assessment of frequency of aerial growth on colonies of nocardioform strains, determined by different methods (Williams *et al*. 1976). Those for *Nocardia caviae*, for example, varied from 9 to 100% frequency of occurrence.

SEM is ideally suited for study of colony morphology as it allows examination of undisturbed growth at high magnification and large depth of focus. Colony development in *Actinomyces* and related species has been studied by Slack & Gerencser (1975) and Locci (1978). The latter worker emphasized the importance of studying undisturbed growth. He noted the occurrence of 'rough' and 'smooth' colonies in the same strain and found the 'mycelial' colonies of *Bacterionema matruchotti* to consist of rods in unbranched chains. Colonies of other bacteria have been studied by SEM including *Diplococcus* and *Streptococcus* (Springer & Roth 1972), *Staphylococcus* (Fass 1973) and various rod-shaped bacteria (Drucker & Whittaker 1971). The stability and diagnostic value of the additional features provided by SEM remain to be evaluated.

The morphology of growth in submerged culture has received less attention from bacterial taxonomists. However, Tresner *et al.* (1967) noted consistent and unusual cell forms in strains of *Streptomyces aureofaciens* grown in AC broth (Difco). These were proposed as an aid to the identification of this species. Similar cell forms are observed in *Stmy. aureofaciens* using SEM (Williams *et al.* 1974*a*).

C. The role of electron microscopy in classification and identification

Detailed studies of the fine structure of actinomycetes and other bacteria clearly help to reinforce or modify existing classifications. However, it is often impracticable to employ fine structure in routine identification. A problem arises when new taxa are described which rely on electron microscopy for their definition and hence also for their identification. Thus *Streptosporangium corrugatum* was distinguished primarily by characters determined by transmission and scanning electron microscopy (Williams & Sharples 1976). Clearly other more accessible characters should be proposed whenever possible.

7. Conclusions

A brief overall assessment of the current value of morphology in actinomycete taxonomy can now be made. Schleifer & Kandler (1972) listed the desirable attributes of a useful taxonomic marker and these, with some modifications, are as follows:

  (i)   The character should be widespread in the taxon to be classified and should correlate with other properties;

 (ii)   It should be determined by as many genes as possible to increase stability and reduce chances of similarities due to convergence;

(iii)   It should reveal derivative sequences, allowing recognition of primitive and advanced states;

(iv)   It should not be too dependent on growth stages and environmental factors;

 (v)   It should be determinable on a routine basis.

Morphological attributes of actinomycetes are usually expressed consistently within a given genus, particularly by the sporoactinomycetes. Information on correlations with other characters is scattered and will be provided only by more studies of phenetic similarity, especially among the sporoactinomycetes. Correlation with wall chemotypes (Lechevalier & Lechevalier 1970) appears to be poor.

Information on the genetical control of sporing in streptomycetes suggests that the second and third conditions can be met, at least by sporoactinomycetes.

Attempts have been made to use morphology in the recognition of evolutionary sequences (e.g. Prauser 1976, 1978). It is, however, more difficult to define objectively derivative sequences using morphology than, for example, peptidoglycan structure. Morphological features are more dependent on environmental factors than is peptidoglycan structure. Nocardioform genera are clearly more variable than sporoactinomycetes which have reproducible features if grown in reasonably standardized conditions.

The feasibility of determining morphological characters clearly depends on the observational technique used. The advantages and problems of the increasing use of electron microscopy have been discussed.

Within the actinomycetes, and the rest of the bacteria, morphological attributes which are distinctive and stable have obvious attractions for the taxonomist. With the rapid development of modern taxonomic methods, however, there is an inherent danger in over-emphasizing the significance of such characters. Thus use of the single criterion of endospore formation for inclusion in the genera *Bacillus* or *Clostridium* groups together strains with considerable genetic diversity (see Chapter 1). Although morphology has played a major role in bacterial taxonomy for many years, more information is needed on its correlation with other taxonomic data. The ideal morphological feature should not only be distinctive and stable but also be predictive of many other properties of the organism.

Part of the work reported here was supported by a Science Research Council research grant GR/A 04309.

8. References

BOUSFIELD, I. J. 1972 A taxonomic study of some coryneform bacteria. *Journal of General Microbiology* 71, 441–455.

BRENNER, D. J. 1973 Deoxyribonucleic acid reassociation in the taxonomy of enteric bacteria. *International Journal of Systematic Bacteriology* 23, 298–307.

BUCHANAN, R. E. & GIBBONS, M. E. 1974 *Bergey's Manual of Determinative Bacteriology*, 8th edn. Baltimore: Williams & Wilkins.

CHATER, K. F. 1972 A morphological and genetic mapping study of white colony mutants of *Streptomyces coelicolor. Journal of General Microbiology* 72, 9–28.

CHATER, K. F. 1975 Construction and phenotypes of double sporulation-deficient mutants in *Streptomyces coelicolor* A3(2). *Journal of General Microbiology* 87, 312–325.

CHATER, K. F. & HOPWOOD, D. A. 1973 Differentiation in actinomycetes. In *Microbial Differentiation* ed. Ashworth, J. M. & Smith, J. E., pp. 143–160. Cambridge: Cambridge University Press.

COLLINS, M. D., PIROUZ, T., GOODFELLOW, M. & MINNIKIN, D. E. 1977 Distribution of menaquinones in actinomycetes and corynebacteria. *Journal of General Microbiology* 100, 221-230.

COWAN, S. T. 1965 Principles and practice of bacterial taxonomy. *Journal of General Microbiology* 39, 143-153.

COWAN, S. T. & LISTON, J. 1974 The mechanism of identification. In *Bergey's Manual of Determinative Bacteriology*, 8th edn. ed. Buchanan, R. E. & Gibbons, N. E. pp. 10-13. Baltimore: Williams & Wilkins.

CROSS, T. 1970 The diversity of bacterial spores. *Journal of Applied Bacteriology* 33, 95-102.

CROSS, T. & ATTWELL, R. W. 1975 Actinomycete spores. In *Spores VI*, pp. 3-14. Washington D.C.: American Society for Microbiology.

CROSS, T. & GOODFELLOW, M. 1973 Taxonomy and classification of the actinomycetes. In *Actinomycetales, Characteristics and Practical Importance*, pp. 11-112, ed. Sykes, G. & Skinner, F. A. London: Academic Press.

CROSS, T., WALKER, P. D. & GOULD, G. W. 1968 Thermophilic actinomycetes producing resistant endospores. *Nature, London* 220, 352-354.

CROSS, T., DAVIES, F. L. & WALKER, P. D. 1971 *Thermoactinomyces vulgaris*. I. Fine structure of the developing endospores. In *Spore Research 1971*, pp. 175-180, ed. Barker, A. N., Gould, G. W. & Wolf, J. London: Academic Press.

CROSS, T., ATTWELL, R. W. & LOCCI, R. 1973 Fine structure of the spore sheath in *Streptoverticillium* species. *Journal of General Microbiology* 75, 421-424.

DIETZ, A. & MATHEWS, J. 1962 Taxonomy by carbon replication. I. An examination of *Streptomyces hygroscopicus*. *Applied Microbiology* 10, 258-263.

DIETZ, A. & MATHEWS, J. 1968 Taxonomy of carbon replication. II. Examination of eight additional cultures of *Streptomyces hygroscopicus*. *Applied Microbiology* 16, 935-941.

DIETZ, A. & MATHEWS, J. 1969 Scanning electron microscopy of selected members of the *Streptomyces hygroscopicus* group. *Applied Microbiology* 18, 694-696.

DIETZ, A. & MATHEWS, J. 1971 Classification of *Streptomyces* spore surfaces into five groups. *Applied Microbiology* 21, 527-533.

DIETZ A. & MATHEWS, J. 1972 Characterization of hairy-spored streptomycetes. *International Journal of Systematic Bacteriology* 22, 173-177.

DIETZ, A. & MATHEWS, J. 1977 Characterization of hairy-spored streptomycetes. II. Twelve additional cultures. *International Journal of Systematic Bacteriology* 27, 282-287.

DRUCKER, D. B. & WHITTAKER, D. K. 1971 Microstructure of colonies of rod-shaped bacteria. *Journal of Bacteriology* 108, 515-525.

ETTLINGER, L., CORBAZ, R. & HÜTTER, R. 1958 Zur Systematik der Actinomyceten. 4. Eine Arteinteilung der Gattung *Streptomyces* Waksman et Henrici: *Archiv für Mikrobiologie* 31, 326-358.

FARINA, G. & BRADLEY, S. G. 1970 Reassociation of deoxyribonucleic acids from *Actinoplanes* and other actinomycetes. *Journal of Bacteriology* 102, 30-35.

FASS, R. J. 1973 Morphology and ultrastructure of staphylococcal L-colonies: light, scanning and transmission electron microscopy. *Journal of Bacteriology* 113, 1049-1053.

GOODFELLOW, M. 1971 Numerical taxonomy of some nocardioform bacteria. *Journal of General Microbiology* 69, 33-80.

GOODFELLOW, M., MORDARSKI, M., SZYBA, K. & PULVERER, G. 1978 Relationships among rhodococci based upon deoxyribonucleic acid reassociation. In *Genetics of the Actinomycetales* ed. Freerksen, E., Tárnok, I. & Thumin, J. H. pp. 231-234. Stuttgart & New York: Gustav Fischer Verlag.

GOTTLIEB, D. 1961 An evaluation of criteria and procedures used in the description and characterisation of the streptomycetes. A co-operative study. *Applied Microbiology* 9, 55-65.

GUTHRIE, R. K., KOCSIS, I. & LASSITER, C. B. 1971 Serological analysis of morphologically and biochemically similar *Streptomyces*. *Applied Microbiology* 21, 643-646.

HEINZEN, R. J. & ENSIGN, J. C. 1975 Effect of growth substrates on morphology of *Nocardia corallina*. *Archiv für Mikrobiologie* 103, 209–217.

HOLMBERG, K. & HALLANDER, H. O. 1973 Numerical taxonomy and laboratory identification of *Bacterionema matruchotii, Rothia dentocariosa, Actinomyces naeslundii, Actinomyces viscosus* and some related bacteria. *Journal of General Microbiology* 76, 43–63.

HOLMBERG, K. & NORD, C. E. 1975 Numerical taxonomy and laboratory identification of *Actinomyces* and *Arachnia* and some related bacteria. *Journal of General Microbiology* 91, 17–44.

HOPWOOD, D. A., WILDERMUTH, H. & PALMER, H. M. 1970 Mutants of *Streptomyces coelicolor* defective in sporulation. *Journal of General Microbiology* 61, 397–408.

HOPWOOD, D. A., CHATER, K. F., DOWDING, J. E. & VIVIAN, A. 1973 Advances in *Streptomyces coelicolor* genetics. *Bacteriological Reviews* 37, 371–405.

HÜTTER, R. 1967 *Systematik der Streptomyceten*. Basel & New York: S. Karger.

JONES, D. 1975 A numerical taxonomic study of coryneform and related bacteria. *Journal of General Microbiology* 87, 52–96.

JONES, D. & SNEATH, P. H. A. 1970 Genetic transfer and bacterial taxonomy. *Bacteriological Reviews* 34, 40–81.

KALAKOUTSKII, L. V. & NIKITINA, E. T. 1977 Natural variability of developmental patterns in *Streptomyces*. *Postepy Higieny I Medycyny Doswiadczalnej* 31, 313–355.

KALAKOUTSKII, L. V., KIRILLOVA, I. P. & KRASSILNIKOV, N. A. 1967 A new genus of the Actinomycetales–*Intrasporangium* gen. nov. *Journal of General Microbiology* 48, 79–85.

KRASSILNIKOV, N. A., NIKITINA, N. I. & KORENJAKO, A. J. 1961 On external features in the taxonomy of actinomycetes. *International Bulletin of Bacteriological Nomenclature and Taxonomy* 11, 133–159.

KÜSTER, E. 1953 Beitrag zur Genese und Morphologie der Streptomycetensporen. *6th International Congress of Microbiology* 1, 114–116.

KÜSTER, E. 1961 Results of comparative study of criteria used in classification of the actinomycetes. *International Bulletin of Bacteriological Nomenclature and Taxonomy* 11, 91–98.

LECHEVALIER, H. A. & LECHEVALIER, M. P. 1969 Ultramicroscopic structure of *Intrasporangium calvum* (Actinomycetales). *Journal of Bacteriology* 100, 522–525.

LECHEVALIER, H. A. & TIKHONENKO, A. S. 1960 Effect of nutritional conditions on surface structure of actinomycete spores. *Mikrobiologiya* 29, 28–39 (English translation).

LECHEVALIER, H. A., SOLOTOROVSKY, M. & McDURMONT, C. I. 1961 A new genus of the Actinomycetales, *Micropolyspora* g.n. *Journal of General Microbiology* 26, 11–18.

LECHEVALIER, H. A., LECHEVALIER, M. P. & HOLBERT, P. E. 1966 Electron microscopic observation of the sporangial structure of strains of Actinoplanaceae. *Journal of Bacteriology* 92, 1228–1235.

LECHEVALIER, H. A., LECHEVALIER, M. P. & GERBER, M. N. 1971 Chemical composition as a criterion in the classification of actinomycetes. *Advances in Applied Microbiology* 14, 47–72.

LECHEVALIER, M. P. 1977 Lipids in bacterial taxonomy–a taxonomist's view. *Critical Reviews in Microbiology* 5, 109–210.

LECHEVALIER, M. P. & LECHEVALIER, H. A. 1970 Chemical composition as a criterion in the classification of aerobic actinomycetes. *International Journal of Systematic Bacteriology* 20, 435–443.

LECHEVALIER, M. P., LECHEVALIER, H. A. & HORAN, A. C. 1973a Chemical characteristics and classification of nocardiae. *Canadian Journal of Microbiology* 19, 965–972.

LECHEVALIER, M. P., LECHEVALIER, H. A. & HEINTZ, C. E. 1973b Morphological and chemical nature of the sclerotia of *Chainia olivacea* Thirumalachar and Sukapure of the order Actinomycetales. *International Journal of Systematic Bacteriology* 23, 157–170.

LECHEVALIER, M. P., DE BIÈVRE, C. & LECHEVALIER, H. 1977 Chemotaxonomy of aerobic actinomycetes: phospholipid composition. *Biochemical Ecology and Systematics* 5, 249–260.

LOCCI, R. 1978 Micromorphological development of *Actinomyces* and related genera. In *Nocardia and Streptomyces* ed. Mordarski, M., Kuryłowicz, W. & Jeljaszewicz, J. pp. 173–180. Stuttgart & New York: Gustav Fischer Verlag.

LUSCOMBE, B. M. & GRAY, T. R. G. 1974 Characteristics of *Arthrobacter* grown in continuous culture. *Journal of General Microbiology* 82, 213–222.

LYONS, A. J. & PRIDHAM, T. G. 1971 *Streptomyces torulosus* sp.n. an unusual knobby-spored taxon. *Applied Microbiology* 22, 190–193.

McVITTIE, A. 1974 Ultrastructural studies on sporulation in wild-type and white colony mutants of *Streptomyces coelicolor*. *Journal of General Microbiology* 81, 291–302.

McVITTIE, A., WILDERMUTH, H. & HOPWOOD, D. A. 1972 Fine structure and surface topography of endospores of *Thermoactinomyces vulgaris*. *Journal of General Microbiology* 71, 367–381.

MINNIKIN, D. E. & GOODFELLOW, M. 1976 Lipid composition in the classification and identification of nocardiae and related taxa. In *The Biology of the Nocardiae* ed. Goodfellow, M., Brownell, G. H. & Serrano, J. A. pp. 160–219. London: Academic Press.

MINNIKIN, D. E., ALSHAMAONY, L. & GOODFELLOW, M. 1975 Differentiation of *Mycobacterium, Nocardia* and related taxa by thin-layer chromatographic analysis of whole-organism methanolysates. *Journal of General Microbiology* 88, 200–204.

MINNIKIN, D. E., GOODFELLOW, M. & ALSHAMAONY, L. 1978 Mycolic acids in the classification of nocardioform bacteria. In *Nocardia and Streptomyces* ed. Mordarski, M., Kuryłowicz, W. & Jeljaszewicz, J. pp. 63–66. Stuttgart & New York: Gustav Fischer Verlag.

MORDARSKA, H., MORDARSKI, M. & GOODFELLOW, M. 1972 Chemotaxonomic characters and classification of some nocardioform bacteria. *Journal of General Microbiology* 79, 77–86.

MORDARSKI, M. & KUDRINA, E. S. 1961 Effect of various nitrogen sources upon the surface of the spore envelope of actinomycetes. *Mikrobiologiya* 30, 79–83 (English translation).

MURRAY, R. G. E. 1968 Microbial structure as an aid to microbial classification and taxonomy. *Spisy (Faculta des Sciences de l'Université J. E. Purkyne Brno)* 43, 249–252.

MURRAY, R. G. E. 1974 A place for bacteria in the living world. In *Bergey's Manual of Determinative Bacteriology* 8th edn. ed. Buchanan, R. E. & Gibbons, N. E. pp. 4–9. Baltimore: Williams & Wilkins.

OKANISHI, M., AKAGAWA, H. & UMEZAWA, H. 1972 An evaluation of taxonomic criteria in streptomycetes on the basis of deoxyribonucleic acid homology. *Journal of General Microbiology* 72, 49–58.

PRAUSER, H. 1976 New nocardioform organisms and their relationship. In *Actinomycetes: The Boundary Micro-organisms* ed. Arai, T. pp. 193–207. Baltimore, London & Tokyo: University Park Press.

PRAUSER, H. 1978 Considerations on taxonomic relations among Gram-positive, branching bacteria. In *Nocardia and Streptomyces*, ed. Mordarski, M., Kuryłowicz, W. & Jeljaszewicz, J. pp. 3–12, Stuttgart & New York: Gustav Fischer Verlag.

PRIDHAM, T. G. 1965 Color and streptomycetes. Report of an International Workshop on determination of color of streptomycetes. *Applied Microbiology* 13, 43–61.

PRIDHAM, T. G., HESSELTINE, C. W. & BENEDICT, R. G. 1958 A guide for the classification of streptomycetes according to selected groups. Placement of strains in morphological sections. *Applied Microbiology* 6, 52–79.

PRIDHAM, T. G. & TRESNER, H. D. 1974 Family VII Streptomycetaceae Waksman and Henrici 1943. In *Bergey's Manual of Determinative Bacteriology*, 8th edn., ed. Buchanan, R. E. & Gibbons, N. E. pp. 747–845. Baltimore: Williams & Wilkins.

SCHLEIFER, K. H. & KANDLER, O. 1972 Peptidoglycan types of bacterial cell walls and their taxonomic implications. *Bacteriological Reviews* 36, 407–477.

SHARPLES, G. P. & WILLIAMS, S. T. 1974 Fine structure of the globose bodies of *Dacty-*

losporangium thailandense (Actinomycetales). *Journal of General Microbiology* **84**, 219-222.

SHARPLES, G. P. & WILLIAMS, S. T. 1976 Development and fine structure of sclerotia and spores of the actinomycete *Chainia olivacea. Microbios* **15**, 37-47.

SHARPLES, G. P., WILLIAMS, S. T. & BRADSHAW, R. M. 1974 Spore formation in the Actinoplanaceae (Actinomycetales). *Archiv für Mikrobiologie* **101**, 9-20.

SHIRLING, E. B. & GOTTLIEB, D. 1968*a* Co-operative description of type cultures of *Streptomyces.* II. Species descriptions from first study. *International Journal of Systematic Bacteriology* **18**, 69-189.

SHIRLING, E. B. & GOTTLIEB, D. 1968*b* Co-operative descriptions of type cultures of *Streptomyces.* III. Additional species descriptions from first and second studies. *International Journal of Systematic Bacteriology* **18**, 279-391.

SHIRLING, E. B. & GOTTLIEB, D. 1969 Co-operative description of type cultures of *Streptomyces.* IV. Species descriptions from the second, third and fourth studies. *International Journal of Systematic Bacteriology* **19**, 391-512.

SHIRLING, E. B. & GOTTLIEB, D. 1972 Co-operative description of type species of *Streptomyces.* V. Additional descriptions. *International Journal of Systematic Bacteriology* **22**, 265-394.

SHIRLING, E. B. & GOTTLIEB, D. 1977 Retrospective evaluation of International Streptomyces Project taxonomic criteria. In *Actinomycetes: The Boundary Microorganisms* pp. 9-41. ed. Arai, T. Baltimore, London & Tokyo: University Park Press.

SLACK, J. M. & GERENCSER, M. A. 1975 *Actinomyces, Filamentous Bacteria: Biology and Pathogenicity.* Minneapolis: Burgess Publishing Co.

SNEATH, P. H. A. 1976 An evaluation of numerical taxonomic techniques in the taxonomy of *Nocardia* and allied taxa. In *The Biology of the Nocardiae.* ed. Goodfellow, M., Brownell, G. H. & Serrano, J. A. pp. 74-101. London: Academic Press.

SPRINGER, E. L. & ROTH, I. L. 1972 Scanning electron microscopy of bacterial colonies. I. *Diplococcus pneumoniae* and *Streptococcus pyogenes. Canadian Journal of Microbiology* **18**, 219-223.

STALEY, T. E. & COLWELL, R. R. 1973 Applications of molecular genetics and numerical taxonomy to the classification of bacteria. *Annual Review of Ecology and Systematics* **4**, 273-300.

SZABO, I. M. & MARTON, M. 1964 Comments on the first results of the international co-operative work on criteria used in characterization of streptomycetes. *International Bulletin of Bacteriological Nomenclature and Taxonomy* **14**, 17-38.

SZABO, I. M. & MARTON, M. 1976 Evaluation of criteria used in the I.S.P. cooperative description of type strains of *Streptomyces* and *Streptoverticillium* species. *International Journal of Systematic Bacteriology* **26**, 105-110.

TAKEO, K. 1976 Existence of a surface configuration on the aerial spore and aerial mycelium of *Micropolyspora. Journal of General Microbiology* **95**, 17-26.

TAKEO, K. & UESAKA, I. 1975 Existence of a simple configuration in the wall surface of *Nocardia* mycelium. *Journal of General Microbiology* **87**, 373-376.

THIEMANN, J. E. 1970 Study of some new genera and species of the Actinoplanaceae. In *The Actinomycetales* ed. Prauser, H. pp. 245-257. Jena: Gustav Fischer.

THIEMANN, J. E., PAGANI, H. & BERETTA, G. 1967 A new genus of the Actinoplanaceae: *Dactylosporangium* gen. nov. *Archiv für Mikrobiologie* **58**, 42-52.

THIRUMALACHAR, M. J. 1955 *Chainia*, a new genus of the Actinomycetales. *Nature, London* **176**, 934-935.

TOYAMA, H., OKANISHI, M. & UMEZAWA, H. 1974 Heterogeneity among whorl-forming streptomycetes determined by deoxyribonucleic acid reassociation. *Journal of General Microbiology* **80**, 507-514.

TRESNER, H. D., DAVIES, M. C. & BACKUS, E. J. 1961 Electron microscopy of *Streptomyces* spore morphology and its role in species differentiation. *Journal of Bacteriology* **81**, 70-80.

TRESNER, H. D., HAYES, J. A. & BACKUS, E. J. 1967 Morphology of submerged growth of streptomycetes as a taxonomic aid. I. Morphological development of *Streptomyces aureofaciens* in agitated liquid media. *Applied Microbiology* **15**, 1185-1191.

WAKSMAN, S. A. 1961 *The Actinomycetes Vol. 2. Classification, Identification and Description of Genera and Species.* Baltimore: Williams & Wilkins.

WILDERMUTH, H., WEHRLI, E. & HORNE, R. W. 1971 The surface structure of spores and aerial mycelium in *Streptomyces coelicolor. Journal of Ultrastructural Research* 35, 168-180.

WILLIAMS, S. T. 1970 Further investigations of actinomycetes by scanning electron microscopy. *Journal of General Microbiology* 62, 67-73.

WILLIAMS, S. T. & DAVIES, F. L. 1967 Use of a scanning electron microscope for the examination of actinomycetes. *Journal of General Microbiology* 48, 171-177.

WILLIAMS, S. T. & SHARPLES, G. P. 1976 *Streptosporangium corrugatum* sp. nov. an actinomycete with some unusual morphological features. *International Journal of Systematic Bacteriology* 26, 45-52.

WILLIAMS, S. T., BRADSHAW, R. M., COSTERTON, J. W. & FORGE, A. 1972 Fine structure of the spore sheath of some *Streptomyces* species. *Journal of General Microbiology* 72, 249-258.

WILLIAMS, S. T., ENTWISTLE, S. & KURYŁOWICZ, W. 1974a The morphology of streptomycetes growing in media used for commercial production of antibiotics. *Microbios* 11, 47-60.

WILLIAMS, S. T., SHARPLES, G. P. & BRADSHAW, R. M. 1974b Spore formation in *Actinomadura dassonvillei* (Brocq-Rousseu) Lechevalier and Lechevalier. *Journal of General Microbiology* 84, 415-419.

WILLIAMS, S. T., SHARPLES, G. P., SERRANO, J. A., SERRANO, A. A. & LACEY, J. 1976 The micromorphology and fine structure of nocardioform organisms. In *The Biology of the Nocardiae* ed. Goodfellow, M., Brownell, G. H. & Serrano, J. A. pp. 102-140. London: Academic Press.

Cell Wall Composition in the Classification and Identification of Coryneform Bacteria

R. M. KEDDIE* AND I. J. BOUSFIELD†

*Department of Microbiology, University of Reading, Reading, Berks, UK

†National Collection of Industrial Bacteria, Torry Research Station, Aberdeen, UK

Contents

1. Introduction

ONE OF the more important aspects of modern microbial taxonomy has been the development of chemotaxonomy, which is simply the use for taxonomic purposes of information obtained from chemical analyses of cells or parts of cells. Chemotaxonomic methods are now making major contributions to classification and identification, particularly in those groups of organisms where the more traditional methods based on form and function have largely failed to provide satisfactory systems.

One of the earliest chemical characters found to be of possible taxonomic value was the cell wall composition of Gram positive bacteria. Over twenty years ago, Cummins & Harris (1956) examined the cell walls of various Gram positive organisms and showed that they did not all have the same qualitative amino acid composition and that, furthermore, the differences that occurred seemed to follow a pattern. In general, the cell wall amino acid composition appeared to be similar for species within a genus, but to differ between genera. Of the relatively few amino acids which occurred, alanine and glutamic acid were always present. Differences were found between genera, however, in that either lysine or diaminopimelic acid (DAP) occurred, and sometimes other amino acids such as aspartic acid and glycine were found as well. Cummins & Harris also showed that there were differences in the sugars present in cell wall hydrolysates, but in this case the differences were often between species of the same genus. This led them to suggest that the cell wall amino acid composition might prove to be a useful taxonomic criterion at the generic level and that the sugar composition might help to distinguish between species of the same genus.

The species examined originally by Cummins & Harris (1956) contained either lysine or DAP in the cell wall and like Work & Dewey (1953), who had previously investigated the distribution of DAP in bacteria, they thought that the presence or absence of this compound might be an important taxonomic character. Support for this idea soon came from further studies of the cell wall composition of various actinomycetes and allied groups: not only was it shown that the presence or absence of DAP was a valuable distinguishing feature but also that organisms in which DAP occurred could be differentiated further according to whether they contained the *meso*- or L-isomer (Hoare & Work 1956, 1957; Cummins & Harris 1958). A few years later, however, it was shown that some plant pathogenic coryneform bacteria contained neither DAP nor lysine in the cell wall but instead contained ornithine or diaminobutyric acid (Perkins & Cummins 1964). Thus the diamino acid present in the cell wall came to be regarded as the major taxonomic character provided by qualitative cell wall analysis and the feeling grew that organisms containing different diamino acids should not be classified in the same genus.

Following these early studies, cell wall composition has become a well-established and widely-used criterion in the classification and identification of Gram positive bacteria. Several workers have used techniques similar to those originally described by Cummins & Harris (1956, 1958) in which qualitative analyses are made of purified cell walls (e.g. for coryneform bacteria see Keddie *et al.* 1966; Robinson 1966*b*; Yamada & Komagata 1970*a*, 1972*a*; Cummins 1971). When considered from a taxonomic viewpoint, however, cell wall analysis has developed in two opposing directions. On the one hand, relatively elaborate, and therefore quite laborious techniques are used to determine the detailed peptidoglycan structure; while on the other, simple so-called 'rapid' methods are used to determine the occurrence of a few, taxonomically useful components.

After the primary structure of the peptidoglycan of bacterial cell walls had been elucidated, Schleifer & Kandler (1967) developed a 'chemical method' which they and their associates have used to determine the peptidoglycan types of a wide range of Gram positive bacteria. This approach provides information of great taxonomic value and indeed has revealed differences between organisms which could not possibly be detected by simple qualitative cell wall analyses. However, because of the specialized techniques involved in determining the peptidoglycan structure, this method cannot be applied readily to large numbers of strains in taxonomic studies—although Schleifer & Kandler (1972) have suggested a 'rapid screening method'—and it obviously has a limited application in routine identification.

Other developments in methods of cell wall analysis have been aimed at simplifying the procedure so that larger numbers of strains can be examined more quickly. Such techniques have the advantage that they can be used in routine identification. These so-called 'rapid' methods involve qualitative analyses either of acid hydrolysates of whole organisms (Becker *et al.* 1964; Murray & Proctor 1965) or of wall preparations obtained by alkali treatment of whole organisms (Boone & Pine 1968; Keddie & Cure 1977). They have been designed mainly to detect those wall components which experience has shown are of most value in classification and identification viz. the diamino acid (which is always present) and certain diagnostic sugars such as arabinose.

The 'whole organism' technique is the simplest and quickest of the 'rapid' methods and it has been of great value in taxonomic studies of the actino-mycetes (see Chapter 9, pp. 192–195, Lechevalier & Lechevalier 1970; Mordarska *et al.* 1972). However, this method is suitable only for detecting components which occur mainly or entirely in the cell wall such as the isomers of diamino-pimelic acid and arabinose. The 'alkali method', on the other hand, can be used for the detection of all the diamino acids which may occur in bacterial walls as well as for taxonomically important sugars (Keddie & Cure 1977).

As mentioned earlier, chemotaxonomic methods have had the greatest impact in those bacterial groups where traditional methods have failed to produce a satisfactory classification. Perhaps nowhere is this more apparent than in the classification of the coryneform bacteria, where in fact little real progress was made until the advent of cell wall analysis. An attempt will be made to illustrate this by looking in turn at each of those genera which can be considered to belong to the group of aerobic coryneform bacteria according to the criteria of Cure & Keddie (1973) and Keddie (1978): i.e. *Corynebacterium, Curtobacterium, Cellulomonas, Arthrobacter, Microbacterium* and *Brevibacterium*. The newly proposed genus *Caseobacter* (Crombach 1978) will also be mentioned briefly. However, the genus *Kurthia* will not be considered, despite its 'tentative' inclusion in the coryneform group in the eighth edition of *Bergey's Manual of Determinative Bacteriology* (Keddie & Rogosa 1974) because it does not have a coryneform morphology.

2. Genera which Contain *Meso*-Diaminopimelic Acid in the Cell-wall Peptidoglycan

A. Corynebacterium

The genus *Corynebacterium* was established by Lehmann & Neumann (1896) to accommodate the diphtheria bacillus; later a few very similar species were added. Because *Cnbc. diphtheriae* and similar species were at that time considered to have very distinctive morphological features and staining reactions, the genus was defined largely in these terms. However, in the years that followed it was realized that morphologically similar organisms occurred in many habitats other than the animal body and, because morphological similarity was then widely held to indicate relatedness, they too were placed in the genus *Corynebacterium* by various investigators. Thus the name *Corynebacterium* was applied not only to animal parasitic species but also to a heterogeneous collection of morphologically similar plant pathogens and to saprophytic species from a wide range of habitats (see Jensen 1952). This practice has continued almost up to the present and thus many species of widely different characteristics now bear the name *Corynebacterium*.

Although some of the earlier investigators (see Conn 1947; Jensen 1952; Clark 1952) felt intuitively that the genus should once more be restricted to *Cnbc. diphtheriae* and very similar animal parasitic species, there was really no way that this could be done with the methods then available other than by resorting largely to habitat relationships. When the chemical composition of the cell wall was introduced as a taxonomic criterion (Cummins & Harris 1956) a means was provided whereby this might be achieved. In the first few years in which these techniques were used it was shown that, with the exception of a few species, the animal parasites were similar to each other in cell wall composition and that they differed for the most part from morphologically similar plant pathogenic and saprophytic species (Cummins & Harris 1956, 1958, 1959; Cummins 1962; Perkins & Cummins 1964). Since then the introduction of further chemotaxonomic methods has provided a means of defining the genus largely in chemical terms. Thus there is now a broad measure of agreement that the genus *Corynebacterium* should be restricted to those organisms which contain *meso*-DAP, arabinose and galactose in the cell wall, which contain relatively short chain mycolic acids (corynomycolic acids) with *ca*. 22–38 carbon atoms (see Barksdale 1970; Schleifer & Kandler 1972; Keddie & Cure 1977, 1978; Minnikin *et al.* 1978) and which have DNA base ratios in the approximate range, 51–59% G+C (see references, Table 1). In addition a majority of species with these chemical attributes are facultatively anaerobic (Cummins *et al.* 1974; Keddie & Cure 1977) and those species examined to date contain dihydrogenated menaquinones with either 8 or 9 isoprene units (see Minnikin *et al.* 1978). This concept of the genus (*Corynebacterium sensu stricto*) is, in general, supported by numerical taxonomic studies (see, for example, Bousfield 1972; Jones 1975).

Corynebacterium, when defined in this way, includes *Cnbc. diphtheriae* and most animal pathogenic and parasitic species, but excludes such species as *Cnbc. pyogenes* and *Cnbc. haemolyticum* which contain lysine in the cell wall peptidoglycan and rhamnose as characteristic wall sugar (Barksdale *et al.* 1957). This restricted concept of the genus also excludes all plant pathogens and many saprophytic species which currently bear the name *Corynebacterium*. However,

TABLE 1

Coryneform taxa which contain meso-*diaminopimelic acid in the cell wall peptidoglycan and some of their chemotaxonomic features*

| Taxon | Arabinose in wall[a] | Mycolic acids[b] | | Major menaquinones[c] | % G+C[d] |
		Present	No. of carbons in chain		
Corynebacterium[e,f] (sensu stricto)	+	+	22–38	MK-8 (H_2)[g] or MK-9 (H_2)	51–59[h] (refs 1–7)
Rhodococcus[i,f]	+	+	34–52	MK-8 (H_2) or MK-9 (H_2)	59–69 (ref. 8)
Caseobacter	+	+[j]	ND[k]	ND	60–67 (ref. 9)
Brevibacterium linens	–	–		MK-8 (H_2)	60–64 (refs 4, 6)
Unassigned species[l]	–	–		ND	ND

+, present; –, absent. [a]See Keddie & Cure (1978) for details. [b]Long-chain 2-alkyl-branched-3-hyroxy acids which have been demonstrated only in coryneform and allied taxa which contain meso-diaminopimelic acid, arabinose and galactose in the cell wall. Data from Minnikin et al. (1978). [c]Data from Yamada et al. (1976); Collins et al. (1977); Goodfellow & Minnikin (1977); Minnikin et al. (1978); Collins et al. (1979). [d]Data from (1) Marmur & Doty (1962); (2) Hill (1966); (3) Abe et al. (1967); (4) Yamada & Komagata (1970b); (5) Bousfield (1972); (6) Crombach (1972); (7) Cummins et al. (1974); (8) Goodfellow & Alderson (1977); (9) Crombach (1978). [e]Bacterionema matruchotii has similar characteristics: see Minnikin et al. (1978). [f]Most members of Corynebacterium sensu stricto do not contain tuberculostearic acid (10-methyloctadecanoic acid) whereas many members of Rhodococcus do (see Minnikin et al. 1978) [g]MK-8, MK-9 etc. indicates the number of isoprene units in the menaquinone; (H_2), (H_4) etc. indicates the number of double bonds hydrogenated. [h]Occasional higher values reported. [i]Usually considered a member of the nocardioform bacteria but included here for comparison with Corynebacterium. [j]Reported to be corynomycolic acids (Crombach 1978). [k]Data not available or incomplete. [l]A heterogeneous collection of mainly un-named strains of uncertain taxonomic position but includes strains presently named Arthrobacter viscosus (ATCC 19584) and Atbc. stabilis (NCIB 10617).

some saprophytes would remain in the genus. Prominent among these are several glutamic acid-producing coryneform species, all 'patent' strains considered by Abe et al. (1967) to belong to the taxon *Corynebacterium glutamicum*. Subsequent chemotaxonomic studies (see Keddie & Cure 1977, 1978; Minnikin et al. 1978) have confirmed that *Cnbc. glutamicum* and its numerous synonyms such as *Cnbc. callunae, Cnbc. herculis, Cnbc. lilium, Brevibacterium divaricatum, Brev. flavum, Brev. roseum* etc. have the characteristics of *Corynebacterium sensu stricto*.

Since the species presently named *Microbacterium flavum* was shown to contain *meso*-DAP, arabinose and galactose in the cell wall (Keddie *et al.* 1966; Robinson 1966b), a considerable amount of different kinds of evidence has been accumulated showing that it should be transferred to the genus *Corynebacterium* (Robinson 1966b; Schleifer 1970; Collins-Thompson *et al.* 1972; Bousfield 1972; Jones 1975; Goodfellow *et al.* 1976; Keddie & Cure 1977; Minnikin *et al.* 1978).

In addition to providing a basis for a new definition of the genus *Corynebacterium*, studies of cell wall composition also gave the first clear indication of a relationship between *Corynebacterium sensu stricto* and the genera *Nocardia* (*sensu stricto*) and *Mycobacterium* (see Cummins & Harris 1958) all of which contain *meso*-DAP, arabinose and galactose in the cell wall. This relationship has been substantiated by subsequent chemotaxonomic studies (see Barksdale 1970; Cummins *et al.* 1974; Minnikin *et al.* 1978). *Rhodococcus*, a genus recently resurrected by Goodfellow & Alderson (1977) to accommodate the so-called 'rhodochrous' complex, also belongs to this group of related genera as does *Bacterionema matruchotii* (see Minnikin *et al.* 1978). Indeed the latter species has the characteristics of *Corynebacterium sensu stricto* (see Pine 1970; Alshamaony *et al.* 1977; Minnikin *et al.* 1978). All of the genera mentioned have a similar cell wall composition but in general can be distinguished by their DNA base composition, fatty acid composition, oxygen relations and by the chain length of their mycolic acids (see Goodfellow & Minnikin 1977; Minnikin *et al.* 1978); but there is some overlap between the genera *Corynebacterium* and *Rhodococcus* (see Keddie & Cure 1978; Minnikin *et al.* 1978).

As was mentioned earlier, the plant pathogenic coryneform bacteria and many saprophytic species currently named *Corynebacterium* do not conform with the description of *Corynebacterium sensu stricto* and should be removed from the genus. However many of them can now be accommodated elsewhere. There has always been a tendency to consider the plant pathogenic coryneform species as a taxonomic entity largely because they are plant pathogens. Indeed, in a recent numerical taxonomic study, Dye & Kemp (1977) concluded that the plant pathogenic coryneform bacteria formed a relatively homogeneous group (only conventional tests were used) which should be classified within a single genus. Although no reference strains of *Corynebacterium sensu stricto* were studied, they nevertheless recommended that the plant pathogens should be retained in the genus *Corynebacterium*, a conclusion which is not supported by the chemotaxonomic data nor by most recent numerical taxonomic studies. The species currently recognized are heterogeneous in cell wall composition and four different groups can be distinguished on the basis of the diamino acid present in the cell wall peptidoglycan. *Corynebacterium fascians* contains *meso*-DAP, arabinose and galactose in the cell wall and for this reason has been considered a possible candidate for the genus *Corynebacterium* (see, e.g. Barksdale 1970).

The results of numerical taxonomic (Jones 1975), lipid (Goodfellow *et al*. 1976; Keddie & Cure 1977; Minnikin *et al*. 1978) and DNA base ratio studies (Yamada & Komagata 1970*b*; Bousfield 1972) indicate, however, that it is a member of the genus *Rhodococcus* (Goodfellow & Alderson 1977). On the other hand *Cnbc. ilicis* contains lysine in the cell wall peptidoglycan (see Keddie & Cure 1978), and its characters conform closely with those of the 'ideal phenotype' of *Atbc. globiformis* (Keddie & Cure 1977). It therefore seems a perfectly legitimate member of the genus *Arthrobacter*, a conclusion supported by numerical taxonomic studies (Bousfield 1972; Jones 1975). A second lysine-containing species, *Cnbc. rathayi* (see Keddie & Cure 1978), was also considered to be a member of the genus *Arthrobacter* by Jones (1975), a view which is supported by the peptidoglycan structure (Schleifer & Kandler 1972). However, most plant pathogenic coryneform bacteria contain either ornithine or 2,4-diaminobutyric acid in the cell wall (Table 2). The taxonomic position of the diaminobutyric acid-containing species, which also include some saprophytes, has not yet been resolved but those containing ornithine have the characters of *Curtobacterium* (see below), a genus proposed by Yamada & Komagata (1972*b*).

Of the many saprophytic *'Corynebacterium'* species which do not conform with the description of *Corynebacterium sensu stricto* some, such as *Cnbc. hydrocarboclastus* and *Cnbc. rubrum*, have the characteristics of the genus *Rhodococcus* (the *'rhodochrous'* taxon, Goodfellow *et al*. 1976; Keddie & Cure 1977). There are also some species, however, whose characteristics overlap those of *Corynebacterium sensu stricto* and the genus *Rhodococcus* and cannot be assigned with confidence to either (see Keddie & Cure 1978; Minnikin *et al*. 1978). Other unassigned species contain lysine, ornithine or diaminobutyric acid as cell wall diamino acid (see Keddie & Cure (1978) for further details).

B. Caseobacter

This genus was proposed (Crombach 1978) to accommodate certain of the so-called 'grey-white' or 'non-orange' cheese coryneform bacteria described originally by Mulder & Antheunisse (1963). The 'grey-white' cheese strains were shown to be heterogeneous in cell wall composition by Keddie & Cure (1977); most of those studied contained *meso*-DAP, arabinose and galactose in the cell wall whereas the remainder contained lysine as wall diamino acid. The genus *Caseobacter* was proposed for the *meso*-DAP-containing strains, representatives of which were stated to contain corynomycolic acids (Crombach 1978). However, the relationship of the proposed new genus to *Corynebacterium* on the one hand, and to *Rhodococcus* on the other, needs to be clarified. Indeed Keddie & Cure (1977) considered that the *meso*-DAP-containing 'grey-white' cheese strains were members of the *'rhodochrous'* complex (genus *Rhodococcus*), a conclusion supported by their DNA base ratios. The fatty acid pro-

TABLE 2

Coryneform taxa which contain diamino acids other than meso-diaminopimelic acid in the cell wall peptidoglycan and some of their chemotaxonomic features

Taxon	Cell wall diamino acid	Peptidoglycan group[a]	Galactose in wall[b]	Major menaquinone[c]	% G+C[d]
Arthrobacter (sensu stricto)	lysine	A	+	MK-9 (H$_2$)[e,f]	59–66[g] (refs 1,2,4–6)
Microbacterium lacticum	lysine	B	+	MK-10, MK-11	69–70[h] (refs 2,5)
'Arthrobacter' (*Atbc. simplex/ Atbc. tumescens*)	L-DAP[i]	A	+	MK-8 (H$_4$)	70–74 (refs 1,2,4)
Cellulomonas	ornithine	A	–	MK-9 (H$_4$)	71–73 (refs 3,5)
Curtobacterium	ornithine	B	(+)[j]	MK-9[k]	67–71 (refs 3,5,7)
'Arthrobacter' (*Atbc. terregens/ Atbc. flavescens*)	ornithine	B (*Atbc. terregens*)	+	ND[l]	69–70 (refs 1,4)
Microbacterium liquefaciens	ornithine	B	–	ND	ND
Unassigned species[m]	DAB[n]	B	ND	MK-10, MK-11 (*Corynebacterium aquaticum*)	64–74[p] (refs 1,2,4–8)

+, present; −, absent. [a] Cross-linkage between positions 3 and 4 (Group A), or positions 2 and 4 (Group B), of two peptide subunits usually by an interpeptide bridge containing one or more amino acids (see Schleifer & Kandler 1972). [b] Other sugars usually present: see Keddie & Cure (1978) for details. [c] Data from Yamada et al. (1976); Minnikin et al. (1978); Collins et al. (1979). Relatively few strains of each taxon have been examined. [d] Data from (1) Skyring & Quadling (1970); (2) Yamada & Komagata (1970b); (3) Yamada & Komagata (1972b); (4) Skyring et al. (1971); (5) Bousfield (1972); (6) Crombach (1972); (7) Starr et al. (1975); (8) Luthy (1974). [e] MK−8, MK−9 etc. indicates the number of isoprene units in the menaquinone; (H₂), (H₄) etc. indicates the number of double bonds hydrogenated. [f] A few species studied have MK−8 and/or MK−9 (see Minnikin et al. 1978 for details). [g] The single strain of Atbe. atrocyaneus has the high value of ca. 70% G+C (Skyring & Quadling 1970; Yamada & Komagata (1970b). The values reported by Bowie et al. (1972) are considerably higher than those of other workers and have not been included. [h] Collins-Thompson et al. (1972) quote the very low values of 63–64% G+C for 2 strains; the values given are those of Yamada & Komagata (1970b) and Bousfield (1972). [i] L-2, 4-diaminopimelic acid. Pitcher (1976) has described unusual skin coryneform bacteria which contain L-DAP, arabinose and galactose. [j] (l), most strains have the indicated reaction. [k] Curtobacterium testaceum differs from all other members of the genus in having MK−11 as the major menaquinone (Yamada et al. 1976). [l] No data. [m] A number of species currently named Corynebacterium but as yet of uncertain taxonomic position. Included are the plant pathogenic species Cnbc. insidiosum, Cnbc. michiganense, Cnbc. sepedonicum and Cnbc. tritici; other species include Cnbc. aquaticum, Cnbc. mediolanum and Cnbc. okanaganae (see Keddie & Cure 1975). [n] 2,4-diaminobutyric acid. [p] Yamada & Komagata (1970b) give the high value 78.1% G+C for Cnbc. insidiosum (ATCC 10253) but Starr et al. (1978) give 72.9 (± 0.9)% G+C for 9 strains including ATCC 10253.

files of a few strains which have been examined resemble, however, those found in members of the genus *Corynebacterium sensu stricto* (T. R. Dando & I. J. Bousfield, unpublished results).

C. Brevibacterium

The genus *Brevibacterium* was proposed by Breed (1953*a,b*) for a number of species of non-spore-forming, Gram positive rods formerly classified in the genus *Bacterium*. It was largely a genus of convenience and one of the few definite statements in the circumscription was that the genus comprised typically short, unbranching rods; no mention was made of a coryneform morphology. It was subsequently shown however, that the type species, *Brev. linens*, had a coryneform morphology and indeed that it had a rod/coccus growth cycle similar to that seen in the genus *Arthrobacter* (Schefferle 1957; Mulder & Antheunisse 1963). Most of the large number of nomenclatural species named *Brevibacterium* have now been shown to have a coryneform morphology.

In view of the vague circumscription, it is hardly surprising that subsequent numerical taxonomic (see Jones 1978) and chemotaxonomic studies (see Fiedler *et al.* 1970; Schleifer & Kandler 1972; Keddie & Cure 1978; Minnikin *et al.* 1978) have shown that the genus *Brevibacterium* is extremely heterogeneous. Because of the morphological similarity of *Brev. linens* to *Atbc. globiformis* it was at one time suggested that *Brev. linens* belonged in the genus *Arthrobacter* (Mulder & Antheunisse 1963). Some earlier numerical taxonomic studies also indicated a relationship between *Brev. linens* and *Atbc. globiformis* and similar species, and indeed it was suggested that *Brev. linens* be transferred to the genus *Arthrobacter* (Da Silva & Holt 1965; Bousfield 1972). The DNA base ratios reported for different strains of *Brev. linens* also lie in the range reported for legitimate *Arthrobacter* spp. (see Tables 1 and 2). Cell wall analysis has shown, however, that *Brev. linens* is quite distinct from *Arthrobacter*; the wall diamino acid is *meso*-DAP, not lysine as in legitimate arthrobacters (see Keddie & Cure 1978). Indeed *Brev. linens* contains a directly cross-linked, Group A peptidoglycan similar to that found in *Corynebacterium* (Schleifer & Kandler 1972), but it does not contain arabinose in the wall polysaccharide (see Keddie & Cure 1978) nor does it contain mycolic acids (see Minnikin *et al.* 1978). Although it was originally reported that *Brev. linens* was characterized by the presence of ribose in cell wall preparations (Keddie & Cure 1977) subsequent and more detailed studies of the wall polysaccharide failed to reveal the presence of this pentose in any of several strains studied (Fiedler & Stackebrandt 1978). However, *Brev. linens* was shown to differ from all other coryneform bacteria studied in containing a glycerol teichoic acid in the wall polysaccharide (Fiedler & Stackebrandt 1978). It seems, therefore, that the report of ribose in the walls of *Brev. linens* strains (Keddie & Cure 1977) resulted from an error in chromatography.

Thus a considerable amount of evidence has been amassed which indicates that *Brev. linens* is a distinct taxonomic entity (see also Jones 1975), which could form the nucleus of a redefined genus *Brevibacterium* as suggested by Yamada & Komagata (1972*b*). Although such a genus would at present contain only the type species, *Brev. linens*, many of the large number of nomenclatural species presently named *Brevibacterium* can now be assigned to other genera. An appreciable number of species contain *meso*-DAP, arabinose and galactose in the cell wall (Schleifer & Kandler 1972; Keddie & Cure 1977) and some are legitimate *Corynebacterium* spp. Essential information is lacking on others and they cannot as yet be assigned to a particular taxon. It is likely, however, that they are either *Corynebacterium* or *Rhodococcus* spp. (see Keddie & Cure 1978; Minnikin *et al.* 1978). Rather fewer species contain lysine in the cell wall and some have been shown to be legitimate *Arthrobacter* spp. (Yamada & Komagata 1972*b*; Schleifer & Kandler 1972; Keddie & Cure 1978) while one, *Brev. fermentans*, is probably an *Oerskovia* sp. (I. J. Bousfield, unpublished results). Also, some ornithine-containing species have been placed in the proposed genus *Curtobacterium* (Yamada & Komagata 1972*b*). Two strains named *Brev. lipolyticum* contain L-DAP in the cell wall (Yamada & Komagata 1972*b*) and both the DNA base ratios (Yamada & Komagata 1970*b*) and the menaquinone system (Yamada *et al.* 1976) are similar to those of *Arthrobacter simplex*. At least one species, *Brev. leucinophagum*, is a Gram negative rod and a possible *Acinetobacter* sp. (Jones & Weitzman 1974). Further information on the cell wall composition and probable taxonomic position of '*Brevibacterium*' species may be obtained in Keddie & Cure (1978).

3. Genera which Contain Diamino Acids Other than *Meso*-Diaminopimelic Acid in the Cell-wall Peptidoglycan

A. Arthrobacter

The genus *Arthrobacter* was originally created for a group of soil bacteria with an unusual morphology in that they were stated to occur as Gram negative rods in young cultures and as Gram positive cocci in older cultures. In addition they were described as aerobic, nutritionally non-exacting and liquefying gelatin slowly (Conn & Dimmick 1947). However, in the ensuing years the genus was broadened by the inclusion of a number of nutritionally exacting species such as *Atbc. terregens* (Lochhead & Burton 1953), *Atbc. citreus* (Sacks 1954), *Atbc. duodecadis* and *Atbc. flavescens* (Lochhead 1958); indeed one of Conn's original cultures of *Atbc. globiformis* was shown to require biotin (Morris 1960; Chan & Stevenson 1962). Thus *Arthrobacter* became a genus of aerobic soil bacteria whose main distinguishing feature was the possession of a growth cycle in which the irregular rods in young cultures were replaced by coccoid forms in older cultures. These coccoid forms, when transferred to fresh medium, produced outgrowths to give irregular rods again and so the cycle was repeated. This 'rod/coccus' growth cycle is not unique, however, to *Arthrobacter* and occurs in some other coryneform bacteria as well as in at least some members of the genus *Rhodococcus*. Thus the considerable emphasis placed on morphological features in the generic circumscription led to much of the confusion that now exists in the classification of the genus *Arthrobacter*.

As early as 1959, Cummins & Harris showed that the few, differently named *Arthrobacter* strains that they examined were heterogeneous in cell wall composition. This heterogeneity has been further emphasized by subsequent studies of cell wall composition (see Keddie & Cure 1978) and peptidoglycan structure (see Schleifer & Kandler 1972), and by studies using numerical taxonomic methods (see Jones 1978), various chemotaxonomic techniques (see Bowie *et al.* 1972; Keddie & Cure 1977; Minnikin *et al.* 1978) and determinations of DNA base ratios (see Skyring & Quadling 1970; Skyring *et al.* 1971).

However, a considerable number of nomenclatural species of *Arthrobacter* contain lysine in the cell wall peptidoglycan and resemble the type species *Atbc. globiformis* in many phenotypic characters (see Keddie & Cure 1978 for list). There is now a considerable measure of agreement that the genus should be restricted to these species and *Atbc. citreus* (ATCC 11624) (Yamada & Komagata 1972b; Schleifer & Kandler 1972; Keddie 1978). The last mentioned species contains lysine as cell wall diamino acid but differs from *Atbc. globiformis* in nutritional and some other respects. *Arthrobacter* in this restricted sense is referred to as *Arthrobacter sensu stricto* in Table 2 in which some of the chemotaxonomic characters mentioned are listed.

Before leaving *Arthrobacter sensu stricto*, however, some comment on the type species *Atbc. globiformis* is appropriate. In *Bergey's Manual of Determinative Bacteriology* (Keddie 1974a), an 'ideal phenotype' was included based on an analysis of some 20 named and un-named strains considered to represent the species *Atbc. globiformis*. Those nomenclatural species conforming closely to this 'ideal phenotype', which contained lysine in the cell wall and which had DNA base ratios similar to that of the type strain of *Atbc. globiformis*, were considered to be synonyms of *Atbc. globiformis*. A few others for which the data were less complete were considered to be possible synonyms. However, despite the very considerable phenotypic resemblance among these differently named species, a similarity confirmed largely by the numerical phenetic study of Skyring *et al.* (1971), more recent data have shown much more heterogeneity than was then apparent. Thus, although these differently named species have cell wall peptidoglycans which have a very similar primary structure, the interpeptide bridges are of many different types (Schleifer & Kandler 1972). Also, while most of these nomenclatural species have DNA base ratios similar to that of the type strain of *Atbc. globiformis,* it is now apparent that some have values that differ by up to 5%: G+C from that of the type strain (Skyring & Quadling 1970; Skyring et al. 1971). It is therefore more appropriate to refer to this assemblage of nomenclatural species as the 'globiformis' group rather than as a single species.

Two pairs of species which bear the name Arthrobacter differ from Arthrobacter sensu stricto in cell wall composition. The first comprises the species Atbc. simplex and Atbc. tumescens which contain L-DAP in the cell wall peptidoglycan (see Keddie & Cure 1978), whose DNA base ratios are substantially higher than that of Atbc. globiformis, and which contain tetrahydrogenated menaquinones with 8 isoprene units (Table 1). In addition the fatty acid profile is unusual in that 10-methyloctadecanoic acid (tuberculostearic acid) occurs in both species, although at much higher levels in Atbc. simplex than in Atbc. tumescens (Minnikin et al. 1978; Collins et al. 1979; T. R. Dando & I. J. Bousfield unpublished results). Two further species, Atbc. terregens and Atbc. flavescens, contain ornithine in the cell wall peptidoglycan (see Keddie & Cure 1978) and have DNA base ratios some 3–5% higher than the type strain of Atbc. globiformis (see Table 1). Numerical phenetic studies also indicate that Atbc. terregens and Atbc. flavescens are distinct from the 'globiformis' group of arthrobacters (Skyring et al. 1971).

It is thus clear that both these pairs of species are distinct from Arthrobacter sensu stricto and should be removed from the genus. At present no suitable alternative location exists for the L-DAP-containing species. However, the characters of Atbc. terregens and Atbc. flavescens conform with those described for the genus Curtobacterium (Yamada & Komagata 1972b) and the peptidoglycan structure described for Atbc. terregens suggests a close affinity with that genus (see Schleifer & Kandler 1972).

In addition to those mentioned a number of so-called Arthrobacter spp. have been described in patent specifications. Most of these contain meso-DAP and arabinose in the cell wall and can be assigned either to Corynebacterium or Rhodococcus (see Keddie & Cure 1978). However, two species Atbc. viscosus and Atbc. stabilis contain meso-DAP in the cell wall but not arabinose (Keddie & Cure 1977) and their taxonomic position is uncertain.

B. Microbacterium

Members of the genus Microbacterium were originally characterized largely by their marked heat resistance, their dairy origin and by their production of small amounts of L(+) lactic acid from glucose (see Breed et al. 1957). Two species were generally recognized, Miba. lacticum and Miba. flavum, while a third, Miba. liquefaciens was considered by many investigators to be merely a gelatin-liquefying variety of Miba. lacticum (see Doetsch & Rakosky 1950; Jayne-Williams & Skerman 1966).

However studies of cell wall composition have shown that the genus Micro-

bacterium was quite heterogeneous. *Microbacterium flavum* was the first saprophytic coryneform organism shown to contain *meso*-DAP and arabinose in the
cell wall (Keddie *et al.* 1966; Robinson 1966*b*) and subsequent studies have
shown that it is a legitimate *Corynebacterium* sp. (see *Corynebacterium*). The
type species *Miba. lacticum*, on the other hand, contains lysine as cell wall
diamino acid (Keddie *et al.* 1966; Robinson 1966*b*) but unlike members of the
genus *Arthrobacter*, which also contain lysine, has a Group B type of peptidoglycan (Schleifer & Kandler 1972). In addition, numerical taxonomic studies
(Jones 1975) and other chemotaxonomic data (see Table 1) indicate that *Miba.
lacticum* is a distinct taxonomic entity. Thus although the genus *Microbacterium*
was considered *incertae sedis* in the eighth edition of *Bergey's Manual of Determinative Bacteriology* (Rogosa & Keddie 1974) more recent evidence indicates
that *Miba. lacticum* could form the nucleus of a redefined genus as suggested by
Jones (1975) and Keddie & Cure (1978).

While *Miba. liquefaciens* is similar to *Miba. lacticum* in some respects the
former species is now considered to represent a distinct taxon (see Robinson
1966*a,b*; Schleifer 1970; Schleifer & Kandler 1972). *Microbacterium liquefaciens*, like *Miba. lacticum*, has a Group B peptidoglycan but the diamino acid
is ornithine, not lysine. Thus the peptidoglycan structure of *Miba. liquefaciens*
suggests a close affinity with the genus *Curtobacterium* but data on such features as DNA base ratios and menaquinone composition are as yet lacking.

A further, more recently described species, *Miba. thermosphactum* (McLean
& Sulzbacher 1953), does not have a coryneform morphology and contains
meso-DAP in the cell wall peptidoglycan (Schleifer & Kandler 1972). Sneath
& Jones (1976) proposed the genus *Brochothrix* to accommodate this species
as *Broc. thermosphacta*.

C. Cellulomonas

Although this genus originally comprised a rather heterogeneous collection of
cellulolytic bacteria (see Bergey *et al.* 1923), the present day concept of *Cellulomonas* as a genus of cellulolytic coryneform bacteria stems from the work
of Clark (1952). The establishment of a separate genus of coryneform bacteria
based to a large extent on the property of cellulolysis was questioned
(see Jensen 1966), but subsequent studies using modern taxonomic methods
have shown that the genus *Cellulomonas*, as revised by Clark (1952), was well-
founded. Thus all the original, authentic cultures of *Cellulomonas* spp. were
shown to contain ornithine in the cell wall peptidoglycan which in addition
had a structure found only in members of this genus (Schleifer & Kandler 1972;
Fiedler & Kandler 1973*a*). Although considerable variation occurs in the sugar
composition of the cell walls—which may militate against the conclusion that
only one species should be recognized, see Keddie (1974*b*)—Keddie *et al.* (1966)

noted that galactose was uniformly absent. This finding was largely confirmed by Fiedler & Kandler (1973a), who, however, recorded a small amount of galactose in the walls of *Celm. biozotea.*

That *Cellulomonas* is a homogeneous and distinct taxon is further supported by both numerical (see Jones 1978) and non-numerical (Yamada & Komagata 1972a,b) taxonomic studies, by the vitamin, nitrogen and carbon nutrition (Keddie *et al.* 1966; Owens & Keddie 1969; Keddie 1974b), and by the menaquinone composition and DNA base ratios (see Table 2).

Some confusion has arisen, however, about the taxonomic position of the species *Celm. fimi.* In two numerical taxonomic studies it was suggested that this species should be removed from the genus *Cellulomonas* (Bousfield 1972; Jones 1975) whereas Keddie & Cure (1977) noted that the cell wall contained ornithine and considered that it should be retained in this genus. The strain of *Celm. fimi* used in all three studies was NCIB 8980 which was supposed to have been derived from the same strain as the authentic co-type, ATCC 484. However, whereas ATCC 484 is cellulolytic, NCIB 8980 is not and differs from the authentic strain in some other respects (I. J. Bousfield, unpublished results). Thus *Celm. fimi* (ATCC 484) is cellulolytic and has the peptidoglycan structure (Fiedler & Kandler 1973a) and other characteristics of the genus. Strain NCIB 8980 is non-cellulolytic, contains ornithine and glycine in the cell wall (Keddie & Cure 1977), has a DNA base ratio of 66% G+C (Bousfield 1972) and may, therefore, be a *Curtobacterium* sp.

A further 'patent' strain named *Celm. cartalyticum* has recently been shown to contain lysine as diamino acid in the cell wall peptidoglycan (Stackebrandt *et al.* 1978), but these authors nevertheless considered that this 'species' should be retained in the genus *Cellulomonas.* However, *Celm. cartalyticum* produces a primary mycelium which fragments later in the growth cycle (I. J. Bousfield, unpublished results) and this, together with the peptidoglycan structure reported for this organism (Stackebrandt *et al.* 1978), suggests that it is a member of the nocardioform genus *Oerskovia.*

D. Curtobacterium

This genus was proposed by Yamada & Komagata (1972b) for a group of coryneform bacteria which contained ornithine in the cell wall but which were considered to be distinct from the genus *Cellulomonas.* Thus, because the cell wall diamino acid was used as a major character in the generic definition, this genus was homogeneous in cell wall composition from the beginning. In addition to containing ornithine in the cell wall, the main features described for the genus *Curtobacterium* were: DNA base ratios in the range 66–71%: G+C; slow and weak acid production from some carbohydrates and all were obligately aerobic.

The species assigned to this proposed new genus included a number of former

Brevibacterium spp. such as *Brev. citreum* and *Brev. albidum* together with *Corynebacterium flaccumfaciens* (2 strains) and *Cnbc. poinsettiae* (one strain), the only two ornithine-containing plant pathogenic coryneform species studied by Yamada & Komagata (1972*b*).

However the numerical taxonomic study of Jones (1975) indicated that *Cnbc. betae* (one strain) should be considered a *Curtobacterium* sp, a conclusion supported by the cell wall composition (see Keddie & Cure 1978) and the DNA base ratios (Starr *et al.* 1975) of a number of strains which have been studied.

Despite a somewhat meagre original description, support for the genus *Curtobacterium* comes from the fact that Schleifer & Kandler (1972) independently grouped together most of the same species on the basis of peptidoglycan structure. Thus those species assigned to *Curtobacterium* have the less common Group B type of peptidoglycan, whereas members of the other ornithine-containing genus of coryneform bacteria, *Cellulomonas*, have a Group A peptidoglycan (Schleifer & Kandler 1972; Fiedler & Kandler 1973*a*; see Table 2). Studies of the menaquinone (and fatty acid) composition, in general, support the genus *Curtobacterium*; but the species *Curtobacterium (Brevibacterium) testaceum* contains menaquinones with 11 isoprene units whereas those of the remaining species contain 9 isoprene units (Yamada *et al.* 1976). The detailed peptidoglycan structure of *Curt. testaceum* has not been determined but the glycan moiety has been shown to be unusual in containing equal amounts of glycolyl and acetyl residues (Uchida & Aida 1977). This evidence together with the somewhat lower DNA base ratio (65.4% G+C, Yamada & Komagata 1970*a*) of *Curt. testaceum* led Yamada *et al.* (1976) to conclude that this species did not belong in the genus *Curtobacterium*.

Yamada & Komagata (1972*b*) also reported that 4 strains named *Brev. helvolum* contained ornithine in the cell wall and assigned them to the genus *Curtobacterium*. On the other hand Schleifer & Kandler (1972) reported that 5 strains named *Brev. helvolum* (including 4 Komagata strains) contained diaminobutyric acid in the cell wall peptidoglycan. A further strain (ATCC 11822) listed in some culture collections as *Brev. helvolum* contains lysine in the cell wall and is a soil isolate originally described by Jensen (1934) under the name *Corynebacterium helvolum*. It is a legitimate *Arthrobacter* sp. of the '*globiformis*' group (see Lochhead 1955; Schleifer & Kandler 1972; Keddie & Cure 1977, 1978). It is thus extremely doubtful that any so-called '*Brev. helvolum*' strains are *Curtobacterium* spp.

4. Conclusion

No attempt has been made here to present a comprehensive review of the cell wall composition of the coryneform bacteria and the reader is referred elsewhere for detailed information on individual strains (Schleifer & Kandler 1972; Keddie

& Cure 1978). However, an attempt has been made to give some idea of the considerable impact that studies of cell wall composition, at all levels of complexity, have had on the taxonomy of this group. They have clearly shown that most coryneform genera were far from homogeneous; but at the same time they have provided a basis for defining these genera more clearly, so that coherent taxa now appear to be emerging from the coryneform complex. Nowhere is this more evident than in the genus *Corynebacterium* itself which, by using a combination of cell wall and other chemotaxonomic features, can now be defined in much more precise terms than was formerly possible. Cell wall studies have also indicated where new genera may be required; for example for L-DAP-containing species such as *Arthrobacter simplex* and *Atbc. tumescens*, and possibly for the diaminobutyric acid-containing coryneform bacteria also. But much more information on a large number of strains is required before the creation of such genera can be contemplated.

Undoubtedly the most important contribution in more recent times has come from studies of the peptidoglycan structure (for details see Schleifer & Kandler 1972; Fiedler & Kandler 1973a,b; Fiedler et al. 1973; Stackebrandt et al. 1978). However, simple cell wall analysis also has an important part to play in taxonomy, largely because of its simplicity and because of the relative ease with which it can be applied to large numbers of strains. But simple qualitative methods must be used with caution in taxonomic studies; taxa whose cell walls contain the same diamino acid and indeed which may have an identical amino acid composition, see Schleifer & Kandler (1972) may nevertheless possess peptidoglycans of quite different structure (see Table 2). By the same token differences in the diamino acid present in the walls of different organisms do not necessarily reflect major differences in peptidoglycan structure (see Schleifer & Kandler 1972). On the other hand some quite different taxa such as *Brevibacterium linens* and *Corynebacterium sensu stricto* have the same peptidoglycan structure (Schleifer & Kandler 1972), but may nevertheless be distinguished by the sugar composition of the cell wall (see Table 1).

Therefore, like other taxonomic criteria, simple cell wall composition (and indeed peptidoglycan structure) should be used in defining taxa only when it can be shown to be concordant with other features, both conventional and chemotaxonomic. Of course one of the main strengths of cell wall composition as a taxonomic character in the coryneform group is the way in which it has been shown to correlate well with other (especially chemical) features, a point also made by Minnikin et al. (1978) in a recent review on lipid composition in the classification of coryneform bacteria.

However, where simple cell wall analysis really comes into its own is in identification. The development of 'rapid' methods (and with it the accumulation of data on a large number of strains) has provided a powerful tool for the primary stage of identification of coryneform bacteria. Indeed, in our opinion, most coryneform bacteria can be identified with some degree of confidence only when the cell wall composition, or at least the diamino acid present, is known.

5. REFERENCES

ABE, S., TAKAYAMA, K. & KINOSHITA, S. 1967 Taxonomical studies on glutamic acid-producing bacteria. *Journal of General and Applied Microbiology* 13, 279-301.

ALSHAMAONY, L., GOODFELLOW, M., MINNIKIN, D. E., BOWDEN, G. H. & HARDIE, J. M. 1977 Fatty and mycolic acid composition of *Bacterionema matruchotii* and related organisms. *Journal of General Microbiology* 98, 205-213.

BARKSDALE, L. 1970 *Corynebacterium diphtheriae* and its relatives. *Bacteriological Reviews* 34, 378-422.

BARKSDALE, W. L., LI, K., CUMMINS, C. S. & HARRIS, H. 1957 The mutation of *Corynebacterium pyogenes* to *Corynebacterium haemolyticum*. *Journal of General Microbiology* 16, 749-758.

BECKER, B., LECHEVALIER, M. P., GORDON, R. E. & LECHEVALIER, H. A. 1964 Rapid differentiation between *Nocardia* and *Streptomyces* by paper chromatography of whole-cell hydrolysates. *Applied Microbiology* 12, 421-423.

BERGEY, D. H., HARRISON, F. C., BREED, R. S., HAMMER, B. W. & HUNTOON, F. M. 1923 *Bergey's Manual of Determinative Bacteriology*, 1st edn. Baltimore: Williams & Wilkins.

BOONE, C. J. & PINE, L. 1968 Rapid method for characterisation of actinomycetes by cell wall composition. *Applied Microbiology* 16, 279-284.

BOUSFIELD, I. J. 1972 A taxonomic study of some coryneform bacteria. *Journal of General Microbiology* 71, 441-455.

BOWIE, I. S., GRIGOR, M. R., DUNCKLEY, G. E., LOUTIT, M. W. & LOUTIT, J. S. 1972 The DNA base composition and fatty acid constitution of some Gram-positive pleomorphic soil bacteria. *Soil Biology and Biochemistry* 4, 397-412.

BREED, R. S. 1953a The *Brevibacteriaceae* fam. nov. of order Eubacteriales. *Riassunti delle Communicazione* VI *Congresso Internazionale di Microbiologia, Roma* 1, 13-14.

BREED, R. S. 1953b The families developed from *Bacteriaceae* Cohn with a description of the family *Brevibacteriaceae* Breed, 1953. In: *Atti del VI Congresso Internazionale di Microbiologia, Roma* 1, 10-15.

BREED, R. S., MURRAY, E. G. D. & SMITH, N. R. (eds) 1957 *Bergey's Manual of Determinative Bacteriology*, 7th edn. Baltimore: Williams & Wilkins.

CHAN, E. C. S. & STEVENSON, I. L. 1962 On the biotin requirement of *Arthrobacter globiformis*. *Canadian Journal of Microbiology* 8, 403-405.

CLARK, F. E. 1952 The generic classification of the soil corynebacteria. *International Bulletin of Bacteriological Nomenclature and Taxonomy* 2, 45-56.

COLLINS, M. D., PIROUZ, T., GOODFELLOW, M. & MINNIKIN, D. E. 1977 Distribution of menaquinones in actinomycetes and corynebacteria. *Journal of General Microbiology* 100, 221-230.

COLLINS, M. D., GOODFELLOW, M. & MINNIKIN, D. E. 1979 Isoprenoid quinones in the classification of coryneform and related bacteria. *Journal of General Microbiology* 110, 127-136.

COLLINS-THOMPSON, D. L., SØRHAUG, T., WITTER, L. D. & ORDAL, Z. J. 1972 Taxonomic consideration of *Microbacterium lacticum, Microbacterium flavum* and *Microbacterium thermosphactum*. *International Journal of Systematic Bacteriology* 22, 65-72.

CONN, H. J. 1947 A protest against the misuse of the generic name *Corynebacterium*. *Journal of Bacteriology* 54, 10.

CONN, H. J. & DIMMICK, I. 1947 Soil bacteria similar in morphology to *Mycobacterium* and *Corynebacterium*. *Journal of Bacteriology* 54, 291-303.

CROMBACH, W. H. J. 1972 DNA base composition of soil arthrobacters and other coryneforms from cheese and sea fish. *Antonie van Leeuwenhoek Journal of Microbiology and Serology* 38, 105-120.

CROMBACH, W. H. J. 1978 *Caseobacter polymorphus* gen. nov., sp. nov., a coryneform bacterium from cheese. *International Journal of Systematic Bacteriology* 28, 354–366.

CUMMINS, C. S. 1962 Chemical composition and antigenic structure of cell walls of *Corynebacterium, Mycobacterium, Nocardia, Actinomyces* and *Arthrobacter. Journal of General Microbiology* 28, 35–50.

CUMMINS, C. S. 1971 Cell wall composition in *Corynebacterium bovis* and some other corynebacteria. *Journal of Bacteriology* 105, 1227–1228.

CUMMINS, C. S. & HARRIS, H. 1956 The chemical composition of the cell wall in some Gram-positive bacteria and its possible value as a taxonomic character. *Journal of General Microbiology* 14, 583–600.

CUMMINS, C. S. & HARRIS, H. 1958 Studies on the cell wall composition and taxonomy of Actinomycetales and related groups. *Journal of General Microbiology* 18, 173–189.

CUMMINS, C. S. & HARRIS, H. 1959 Taxonomic position of *Arthrobacter. Nature, London* 184, 831–832.

CUMMINS, C. S., LELLIOTT, R. A. & ROGOSA, M. 1974 'Genus *Corynebacterium*', In *Bergey's Manual of Determinative Bacteriology* 8th edn. ed. Buchanan, R. E. & Gibbons, N. E. Baltimore: Williams & Wilkins.

CURE, G. L. & KEDDIE, R. M. 1973 Methods for the morphological examination of aerobic coryneform bacteria. In *Sampling–Microbiological Monitoring of Environments*, Society for Applied Bacteriology Technical Series No. 7, ed. Board, R. G. & Lovelock, D. W. pp. 123–135, London: Academic Press.

DA SILVA, G. A. N. & HOLT, J. G. 1965 Numerical taxonomy of certain coryneform bacteria. *Journal of Bacteriology* 90, 921–927.

DOETSCH, R. N. & RAKOSKY, J. 1950 Is there a *Microbacterium liquefaciens? Bacteriological Proceedings* (G16) p. 38.

DYE, D. W. & KEMP, W. J. 1977 A taxonomic study of plant pathogenic *Corynebacterium* species. *New Zealand Journal of Agricultural Research* 20, 563–582.

FIEDLER, F. & KANDLER, O. 1973*a* Die Murientypen in der Gattung *Cellulomonas* Bergey *et al. Archiv für Mikrobiologie* 89, 41–50.

FIEDLER, F. & KANDLER, O. 1973*b* Die Aminosäuresequenz von 2,4-Diaminobuttersäure enthaltenden Mureinen bei verschiedenen coryneformen Bakterien und *Agromyces ramosus. Archiv für Mikrobiologie* 89, 51–66.

FIEDLER, F. & STACKEBRANDT, E. 1978 Taxonomical studies on *Brevibacterium linens. Abstracts of the XII International Congress of Microbiology, München* (C45) p. 96.

FIEDLER, F., SCHLEIFER, K. H., CZIHARZ, B., INTERSCHICK, E. & KANDLER, O. 1970 Murein types in *Arthrobacter*, brevibacteria, corynebacteria and microbacteria. *Publications de la Faculté des Sciences de l'Université J. E. Purkyne, Brno* 47, 111–122.

FIEDLER, F., SCHLEIFER, K. H. & KANDLER, O. 1973 Amino acid sequence of the threonine-containing mureins of coryneform bacteria. *Journal of Bacteriology* 113, 8–17.

GOODFELLOW, M. & ALDERSON, G. 1977 The actinomycete–genus *Rhodococcus* : a home for the '*rhodochrous*' complex. *Journal of General Microbiology* 100, 99–122.

GOODFELLOW, M. & MINNIKIN, D. E. 1977 Nocardioform bacteria. *Annual Review of Microbiology* 31, 159–180.

GOODFELLOW, M., COLLINS, M. D. & MINNIKIN, D. E. 1976 Thin-layer chromatographic analysis of mycolic acid and other long-chain components in whole-organism methanolysates of coryneform and related taxa. *Journal of General Microbiology* 96, 351–358.

HILL, L. R. 1966 An index to deoxyribonucleic acid base compositions of bacterial species. *Journal of General Microbiology* 44, 419–437.

HOARE, D. S. & WORK, E. 1956 Distribution and metabolism of the stereoisomers of diaminopimelic acid in certain bacteria. *Journal of General Microbiology* 15, xiii.

HOARE, D. S. & WORK, E. 1957 The stereoisomers of α,ϵ-diaminopimelic acid 2. Their distribution in the bacterial order Actinomycetales and in certain Eubacteriales. *Biochemical Journal* 65, 441–447.

JAYNE-WILLIAMS, D. J. & SKERMAN, T. M. 1966 Comparative studies on coryneform bacteria from milk and dairy sources. *Journal of Applied Bacteriology* 29, 72-92.

JENSEN, H. L. 1934 Studies on saprophytic mycobacteria and corynebacteria. *Proceedings of the Linnean Society of New South Wales* 59, 19-61.

JENSEN, H. L. 1952 The coryneform bacteria. *Annual Review of Microbiology* 6, 77-90.

JENSEN, H. L. 1966 Some introductory remarks on the coryneform bacteria. *Journal of Applied Bacteriology* 29, 13-16.

JONES, D. 1975 A numerical taxonomic study of coryneform and related bacteria. *Journal of General Microbiology* 87, 52-96.

JONES, D. 1978 An evaluation of the contribution of numerical taxonomy to the classification of the coryneform bacteria. In *Special Publications of the Society for General Microbiology* I. *Coryneform Bacteria*. ed. Bousfield, I. J. & Callely, A. G. pp. 13-46. London: Academic Press.

JONES, D. & WEITZMAN, P. D. J. 1974 Reclassification of *Brevibacterium leucinophagum* Kinney and Werkman as a Gram-negative organism, probably in the genus *Acinetobacter*. *International Journal of Systematic Bacteriology* 24, 113-117.

KEDDIE, R. M. 1974a. 'Genus *Arthrobacter*'. In *Bergey's Manual of Determinative Bacteriology*, 8th edn. ed. Buchanan, R. E. & Gibbons, N. E. Baltimore: Williams & Wilkins.

KEDDIE, R. M. 1974b 'Genus *Cellulomonas*'. In *Bergey's Manual of Determinative Bacteriology*, 8th edn. ed. Buchanan, R. E. & Gibbons, N. E. Baltimore: Williams & Wilkins.

KEDDIE, R. M. 1978 What do we mean by coryneform bacteria? In: *Special Publications of the Society for General Microbiology* I. *Coryneform bacteria*. ed. Bousfield, I. J. & Callely, A. G. pp. 1-12. London: Academic Press.

KEDDIE, R. M. & CURE, G. L. 1977 The cell wall composition and distribution of free mycolic acids in named strains of coryneform bacteria and in isolates from various natural sources. *Journal of Applied Bacteriology* 42, 229-252.

KEDDIE, R. M. & CURE, G. L. 1978 Cell wall composition of coryneform bacteria. In: Special Publications of the Society for General Microbiology I. *Coryneform Bacteria* ed. Bousfield, I. J. & Callely, A. G. pp. 47-84. London: Academic Press.

KEDDIE, R. M. & ROGOSA, M. 1974 'Genus *Kurthia*'. In *Bergey's Manual of Determinative Bacteriology*, 8th edn. ed. Buchanan R. E. & Gibbons, N. E. Baltimore: Williams & Wilkins.

KEDDIE, R. M., LEASK, B. G. S. & GRAINGER, J. M. 1966 A comparison of coryneform bacteria from soil and herbage: cell wall composition and nutrition. *Journal of Applied Bacteriology* 29, 17-43.

LECHEVALIER, M. P. & LECHEVALIER, H. 1970 Chemical composition as a criterion in the classification of aerobic actinomycetes. *International Journal of Systematic Bacteriology* 20, 435-443.

LEHMANN, K. B. & NEUMANN, R. 1896 *Atlas und Grundriss der Bakteriologie und Lehrbuch der speciellen bakteriologischen Diagnostik*. 1st edn. München: J. F. Lehmann.

LOCHHEAD, A. G. 1955 *Brevibacterium helvolum* (Zimmermann) comb. nov. *International Bulletin of Bacteriological Nomenclature and Taxonomy* 5, 115-119.

LOCHHEAD, A. G. 1958 Two new species of *Arthrobacter* requiring respectively vitamin B12 and the terregens factor. *Archiv für Mikrobiologie* 31, 163-170.

LOCHHEAD, A. G. & BURTON, M. O. 1953 An essential bacterial growth factor produced by microbial synthesis. *Canadian Journal of Botany* 31, 7-22.

LUTHY, P. 1974 *Corynebacterium okanaganae*, an entomopathogenic species of the Corynebacteriaceae. *Canadian Journal of Microbiology* 20, 791-794.

McLEAN, R. A. & SULZBACHER, W. L. 1953 *Microbacterium tnermosphactum*, spec nov; a nonheat resistant bacterium from fresh pork sausage. *Journal of Bacteriology* 65, 428-433.

MARMUR, J. & DOTY, P. 1962 Determination of the base composition of deoxyribonucleic acid from its thermal denaturation temperature. *Journal of Molecular Biology* 5, 109-118.

MINNIKIN, D. E., GOODFELLOW, M. & COLLINS, M. D. 1978 Lipid composition in the classification and identification of coryneform and related taxa. In *Special Publications of the Society for General Microbiology* I. *Coryneform Bacteria*. ed. Bousfield, I. J. & Callely, A. G. pp. 85–160. London: Academic Press.
MORDARSKA, H., MORDARSKI, M. & GOODFELLOW, M. 1972 Chemotaxonomic characters and classification of some nocardioform bacteria. *Journal of General Microbiology* 71, 77–86.
MORRIS, J. G. 1960 Studies on the metabolism of *Arthrobacter globiformis*. *Journal of General Microbiology* 22, 564–582.
MULDER, E. G. & ANTHEUNISSE, J. 1963 Morphologie, physiologie et écologie des *Arthrobacter*. *Annales de l'Institut Pasteur* 105, 46–74.
MURRAY, I. G. & PROCTOR, A. G. J. 1965 Paper chromatography as an aid to the identification of *Nocardia* species. *Journal of General Microbiology* 41, 163–167.
OWENS, J. D. & KEDDIE, R. M. 1969 The nitrogen nutrition of soil and herbage coryneform bacteria. *Journal of Applied Bacteriology* 32, 338–347.
PERKINS, H. R. & CUMMINS, C. S. 1964 Ornithine and 2,4-diaminobutyric acid as components of the cell walls of plant pathogenic corynebacteria. *Nature, London* 201, 1105–1107.
PINE, L. 1970 Classification and phylogenetic relationship of microaerophilic actinomycetes. *International Journal of Systematic Bacteriology* 20, 445–474.
PITCHER, D. G. 1976 Arabinose with L-diaminopimelic acid in the cell wall of an aerobic coryneform organism isolated from human skin. *Journal of General Microbiology* 94, 225–227.
ROBINSON, K. 1966a Some observations on the taxonomy of the genus *Microbacterium*. I. Cultural and physiological reactions and heat resistance. *Journal of Applied Bacteriology* 29, 607–615.
ROBINSON, K. 1966b Some observations on the taxonomy of the genus *Microbacterium*. II. Cell wall analysis, gel electrophoresis and serology. *Journal of Applied Bacteriology* 29, 616–624.
ROGOSA, M. & KEDDIE, R. M. 1974 'Genus *Microbacterium*'. In *Bergey's Manual of Determinative Bacteriology*, 8th edn. ed. Buchanan, R. E. & Gibbons, N. E. Baltimore: Williams & Wilkins.
SACKS, L. E. 1954 Observations on the morphogenesis of *Arthrobacter citreus*, spec. nov. *Journal of Bacteriology* 67, 342–345.
SCHEFFERLE, H. E. 1957 *An investigation of the microbiology of built-up poultry litter*. Ph.D. Thesis, University of Edinburgh, U. K.
SCHLEIFER, K. H. 1970 Die Mureintypen in der Gattung *Microbacterium*. *Archiv. für Microbiologie* 71, 271–282.
SCHLEIFER, K. H. & KANDLER, O. 1967 Zur chemischen Zusammensetzung der Zellwand der Streptokokken I. Die Aminosäuresequenz des Mureins von *Str. thermophilus* und *Str. faecalis*. *Archiv für Mikrobiologie* 57, 335–364.
SCHLEIFER, K. H. & KANDLER, O 1972 Peptidoglycan types of bacterial cell walls and their taxonomic implications. *Bacteriological Reviews* 36, 407–477.
SKYRING, G. W. & QUADLING, C. 1970 Soil bacteria: a principal component analysis and guanine-cytosine contents of some arthrobacter-coryneform soil isolates and of some named cultures. *Canadian Journal of Microbiology* 16, 95–106.
SKYRING, G. W., QUADLING, C. & ROUATT, J. W. 1971 Soil bacteria: principal component analysis of physiological descriptions of some named cultures of *Agrobacterium*, *Arthrobacter* and *Rhizobium*. *Canadian Journal of Microbiology* 17, 1299–1311.
SNEATH, P. H. A. & JONES, D. 1976 *Brochothrix*, a new genus tentatively placed in the family Lactobacillaceae. *International Journal of Systematic Bacteriology* 26, 102–104.
STACKEBRANDT, E., FIEDLER, F. & KANDLER, O. 1978 Peptidoglycantyp und Zusammensetzung der Zellwandpolysaccharide von *Cellulomonas cartalyticum* und einigen coryneformen Organismen. *Archives of Microbiology* 117, 115–118.
STARR, M. P., MANDEL, M. & MURATA, N. 1975 The phytopathogenic coryneform

bacteria in the light of DNA base composition and DNA-DNA segmental homology. *Journal of General and Applied Microbiology* **21**, 13-26.

UCHIDA, K. & AIDA, K. 1977 Acyl type of bacterial cell wall: its simple identification by colorimetric method. *Journal of General and Applied Microbiology* **23**, 249-260.

WORK, E. & DEWEY, D. L. 1953 The distribution of α,ϵ-diaminopimelic acid among various micro-organisms. *Journal of General Microbiology* **9**, 394-406.

YAMADA, K. & KOMAGATA, K. 1970a Taxonomic studies on coryneform bacteria II. Principal amino acids in the cell wall and their taxonomic significance. *Journal of General and Applied Microbiology* **16**, 103-113.

YAMADA, K. & KOMAGATA, K. 1970b Taxonomic studies on coryneform bacteria III. DNA base composition of coryneform bacteria. *Journal of General and Applied Microbiology* **16**, 215-224.

YAMADA, K. & KOMAGATA, K. 1972a Taxonomic studies on coryneform bacteria IV. Morphological, cultural, biochemical and physiological characteristics. *Journal of General and Applied Microbiology* **18**, 399-416.

YAMADA, K. & KOMAGATA, K. 1972b Taxonomic studies on coryneform bacteria V. Classification of coryneform bacteria. *Journal of General and Applied Microbiology* **18**, 417-431.

YAMADA, Y., INOUYE, G., TAHARA, Y. & KONDO, K. 1976 The menaquinone system in the classification of coryneform and nocardioform bacteria and related organisms. *Journal of General and Applied Microbiology* **22**, 203-214.

Lipid Composition in the Classification and Identification of Acid-fast Bacteria

D. E. MINNIKIN AND M. GOODFELLOW

Departments of Organic Chemistry and Microbiology, The University, Newcastle upon Tyne, UK

Contents

1. Introduction

A. Historical aspects

MOST ACID-FAST bacteria are currently classified in the genera *Mycobacterium*, *Nocardia* and *Rhodococcus*. The actinomycetes included in these taxa have had a long and complex taxonomic history which will only be mentioned briefly as detailed reviews are available (Cross & Goodfellow 1973; Bousfield & Goodfellow 1976; Lechevalier 1976; Goodfellow & Minnikin 1977, 1978). The difficulties inherent in the classification and identification of acid-fast bacteria do, however, go back to the very earliest classifications of these and related bacteria.

Nocard (1888) pointed out that the actinomycete subsequently known as *Nocardia farcinica* resembled *Corynebacterium diphtheriae* in laboratory culture and Lehmann & Neumann (1896) noted the similarity between *Corynebacterium diphtheriae* and *Mycobacterium tuberculosis*. Jensen (1931) recognized these relationships when he classified the genera *Corynebacterium*, *Mycobacterium* and *Proactinomyces (Actinomyces* and *Nocardia)* in the family Proactinomycetaceae (Lehmann & Neumann 1927). Subsequently, the label *Proactinomyces* fell into disuse with the establishment of the priority of the names *Actinomyces* and *Nocardia* for the anaerobic and aerobic proactinomycetes respectively (Waksman & Henrici 1943).

Jensen distinguished proactinomycetes from corynebacteria and mycobacteria by their ability to produce an "initial mycelium" and from streptomycetes and related actinomycetes by their inability to form spores. These morphological divisions were not to stand the test of time as many actinomycetes were found to vary in their morphological properties. Thus, in 1953 Jensen noted that nocardiae could occasionally be less mycelial than mycobacteria and that some mycobacteria formed rudimentary aerial hyphae. Most actinomycetes can now be divided into two broad morphological groups, the nocardioform and sporoactinomycetes (Prauser 1967, 1978; Prauser & Bergholz 1974). Nocardioform actinomycetes are branching bacteria that reproduce by fragmentation of all, or more or less accidentally involved, parts of their hyphae into bacilli and coccoid elements. In contrast, sporoactinomycetes reproduce by spores formed on or in definite parts of the mycelium (see Chapter 7). While sporoactinomycetes and nocardioform bacteria can usually be distinguished there are many transitions between the latter and coryneform bacteria. The coryneforms can be defined as aerobic, Gram positive, non-acid-fast, non-spore-forming, rod-shaped bacteria reproducing by snapping or bending type cell division (Yamada & Komagata 1972).

It is generally recognized that morphology has been and can be important in differentiating actinomycete taxa, but over-reliance on morphological properties has led to much confusion in the classification of coryneform and nocardioform bacteria (Williams *et al.* 1976, see Chapter 7; Keddie 1978). Morphological differences amongst the nocardioform-coryneform bacteria are ones of degree rather than kind and morphological properties should not be given undue emphasis in the classification and identification of these bacteria (Goodfellow 1973; Bousfield & Goodfellow 1976). The production of nocardioform mutants from a coryneform bacterium (Jičinská 1973) highlights the problems of dealing with morphological characters of these organisms. Indeed, improvements in the classification and identification of nocardioform and coryneform bacteria had to await the application of new approaches such as those based on numerical and chemical methods.

B. Numerical taxonomy

It is now accepted that natural or general purpose classifications should be based upon different kinds of taxonomic criteria. A reliable and relatively quick way of detecting centres of variation in poorly defined taxa is to examine many strains for a large number of equally weighted characters and classify them on the basis of shared similarities (see Chapter 4). Most other modern taxonomic methods cannot readily handle large numbers of strains but can be used to examine representatives of phena defined in numerical phenetic surveys. Thus, data from chemical, genetical and serological investigations can be used to evaluate and extend classifications based upon numerical taxonomic techniques. Work based upon this strategy has helped to clarify the subgeneric classification of *Mycobacterium* and *Nocardia* and gone someway towards establishing relationships between these genera and allied taxa (Sneath 1976; Goodfellow & Minnikin 1978; Saito *et al.* 1977; Wayne *et al.* 1978).

Numerical phenetic surveys have led to improved speciation in the genus *Mycobacterium* (Wayne *et al.* 1971; Kubica *et al.* 1972; Meissner *et al.* 1974) where 30 species are currently recognized (Runyon *et al.* 1974). Numerical methods have also been applied with good effect in the classification and identification of *Nocardia* and related actinomycetes (see Goodfellow & Minnikin 1977, 1978; Goodfellow & Schaal 1979) but unfortunately the advances reported in these papers came too late to be included in the current edition of *Bergey's Manual of Determinative Bacteriology* (Buchanan & Gibbons 1974) where the section on the family Nocardiaceae (McClung 1974) does not reflect adequately present views. *Nocardia sensu* Waksman (1961) falls into a number of aggregate clusters which have been given generic status mainly on the basis of data derived from numerical, chemical and morphological investigations.

1. *Nocardia sensu stricto* (Cross & Goodfellow 1973) contains the well established taxa *Nrda. asteroides*, *Nrda. brasiliensis* and *Nrda. otitidis-caviarum (Nrda. caviae)*, and the less intensively studied *Nrda. amarae*, *Nrda. carnea*, *Nrda. salmonicida*, *Nrda. transvalensis* and *Nrda. vaccinii* (Goodfellow & Minnikin 1977; Gordon *et al.* 1978). *Nocardia amarae*, *Nrda. brasiliensis* and *Nrda. otitidis-caviarum* are good taxospecies (Lechevalier & Lechevalier 1974; Sneath 1976) but *Nrda. asteroides* is heterogeneous (Goodfellow & Minnikin 1977, 1978). The status of the type species *Nrda. farcinica* is not clear and will be considered later.

2. The monotypic genus *Rothia* (Georg & Brown 1967), created for strains previously classified as *Nrda. dentocariosus* and *Nrda. salivae*, forms a phenon distinct from clusters containing aerobic and anaerobic actinomycetes (Holmberg & Hallander 1973).

3. The genus *Rhodococcus* (Goodfellow & Alderson 1977) was resurrected for actinomycetes previously classified as *Gordona*, *Jensenia*, *'Mycobacterium' rhodochrous*, and the *'rhodochrous'* complex (Tsukamura 1971; Bousfield & Goodfellow 1976). The genus contains 10 species including the type species, *Rhod. rhodochrous*.

4. The genus *Oerskovia* (Prauser *et al.* 1970), erected for strains previously known as *Nocardia turbata*, forms a tight cluster distinct from phena equated with *Nocardia* and *Rhodococcus* (Goodfellow 1971). *Oerskovia xanthineolytica* was added subsequently to the genus (Lechevalier 1972) which may also provide a suitable niche for *Nocardia cellulans* (Goodfellow 1971; Jones 1975; Minnikin *et al.* 1979*a*).

5. The genus *Actinomadura* (Lechevalier & Lechevalier 1970*a*) was proposed for actinomycetes classified as *Nrda. dassonvillei, Nrda. madurae* and *Nrda. pelletieri*. These taxa form loosely related clusters (Tsukamura 1969; Alderson & Goodfellow 1979; Goodfellow *et al.* 1979) all of which can be sharply distinguished from phena corresponding to *Nocardia sensu stricto, Oerskovia* and *Rhodococcus* (Goodfellow 1971). Since 1970, many additional *Actinomadura* spp. have been described primarily on the basis of minor morphological characters that have proved unreliable in other taxa (Lacey *et al.* 1978). Meyer (1976) proposed the genus *Nocardiopsis* for *Actinomadura dassonvillei* strains.

6. Early numerical phenetic investigations indicated the equivocal position of the taxon *Gordona aurantiaca* in the genus *Gordona* (Tsukamura 1974; 1975). Representatives of *Grda. aurantiaca*, some of which are acid-fast, have been found to form a tight phenon equivalent in rank to clusters corresponding to the genera *Nocardia sensu stricto* and *Rhodococcus* (Goodfellow *et al.* 1978; Tsukamura *et al.* 1979). Goodfellow and his colleagues considered that *Grda. aurantiaca* strains should be provisionally recognized as the '*aurantiaca*' taxon and further comparative studies done to determine their taxonomic status.

Numerical phenetic surveys have also highlighted the heterogeneity of the genus *Corynebacterium* (Jones 1978) and there is good evidence that the taxon *Corynebacterium sensu stricto* should be reserved for *Cnbc. diphtheriae*, certain animal pathogens and some saprophytic species notably *Cnbc. glutamicum* (Keddie 1978; Minnikin *et al.* 1978*c*; Chapter 8). It appears that *Corynebacterium sensu stricto* is more closely related to *Mycobacterium, Nocardia* and *Rhodococcus* than to other coryneform taxa (Barksdale 1970; Jones 1978; Minnikin *et al.* 1978*c*).

Numerical taxonomic methods have led to dramatic improvements in the classification of acid-fast bacteria but have been less successful in highlighting good characters for the differentiation of *Corynebacterium sensu stricto, Mycobacterium, Nocardia sensu stricto, Rhodococcus,* the '*aurantiaca*' taxon and related taxa. Chemical characters are, however, proving to be of particular value in the identification of actinomycete and coryneform genera (Minnikin *et al.* 1978*c*; Goodfellow & Schaal 1979; Chapter 8).

C. *Chemotaxonomy*

Chemical methods are becoming well-established in bacterial taxonomy and are providing good characters for the classification and identification of actinomycetes and coryneform bacteria (Lechevalier *et al.* 1971*a*; Minnikin & Goodfellow 1976; Goodfellow & Minnikin 1977; Lechevalier 1977; Minnikin *et al.* 1978*c*; Keddie & Cure 1978). The introduction of wall sugar and amino acid analysis (Chapter 8) led to a reappraisal of the classification of actinomycetes and coryneform bacteria and valuable data have also been provided by deoxyribo-

TABLE 1

Distribution of some non-lipid chemical characters in acid fast and related bacteria

Wall chemotype*	Major distinguishing wall constituents†	Taxon	%G+C‡
I	L-DAP, glycine	*Streptomyces*	69–77
IIIA	*meso*-DAP, madurose §	*Actinomadura*	65–77
IIIB	*meso*-DAP	*Nocardiopsis*	ND
		Actinopolyspora	64
		Bacterionema	55–57
		Corynebacterium sensu stricto	48–59
		Micropolyspora brevicatena	66–68
		Micropolyspora faeni	ND
		Mycobacterium	62–70
		Nocardia sensu stricto	64–68
IV	*meso*-DAP, arabinose, galactose	*Nocardia autotrophica*	ND
		Nocardia mediterranea	ND
		Nocardia orientalis	ND
		Pseudonocardia	79.3
		Rhodococcus	59–69
		Saccharomonospora	69–75
		Saccharopolyspora	ND
		'aurantiaca' taxon	ND
	meso-DAP, galactose	*Nocardia aerocolonigenes*	ND
VI	Lysine, aspartic acid, galactose	*Oerskovia*	70–75
		Rothia	65–70

*Wall chemotypes and data according to Lechevalier & Lechevalier (1970*b*); Gochnauer *et al.* (1975); Goodfellow & Minnikin (1977, 1978); Goodfellow *et al.* (1978); Gordon *et al.* 1978). † All wall preparations contain major amounts of alanine, glutamic acid, glucosamine and muramic acid; DAP, diaminopimelic acid. ‡Data from Bradley *et al.* (1973); Gochnauer *et al.* (1975); Goodfellow & Minnikin (1977, 1978); Lechevalier *et al.* (1971*a*); Tsyganov *et al.* (1966); Yamaguchi (1967); ND, not determined. § Madurose only in whole cells. ‖Detailed structure of peptidoglycan different to *Nocardia sensu stricto* (Bordet *et al.* 1972).

nucleic acid base determinations (Table 1) and nucleic acid reassociation studies (Bradley & Mordarski 1976; Mordarski *et al.* 1978*a,b*). Lipid analyses are also proving to be of value in bacterial classification (Goldfine 1972; Shaw 1974; Lechevalier 1977) and have been found to be effective for the classification of actinomycetes and coryneform bacteria (Minnikin & Goodfellow 1976; Lechevalier 1977; Minnikin *et al.* 1978*c*).

When considered from a taxonomic view point wall analyses have developed in two contrasting ways (Chapter 8). Thus, relatively elaborate and laborious techniques are used to determine the peptidoglycan type (Schleifer & Kandler 1972) while, on the other hand, simple 'rapid' methods are applied to detect the distribution of a few, taxonomically useful, wall components. Elucidation of the primary structure of the peptidoglycan of bacterial walls provides more taxonomically useful information than that obtained using simple qualitative methods but the more detailed analyses require the use of specialized techniques that are not readily applicable to a large number of strains. The 'rapid' qualitative methods not only make it possible to study large numbers of bacteria but also provide a powerful tool for the primary identification of actinomycetes and related bacteria (Lechevalier 1968; Berd 1973; Staneck & Roberts 1974; Goodfellow & Schaal 1979).

Simple qualitative analyses of the wall amino-acid and sugar content of actinomycetes and some coryneform bacteria have shown that these organisms can be classified into several large groups or wall chemotypes on the basis of the limited distribution of certain major components of the peptidoglycan layer (Lechevalier *et al.* 1966; Lechevalier & Lechevalier 1970*b*; Lechevalier 1976). On the basis of such data many strains formerly assigned to the genus *Nocardia* have been transferred to *Actinomadura, Oerskovia* and *Rothia* (Table 1). Simple wall composition analyses do not, however, distinguish between *Corynebacterium sensu stricto, Mycobacterium, Nocardia sensu stricto, Rhodococcus* and the '*aurantiaca*' taxon or between them and certain sporoactinomycete taxa (Table 1). Strains belonging to these taxa all have a wall chemotype IV, that is, they contain major amounts of *meso*-diaminopimelic acid (*meso*-DAP), arabinose and galactose.

Structural variations in the peptidoglycans of wall chemotype IV organisms may, however, be of value in classification. Representatives of *Mycobacterium* and *Nocardia sensu stricto* have N-glycol substituents on the peptidoglycan muramic acid rather than the more usual N-acetyl group (Lederer *et al.* 1975). A relatively simple procedure for detecting the presence of N-glycolation of peptidoglycans has been devised by Uchida & Aida (1977). Glycolic acid was detected in the expected amounts in hydrolysates of walls of *Myco. phlei, Myco. tuberculosis, Nrda. asteroides, Nrda. erythropolis* and *Cnbc. equi.* The latter two species are possibly members of the genus *Rhodococcus* (Goodfellow & Alderson 1977) but a strain labelled '*Mycobacterium rhodochrous*' and another, *Cnbc. diphtheriae*, produced only acetic acid on hydrolysis of their walls. The peptidoglycan of *Nrda. mediterranea* has been shown to be N-acetylated and it also differs in other details from those of *Nrda. otitidis-caviarum (caviae)* and *Nrda. asteroides* (Bordet *et al.* 1972).

Lipid analyses also have potential for discriminating between various acid-fast and other bacteria of wall chemotype IV. Representatives of *Mycobacterium, Nocardia sensu stricto, Rhodococcus, Corynebacterium sensu stricto* and a number of related taxa contain characteristic high molecular weight 3-hydroxy-2-branched acids, mycolic acids. These molecules have not been found so far in any other living organisms. As mycolic acids vary widely in structure a variety of techniques have been developed to recognize the different types. Studies on the more widely distributed non-hydroxylated fatty acids also provide valuable chemotaxonomic data and analyses of isoprenoid quinones and polar lipids give additional discriminatory characters. Mycolic acid-containing bacteria also usually produce the so called 'cord factor' lipids which are dimycolates of trehalose and whose detailed structural analysis should yield further useful data. Mycobacteria also contain a variety of very characteristic complex lipids and long chain components of particular systematic potential (Asselineau & Asselineau 1978a,b).

The characteristic acid-fast properties (resistance to decolourization with acidic ethanol after staining with, for example, fuchsin) of mycobacteria and related organisms, have been discussed at length by Barksdale & Kim (1977). A recent study (Goren et al. 1978) led to the conclusion that the lipid barrier of the cell wall mycolyl-arabinogalactan effectively hinders the penetration of the bleaching acid. The different degrees of acid-fastness encountered may therefore be related to variations in the chemical composition of the mycolyl-arabinogalactan layer.

2. Lipids Having Chemotaxonomic Potential

A. Mycolic acids

Mycolic acids are long-chain 3-hydroxy 2-branched acids found only in bacteria having an arabinogalactan in their walls and peptidoglycans based on meso-DAP. Bacteria having this so-called wall chemotype IV (Lechevalier & Lechevalier 1970b; Lechevalier 1977; Goodfellow & Minnikin 1978) and mycolic acids are found mainly in the genera Mycobacterium, Nocardia, Rhodococcus and Corynebacterium (Asselineau 1966; Etémadi 1967a,b; Minnikin & Goodfellow 1976; Lechevalier 1977; Minnikin et al. 1978b,c; Asselineau & Asselineau 1978b). In the following sections procedures for the investigation of mycolic acids will be outlined followed by a discussion of the significance of the distribution of mycolic acid types.

Mycolic acids were originally identified as an unsaponifiable fraction obtained from Mycobacterium tuberculosis (Stodola et al. 1938) and alkaline degradation has continued to be used as a reliable means for their isolation (Asselineau 1966; Etémadi 1967a,b). Alkaline reagents may, however, cause isomerization of mycolates (Minnikin & Polgar 1966a; Etémadi 1967c) and acid methanolysis is a convenient alternative (Minnikin et al. 1975; Goodfellow et al. 1976; Minnikin et al. 1980) but the possibility of acid degradation of components containing cyclopropane rings must be borne in mind (Minnikin 1972).

The introduction of thin-layer chromatography (t.l.c.) analysis of mycolic acid esters (Lanéelle 1963) enabled the true complexity of the natural mixtures to be assessed clearly for the first time. Mycolic acids from nocardioform bacteria and corynebacteria gave single spots on thin-layer chromatography (Lanéelle et al. 1965a; Bordet & Michel 1969; Maurice et al. 1971; Goodfellow et al. 1976;

Minnikin *et al.* 1975, 1980) but those from strains of *Mycobacterium*, so far examined, produced multispot patterns (Lanéelle 1963; Etémadi 1967*a*; Minnikin & Polgar 1967*a,b*; Minnikin *et al.* 1975, 1980). Single spots corresponding to mycolic esters were also found in methanolysates of representatives of the '*aurantiaca*' taxon (Goodfellow *et al.* 1978). Argentation t.l.c. may be used to resolve further mixtures of mycolic esters containing double bonds (Etémadi 1967*a*; Bordet & Michel 1969; Minnikin *et al.* 1974).

An alternative approach to the t.l.c. analysis of mycolic acids involving separation of the free acids by unidimensional multiple chromatography has been suggested (Michalec & Mára 1975; Michalec *et al.* 1975). This procedure gave one minor and two major spots for the mycolic acids of *Mycobacterium bovis*, BCG and further studies will be necessary before the results can be compared with those obtained by the more usual chromatography of mycolic acid methyl esters.

Thin-layer chromatography has also been used to detect the presence of characteristic lipids found only in ethanol-diethyl ether (1:1, by vol.) extracts of nocardiae and related bacteria (Mordarska 1968; Mordarska & Mordarski 1969; Mordarska & Réthy 1970; Mordarska *et al.* 1972). These so-called LCN-A lipids were shown to be free mycolic acids (Goodfellow *et al.* 1973) and their continued study has proved valuable in the classification of nocardioform and coryneform bacteria (Minnikin *et al.* 1974; Minnikin & Goodfellow 1976; Alshamaony *et al.* 1976*a,b*; Keddie & Cure 1977; Minnikin *et al.* 1978*b,c*). Free mycolic acids are not usually detected in ethanol-diethyl ether extracts of mycobacteria because the mycolic acids from these organisms are insoluble in this solvent mixture.

The characteristic high insolubility of mycobacterial mycolic acids in alcoholic solvents has been developed into a procedure for distinguishing mycobacteria from other bacterial taxa (Kanetsuna & Bartoli 1972). Lipid extracts produced by saponification are dissolved in diethyl ether and an equal amount of ethanol added. Copious white precipitates are produced only by mycobacterial extracts. The procedure was expanded by Hecht & Causey (1976) to include the analysis, by t.l.c., of the supernatant remaining after the mycolic acid precipitation process. The mycolic acids from mycobacteria were found only in the precipitates produced by addition of ethanol to ethereal solutions but those from nocardioform bacteria and corynebacteria were detected on t.l.c. of the solution remaining after attempted precipitation. This method, as well as identifying members of the genus *Mycobacterium,* allows representatives of the other mycolic acid-containing taxa to be partially characterized. It is, however, incorrect to regard the component observed on t.l.c. of the extracts of nocardioform and corynebacteria as being equivalent to the characteristic lipid LCN-A. Hecht & Causey (1976) in fact studied the *total* mycolic acids but lipid LCN-A corresponds only to ethanol-diethyl ether extractable *free* mycolic acids. The mycolic acids in free lipids have been found in one case to differ from those in whole-organism hydrolysates (Ioneda *et al.* 1970), a subsequent study (Lechevalier & Lechevalier 1974) did not, however, lend support to this finding.

Partial characterization of mycolic acids may be made by pyrolysis gas chromatography of their methyl esters (Etémadi 1964, 1967*d*; Lechevalier *et al.* 1971*b*) by which process long-chain fatty acid esters are relased and analysed as shown.

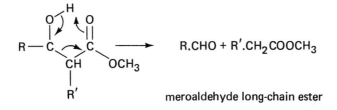

$$R.CHO + R'.CH_2COOCH_3$$

meroaldehyde long-chain ester

For relatively low molecular weight mycolates the meroaldehydes (Morgan & Polgar 1957) may also be analysed (Etémadi 1967*d*; Promé *et al.* 1976). Mycolic esters may also be pyrolysed under vacuum, the esters and meroaldehydes isolated and studied separately (Etémadi 1967*d*; Minnikin & Polgar 1967*a,b*).

Mass spectrometry of mycolic esters allows determinations of overall size, degree of unsaturation and the size of the chain in 2-position to be made (Etémadi 1967*a,b,d*; Maurice *et al.* 1971; Minnikin *et al.* 1974; Alshamaony *et al.* 1976*a,b*; Minnikin & Goodfellow 1976). Combined gas chromatography mass spectrometry of trimethylsilyl ethers of mycolic esters enables separation according to molecular size to be made and structural details of the separated components to be determined (Batt *et al.* 1971; Yano & Saito 1972; Yano *et al.* 1972, 1978).

The location of cyclopropane rings in mycobacterial mycolates using mass spectrometry alone is unreliable (Puzo & Promé 1973; Minnikin 1972, 1978) and derivatives or degradation products should be analysed (Minnikin & Polgar 1967*a,b*; Minnikin 1972, 1978; Promé 1968; Asselineau *et al.* 1969*a*).

High performance liquid chromatography (h.p.l.c.) has been shown to be an extremely powerful tool for the resolution of mycolic acids as their *p*-bromo-phenacyl esters (Steck *et al.* 1978; Qureshi *et al.* 1978). Using adsorption chromatography the mycolic esters from *Myco. bovis*, *Myco. smegmatis* and *Myco. tuberculosis* were resolved into several fractions whose homologous components were then separated on a reverse-phase column (Steck *et al.* 1978). The separated components were shown by mass spectrometry to be mycolic esters but no further characterization was made. In an independent study Qureshi *et al.* (1978) studied one component of the mycolates of *Myco. tuberculosis* H37Ra

TABLE 2

Mycolic acids of representative nocardioform, coryneform and related bacteria*

Taxon	Total number of carbons	Number of double bonds	Acid released on pyrolysis†	
Bacterionema matruchotii	30–36	0,1,2	14:0, 16:0, 18:0	Alshamaony *et al.* (1977)
Corynebacterium bovis	22–32	0,1,2	8:0, 10:0	
Corynebacterium spp.‡	26–38	0,1,2	14:0–18:0 (14:1–18:1)	
'Arthrobacter' roseoparaffinus, *'Brevibacterium' paraffinolyticum*, *'Brev.' sterolicum*, *Cnbc. equi*, *Cnbc. hydrocarboclastus*,	30–48	0,1,2	10:0–18:0	Minnikin *et al.* (1978c) and unpublished results
Cnbc. fascians, *Cnbc. rubrum*	38–52	0,1,2	10:0–16:0	
Rhodococcus coprophilus, *Rhod. equi*, *Rhod. erythropolis*, *Rhod. rhodnii*, *Rhod. rhodochrous*, *Rhod. ruber*	34–52	0–3	12:0–16:0	
Rhod. bronchialis, *Rhod. corallinus*	52–66	1–4	16:0, 18:0	Alshamaony *et al.* (1976a,b)
Rhod. terrae, *Rhod. rubropertinctus*	38–64	1–4	12:0–18:0	Minnikin *et al.* (1978b) and unpublished results
Nocardia asteroides, *Nrda. brasiliensis*, *Nrda. otitidis-caviarum (caviae)*	46–60	0–3	12:0–18:0	

Nocardia amarae	ND	ND	16:1, 18:1	Lechevalier & Lechevalier (1974)
Micropolyspora brevicatena	46–56	ND	ND	Collins et al. (1977)
'aurantiaca' taxon	68–74	1–5	20:0, 20:1, 22:0, 22:1	Goodfellow et al. (1978)
'Mycobacterium rhodochrous'§	28–49	0,1,2	10:0–16:0	
Nocardia lutea	36–51	0, 1, 2,	12:0–16:0	
Nrda. asteroides, Nrda. blackwellii, Nrda. gardneri, Nrda. mexicana, Nrda. eppingerii	44–58	0–4	14:0, 16:0	Yano et al. (1978)
Nrda. polychromogenes	54–66	1–5	16:0, 18:0	
'Gordona' bronchialis	56–68	1–5	16:0, 18:0	

ND, not determined. *For data on other strains see reviews by Minnikin & Goodfellow (1976), Goodfellow & Minnikin (1977), Lechevalier (1977) and Minnikin et al. (1978b,c). † Abbreviations exemplified by: 14:0, tetradecanoate; 16:1, hexadecenoate. ‡ Including strains bearing the labels 'Arthrobacter', 'Brevibacterium' and 'Microbacterium' (Minnikin et al. 1978c). § Including strains labelled Nrda. erythropolis, Nrda. rubra and Nrda. corallina.

by reverse-phase h.p.l.c. The dicyclopropyl ester investigated was separated into four major homologous components which were then subjected to structural analysis. Reverse-phase h.p.l.c. therefore, enables homologous components of mycolates, too large for simple gas chromatography, to be resolved and prepared for structural analysis.

In considering the distribution of mycolic acids it should first be noted that certain bacteria having a wall chemotype IV do not contain these acids. This category includes representatives of *Pseudonocardia, Saccharomonospora, Saccharopolyspora, Micropolyspora caesia, Micropolyspora faeni, Nocardia autotrophica (Nocardia coeliaca), Nocardia lurida, Nocardia orientalis* and *Nocardia mediterranea* (Yano *et al.* 1969; Pommier & Michel 1973; Kroppenstedt & Kutzner 1976, 1978; Lechevalier *et al.* 1977; Goodfellow & Minnikin 1977). *Actinopolyspora halophila* also has a wall chemotype IV, is acid-fast, but apparently does not contain mycolic acids (Gochnauer *et al.* 1975). Organisms labelled *Nocardia aerocolonigenes* do not contain mycolic acids but they also lack arabinose in their walls (Lechevalier *et al.* 1977; Gordon *et al.* 1978).

Relatively simple homologous mixtures of saturated and unsaturated mycolic acids (I) are found in representatives of *Bacterionema, Corynebacterium, Nocardia, Rhodococcus, Micropolyspora brevicatena* and the 'aurantiaca' taxon. Structural features of these acids are summarized in Table 2; more detailed surveys have been compiled (Minnikin & Goodfellow 1976; Goodfellow & Minnikin 1977; Minnikin *et al.* 1978b,c). Uncharacterized mycolates have been noted in extracts of *Micropolyspora fascifera* (Soina & Agre 1976), *Nrda. carnea, Nrda. transvalensis* and *Nrda. vaccinii* (Lechevalier & Lechevalier 1974; Gordon *et al.* 1978). In addition to mycolic acids certain nocardioform actinomycetes and corynebacteria sometimes contain long-chain ketones or alcohols having a close structural relationship to mycolic acids (Pudles & Lederer 1954; Bordet & Michel 1964, 1969; Lanéelle *et al.* 1965a; Lacave *et al.* 1967; Minnikin & Goodfellow 1976).

$$\underset{\substack{| \\ HO}}{}\;\;\underset{\substack{| \\ COOH}}{}$$

$$CH_3.(CH_2)_l\;[CH=CH.(CH_2)_m]_n.\;\underset{\substack{| \\ HO}}{CH}.\underset{\substack{| \\ COOH}}{CH}.(CH_2)_x.\;[CH=CH.(CH_2)_y]_z.CH_3$$

$$l, m, x, y\; all > 1;\qquad 0 \leqslant n \leqslant 4;\qquad x = 0\; or\quad z = 0\; or\; 1$$

Compound I

The mycolic acids of mycobacteria, studied so far, are much more complicated in character; components having oxygen functions in addition to the 3-hydroxy acid system are usually found. Because of this complexity no satis-

TABLE 3

Patterns of mycobacterial mycolic acids and related long-chain compounds*

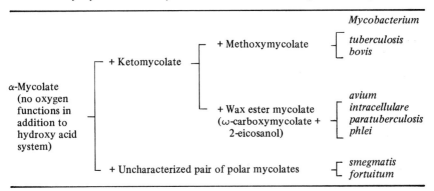

*After Minnikin *et al.* (1980).

factory systematic studies have yet been completed. A preliminary survey, using specially-developed t.l.c. analysis of whole-organism methanolysates (Minnikin *et al.* 1980), has shown that the superficial patterns of mycolic acids and other long-chain components from representative mycobacteria fall into three main groups as summarized in Table 3. The separation of the components detected by t.l.c. is probably due to differences in the nature and number of their oxygen functions but hydrocarbon skeletal differences would not be readily revealed.

The least polar mycolates from mycobacteria, investigated so far, show several structural variations. *Mycobacterium smegmatis,* for example, produce a family of related C_{76} to C_{82} mycolic acids (IIa–c) occurring with lesser amounts of lower molecular weight C_{66} and C_{68} diunsaturated acids (IIa, n = 7; $\neq$ 15, 17) and C_{62} monounsaturated acids (IId) (Etémadi 1967a,b; Etémadi *et al.* 1967). Essentially the same diunsaturated acid (IIa) and a closely related dicyclopropyl acid were isolated from *Myco. phlei* (Lamonica & Etémadi 1967b).

Dicyclopropyl mycolic acids are, in several instances, found to be the only type of least polar mycolates encountered. *Mycobacterium tuberculosis* has been studied in the greatest detail and the essential structure (III) has become established for the major components (Minnikin & Polgar 1967a; Asselineau *et al.* 1969a; Gensler & Marshall 1977) though Qureshi *et al.* (1978) have found a different acid (III, 1 = 17, 19, m = 10, n = 17, 19) in the H37Ra strain. The positions of the cyclopropane rings in the least polar mycolate from *Myco. paratuberculosis* have been determined and the structure (IV) proposed (Lanéelle & Lanéelle 1970). The same overall structure (IV) had been found previously for an acid from *Myco. avium* (Lamonica & Etémadi 1967a) but some evidence of a methyl branch was advanced (Minnikin & Polgar 1967a). The least polar mycolic

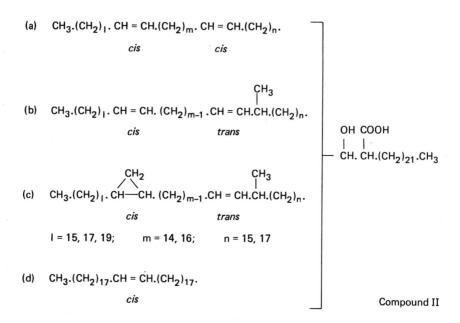

acids from *Myco. kansasii* (Etémadi 1967*a,b*), *Myco. marianum (scrofulaceum)* (Bruneteau & Michel 1968) and *Myco. leprae* (Etémadi & Convit 1974) were also suggested to have the structure (IV); *Myco. kansasii* also contained related long-chain ketones (Miquel *et al.* 1966). A similar dicyclopropyl mycolic acid (IV) was isolated from the least polar mycolates of *Myco. phlei* (Lamonica & Etémadi 1967*b*).

$$CH_3.(CH_2)_l \overset{\overset{\displaystyle CH_2}{\diagup\diagdown}}{CH-CH}.(CH_2)_m. \overset{\overset{\displaystyle CH_2}{\diagup\diagdown}}{CH-CH}.(CH_2)_n. \overset{\overset{\displaystyle OH}{|}}{CH}.\overset{\overset{\displaystyle COOH}{|}}{CH}.(CH_2)_{23}.CH_3$$

         cis   *cis*

$l = 19$;   m = 14, n = 11, 13

                           Compound III

$$CH_3.(CH_2)_l. \overset{\overset{\displaystyle CH_2}{\diagup\diagdown}}{CH-CH}.(C_mH_{2m}). \overset{\overset{\displaystyle CH_2}{\diagup\diagdown}}{CH-CH}.(C_nH_{2n}). \overset{\overset{\displaystyle OH}{|}}{CH}.\overset{\overset{\displaystyle COOH}{|}}{CH}.(CH_2)_{21}.CH_3$$

         cis   *cis*

$l = 17$, m = 14, n = 17

                           Compound IV

Ketomycolates are found in extracts of many mycobacterial species (Table 3) (Minnikin *et al.* 1980) but only those from *Myco. tuberculosis* have been studied in detail (Minnikin & Polgar 1967*b*) and found to be mixtures of two closely related homologous series (V, main components). Complex mixtures of saturated monounsaturated and cyclopropyl ketomycolates have been partially characterized from *Myco. phlei* (Kusamran *et al.* 1972). It has been reported that ketomycolates were present in *Myco. smegmatis* (Etémadi 1967*b*) but in a recent survey (Minnikin *et al.* 1980) this organism (Table 3) did not contain components co-chromatographing with ketomycolates from *Myco. tuberculosis*. Methoxymycolates are considered to be derived biosynthetically from ketomycolates or their precursors (Etémadi 1967*a,b*) and for *Myco. tuberculosis* the structures (VI), for the main components, have been determined (Minnikin & Polgar 1967*b*). Keto and methoxymycolates were suggested to be present in *Myco. kansasii* (Etémadi 1967*b*) and a ketomycolate was isolated from *Myco. leprae* (Etémadi & Convit 1974).

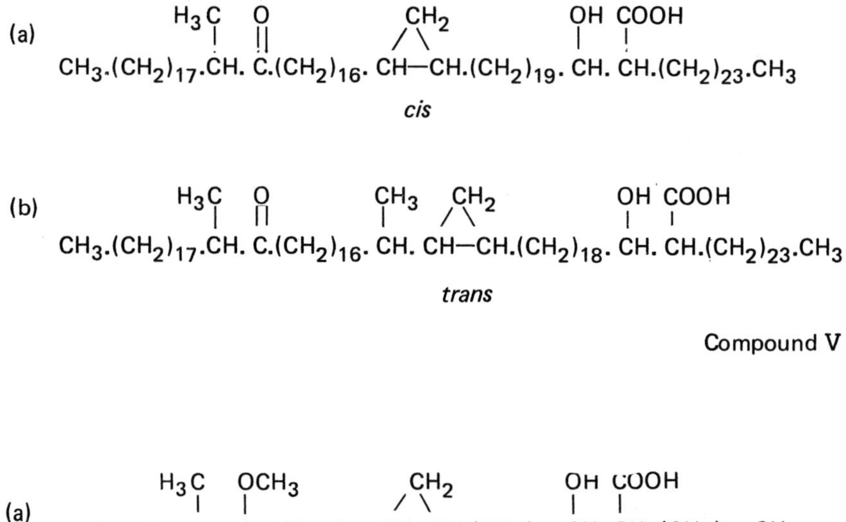

(a)

$$\underset{\text{cis}}{CH_3.(CH_2)_{17}.CH.\overset{\overset{\displaystyle H_3C}{|}}{\underset{\overset{\displaystyle \|}{O}}{C}}.(CH_2)_{16}.\overset{\overset{\displaystyle CH_2}{\diagup\diagdown}}{CH-CH}.(CH_2)_{19}.CH.\overset{\overset{\displaystyle OH}{|}}{\underset{}{CH}}.\overset{\overset{\displaystyle COOH}{|}}{(CH_2)_{23}.CH_3}}$$

(b)

$$\underset{\text{trans}}{CH_3.(CH_2)_{17}.CH.\overset{\overset{\displaystyle H_3C}{|}}{\underset{\overset{\displaystyle \|}{O}}{C}}.(CH_2)_{16}.CH.\overset{\overset{\displaystyle CH_3}{|}}{}\overset{\overset{\displaystyle CH_2}{\diagup\diagdown}}{CH-CH}.(CH_2)_{18}.CH.CH.(CH_2)_{23}.CH_3}$$

Compound V

(a)

$$\underset{\text{cis}}{CH_3.(CH_2)_{17}.CH.\overset{\overset{\displaystyle H_3C}{|}}{}\overset{\overset{\displaystyle OCH_3}{|}}{CH}.(CH_2)_{16}.\overset{\overset{\displaystyle CH_2}{\diagup\diagdown}}{CH-CH}.(CH_2)_{17}.CH.CH.(CH_2)_{23}.CH_3}$$

(b)

$$\underset{\text{trans}}{CH_3.(CH_2)_{17}.CH.\overset{\overset{\displaystyle H_3C}{|}}{}\overset{\overset{\displaystyle OCH_3}{|}}{CH}.(CH_2)_{16}.CH.\overset{\overset{\displaystyle H_3C}{|}}{}\overset{\overset{\displaystyle CH_2}{\diagup\diagdown}}{CH-CH}.(CH_2)_{16}.CH.CH.(CH_2)_{23}.CH_3}$$

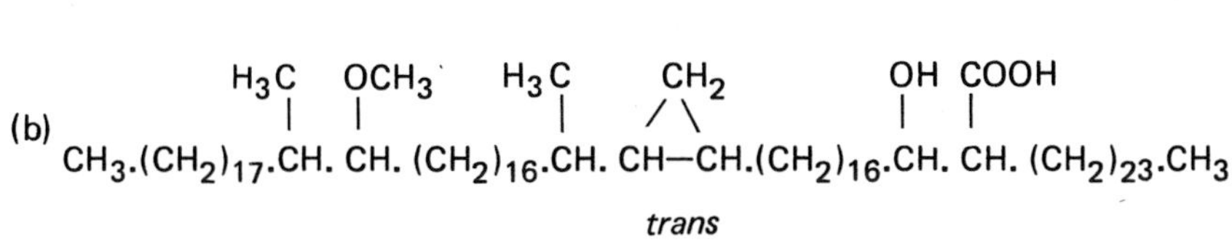

Compound VI

Since ω-carboxymycolates and 2-eicosanol (VII) were found to occur together it was proposed (Etémadi 1967a,b) that these components might result from hydrolysis of a wax ester mycolate formed by biological oxidation of a ketomycolate or its precursor. Support for this hypothesis was provided by Lanéelle & Lanéelle (1970) who isolated a ketomycolate (VIII) and a wax ester mycolate (IX) from *Myco. paratuberculosis* and a similar wax ester mycolate has been characterized from *Myco. phlei* (Promé *et al.* 1976). The ω-carboxymycolate of *Myco. phlei* consist of complex mixtures of monounsaturated (X) and cyclopropyl acids (Markovits *et al.* 1966; Kusamran *et al.* 1972), and similar monounsaturated acids were isolated from walls of *Myco. marianum (scrofulaceum)* (Bruneteau & Michel 1968).

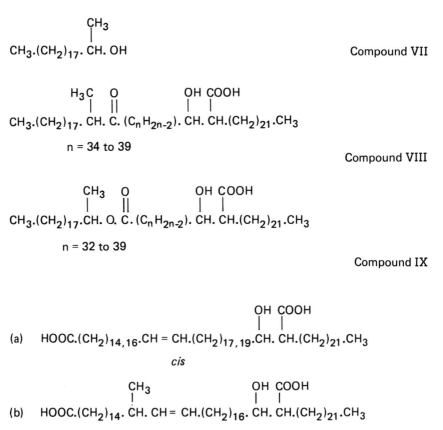

$$CH_3.(CH_2)_{17}.\overset{\overset{\displaystyle CH_3}{|}}{CH}.OH$$

Compound VII

$$CH_3.(CH_2)_{17}.\overset{\overset{\displaystyle H_3C}{|}}{CH}.\overset{\overset{\displaystyle O}{||}}{C}.(C_nH_{2n-2}).\overset{\overset{\displaystyle OH}{|}}{CH}.\overset{\overset{\displaystyle COOH}{|}}{CH}.(CH_2)_{21}.CH_3$$

n = 34 to 39

Compound VIII

$$CH_3.(CH_2)_{17}.\overset{\overset{\displaystyle CH_3}{|}}{CH}.O.\overset{\overset{\displaystyle O}{||}}{C}.(C_nH_{2n-2}).\overset{\overset{\displaystyle OH}{|}}{CH}.\overset{\overset{\displaystyle COOH}{|}}{CH}.(CH_2)_{21}.CH_3$$

n = 32 to 39

Compound IX

(a) $HOOC.(CH_2)_{14,16}.CH = CH.(CH_2)_{17,19}.\overset{\overset{\displaystyle OH}{|}}{CH}.\overset{\overset{\displaystyle COOH}{|}}{CH}.(CH_2)_{21}.CH_3$

cis

(b) $HOOC.(CH_2)_{14}.\overset{\overset{\displaystyle CH_3}{|}}{CH}.CH = CH.(CH_2)_{16}.\overset{\overset{\displaystyle OH}{|}}{CH}.\overset{\overset{\displaystyle COOH}{|}}{CH}.(CH_2)_{21}.CH_3$

trans

Compound X

The detailed differences in mycolic acid types have great potential in the classification and identification of mycobacteria and related taxa but routine systematic procedures are not yet available. The mycolic acids of mycobacteria as a whole may be distinguished from all others by their relative insolubility in ethanol-diethyl ether (Kanetsuna & Bartoli 1972; Hecht & Causey 1976). The insolubility in ethanolic solvents of mycobacterial mycolates is attributable to their higher molecular weights but the fact that they apparently never contain more than two points of unsaturation in the molecule may also be important. Mycolic acids from representatives of *Corynebacterium*, *Nocardia* and *Rhodococcus* are relatively much more unsaturated as well as being smaller in size and both these factors would be expected to increase their solubility in polar solvents such as ethanol. Indeed in the mycolic acids of bacteria other than mycobacteria there appears to be a correlation between molecular size and degree of unsaturation (Minnikin *et al.* 1974) as can be discerned by inspection of the data in Table 2. This phenomenon may represent an attempt to maintain the fluidity of the bacterial envelope within particular limits. The upper extreme in this progression is represented by the mycolic acids of the '*aurantiaca*' taxon having up to 74 carbons and five double bonds. There is a sharp discontinuity between these highly unsaturated acids and those of *Mycobacterium* which are only slightly larger but never more than diunsaturated.

Pyrolysis gas chromatography of mycolic acid methyl esters also allows mycobacteria to be distinguished from other mycolic acid-containing taxa. Systematic studies have shown that mycobacterial mycolates release C_{22} to C_{26} esters but the mycolates from representatives of *Bacterionema*, *Corynebacterium*, *Nocardia*, *Rhodococcus* and *Micropolyspora brevicatena* release esters in the range C_{12} to C_{18} (Etémadi 1967*d*; Lechevalier 1977; Lechevalier *et al.* 1971*b*, 1973; Goodfellow & Minnikin 1977). Pyrolysis gas chromatography of mycolates has been of value in the characterization of two particular actinomycete taxa. Representatives of *Nrda. amarae* contained mycolic acids whose esters released C_{16} and C_{18} monounsaturated esters (Lechevalier & Lechevalier 1974) and members of the '*aurantiaca*' taxon are characterized by the release of C_{20} and C_{22} monounsaturated esters from their mycolates (Goodfellow *et al.* 1978).

B. Non-hydroxylated fatty acids

Relatively simple fatty acids having from 12 to around 20 carbon atoms are found in all acid-fast and related bacteria but, in certain mycobacteria, acids having longer chains are encountered. These latter acids have not been systematically investigated but such is the variation of the structural types that their study would be very profitable.

The best-studied examples of the high molecular weight non-hydroxylated fatty acids are the multimethyl branched mycolipenic (phthienoic) and myco-ceranic (mycocerosic) acids (Asselineau 1966; Polgar 1971). Mycoceranic (myco-cerosic) acids (XI) are found both in *Myco. tuberculosis* and *Myco. bovis* BCG but the latter does not contain phthienoic (mycolipenic) acids (XII) (Bennet & Asselineau 1963). Mycolipanolic acids (XIII), closely related to mycolipenic acids (XII), have been isolated from *Myco. tuberculosis* (Coles & Polgar 1968). Related multimethyl branched acids, the phthioceranic acids (XIV) which in-clude hydroxyphthioceranates (XIVb), have been characterized from the acyl components of a trehalose sulphate lipid from virulent tubercle bacilli (Goren *et al.* 1971, 1976).

$$\text{(a)} \quad CH_3.(CH_2)_{21}.\underset{\underset{D}{|}}{\overset{\overset{CH_3}{|}}{C}}H.CH_2.\underset{\underset{D}{|}}{\overset{\overset{CH_3}{|}}{C}}H.CH_2.\underset{\underset{D}{|}}{\overset{\overset{CH_3}{|}}{C}}H.CH_2.\underset{\underset{D}{|}}{\overset{\overset{CH_3}{|}}{C}}H.COOH$$

$$\text{(b)} \quad CH_3.(CH_2)_{21}.\underset{\underset{D}{|}}{\overset{\overset{CH_3}{|}}{C}}H.CH_2.\underset{\underset{D}{|}}{\overset{\overset{CH_3}{|}}{C}}H.CH_2.\underset{\underset{D}{|}}{\overset{\overset{CH_3}{|}}{C}}H.COOH \qquad \text{Compound XI}$$

$$CH_3.(CH_2)_{17}.\underset{\underset{L}{|}}{\overset{\overset{CH_3}{|}}{C}}H.CH_2.\underset{\underset{L}{|}}{\overset{\overset{CH_3}{|}}{C}}H.CH{=}\overset{\overset{CH_3}{|}}{C}.COOH \qquad \text{Compound XII}$$

$$CH_3.(CH_2)_{17}.\underset{\underset{L}{|}}{\overset{\overset{CH_3}{|}}{C}}H.CH_2.\underset{\underset{L}{|}}{\overset{\overset{H_3C}{|}}{C}}H.\overset{\overset{OH}{|}}{C}H.\overset{\overset{CH_3}{|}}{C}H.COOH \qquad \text{Compound XIII}$$

$$\text{(a)} \quad CH_3.(CH_2)_{14}.CH_2.\underset{\underset{CH_3}{|}}{(CH.CH_2)_n}.\underset{\underset{CH_3}{|}}{CH}.COOH$$

$$n = 4 - 9$$

$$\text{(b)} \quad CH_3.(CH_2)_{14}.\underset{\underset{OH}{|}}{CH}.\underset{\underset{CH_3}{|}}{(CH.CH_2)_n}.\underset{\underset{CH_3}{|}}{CH}.COOH$$

$$n = 2, 4 - 9 \qquad\qquad\qquad \text{Compound XIV}$$

Series of complex 'mycobacteric' acids having 40–68 carbons and functional groups similar to those in mycolic acids were partially characterized from *Myco. tuberculosis*, *Myco. bovis* BCG and *Myco. phlei* (Promé *et al.* 1966); these acids may be related to those isolated by Takayama *et al.* (1975) in studies of mycolic acid biosynthesis. The phleic acids (XV) are polyunsaturated acids of unusual structure and occur as esters of trehalose only in *Myco. phlei* (Asselineau & Montrozier 1976; Asselineau *et al.* 1969*b*; Asselineau & Asselineau 1978*b*).

$$CH_3.(CH_2)_m.(CH\!=\!CH.CH_2.CH_2)_n.COOH$$

$$m = 14, n = 4\text{-}6; \qquad m = 12, n = 5, 6 \qquad\qquad \text{Compound XV}$$

Mycolic acid-containing bacteria contain relatively simple mixtures of straight-chain and unsaturated fatty acids and tuberculostearic (10-methyloctadecanoic) acid is widely encountered (Tables 4 and 5). Organisms lacking mycolic acids but containing major amounts of arabinose and galactose in their walls (chemotype IV) have different fatty acid profiles consisting mainly of branched-chain iso and anteiso acids (Table 6).

The great majority of mycobacteria studied have fairly uniform overall fatty acid patterns (Table 4) but detailed differences between species may be of value in classification (Cattaneo *et al.* 1965; Lucchesi *et al.* 1967). The presence of a specific component, tentatively identified as a 2-methyl branched acid, in *Myco. kansasii* is particularly worthy of mention (Thoen *et al.* 1971*b*, 1972) and the suggested absence of tuberculostearic acid in *Myco. abscessus* (Lechevalier *et al.* 1977) and *Mycobacterium* sp. Lausanne 1369 (Lucchesi *et al.* 1967) is potentially valuable (Table 4). In a detailed study Campbell & Naworal (1969) gave evidence for the presence of iso and anteiso branched acids in *Myco. phlei*.

TABLE 4

Fatty acid profiles of mycobacteria

Taxon *Mycobacterium* Species	Straight-chain	Unsaturated	Tuberculo-stearic	Iso and anteiso
aquae (gordonae)[1], *avium*[1-7], *bovis*[4,3], *bovis* BCG[9], *fortuitum*[1,4], *intracellulare* (battey)[1,7], *kansasii*[1,7,10,11], *marianum*[1], *marinum*[11], *phlei*[1,12-15], *scrofulaceum*[1,7], *smegmatis (butyricum)*[1,4,8,12,14,16], *tuberculosis*[1-5,12,17,18], *vaccae*[19,20], spp. (various)[1,21], sp. 1217[22], sp. 607[2,5], and sp. 4524[4]	+	+	+	−
abscessus[4] and sp. Lausanne 1369[1]	+	+	−	−
phlei[23]	+	+	+	+

[1] Cattaneo *et al.* (1964, 1965) Lucchesi *et al.* (1967); [2] Chandramouli & Venkitasubramanian (1973); [3] Alberghina *et al.* (1977); [4] Lechevalier *et al.* (1977); [5] Khuller & Subrahmanyam (1974); [6] Thoen *et al.* (1971a); [7] Javora & Bacilek (1972); [8] Hung & Walker (1970) Walker *et al.* (1970); [9] Mára *et al.* (1975) Portelance & Asselineau (1970); [10] Thoen *et al.* (1971b); [11] Thoen *et al.* (1972); [12] Okuyama *et al.* (1967); [13] Brennan & Ballou (1967); [14] Dhariwal *et al.* (1976); [15] Dhariwal *et al.* (1977); [16] Weir *et al.* (1972); [17] Subrahmanyam (1965); [18] Bennet & Asselineau (1970); [19] Dunlap & Perry (1967, 1968); [20] King & Perry (1975); [21] Kroppenstedt & Kutzner (1976, 1978); [22] Lanéelle *et al.* (1965b), and [23] Campbell & Naworal (1969).

TABLE 5

Fatty acid profiles of mycolic acid-containing organisms other than Mycobacterium

Organism	Straight chain	Unsaturated	Tuberculo-stearic	Iso and anteiso
Bacterionema matruchotii[1], Brevibacterium lactofermentum[2], Brev. thiogenitalis[3], Corynebacterium alkanolyticum[4], Cnbc. diphtheriae[1,5-8], Cnbc. ovis[5], Cnbc. xerosis[1,9], Microbacterium ammoniaphilum[10], Nocardia convoluta[11], Nrda. kirovani[12], Nrda. opaca[13], Nrda. rubra[7], Nocardia sp. P48[13]	+	+	—	—
Cnbc. bovis[8], Cnbc. equi[14], Micropolyspora brevicatena[8,14], 'Mycobacterium' rhodochrous[14-16], Nrda. amarae[8], Nrda. asteroides[8,14,17,18], Nrda. brasiliensis[8,11,14,16,19], Nrda. caviae[8,14,18], Nrda. erythropolis[18], Nrda. farcinica[18,20], Nrda. pellegrino[16,19], Nrda. polychromogenes[21], Nrda. rubropertincta[19], Nocardia sp. C1CM 7060[19], Nocardia sp. OC2A[22], 'aurantiaca' 'taxon[23]	+	+	+	—
Brev. ammoniagenes[13], Cnbc. diphtheriae[24], Cnbc. equi[25], Cnbc. fascians[13], Cnbc. pseudodiphtheriticum[25], Nrda. corallina[11,13], Nrda. farcinica[13], Nrda. ruber[11]	+	+	—	+
Chbc. equi[24], Cnbc. ovis[24], Micr. brevicatena[26]	+	—	—	+
Nrda. asteroides[11,27], Nrda. brasiliensis[19,27], Nrda. globerulus[11], Nrda. minimus[11], Nrda. ruber[11]	+	+	+	+

[1] Alshamaony et al. (1977); [2] Takinami et al. (1968); [3] Okazaki et al. (1968); [4] Kikuchi & Nakao (1973a); [5] Asselineau (1961); [6] Fulco et al. (1964); [7] Blaschy & Zimmermann (1971); [8] Lechevalier et al. (1977); [9] Drucker & Owen (1973); [10] Shibukawa et al. (1973); [11] Tsyganov et al. (1973); [12] Guinand et al. (1970) Vacheron et al. (1972); [13] Bowie et al. (1972); [14] Kroppenstedt & Kutzner (1976, 1978); [15] Whiteside et al. (1971); [16] Lanéelle et al. (1965a); [17] Farshtchi & McClung (1970); [18] Pommier & Michel (1973); [19] Ballio & Barcellona (1968); [20] Asselineau et al. (1969c), Chamoiseau & Asselineau (1970); [21] Yano et al. (1968); [22] Dunlap & Perry (1967); [23] Goodfellow et al. (1978); [24] Brennan & Lehane (1971); [25] Tadayon & Carroll (1971); [26] Guzeva et al. (1973), and [27] Bordet & Michel (1963).

TABLE 6

Fatty acid profiles of wall chemotype IV and related bacteria lacking mycolic acids

Taxon	Straight-chain	Unsaturated	Tuberculo-stearic	Iso and anteiso
Nocardia autotrophica (coeliaca)[1-3], *Pseudonocardia thermophila*[2]	+	±	±	+
Nrda. lurida[3-4], *Nrda. rugosa*[4], *Saccharomonospora*[3]	+	+	–	+
Actinopolyspora halophila[5], *Micropolyspora caesia*[3], *Micr. faeni*[2,3], *Micropolyspora* spp.[6], *Nrda. gardneri*[7], *Nrda. leishmanii*[8], *Nrda. mediterranea*[8,9], *Nrda. orientalis*[2]	+	–	–	+

[1] Yano *et al.* (1969); [2] Lechevalier *et al.* (1977); [3] Kroppenstedt & Kutzner (1976, 1978); [4] Bordet & Michel (1963); [5] Gochnauer *et al.* (1975); [6] Guzeva *et al.* (1973); [7] Lanéelle *et al.* (1968*l*); [8] Yano *et al.* (1970) and [9] Pommier & Michel (1973).

Iso and anteiso fatty acids are also occasionally found in other mycolic acid-containing organisms but the presence or absence of tuberculostearic acid may be valuable in the classification of these bacteria (Table 5). Organisms lacking mycolic acid may possibly be distinguished by their content of unsaturated and methyl-branched acids (Table 6). The relative proportions of iso and anteiso acids have been shown to be valuable in the classification of members of the genus *Bacillus* (Kaneda 1977) but the lack of sufficient systematic data does not allow positive conclusions to be made at present for the organisms listed in Table 6. The interesting observation (Kroppenstedt & Kutzner 1978) that *Nrda. autotrophica (coeliaca)* has significant amounts of 10-methylheptadecanoate but no 10-methyloctadecanoate (tuberculostearate) may be of value in future studies.

A procedure combining the gas chromatographic analysis of fatty acid methyl esters and methyl glycoside trifluoroacetates, produced after acid methanolysis of mycobacteria, has been developed by Larsson & Mårdh (1976). The fatty acid components were not identified. The patterns for *Myco. tuberculosis* and *Myco. bovis* BCG were similar but distinct profiles were obtained for extracts of *Myco. avium* and *Myco. kansasii*. A very rapid small-scale method involving gas chromatography of methanolic tetramethylammonium hydroxide digests of mycobacteria and rhodococci also gave characteristic profiles whose components were considered to be mainly fatty acid methyl esters formed by pyrolysis of the tetramethylammonium salts (Ohashi *et al.* 1977). *Mycobacterium tuberculosis* was distinguished by the presence of a component eluting very slowly and eight other mycobacterial species and the rhodococci had detailed differences in their profiles which may be useful for identification.

C. Isoprenoid quinones

Isoprenoid quinones are characteristic components of the membranes of most bacteria and those of acid-fast and related bacteria are exclusively menaquinones (XVI). The chemotaxonomic potential of menaquinones lies in the variation of the numbers of isoprene units and hydrogenated double bonds (Minnikin & Goodfellow 1976; Goodfellow & Minnikin 1977; Minnikin *et al.* 1978*a,c*). The distribution of menaquinones in acid-fast and related organisms is summarized in Table 7.

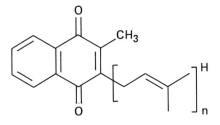

Compound XVI

Representatives of the '*aurantiaca*' taxon are the only mycolic acid-containing organisms whose menaquinones are fully unsaturated (Table 7) (Goodfellow *et al.* 1978). Mycobacteria examined so far have menaquinones with one of the nine isoprene units hydrogenated, abbreviated as MK-$9(H_2)$. Other acid-fast and related bacteria having this menaquinone system are *Cnbc. bovis* and a range of organisms bearing assorted labels (Table 6) but probably having a close relationship to *Cnbc. glutamicum* (Minnikin *et al.* 1978*c*). The rhodococci having MK-$9(H_2)$, *Rhodococcus bronchialis, Rhod. corallinus* and *Rhod. rubropertinctus (Nocardia rubropertincta)* (Table 7), also have relatively large mycolic acids (Minnikin *et al.* 1978*b*). Organisms having MK-$8(H_2)$ as major isoprenoid quinone may probably be divided between species of *Corynebacterium*, related to *Cnbc. diphtheriae*, and *Rhodococcus* (Table 7); certain strains containing comparable amounts of MK-$8(H_2)$ and MK-$9(H_2)$ (Table 7) are more difficult to classify.

Nocardia sensu stricto is clearly distinguished from most other mycolic acid-containing bacteria by the presence of tetrahydrogenated menaquinones (Table 7). *Micropolyspora brevicatena*, however, contains MK-$8(H_4)$ and mycolic acids but *Nrda. autotrophica (coeliaca)* lacks mycolic acids. The classification of *Streptomyces (Nocardia) gardneri* requires further study since the type strain has MK-$8(H_4)$ (Table 7) (Yamada *et al.* 1977) and an unspecified strain has mycolic acids (Yano *et al.* 1978). More heavily hydrogenated menaquinones are found in certain organisms having wall chemotype IV but lacking mycolic acids.

TABLE 7

Major menaquinones of acid-fast and related bacteria

Organisms	Major menaquinone*
'aurantiaca' taxon[1]	MK-9
Arthrobacter albidus[2] , *Brevibacterium ammoniagenes*[2],[3] , *Brev. divaricatum*[2], *Brev. flavum*[2], *Brev. helvolum*[3], *Brev. immariophilum*[2], *Brev. lactofermentum*[2], *Brev. roseum*[2], *Brev. saccharolyticum*[2], *Brev. thiogenitalis*[4], *Corynebacterium acetoacidophilum*[2], *Cnbc. bovis*[2],[6], *Cnbc. callunae*[2], *Cnbc. createnovorans*[5], *Cnbc. glutamicum*[3],[6], *Cnbc. herculis*[2], *Cnbc. lilium*[2],[3], *Cnbc. melassecola*[2], *Cnbc. xerosis*[3], *Microbacterium ammoniaphilum*[2], *Mycobacterium avium*[6],[7], *Myco. bovis*[6], *Myco. fortuitum*[7], *Myco. intracellulare*[6], *Myco. johnei*[6], *Myco. phlei*[3],[5],[7],[8], *Myco. smegmatis*[6] *(butyricum*[7]*), Myco. tuberculosis*[3],[5],[6],[7], *'Nrda'. farcinica (Myco. farcinogenes)*[6], *'Nrda.' rubropertincta*[9], *Rhodococcus ('Gordona') bronchialis*[6], *Rhod. corallinus ('Gordona' rubra)*[6]	MK-9(H$_2$)
Arthrobacter variabilis[2], *Bacterionema matruchotii*[10], *Brevibacterium stationis*[2], *Microbacterium flavum*[3], *Nocardia convoluta*[9], *Nrda. polychromogenes*[9]	MK-8,9(H$_2$)
Arthrobacter roseoparaffinus[2], *Brevibacterium paraffinolyticum*[2], *Brev. sterolicum*[2], *Brev. vitarumen*[4], *Corynebacterium diphtheriae*[2],[3],[7],[11], *Cnbc. equi*[2],[3], *Cnbc. fascians*[2],[3],[6], *Cnbc. flavidum*[6], *Cnbc. hoagii*[2], *Cnbc. hydrocarboclastus*[2], *Cnbc. minutissimum*[2], *Cnbc. murium*[2], *Cnbc. mycetoides*[2], *Cnbc. pseudodiphtheriticum*[2], *Cnbc. pseudotuberculosis*[6], *Cnbc. renale*[2],[6], *Cnbc. rubrum*[2],[7], *Cnbc. ulcerans*[2],[6], *Cnbc. xerosis*[2],[6], *Microbacterium flavum*[2], *Nocardia butanica*[9], *Nrda. calcarea*[9], *Nrda. corallina*[9], *Nrda. erythropolis*[9], *Nrda. globerula*[9], *Nrda. lutea*[9], *Nrda. opaca*[9], *Nrda. restricta*[9], *Nrda. rubra*[9], *Nrda. rugosa*[9], *'rhodochrous'* strains[2],[6] *(Rhodococcus erythropolis, Rhod. rhodochrous, Rhod. ruber*[12]*)*	MK-8(H$_2$)
Micropolyspora brevicatena[6], *Nocardia asteroides*[3],[6],[9] *(farcinica*[9]*), Nrda. bostroemi*[9], *Nrda. brasiliensis*[3],[6],[9], *Nrda. coeliaca*[3],[9], *Nrda. mexicana*[9], *Nrda. otitidis-caviarum (caviae)*[6],[9], *Nrda. transvalensis*[6], *Nrda. vaccinii*[6], *Streptomyces gardneri*[9]	MK-8(H$_4$)
Micropolyspora faeni[6], *Nocardia leishmanii*[9]	MK-9(H$_4$,H$_6$)

*Abbreviations exemplified by MK-9(H$_2$): menaquinone with 9 isoprene units having one double bond hydrogenated.
[1] Goodfellow *et al.* (1978); [2] Minnikin *et al.* (1978c), Collins *et al.* (1979); [3] Yamada *et al.* (1976); [4] Kanzaki *et al.* (1974); [5] Dunphy *et al.* (1971); [6] Collins *et al.* (1977); [7] Beau *et al.* (1966); [8] Gale *et al.* (1963), Campbell & Bentley (1968); [9] Yamada *et al.* (1977); [10] Unpublished results of M. D. Collins, M. Goodfellow and D. E. Minnikin, see also Goodfellow & Minnikin (1977); [11] Scholes & King (1965) and [12] Goodfellow & Alderson (1977).

D. Polar lipids

Polar lipids are amphipathic molecules that interact with other such molecules and proteins to form the permeability barrier of the plasma membrane. All free-living organisms must contain polar lipids and their extraction with neutral organic solvents and analysis by t.l.c. gives clear patterns of diagnostic potential (Minnikin *et al.* 1977*a,b*, 1979). Phospholipids are the most common polar lipid types but glycolipids and certain amino acid amines are also encountered.

The phospholipids of acid-fast and related bacteria are derivatives of phosphatidic acid (XVII, Y = H), the most common being phosphatidylglycerol (PG) (XVII, Y = glycerol), diphosphatidylglycerol (DPG) (XVII, Y = phosphatidylglycerol), phosphatidylethanolamine (PE) (XVII, Y = ethanolamine) and phosphatidylinositol (PI, XVII, Y = inositol). Methylated derivatives of PE, such as phosphatidylmonomethylethanolamine (PME) (XVII, Y = N-methylethanolamine) and phosphatidylcholine (PC) (XVII, Y = choline) are less widely encountered. The above lipids are widespread in many bacteria and other living organisms but certain actinomycetes and coryneform bacteria contain very characteristic lipids, the phosphatidylinositol mannosides (PIM's) (XVIII). The most common PIMs are mono- and di-acyl dimannosides (XVIII, n = 1) but pentamannosides (XVIII, n = 4) have been found in mycobacteria.

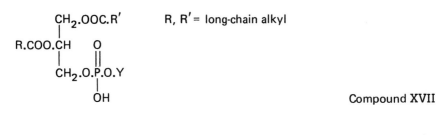

Compound XVII

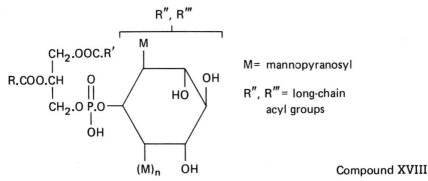

Compound XVIII

The glycolipids of mycolic acid-containing bacteria and related organisms have not been the subject of detailed systematic studies. Diglycosyl diacylglycerols (XIX) have been characterized from extracts of many Gram positive bacteria (Shaw 1975) but they have not been characterized from acid-fast bacteria although these organisms do contain many unidentified glycolipids

TABLE 8
Polar lipids of mycobacteria

Taxon	DPG*	PG	PE	PI	MM	DM	PM	?	GLY	Other	
Mycobacterium											
abscessus	+		+	+				+			Lechevalier et al. (1977)
aurum	+	+	+	+		+			+	+†	Minnikin et al. (1977b)
											Chandramouli & Venkitasubramanian (1974)
											Alberghina (1976)
avium	+		+		+		+	+			Lechevalier et al. (1977)
bovis	+		+	+				+			Lechevalier et al. (1977)
											Akamatsu & Nojima (1965); Okuyama et al. (1967); Walker et al. (1970)
bovis BCG	+		+					+		+‡	Portelance & Asselineau (1970)
											Asselineau & Portelance (1974)
	+	+	+	+	+	+			+		Brennan et al. (1970)
											Khuller et al. (1972)
					+		+			+‡	Promé et al. (1969)
chitae	+	+	+	+	+	+			+	+†	Minnikin et al. (1977b)
diernhoferi	+	+	+	+	+	+			+		
duvalii	+	±	+	±		+			+	+†	
fortuitum	+		+	+			+		+		Lechevalier et al. (1977)
											Vilkas & Rojas (1964), Vilkas et al. (1968)

Species	1	2	3	4	5	6	7	8	9	References
gallinarum	+	+	+	+	+	+			+†	Minnikin et al. (1977b)
gilvum	+	+	+	+	+	+			+†	Minnikin et al. (1977b)
kansasii	+	+	+		+	+	+			Akamatsu & Nojima (1965); Akamatsu et al. (1966, 1967); Okuyama et al. (1967); Brennan & Ballou (1967)
phlei	+	+		+	+		+			Komura et al. (1975)
	+	±	+	+	+	+		+	±‡	Dhariwal et al. (1975, 1976, 1977, 1978)
rhodesiae	+	+	+	+	+		+			
salmoniphilum	+	+	+	+	+		+		+†	Minnikin et al. (1977b)
scrofulaceum	+	±	+	+	+	+		+	±†	
smegmatis	+	+	+	+		+				Okuyama et al. (1967)
(butyricum)	+	+	+	+	+	+				La Belle & Walker (1972); Chandramouli & Venkitasubramanian (1974); Dhariwal et al. (1976); Lechevalier et al. (1977)
thermoresistible	+		+	+		+	+		+†	Minnikin et al. (1977b)
	+		+	+		+				Okuyama et al. (1967); Banerjee et al. (1974); Alberghina (1976)
tuberculosis	+	+	+	+	+		+			Pangborn & McKinney (1966), Pangborn (1968); Sasaki (1975); Chandramouli & Venkitasubramanian (1974); Komura et al. (1975); Lechevalier et al. (1977)
sp. 607	+	+	+	+	+		+		+§	Khuller & Subrahmanyam (1968, 1970); Khuller et al. (1972); Chandramouli & Venkitasubramanian (1974); Banerjee et al. (1974)
sp. P₆	+	+	+	+	+					Oka et al. (1968a,b); Motomiya et al. (1969); Khuller & Brennan (1972a)
sp. NCTC 4524 (Nocardia farcinica)	+	+	+	+		+				Lechevalier et al. (1977)
'Nocardia' farcinica	+	+				+				Asselineau et al. (1969c)

*Abbreviations: DPG, diphosphatidylglycerol; PG, phosphatidylglycerol; PE, phosphatidylethanolamine; PI, phosphatidylinositol; PIM, phosphatidylinositol mannoside (MM, monomannoside; DM, dimannoside; PM, pentamannoside and/or similar polymannoside; ?, uncharacterized); GLY, glycolipid. †Lipids giving no positive reaction with specific spray reagents for lipid phosphorus, carbohydrates or vicinal glycols. ‡Acylated ornithine amide §Ornithine ester of PG.

TABLE 9

Polar lipids of mycolic acid-containing organisms other than Mycobacterium

Taxon	DPG*	PG	PE	PI	PIM MM	PIM DM	PIM ?	GLY	Other	Reference
'aurantiaca' taxon	+	+	±	+	+			+		Minnikin et al. (1977b)
Bacterionema matruchotii	+	+		+	+			+		Komura et al. (1975)
Brevibacterium ammoniagenes	+	+	Tce	+			+			
Corynebacterium										
alkanolyticum	+	+	+	+					+†	Kikuchi & Nakao (1973b)
bovis	+	+	+	+	+	+			+‡	Lechevalier et al. (1977)
	+	+				+			+§	Minnikin et al. (1977b)
										Asselineau (1961)
							+			Gomes et al. (1966)
diphtheriae	+	+	Tce	+		+				Brennan & Lehane (1971)
	+	+		+			+			Komura et al. (1975)
	±	+		+			+		±‡	Lechevalier et al. (1977)
	+	+	Tce	+		+				Brennan & Lehane (1971)
equi	+	+	+	+			+			Komura et al. (1975)
fascians	+	+		+			+			
lilium	+	+	Tce	+			+			Lacave et al. (1967)
ovis	+	+		+	+					Brennan & Lehane (1971)
xerosis	+	+	+	+	+	+			+§	Minnikin et al. (1977b)
	+		+	+						Shibukawa et al. (1970)
Microbacterium										
ammoniaphilum	+								+‖	Komura et al. (1975)
flavum	+	+	Tce	+			+			Lechevalier et al. (1977)
Micropolyspora brevicatena	+	Tce	+	+			+			Pommier & Michel (1973)
Nocardia										
amarae	+	+	+	+			+	+		Komura et al. (1975)
asteroides	+	+	+	+	+					Khuller (1976, 1977a)
	+	±	+	+	+			+		Minnikin et al. (1977b)

Rhodococci

Taxon									References
brasiliensis	+		+	+	+			+	Lanéelle et al. (1965a)
	+	+	+	+	+			+	Komura et al. (1975)
	+	±	+	+	+			+	Lechevalier et al. (1977)
farcinica	+	±	+	+	+	+			Minnikin et al. (1977b)
	+	+	+	+	+				Pommier & Michel (1973)
	+	+	+	+	+	+	+	+	Komura et al. (1975)
otitidis-caviarum	+	±	+	+	+	±		+	Pommier & Michel (1972, 1973, 1976)
(caviae)	+	±	+	+	+			±	Komura et al. (1975)
	+		+	+	+			+	Minnikin et al. (1977b)
	+	+	+	+	+				Lechevalier et al. (1977)
	+	+	+	+	+	+			Kataoka & Nojima (1967)
polychromogenes	+		+	+	+	+			Yano et al. (1968)
			+	+	+				++† Khuller & Brennan (1972b); Khuller (1977a,b); Khuller & Trana (1977)
'Gordona'	+		+	+	+		+	+	
bronchialis	+	±	+	+	+		+	+	
rubra	+		+	+	+		+	+	Minnikin et al. (1977b)
terrae	+	±	+	+	+		+	+	
'Jensenia canicruria'	+	+	+	+	+		+		
'Nocardia'	+	+	+	+	+	+			Komura et al. (1975)
calcarea	+		+	+	+		+	+	
corallina	+	+	+	+	+	+			Minnikin et al. (1977b); Kobayashi et al. (1971); Komura et al. (1973)
erythropolis	+	±	+	+	+	+			Pommier & Michel (1973)
	+	+	+	+	+		+	+	Komura et al. (1975)
opaca	+	+	+	+	+		+	+	Lanéelle et al. (1965a)
pellegrino	+		+	+	+		+		Whiteside et al. (1971)
rhodochrous	+	+	+	+	+		+		Lechevalier et al. (1977)
'rhodochrous'									
sp. R43	+	+	+	+	+		+	+	
sp. N146	+		+	+	+				+§ Minnikin et al. (1977b)

*Abbreviations: DPG, diphosphatidylglycerol; PG, phosphatidylglycerol; PE, phosphatidylethanolamine; PI, phosphatidylinositol; PIM, phosphatidylinositol mannoside (MM, monomannoside; DM, dimannoside; PM, pentamannoside and/or similar polymannoside; ?, uncharacterized); GLY, glycolipid; Tce, trace. †Unidentified phospholipids. ‡Acylated phosphatidylglycerol. §Lipids giving no positive reaction with specific spray reagents for lipid phosphorus, carbohydrates or vicinal glycols. ||Possibly phosphatidylcholine.

(Minnikin *et al.* 1977*b*). An unusual acidic glycolipid, identified as 2-0-octanoyl-2′,3-di-0-decanoyl-6-0-succinyl-α,α′-D-trehalose, has been isolated from the lipids of *Myco. paraffinicum* (Batrakov *et al.* 1977). Phosphorus-free ornithine and lysine amides are common in the lipids of streptomycetes (Minnikin & Goodfellow 1976; Batrakov & Bergelson 1978) but the only report of such lipids in acid-fast bacteria is the isolation of the acylated ornithine-amide (XX) from *Myco. bovis* BCG (Promé *et al.* 1969; Portelance & Asselineau 1970).

$$
\begin{array}{l}
\text{Hexose} \!-\!\!\!-\!\! \text{Hexose} \!-\!\!\!-\!\! O.CH_2 \\
\qquad\qquad\qquad\qquad\;\; | \\
\qquad\qquad\qquad\qquad\; CH.OOC.R' \\
\qquad\qquad\qquad\qquad\;\; | \\
\qquad\qquad\qquad\qquad\; CH_2.OOC.R'
\end{array}
$$

Compound XIX

$$
CH_3.(CH_2)_{14}.\underset{\underset{\textstyle OH}{|}}{CH}.CH_2.CO.NH.\underset{\underset{\textstyle (CH_2)_3.NH_2}{|}}{CH}.CO.O.\,CH_2.CH_2.O.CO\,(CH_2)_8.\underset{\underset{\textstyle CH_3}{|}}{CH}.(CH_2)_7.CH_3
$$

Compound XX

The polar lipids of mycobacteria, summarized in Table 8, appear quite uniform in their overall patterns. DPG, PE and PI are usually encountered but the close chromatographic mobility of the latter to phosphatidylinositol dimannosides (PIDMs) (Pangborn & McKinney 1966; Minnikin *et al.* 1977*b*) may explain why its presence was not included in several reports. The difficulty in separating PI from PIDMs may also explain (Khuller & Brennan 1972*a,b*) why phosphatidylinositol monomannosides have been described in several studies (Table 8). It appears from recent studies that mono- and diacylated PIDMs are the most common PIMs with components having up to five or six mannosides also expected in mycobacteria. Many unidentified glycolipids have been detected in mycobacteria (Minnikin *et al.* 1977*b*); the relationship of these to the diacylated trehaloses from *Myco. fortuitum* (Vilkas & Rojas 1964; Vilkas *et al.* 1968) requires further study.

The polar lipids of other mycolic acid-containing bacteria (Table 9) resemble those of mycobacteria in their general composition but certain potentially valuable differences should be noted. Representatives of *Corynebacterium sensu stricto* appear to lack PE but usually contain PG thus possibly allowing these organisms to be distinguished from nocardiae and rhodococci (Komura *et al.* 1975; Minnikin *et al.* 1977*b*; Lechevalier *et al.* 1977). In one of these studies (Minnikin *et al.* 1977*b*), however, the presence of PE and absence of PG in

TABLE 10

Polar lipids of wall chemotype IV and related bacteria lacking mycolic acids

Taxon	DPG*	PG	PE	PME	PC	PI	PIM MM	PIM DM	PIM ?	GLY	Other	
Actinopolyspora halophila	+	+			+					+	+†	Gochnauer et al. (1975)
Micropolyspora												
faeni	+	+	+	+	+	+						⎫ Lechevalier et al. (1977)
sp. Y39	+	+	±	+	+	+			±			⎭
Nocardia												
autotrophica (coeliaca)	+	+	+		+	+	+			+		⎫ Yano et al. (1969)
	+	+	+		+	+		+		+		⎭ Khuller & Brennan (1972b)
gardneri	+		+			+			+			Lanéelle et al. (1968)
leishmanii	+	+	+			+	+					Yano et al. (1970)
mediterranea	+	+	+				+					Pommier & Michel (1973)
orientalis	+		+		+	+			±		+‡	Lechevalier et al. (1977)
rugosa	+	+	+		+	+			+			Komura et al. (1975)
Pseudonocardia thermophila	+		+		+	+			±			Lechevalier et al. (1977)

*Abbreviations: DPG, diphosphatidylglycerol; PG, phosphatidylglycerol; PE, phosphatidylethanolamine; PME, phosphatidylmonomethylethanolamine; PC, phosphatidylcholine; PI, phosphatidylinositol; PIM, phosphatidylinositol mannoside (MM, monomannoside; DM, dimannoside; PM, pentamannoside and/or similar polymannoside;?, uncharacterized): GLY, glycolipid. † Unidentified phospholipids. ‡ Minor unknown component.

nocardiae and rhodococci was not found to be totally reliable. PIDMs appear to be the major PIMs of nocardiae, rhodococci and corynebacteria but a preliminary report (Khuller 1977a) suggests that, as in mycobacteria, more glycosylated PIMs may also be present. No significant differences have been found between the polar lipids of rhodococci and nocardiae, though two strains of 'Gordona rubra' (Rhodococcus corallinus) appeared to have a potentially simple diagnostic pattern (Minnikin et al. 1977b).

The polar lipids of actinomycetes with a wall chemotype IV but lacking mycolic acids show wide variations in composition (Table 10). PI is found in most organisms but the occurrence of PIMs seems variable. The most significant point is the presence of phosphatidylcholine (PC) and phosphatidylmonomethylethanolamine (PME) in Nrda. autotrophica and in some micropolysporae lacking mycolic acids (Table 10). A single systematic study of these bacteria would show the true value of polar lipid analyses in their classification.

E. Mycosides and peptidolipids

The term 'mycoside' was proposed (Smith et al. 1960a) for "type-specific glycolipids of mycobacterial origin which had been discovered by infrared spectroscopy of chromatographically-fractionated ethanol-diethyl ether extracts" (Smith et al. 1957, 1960b). Subsequent studies have shown that mycosides fall into two distinct chemical classes, phenol-phthiocerol glycosides and peptidoglycolipids (for reviews see Randall & Smith 1964; Asselineau 1964, 1966; Lederer 1967; Goren 1972; Gastambide-Odier 1974; Lederer et al. 1975; Barksdale & Kim 1977; Asselineau & Asselineau 1978a). Peptidolipids lacking sugars have been characterized from a limited number of mycobacteria and nocardiae (Lanéelle et al 1965c; Barber et al 1965; Guinand & Michel 1966). .

Mycosides that lack amino acids are phenolic glycosides of phenol-phthiocerol (XXI, R = R' = H), their structures and distribution are shown in Table 11. Minor variants of the mycosides from Myco. kansasii and Myco. marinum, having mycolic acids esterified to sugars, have also been isolated (Gastambide-Odier 1973). A rapid method, involving t.l.c. of petroleum ether extracts, has been developed for the recognition of mycosides A and B from Myco. kansasii and Myco. bovis BCG, respectively (Koul & Gastambide-Odier 1977). It is notable that mycosides B, characteristic of Myco. bovis, are not found in Myco. tuberculosis (Asselineau & Asselineau 1978a).

$$R\,C - \underset{}{\text{\Large\bigcirc}} - (CH_2)_n.\underset{OR'}{CH}.CH_2.\underset{OR'}{CH}.(CH_2)_4.\underset{CH_3}{CH}.\overset{OCH_3}{CH}.CH_2.CH_3$$

Compound XXI

TABLE 11

Structure and distribution of mycosides based on phenol-phthiocerol

| Taxon | Mycoside | Phenol-phthiocerol (XXI) | | | |
		n	R'	R	
Mycobacterium *kansasii*	A	16,17,18	12:0–18:0, tuberculostearate, mycocerosate	Trisaccharide containing 2 and 3-0-methyl fucose and rhamnose and 2,4 di-0-methyl rhamnose	Gastambide-Odier *et al.* (1965); Gastambide-Odier & Sarda (1970; Gastambide-Odier & Villé (1970).
bovis	B	14,15,16, 17,18	16:0, mycocerosate	2-0-methyl-D-rhamnose	Demarteau-Ginsburg & Lederer (1963); Gastambide-Odier *et al.* (1965); Gastambide-Odier & Sarda (1970).
marinum	G	18,19,20, 21,22	mycocerosates	2-0-methyl-L-rhamnose	Navalkar *et al.* (1965); Sarda & Gastambide-Odier (1967); Villé & Gastambide-Odier (1970).

TABLE 12

Structure and distribution of mycobacterial peptidoglycolipids (mycosides C)

Taxon	Mycoside	Peptidoglycolipid structure (XXII)								
		p	q	r	s	:sc	Deoxy-L-talose	L-rhamnose	
Mycobacterium									
avium	C_2	12	0	1	1	H	+	3,4-di-OMe	Voiland *et al.* (1971)
scrofulaceum	C_s	28,29,30,31	0	0	1	OCH_3	+	3,4-di-OMe	Vilkas *et al.* (1968b)
butyricum (smegmatis)	C_{b1}	22	1	0	1	OCH_3	+	2,3,4-tri-OMe	Vilkas & Lederer (1968)
sp. 1217	C_{1217}	24	0	0	1	OH	+	2,3,4-tri-OMe	Lanéelle & Asselineau (1968)
farcinogenes	C_{378}	24	0	1	0	OCH_3	−	Di-OMe	Lanéelle *et al.* (1971)

Characteristic mycobacterial peptidoglycolipids, collectively termed 'mycosides C', are found as complex families of lipids of which only a few individuals have been characterized in detail. The structures determined so far correspond to variations on the general structure XXII and representative examples are given in Table 12; the recent review by Asselineau & Asselineau (1978a) contains additional data. Further comparisons of the similarity of mycosides C_s (Table 12) and C_m from Myco. scrofulaceum and Myco. marianum (Chaput et al. 1962), respectively, should be made since Myco. marianum is considered to be a synonym of Myco. scrofulaceum (Barksdale & Kim 1977). Likewise Myco. butyricum is a synonym of Myco smegmatis (Barksdale & Kim 1977) so that mycosides (C_{b1} (Table 12) and C_{sm} (Furuchi & Tokunaga 1972) should be closely compared. The isolated report (Nacasch & Vilkas 1967) that mycosides C (XXII) can co-occur with those based on phenol-phthiocerol (XXI) in the same organism (Myco. kansasii) requires confirmation (Gastambide-Odier 1974).

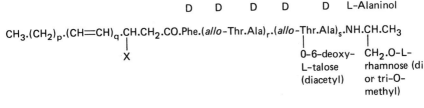

Compound XXII

A peptidolipid, termed mycoside D, has been partially characterized from the lipids of a scotochromogenic Mycobacterium sp. P5 (Fregnan et al. 1962) and it appears to be closely related to the C-mycosides (MacLennan 1962). The distribution of mycosides C and D, distinguishable by their infra-red spectra, was found to be of possible value in the classification of mycobacteria classified in Runyon's (1959) Groups II and III (Navalkar et al. 1964; Walker et al. 1967). Certain Group III mycobacteria contained C-mycosides but D-mycosides were only found in the scotochromogenic Group II organisms. Mycosides, however, were absent in a substantial proportion of the strains in one of these studies (Navalkar et al. 1964); it had previously been shown that a rough mutant of Mycobacterium sp. P5 lacked mycosides (Fregnan et al. 1962).

A crude peptidoglycolipid isolated from the capsule of Myco. lepraemurium (Draper & Rees 1973), on hydrolysis, released alanine, allo-threonine, phenylalanine, valine, leucine, alaninol, 6-deoxytalose and another methyl pentose but t.l.c. of the intact lipid distinguished it from the C-mycosides of Myco. avium (Draper 1974). Myco. fortuitum was found to contain an uncharacterized mycoside whose infra-red spectrum resembled those of acylated trehaloses (Fregnan *et al.* 1961); the relationship of this mycoside to trehalose mycolates (cord-factors) and other trehalose esters from *Myco. fortuitum* (Vilkas & Rojas 1964; Vilkas *et al.* 1968*a*) is in need of clarification.

Carbohydrate-free peptidolipids are very limited in their distribution and indeed appear to be confined to individual strains rather than whole species. A single strain of *Nrda. asteroides*, ATCC 9969, produces a mixture of characteristic peptidolipids of which 'peptidolipin NA', having the cyclic structure (XXIII), is the main component (Guinand & Michel 1966). Peptidolipids having linear structures (XXIV, XXV) have been characterized from the lipids of *Myco. johnei (paratuberculosis)* (Lanéelle *et al.* 1965*c*; Lanéelle 1969); and a strain of *Myco. smegmatis* (Barber *et al.* 1965), respectively. An incompletely characterized peptidolipid, containing leucine, phenylalanine, alanine, glycine, glutamic acid, serine, threonine and a long-chain hydroxy acid was isolated from *Myco. rubrum* (Krasil'nikov & Koronelli 1971).

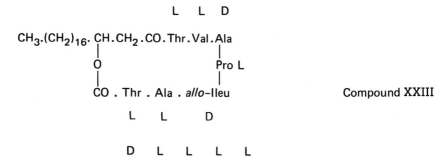

Traditional procedures for the analysis of mycobacterial peptidolglycolipids (mycosides C) are relatively complicated involving solvent and chromatographic fractionation and detailed chemical analyses, though infra-red spectroscopy enables the general type of mycoside to be determined (Smith *et al.* 1957, 1960*a,b*; Walker *et al.* 1967). A new approach has been made by Brennan *et al.* (1978) in which t.l.c. of mildly saponified total free lipids allows the ready analysis of the deacetylated peptidoglycolipids free from overlapping conventional complex lipids. Representatives of the various serotypes of the *Myco. avium–Myco. intracellulare–Myco. scrofulaceum* complex were all found to have very similar families of up to six apolar deacetylated C-mycosidic peptidoglycolipids. These apolar lipids probably correspond to the mycosides of the type summarized in Table 12 and indicate that the structures of only selected components have been characterized so far. It was possible, however, to distinguish between all the 17 serotypes by the chromatographic mobility of, in most cases, single polar peptidoglycolipids and six unknown isolates were identified. These polar glycolipids are derived from the surface (Schaefer) antigens of these mycobacteria and are based on multiglycosylated C-mycoside cores; the apolar C-mycosides are devoid of serological activity (Brennan *et al.* 1978).

Characteristic orcinol-positive lipids identified by t.l.c. of ethanol-diethyl ether extracts of mycobacteria (Marks & Szulga 1965) are believed to be closely related to the C-mycoside based peptidoglycolipids studied by Brennan *et al.* (1978). The original procedure of Marks & Szulga (1965) has been systematically applied to the classification and identification of clinically-significant mycobacteria and the results are illustrated in Table 13. In order to obtain the maximum amount of diagnostic information it was necessary to use single and two-dimensional thin-layer chromatographic analysis and the use of a second one-dimensional system was also valuable (Table 12). Good correlations between lipid patterns and serotypes of *Myco. avium, Myco. intracellulare* and *Myco. scrofulaceum* were obtained (Marks *et al.* 1971), the patterns obtained being more complex than those of the deacetylated polar peptidoglycolipids analysed by Brennan *et al.* (1978). A particular achievement of this thin-layer chromatographic procedure was the recognition of a new pathogen, *Myco. szulgai* (Marks *et al.* 1972). The procedure was also successful in giving useful patterns for the other mycobacteria listed in Table 13 but no consistent patterns were recorded for *Myco. tuberculosis* or *Myco. bovis* (Jenkins & Marks 1973). Two extensive co-operative numerical taxonomic studies of mycobacteria have included the analysis of these characteristic polar glycolipids (Kubica *et al.* 1972; Meissner *et al.* 1974).

The analysis of these characteristic orcinol-positive mycobacterial lipids has been used in other laboratories. In a study of rapidly growing mycobacteria Sehrt *et al.* (1975) found characteristic lipid patterns for *Myco. borstelense, Myco. chelonei (abscessus), Myco. phlei, Myco. runyonii* and *Myco. smegmatis,* and the *Myco. fortuitum* complex was divisable into a group containing representatives of *Myco. fortuitum, Myco. giae, Myco. peregrinum* and *Myco. thamnopheos* and another containing *Myco. minetti, Myco. salmoniphilum* and other strains of *Myco. fortuitum.* The patterns and spot colours were not exactly as described by Marks and co-workers (Table 13); it was notable that *Myco. borstelense* and *Myco. chelonei (abscessus)* had very different patterns in contrast to the results of Jenkins *et al.* (1971). In a further study (Sehrt *et al.* 1976),

TABLE 13

Characteristic polar orcinol-positive lipids detected on thin-layer chromatography of mycobacterial extracts[1]

Taxon	Serotype	No. and RFs of spots in 1 D systems[2]		No. of spots in 2 D		Spot colour
		1	3	Mobile[3]	Immobile	
Mycobacterium avium	1	2(0.8,0.9)	3(0.6-0.7)	Two	Two	Golden brown
	2	3(0.85-0.95)	3(0.5-0.6)	Three	Two	
	3	1(0.65)	1(0.05)	ND	ND	
intracellulare[5]	Various	1 or 2(0.7-1.0)	1-3(0.05-0.6)	ND	ND	
scrofulaceum[5]	Various	2 or 3(0.6-0.95)	1 or 2 (0.1-0.2)	Two-Five	Two	
xenopei	—	1(0.95)	1(0.85)	ND	ND	Pale yellow
gordonae	—	2(0.6)	1(0.05)	Two	Two	Green or greenish-brown
aquae	—	3(0.3-0.5)	ND	Three	Three	Green, brown or intermediate
kansasii	—	3(0.3-0.5)	ND	Three	Three	Bluish-grey (lowest brown-tinged)
szulgai	—	1(0.1)	ND	One	One	Dark brown
fortuitum	—	4(0.7-0.9)	2(0.4,0.9)	Two	Two	Purple
peregrinum	—	2(0.8,0.95)	4(0.2-0.6)	Four	Two	Orange
abscessus[6]	—	3(0.95-1.0)	4(0.3-0.95)	Three	Two	Orange and pink
borstelense[6]	—	3(0.9-1.0)	4(0.3-0.9)	Three	Two	Orange and pink
marinum	—	2(0.5-0.6)	ND	Three	Two	Dense brown

[1] Data from Marks & Szulga (1965); Szulga et al. (1966); Birn et al. (1967); Jenkins et al. (1971, 1972); Marks et al. (1971); Jenkins & Marks (1973). [2] Solvents for one-dimensional (1D) thin-layer chromatography (1) Propan-1-ol: water: 0.880 ammonia; 75:22:3. (3) Propan-1-ol di-*iso*-propyl ether: water; 45:45:10. [3] Solvent 1 from 1D system used is first direction of two-dimensional (2D) system; butan-1-ol: acetic acid: water; 60:20:20 used in second dimension. Mobile and immobile refer to the second dimension. [4] Developed by spraying finely with a mixture of two volumes 60% H_2SO_4 to one of 0.1% aqueous orcinol followed by heating on a hot plate at 120–150° C. [5] The patterns obtained correlated well with serotypes but did not fall into clear groups similar to those of *Myco. avium*. [6] The two closely related patterns differed slightly but significantly in the mobility and order of spots.

lipid patterns did not allow *Myco. avium* and *Myco. intracellulare* to be distinguished but again there was good correlation between lipid analysis and serological typing. Timbol & Shimizu (1977) found similar lipid patterns in *Myco. avium, Myco. intracellulare* and *Myco. scrofulaceum* and were able to assign some unknown isolates to *Myco. gordonae*. A number of '*avium*-like' mycobacteria were screened (Boughton & Bernstad 1974) by use of the procedure of Marks *et al.* (1971).

A new procedure for the differentiation of mycobacterial species by t.l.c. of ethanol-diethyl ether extracts of cells, previously incubated with ^{35}S-methionine, has been introduced by Tsukamura & Mizuno (1975). In most cases characteristic patterns were found for each of the many species examined and it was shown later that *Nocardia* could be differentiated from *Rhodococcus* (Tsukamura & Mizuno 1978). The nature of the radioactive spots recorded in this procedure is unknown but it should be noted that the developing solvent for t.l.c. (propanol : butanol : water : ammonia, 57 : 20 : 20 : 3, by vol.) was used previously by Marks *et al.* (1971) for the analysis of their characteristic orcinol-positive lipids.

F. Acylated trehaloses including sulpholipids

Glycolipids based on trehalose are encountered in a variety of structural forms in the lipids of acid-fast and related bacteria and the subject has been comprehensively reviewed (Asselineau & Asselineau 1978*a,b*). The best-known examples are the dimycolic esters of trehalose, the so-called 'cord-factors', but esters of trehalose with medium chain, phleic (XV) or phthioceranic (XIV) acids are also encountered. The medium chain fatty acid esters of trehalose characterized from *Myco. fortuitum* (Vilkas & Rojas 1964; Vilkas *et al.* 1968*a*) may be regarded as polar lipids (Table 8) and are not considered further here. The trehalose phleates (Asselineau *et al.* 1969*b*) and sulphated trehalose phthioceranates (Goren *et al.* 1971, 1976) are, however, much more characteristic and their structures and distribution will be discussed.

The name 'cord factor' was applied by Bloch (1950) to a lipid material obtained from 'cord-forming' strains of tubercle bacilli. Purification of components of this crude material led to the isolation of a glycolipid (Noll 1956), toxic for mice, shown to be 6,6' dimycoloyl α,α'-D-trehalose (XXVI). Subsequent studies have failed to show that this material has a role in the cord-forming ability of virulent strains but the name cord-factor has continued to be applied to all dimycolates of trehalose. Cord-factors have been isolated from every mycobacterial species specifically examined and indeed it appears likely that they will be found in all mycolic acid-containing bacteria [for reviews see Goren (1972, 1975), Lederer (1976), Asselineau & Asselineau (1978*b*)] .

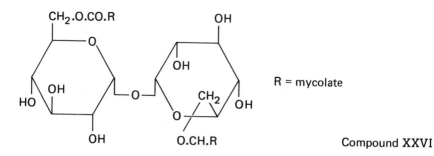

R = mycolate

Compound XXVI

As it is expected that cord factors are present in all mycolic acid-containing organisms the value in classification and identification lies mainly in their structural details. The hydrophilic sugar portion of cord-factors (XXVI) is invariable so that structural differences will depend on the mycolic acid composition of these glycolipids. It is only in recent years that cord-factors have been analysed in any detail and the results of these studies are summarized in Table 14.

The separation of individual cord-factors according to their mycolic acid content was greatly facilitated by the use of t.l.c. of their trimethylsilyl ether derivatives, a procedure introduced by Promé et al. (1976). In addition to the three different trehalose dimycolates characterized from *Myco. phlei* by Promé *et al.* (1976) two trehalose monomycolates containing the same mycolic acids were also isolated. As well as the wax-ester mycolates (IX) and mycolates without oxygen-functions in addition to the 3-hydroxy acid system (α-mycolates) (IIa and/or IV), *Myco. phlei* is expected to contain keto-mycolates (VIII) (Table 3; Minnikin *et al.* 1979b) so that it is very significant that only two particular mycolic acids are found in the cord-factor lipid. A more extensive, but less detailed, study by Strain *et al.* (1977) showed that for *Myco. bovis* AN-5 and *Myco. tuberculosis* all combinations of the three mycolic acid types were found in the cord-factors but for *Myco. bovis* BCG methoxymycolates (VI) were absent (Table 14). The cord-factors of *Myco. avium* and *Myco. phlei* gave similar thin-layer patterns to that determined for *Myco. phlei* by Promé *et al.* (1976) but the mycolic acids were not characterized.

In an earlier study (Adam *et al.* 1967), involving mass spectrometry of peracetylated cord-factors and the constituent mycolic acid esters, a methoxymycolate (VI) was partially characterized from the cord-factor of *Myco. tuberculosis* but a BCG strain had dicyclopropyl (III) and ketomycolates (V) but apparently no methoxymycolates. The cord-factor of *Myco. butyricum*, a synonym of *Myco. smegmatis*, gave a diunsaturated mycolate lacking a cyclopropane ring (Adam *et al.* 1967) in contrast to the structure of the acid (IIc) from the cord-factor (Table 14) of *Myco. smegmatis* (Mompon *et al.* 1978). Detailed analyses of mycobacterial cord-factor lipids appear, therefore, to have considerable potential in studies of mycobacterial classification.

TABLE 14
Structural features of trehalose dimycolates (cord-factors)

Taxon	Cord-factors*	Mycolic acid composition†	
Mycobacterium *phlei*	CS_A CS_B CS_C	di-α‡-(IIa,IV) α(IIa,IV), wax ester (IX) di-wax ester (IX)	Promé *et al.* (1976)
bovis AN-5 *tuberculosis* (Aoyama B and Peurois)	I II III IV V VI	di-α(III) α(III), methoxy (VI) α(III), keto (V) di-methoxy (VI) methoxy (VI), keto (V) di-keto (V)	Strain *et al.* (1977)
bovis BCG	I III VI	di-α (III) α (III), keto (V) di-keto (V)	
avium *phlei*	I III VI	di-α (IIa,IV) α (IIa,IV), γ § (VIII or IX) di-γ (VIII or IX)	
smegmatis	—	di-α (II‖)	Mompon *et al.* (1978)

*Components of total cord factors separated by thin-layer chromatography of trimethylsilyl ether derivatives. † Numbers in brackets indicate that structures are similar to, but not necessarily identical to, those given previously in the section on mycolic acids. ‡ Mycolic acids with no additional oxygen functions in addition to 3-hydroxy acid system. § Not determined whether these 'γ'-mycolic acids were of keto (VIII) or wax-ester (IX) type. ‖ Structural assignment based on spectroscopic analysis of the total mycolates released on hydrolysis.

A limited number of dimycolyl trehaloses have been characterized from mycolic acid-containing taxa other than mycobacteria (Lederer 1976; Minnikin & Goodfellow 1976; Minnikin *et al.* 1978c; Asselineau & Asselineau 1978b). The cord-factors from *Cnbc. diphtheriae* (Ioneda *et al.* 1963; Senn *et al.* 1967) and *Brev. vitarumen* (Lanéelle & Asselineau 1977), probably a species of *Corynebacterium sensu stricto* (Minnikin *et al.* 1978c), contained mycolic acids typical of corynebacteria (Table 2). A *'rhodochrous'* strain also contained a cord-factor having mycolic acids (C_{38}–C_{46}) similar to those in whole organisms but the mycolic acids from the cord-factor of *Nrda. asteroides* had lower molecular weights (C_{32}–C_{36}) than those from the walls (C_{50}–C_{58}) (Ioneda *et al.* 1970). Trehalose dimycolates and monomycolates were isolated from several strains of nocardiae (Yano *et al.* 1971) but the structures of the mycolic acid residues were not determined. A diester of trehalose, having mycolic acids (C_{28}–C_{38}) in undetermined positions, was isolated from the emulsion layer of *'Arthrobacter' paraffineus* grown on hydrocarbons and a range of related organisms produced similar lipids (Suzuki *et al.* 1969). Further systematic studies will be necessary to assess the value of cord-factor structures in the classification of nocardioform actinomycetes and corynebacteria.

Petroleum ether extracts of *Myco. phlei* and *Myco. smegmatis* were found (Asselineau *et al.* 1969*b*, 1972) to contain a complex mixture of esters of trehalose with polyunsaturated phleic acids (XV). Purification of all the individual components of the mixture was not achieved but the least polar and most abundant compound from *Myco. phlei* was shown to be an octaphleate of trehalose. The trehalose phleates were not formed when the bacteria were grown in a dispersed state and their occurrence was related to the degree of clumping.

Sulphur-containing lipids were discovered in extracts of virulent tubercle bacilli (Middlebrook *et al.* 1959) and were later characterized as acylated trehalose sulphates (Goren 1970*a,b*; Goren *et al.* 1971, 1976). The structure (XXVII) of the principal component SL-I of *Myco. tuberculosis* H37 Rv has been established and the gross structure of four closely related trehalose sulphates determined (Goren *et al.* 1976). The original correlation of virulence of tubercle bacilli with sulpholipid content (Middlebrook *et al.* 1959) was supported by a further extensive study (Gangadharam *et al.* 1963). In a more recent study it was found that other strongly acidic lipids, such as phospholipids, may also be implicated in virulence (Goren *et al.* 1974*a*) and it was also demonstrated that these sulphatides enhanced the toxicity of cord-factors (Kato & Goren 1974). Uncharacterized sulpholipids were synthesized by *Myco. avium* undergoing cell division (McCarthy 1976).

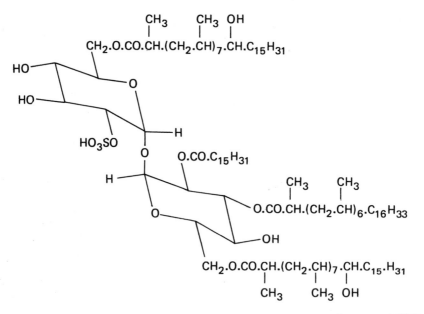

Compound **XXVII**

G. Phthiocerols and related compounds

The phthiocerols are a family of long-chain β-diols characterized from *Myco. tuberculosis* and *Myco. bovis* (Asselineau 1966; Asselineau & Asselineau 1978*a*). Phthiocerol A (XXVIII) and phthiocerol B (XXIV) are methoxylated diols (Minnikin & Polgar 1966*b*) and phthiodiolone A (XXX) (Minnikin & Polgar 1967*c*) and phthiotriol A (XXXI) (Gieusse & Lanéelle 1968; Welby-Gieusse & Tocanne 1970) are closely related. The phthiocerols, including phthiodiolone and phthiotriol, occur naturally as diesters with mycocerosic (mycoceranic) acids (XI). These dimycocerosates of phthiocerol, commonly abbreviated as DIM and characterized by Noll (1957), were also detected by infra-red spectroscopy of chromatographically-fractionated lipid extracts of *Myco. tuberculosis* and *Myco. bovis* (Randall & Smith 1964) but a report of the presence of DIM in *Myco. marinum* (Navalkar *et al.* 1965) requires confirmation. Phthiocerol dimycocerosate was found in 40 clinical isolates of *Myco. tuberculosis* but a mutant of *Myco. tuberculosis* H37 Rv lacking DIM was markedly attenuated (Goren *et al.* 1974*b*) thus suggesting a possible role in virulence. A rapid method for the detection of DIM in *Myco. bovis* BCG involving t.l.c. of petroleum ether extracts has been developed by Koul & Gastambide-Odier (1977).

$$CH_3.(CH_2)_{20,22}.\overset{\overset{\displaystyle OH}{|}}{CH}.CH_2.\overset{\overset{\displaystyle OH}{|}}{CH}.(CH_2)_4.\overset{\overset{\displaystyle H_3C}{|}}{CH}.\overset{\overset{\displaystyle OCH_3}{|}}{CH}.CH_2CH_3$$

Compound **XXVIII**

$$CH_3.(CH_2)_{20,22}.\overset{\overset{\displaystyle OH}{|}}{CH}.CH_2.\overset{\overset{\displaystyle OH}{|}}{CH}.(CH_2)_4.\overset{\overset{\displaystyle H_3C}{|}}{CH}.\overset{\overset{\displaystyle OCH_3}{|}}{CH}.CH_3$$

Compound **XXIX**

$$CH_3.(CH_2)_{20,22}.\overset{\overset{\displaystyle OH}{|}}{CH}.CH_2.\overset{\overset{\displaystyle OH}{|}}{CH}.(CH_2)_4.\overset{\overset{\displaystyle H_3C}{|}}{CH}.\overset{\overset{\displaystyle O}{||}}{C}.CH_2.CH_3$$

Compound **XXX**

$$CH_3.(CH_2)_{20,22}.\overset{\overset{\displaystyle OH}{|}}{CH}.CH_2.\overset{\overset{\displaystyle OH}{|}}{CH}.(CH_2)_4.\overset{\overset{\displaystyle H_3C}{|}}{CH}.\overset{\overset{\displaystyle OH}{|}}{CH}.CH_2CH_3$$

Compound **XXXI**

Certain characteristic mycosides (Table 11) have structures based on a 'phenol-phthiocerol' (XXI) and this compound also forms the core of a so-called attenuation indicator (AI) lipid found in attenuated strains of tubercle bacilli (Goren *et al.* 1974*b*). This lipid is a phenol-phthiocerol (XXI) methylated on the phenol function and acylated on the β-diol with palmitic, stearic and mycocerosic acids and it was not detected in any of the more virulent strains.

H. Other lipids

Two other categories of compounds, mycobactins and carotenoid pigments, are also lipid soluble but will not be considered in detail here. Studies on the mycobactins and related iron-binding compounds have been reviewed extensively (Snow 1970; Ratledge 1976; Ratledge & Patel 1976). It will suffice to note here that detailed structural differences have been found between the mycobactins from various species of mycobacteria and that three characteristic types of 'nocobactins' are found in nocardiae. 'Gordona' rubra produced a nocobactin-like compound but eight other rhodococci (Goodfellow & Alderson 1977) gave no lipid soluble iron-binding compounds (Ratledge & Patel 1976). The future value of the analysis of mycobactins and related compounds would appear to lie in differentiating between closely related species and also excluding taxa which are unable to synthesize such molecules, though cultivation and analytical methods require further development (Ratledge & Patel 1976).

Similar limitations are also applicable to analyses of carotenoids, the variable extent of the pigmentation and instability and complexity of the pigments posing analytical problems. The pigmentation of mycobacteria and related organisms is important in taxonomic studies (Runyon 1965) but, as pointed out by Tárnok & Tárnok (1971), it is difficult to obtain a general view of the distribution of carotenoid types as effective systematic studies have not been performed. The latter authors have demonstrated by use of t.l.c. and spectrophotometry that β-carotene is the most common pigment of chromogenic mycobacteria but it was absent from certain strains. Species-specific patterns were obtained, however, for the chromogenic mycobacteria investigated (Tárnok & Tárnok 1970, 1971) with the exceptions that the pairs *Myco. kansasii/Myco. marinum* and *Myco. aquae/Myco. scrofulaceum* could not be separated. The patterns of pigments from nocardiae (Röhrscheidt & Tárnok 1972), including organisms now classified in the genus *Rhodococcus* (Goodfellow & Alderson 1977), had taxonomic potential but were different to those of mycobacteria particularly in the absence of β-carotene and related carotenoids. Further information on the distribution of carotenoids in mycobacteria and related organisms may be found by reference either to general reviews on microbial carotenoids (Goodwin 1972; Liaaen-Jensen & Andrewes 1972) or to reviews on mycobacteria (Ratledge 1976; Barksdale & Kim 1977).

3. Impact on Classification and Identification

Traditional taxonomic methods based on form and function largely failed to produce a workable classification of acid-fast bacteria and closely related strains nor did they provide good characters for identification. Chemical analyses of wall amino acids, sugars and peptidoglycan structure have only served to emphasize the close relationship that exists between *Mycobacterium, Nocardia, Rhodococcus* and *Corynebacterium sensu stricto,* though the presence of N-glycolylmuramic acid in the peptidoglycan of mycobacteria and nocardiae compared with the more usual N-acetylmuramic acid is a potentially useful diagnostic character (Lederer *et al.* 1975; Uchida & Aida 1977). Analyses of deoxyribonucleic acid base composition in most instances do not provide clear separations of acid-fast bacteria (Table 1) while DNA reassociation assays are effective mainly in the definition of species (Mordarski *et al.* 1978*a,b*). In contrast, lipid analyses, as outlined in previous sections, have great potential in clarifying the classification, and providing methods for the identification, of acid-fast and related bacteria.

Sufficient systematic studies have now been performed to make it possible to propose a provisional co-ordinated scheme for the use of lipid analysis in the classification and identification of bacteria having major amounts of *meso*-DAP, arabinose and galactose in their walls (Table 15); a more detailed plan for mycobacteria is shown in Table 16. Thus, bacteria with a wall chemotype IV (Lechevalier & Lechevalier 1970*b*) can be classified into several groups according to their lipid composition as shown in Table 15. In some instances these groups correspond to established genera such as *Mycobacterium* and *Nocardia* but further comparative studies are required to determine the status of other groups, for example, '*Nocardia*' *autotrophica* and '*Nocardia*' *mediterranea.*

The presence or absence of mycolic acids separates wall chemotype IV bacteria into two broad groups (Table 15) and the fatty acids of those lacking mycolic acids contain higher proportions of branched chain iso and anteiso acids. The fatty acids of the mycolic acid-containing strains are usually simple mixtures of straight-chain unsaturated and 10-methyloctadecanoic (tuberculostearic) acids though *Cnbc. diphtheriae* and its close relatives lack this latter component.

The complexity of mycobacterial mycolic acids, their characteristic insolubility in alcoholic solvent mixtures and the greater size of the long-chain esters released on pyrolysis clearly distinguish members of the genus *Mycobacterium* from related taxa (Tables 3 and 15). Variations in the size and structure of non-mycobacterial mycolic acids are also useful in classification and, in combination with menaquinone composition, groups approximating to genera are obtained. The phospholipid composition of mycolic acid containing taxa is relatively homogeneous but the detailed analysis of polar glycolipids and

TABLE 15

Diagnostic lipids in the classification and identification of acid-fast and related bacteria

Long-chain compounds*		Mycolic acid solubility†	Ester released on pyrolysis GC of mycolate‡	Major menaquinone§
Iso and anteiso acids	Mycolates absent	−	−	ND ND ND ND ND MK-8(H$_4$) ND MK-9(H$_4$)
Straight-chain unsaturated and tuberculostearic acids	Complex mycolates	Insoluble	22:0–26:0	MK-9(H$_2$)
	Single mycolate	ND	20:0, 20:1, 22:0, 22:1	MK-9
			16:1, 18:1	ND
				MK-8(H$_4$)
		Soluble	12:0, 18:0	MK-8(H$_4$) MK-8(H$_2$)
				MK-9(H$_2$)
			8:0–10:0	MK-8,9(H$_2$) MK-9(H$_2$)
Straight-chain and unsaturated acids only	Single mycolate	Soluble	14:0–18:0 (14:1–18:1)	MK-8,9(H$_2$)

phosphatidylinositol mannosides is worthy of further study (Minnikin *et al.* 1977*b*).

The *'aurantiaca'* taxon is especially well defined by its lipid composition as it contains fully unsaturated MK-9 menaquinones and characteristic very unsaturated mycolic acids. *Nocardia amarae* can be distinguished from other members of *Nocardia sensu stricto* by having mycolic acids with unsaturated branches in 2-position and is in need of further study. The same applies to *Micr. brevicatena* which is chemically very similar to *Nocardia sensu stricto* but quite distinct from *Micr. faeni.* The genera *Corynebacterium sensu stricto* and *Rhodococcus* accommodate a heterogeneous range of bacteria united by their content of dihydrogenated menaquinones and relatively simple mycolic acids ranging in size from 22 to 64 carbon atoms. Using lipid analyses alone it is difficult to separate representatives of *Corynebacterium sensu stricto* and *Rhodococcus.* However, true

Table 15 Part 2

Genus	Mycolate size (no. of carbons)‖	Phospholipids¶	Species
Actinopolyspora	–	PG, PC	*halophila*
Saccharomonospora	–	ND	*viridis*
'Nocardia'	–	PE, PI	*orientalis*
'Nocardia'	–	PG, PE, PIM	*mediterranea*
'Nocardia'	–	PG, PE, PI, PIM	*rugosa*
'Nocardia'	–	PE, PME, PC, PI, PIM	*autotrophica*
Pseudonocardia	–	PE, PME, PC, PI, PIM	*thermophila*
'Micropolyspora'	–	PG, PME, PC, PI	*faeni*
Mycobacterium	60–90	PE, PI, PIM	spp.
'Gordona'	68–74	PE, PI, PIM	*aurantiaca*
Nocardia	ND	PE, PI, PIM	*amarae*
Micropolyspora	46–56	PE, PI, PIM	*brevicatena*
Nocardia	46–60	PE, PI, PIM	*asteroides, brasiliensis* and *otitidis-caviarum*
Rhodococcus	34–52	PE, PI, PIM	*coprophilus, equi, erythropolis, rhodnii, rhodochrous* and *ruber*
Rhodoccocus	48–64	PE, PI, PIM	*bronchialis, corallinus* and *terrae*
Rhodococcus	38–64	ND	*rubropertinctus*
Corynebacterium	22–32	PG, PI, PIM	*bovis*
Corynebacterium	26–38	PG, PI, PIM	spp.

ND, not determined. *For data see Tables 2–6. †In ethanol-diethyl ether (Kanetsuna & Bartoli 1972; Hecht & Causey 1976).‡Esters detected by pyrolysis gas chromatography (GC). See Table 2; Etémadi 1967d; Lechevalier et al. 1971b; 1973 for data. Abbreviations exemplified by 18:0, octadecanoate; 18:1 octadecenoate. §For data see Table 7. Abbreviations exemplified by MK-9, menaquinone having 9 isoprene units; MK-9(H₄), menaquinone having two of the nine isoprene units hydrogenated. ‖For data see Table 2 and text for information on mycobacteria. ¶Diagnostic phospholipids: phosphatidylcholine (PC), phosphatidyl-ethanolamine (PE), phosphatidylmonomethylethanolamine (PME), phosphatidylinositol (PI), and phosphatidylinositol mannosides (PIM) occurring in addition to diphosphatidylglycerol.

corynebacteria usually have high proportions of phosphatidylglycerol but lack phosphatidylethanolamine whereas for rhodococci the reverse is the case. Again, corynebacteria usually lack tuberculostearic acid but in a single study (Lechevalier et al. 1977) this acid was found in *Cnbc. bovis,* strains of which also contain characteristic low molecular weight mycolic acids. Further comparative studies are required to determine the exact relationships between true coryne-

TABLE 16

Diagnostic lipids in the classification and identification of mycobacteria

Mycolic acid patterns*		Species	Characteristic glycolipids		Other lipids§
Whole cells	Cord factor		Mycoside†	TLC pattern‡	
α, keto	ND	*Mycobacterium* *leprae*	ND	ND	
α, methoxy, keto	α, methoxy, keto (V)	*tuberculosis*‖ { V	–	–	TS ⎫
	α, keto (A)	{ A	–	–	AI ⎬ ML ⎫
	α, keto	*bovis*	Ph(B)	–	⎬ DIM
	ND	*bovis* BCG	Ph(A)	–	ML ⎭
ND	ND	*kansasii*	Ph(A)	+	
ND		*marinum*	Ph(G)	+	

Mycolates		Species	Ph (XXI)		
α, keto, wax ester	α, wax ester	*phlei*	−	+	Trehalose phleate
	ND	*paratuberculosis*	ND	ND	Peptidolipid (XXIV)
	α, keto or wax ester	*avium*	C_2	+	
	ND	*intracellulare*	ND		
ND		*scrofulaceum*	C_S	+	
		xenopei	ND	ND	
	ND	*lepraemurium*	+	+	
		peregrinum	∼ND	+	Peptidolipid (XXV)
α, 2 polar mycolates	ND	*fortuitum*	−	+	Trehalose phleate
	α	*smegmatis*	C_{p1}	ND	
		farcinogenes	C_{378}	ND	
		sp. 1217	C_{1217}	+	
		abscessus	ND	+	
ND	ND	*aquae*	ND	+	
		borstelense	ND	+	
		gordonae	ND	+	
		szulgai	ND	+	

*For data and abbreviations (Tables 3 and 13). †For structures of mycosides based on phenol-phthiocerol (Ph) (XXI) see Table 11, for mycosides 'C' see Table 12 and for the mycoside of *Myco. lepraemurium* see Draper & Rees 1973 and Draper 1974. ‡Data from Table 13 and Sehrt *et al.* 1975, 1976. §Abbreviations: TS, acylated trehalose sulphate (XXVII); AI, attenuation indicator lipid based on phenol-phthiocerol (XXI); ML, mycolipenic (phthienoic) acid (XII); DIM, dimycoserate (XI) of phthiocerol (XXVIII). ‖Virulent (V) and attenuated (A) strains of *Myco. tuberculosis.*

bacteria, rhodococci and representatives of the recently described *Caseobacter* (Crombach 1978).

Actinomycetes with a wall chemotype IV but lacking mycolic acids are also in need of intensive study. *Nocardia autotrophica* has the same menaquinones as *Nocardia sensu stricto* but very different phospholipids and fatty acids. *Pseudonocardia thermophila* has been reported to have lipids similar to those of *Nrda. autotrophica* (Lechevalier et al. 1977). The fragmentary data for the other mycolic acid free bacteria mentioned in Table 15 needs to be consolidated before their affinities and taxonomic status can be considered profitably.

The potential for the use of lipid analysis in the classification and identification of mycobacteria is enormous, (Table 16) but despite half a century of study the overall lipid patterns of all species are not clearly defined. The fatty acid, phospholipid and menaquinone composition of mycobacteria appear to be superficially uniform but detailed differences may be valuable. The analysis of overall mycolic acid patterns (Table 3 and Table 16) appear to have the potential to separate mycobacteria into broad groups (Minnikin et al. 1980) but extensive systematic studies are lacking. Exact determinations of mycolic acid composition may then lead to further refinements.

The detailed characterization of trehalose mycolates (Table 14) may also provide valuable data for the classification of mycobacteria. The most powerful procedure currently under development appears to be the analysis of species and serotype specific mycosides and other characteristic glycolipids (Tables 11, 12, 13 and 16). The empirical chromatographic patterns used routinely by Jenkins and co-workers (Table 13) for identification of atypical mycobacteria may be effectively related to mycoside structure (Table 12) by the new t.l.c. procedure of Brennan et al. (1978) to yield a powerful technique. Analyses of other characteristic lipids such as acylated trehalose sulphates and phenol-phthiocerol derivatives (Goren et al. 1974a,b) are even sensitive enough to distinguish virulent strains of *Myco. tuberculosis* from attenuated varieties.

The application of lipid analysis proved particularly effective in the identification of bacteria responsible for cases of bovine farcy in zebu cattle (Chamoiseau 1969). This actinomycete was found to produce mycolic acids whose esters on pyrolysis gas chromatography gave a single peak corresponding to a C_{24} ester (Chamoiseau & Asselineau 1970) and as a result was considered to belong to the genus *Mycobacterium*. This discovery was of particular interest as the causal agent of bovine farcy was considered previously to be *Nrda. farcinica*, the type of species of the genus *Nocardia* (Nocard 1888; Trevisan 1889). Lipid data have also helped to highlight that the name *Nrda. farcinica* covers at least two groups of actinomycetes. Strains, such as *Nrda. farcinica* NCTC 4524, have properties in common with mycobacteria and should be reclassified as *Myco. farcinogenes* var. *senegalense* (Chamoiseau 1973; Lanéelle et al. 1971; Magnusson 1976). Other strains, including *Nrda. farcinica* ATCC 3318, fall within the range of variation covered by the *Nrda. asteroides* complex (Tsukamura 1969; Tsukamura et al.

1979; Schaal & Reutersberg 1978) but an acceptable name for this taxon awaits the resolution of the nomenclatural confusion surrounding the epithet *Nrda. farcinica.*

Thanks are due to M. D. Collins, A.B. Caldicott and I. G. Hutchinson who have recently worked in close collaboration with the authors on the present subject; M.D.C. was the recipient of a Luccock Scholarship (Medical Scholarships and Research Committee, Faculty of Medicine, University of Newcastle upon Tyne) and A.B.C. and I.G.H. were supported by Medical Research Council grant G974/522/S.

4. References

ADAM, A., SENN, M., VILKAS, E. & LEDERER, E. 1967 Spectrométrie de masse de glycolipides. 2. Diesters de tréhalose naturels et synthétiques. *European Journal of Biochemistry* 2, 460–468.

AKAMATSU, Y. & NOJIMA, S. 1965 Separation and analyses of the individual phospholipids of mycobacteria. *Journal of Biochemistry* 57, 430–439.

AKAMATSU, Y., ONO, Y. & NOJIMA, S. 1966 Phospholipid patterns in subcellular fractions of *Mycobacterium phlei*. *Journal of Biochemistry* 59, 176–182.

AKAMATSU, Y., ONO, Y. & NOJIMA, S. 1967 Studies on the metabolism of phospholipids in *Mycobacterium phlei*. I. Difference in turnover rates of individual phospholipids. *Journal of Biochemistry* 61, 96–102.

ALBERGHINA, M. 1976 Relationship between lipid composition and antibiotic-resistance to isoniazid, streptomycin and p-amino-salicyclic acid, ethambutol and rifampicin in mycobacteria. *Italian Journal of Biochemistry* 25, 127–151.

ALDERSON, G. & GOODFELLOW, M. 1979 Classification and identification of actinomycetes causing mycetoma. *Postepy Higieny i Medycyny Doswiadczalnej* 33, 109–124.

ALSHAMAONY, L., GOODFELLOW, M. & MINNIKIN, D. E. 1976a Free mycolic acids in the classification of *Nocardia* and the '*rhodochrous*' complex. *Journal of General Microbiology* 92, 188–199.

ALSHAMAONY, L., GOODFELLOW, M., MINNIKIN, D. E. & MORDARSKA, H. 1976b Free mycolic acids as criteria in the classification of *Gordona* and the '*rhodochrous*' complex. *Journal of General Microbiology* 92, 183–187.

ALSHAMAONY, L., GOODFELLOW, M., MINNIKIN, D. E., BOWDEN, G. H. & HARDIE, J. M. 1977 Fatty and mycolic acid composition of *Bacterionema matruchotii* and related organisms. *Journal of General Microbiology* 98, 205–213.

ASSELINEAU, C. & ASSELINEAU, J. 1978a Lipides spécifiques des mycobactéries. *Annales de Microbiologie (Institut Pasteur)* 129A, 49–69.

ASSELINEAU, C. & ASSELINEAU, J. 1978b Trehalose-containing glycolipids. *Progress in the Chemistry of Fats and Other Lipids* 16, 59–99.

ASSELINEAU, C. & MONTROZIER, H. L. 1976 Étude du processus de biosynthèse des acides phléiques, acides polyinsaturés synthétisés par *Mycobacterium phlei*. *European Journal of Biochemistry* 63, 509–518.

ASSELINEAU, C., MONTROZIER, H. & PROMÉ, J. C. 1969a Structure des acides α-mycoliques isolés de la souche Canetti de *Mycobacterium tuberculosis*. *Bulletin de la Société Chimique de France* 592–596.

ASSELINEAU, C., MONTROZIER, H. & PROMÉ, J. C. 1969b Présence d'acides polyinsatures dans une bactérie: isolement, à partir des lipides de *Mycobacterium phlei*, d'acide hexatriacontapentaène–4,8,12,16,20 oique et d'acides analogues. *European Journal of Biochemistry* 10, 580–584.

ASSELINEAU, C., MONTROZIER, H. L., PROMÉ, J. C., SAVAGNAC, A. M. & WELBY, M. 1972 Étude d'un glycolipide polyinsaturé synthétisé par *Mycobacterium phlei*. *European Journal of Biochemistry* 28, 102–109.

ASSELINEAU, J. 1961 Sur quelque applications de la chromatographie en phase gazeuse a l'étude d'acides gras bacteriens. *Annales de l'Institut Pasteur, Paris* 100, 109–119.

ASSELINEAU, J. 1964 Glycolipides et peptido-glycolipides des mycobactéries. *Zentralblatt für Bakteriologie, Parasitenkunde, Infektionskrankheiten und Hygiene. I. Abteilung* 194, 157–176.

ASSELINEAU, J. 1966 *The Bacterial Lipids.* Paris: Hermann.

ASSELINEAU, J. & PORTELANCE, V. 1974 Comparative study of the free lipids of eight BCG daughter strains. *Recent Results in Cancer Research* 47, 214–220.

ASSELINEAU, J., LANÉELLE, M. A. & CHAMOISEAU, G. 1969 De l'etiologie du farcin de zébus tchadiens nocardiose ou mycobactériose. II. Composition lipidique. *Revue d'élevage et de Mèdecine Vétérinaire des Pays Tropicaux* 22, 205–209.

BALLIO, A. & BARCELLONA, S. 1968 Relations chimiques et immunologiques chez les actinomycétales. 1. Les acides gras de 43 souches d'actinomycètes aérobies. *Annales de l'Institut Pasteur, Paris* 114, 121–137.

BANERJEE, B., JAIN, S. K. & SUBRAHMANYAN, D. 1974 Separation of phosphatidyl inositomannosides of mycobacteria. *Journal of Chromatography* 94, 342–344.

BARBER, M., JOLLES, P., VILKAS, E. & LEDERER, E. 1965 Determination of amino acid sequences in oligopeptides by mass spectrometry. I. The structure of fortuitine, an acyl nonapeptide methyl ester. *Biochemical and Biophysical Research Communications* 18, 469–473.

BARKSDALE, L. 1970 *Corynebacterium diphtheriae* and its relatives. *Bacteriological Reviews* 34, 378–422.

BARKSDALE, L. & KIM, K-S. 1977 *Mycobacterium. Bacteriological Reviews* 41, 217–372.

BATRAKOV, S. G. & BERGELSON, L. D. 1978 Lipids of the streptomycetes. Structural investigation and biological interrelation. *Chemistry and Physics of Lipids* 21, 1–29.

BATRAKOV, S. G., ROSYNOV, B. V., KORONELLI, T. V., KOZHUHOVA, R. A. & BERGELSON, L. D. 1977 The lipids of mycobacteria. I. Unusual trehalose derivative from *Mycobacterium paraffinicum. Bioorganicheskaya Khimiya* 3, 55–67.

BATT, R. D., HODGES, R. & ROBERTSON, J. G. 1971 Gas chromatography and mass spectrometry of the trimethylsilyl ether methyl ester derivatives of long chain hydroxy acids from *Nocardia corallina. Biochimica et Biophysica Acta* 239, 368–373.

BEAU, S., AZERAD, R. & LEDERER, E. 1966 Isolement et caracterisation des dihydromenaquinones des myco- et corynebactéries. *Bulletin de la Société de Chimie Biologique* 48, 569–581.

BENNET, P. & ASSELINEAU, J. 1963 Nature des acides gras des graisses de la souche bovine B.C.G. de *Mycobacterium tuberculosis. Bulletin de la Société de la Chimie Biologique* 45, 1379–1393.

BENNET, P. & ASSELINEAU, J. 1970 Influence de l'age sur la teneur en acides gras a chaine ramifiée du bacille tuberculeux. *Annales de l'Institut Pasteur, Paris* 118, 324–329.

BERD, D. 1973 Laboratory identification of clinically important actinomycetes. *Applied Microbiology* 25, 665–681.

BIRN, K. J., SCHAEFER, W. B., JENKINS, P. A., SZULGA, T. & MARKS, J. 1967 Classification of *Mycobacterium avium* and related opportunist mycobacteria met in England and Wales. *Journal of Hygiene, Cambridge* 65, 575–589.

BLASCHY, H. & ZIMMERMANN, W. 1971 Gas-chromatographical investigations on fatty acids of various kinds of bacteria. *Zentralblatt für Bakteriologie, Parasitenkunde, Infektionskrankheiten und Hygiene. I. Abteilung Originale* 218, 468–477.

BLOCH, H. 1950 Studies on the virulence of tubercle bacilli. Isolation and biological properties of a constituent of virulent organisms. *Journal of Experimental Medicine* 91, 197–217.

BORDET, C. & MICHEL, G. 1963 Étude des acides gras isolés de plusieurs espèces de *Nocardia. Biochimica et Biophysica Acta* 70, 613–626.

BORDET, C. & MICHEL, G. 1964 Isolement d'un nouvel alcool, le 16-hentriaconatanol a partir des lipides de *Nocardia brasiliensis. Bulletin de la Société de Chimie Biologique* 46, 1101–1112.

BORDET, C. & MICHEL, G. 1969 Structure et biogenèse des lipides à haut poids moléculaire de *Nocardia asteroides. Bulletin de la Société de Chimie Biologique* 51, 527–548.

BORDET, C., KARAHJOLI, M. GATEAU, O. & MICHEL, G. 1972 Cell walls of nocardiae and related actinomycetes: identification of the genus *Nocardia* by cell wall analyses. *International Journal of Systematic Bacteriology* 22, 251-259.

BOUGHTON, E. & BERNSTAD, S. 1974 Screening by thin-layer chromatography of '*avium*-like' mycobacteria isolated from animals in Sweden. *Zentralblatt fuer Veterinaermedizin* B 21, 171-175.

BOUSFIELD, I. J. & GOODFELLOW, M. 1976 The '*rhodochrous*' complex and its relationships with allied taxa. In *The Biology of the Nocardiae* ed. Goodfellow, M., Brownell, G. H. & Serrano, J. A. pp. 39-65. London: Academic Press.

BOWIE, I. S., GRIGOR, M. R., DUNCKLEY, G. C., LOUTIT, M. W. & LOUTIT, J. S. 1972 The DNA base composition and fatty acid constitution of some Gram-positive pleomorphic soil bacteria. *Soil Biology and Biochemistry* 4, 397-412.

BRADLEY, S. G. & MORDARSKI, M. 1976. Association of polydeoxyribonucleotides of deoxyribonucleic acids from nocardioform bacteria. In *The Biology of the Nocardiae* ed. Goodfellow, M., Brownell, G. H. & Serrano, J. A. pp. 310-336. London: Academic Press.

BRADLEY, S. G., BROWNELL, G. H. & CLARK, J. 1973 Genetic homologies among nocardiae and other actinomycetes. *Canadian Journal of Microbiology* 19, 1007-1014.

BRENNAN, P. & BALLOU, C. E. 1967 Biosynthesis of mannophosphoinositides by *Mycobacterium phlei. Journal of Biological Chemistry* 242, 3046-3056.

BRENNAN, P. J. & LEHANE, D. P. 1971 The phospholipids of corynebacteria. *Lipids* 6, 401-409.

BRENNAN, P. J., ROONEY, S. A. & WINDER, F. G. 1970 The lipids of *Mycobacterium tuberculosis* BCG: Fractionation, composition, turnover and the effects of isoniazid. *Irish Journal of Medical Science* 3, 371-390.

BRENNAN, P. J., SOUHRADA, M., ULLOM, B., McCLATCHY, J. K. & GOREN, M. B. 1978 Identification of atypical mycobacteria by thin-layer chromatography of their surface antigens. *Journal of Clinical Microbiology* 8, 374-379.

BRUNETEAU, M. & MICHEL, G. 1968 Structure d'un dimycolate d'arabinose isole de *Mycobacterium marianum. Chemistry and Physics of Lipids* 2, 229-239.

BUCHANAN, R. E. & GIBBONS, N. E. (Eds.) 1974 *Bergey's Manual of Determinative Bacteriology*, 8th edn. Baltimore: Williams & Wilkins.

CAMPBELL, I. M. & BENTLEY, R. 1968 Inhomogeneity of vitamin K_2 in *Mycobacterium phlei. Biochemistry* 7, 3323-3327.

CAMPBELL, I. M. & NAWORAL, J. 1969 Composition of the saturated and monounsaturated fatty acids of *Mycobacterium phlei. Journal of Lipid Research* 10, 593-598.

CATTANEO, C., LUCCHESI, M., de RITIS, G. C., PIETROPAOLO, C., ROSSI, P. & FRISANI, G. 1964 Separazione gas cromatografica degli acidi grassi (C12–C20) presenti nei lipidi micobatterici. *Annali-Instituto "Carlo Forlanini"* 24, 353-361.

CATTANEO, C., DE RITIS, G. C., LUCCHESI, M., ROSSI, P. & FERRARI, S. 1965 Analisi gas cromatografica degli acidi grassi (C11–C20) presenti nei micobatteri. *Annali-Instituto "Carlo Forlanini"* 25, 349-388.

CHAMOISEAU, G. 1969 De l'etiologie du farcin de zébus tchadiens: nocardiose ou mycobacteriose? I. Étude bactériologique et biochimique. *Revue d'Élevage et de Médecine Vétérinaire des Pays Tropicaux* 22, 195-204.

CHAMOISEAU, G. 1973 *Mycobacterium farcinogenes* agent causal du farcin du boeuf en Afrique. *Annales de Microbiologie (Institut Pasteur)* 124A, 215-222.

CHAMOISEAU, G. & ASSELINEAU, J. 1970 Examen des lipides d'une souche de *Nocardia farcinica*: présence d'acides mycoloques. *Compte Rendu Hebdomadaire des Séances de l'Académie des Sciences* 270D, 2603-2604.

CHANDRAMOULI, V. & VENKITASUBRAMANIAN, T. A. 1973 Effect of age on the fatty acids C_{14} to C_{19} of mycobacteria. *American Review of Respiratory Diseases* 108, 387-390.

CHANDRAMOULI, V. & VENKITASUBRAMANIAN, T. A. 1974 Effect of age on the lipids of mycobacteria. *Indian Journal of Chest Diseases, Supplement* 16, 199-207.

CHAPUT, M., MICHEL, G. & LEDERER, E. 1962 Structure du mycoside C_m, peptidoglycolipide de *Mycobacterium marianum. Biochimica et Biophysica Acta* 63, 310-326.

COLES, L. & POLGAR, N. 1968 The mycolipanolic acids. *Journal of the Chemical Society C,* 1541-1544.

COLLINS, M. D., PIROUZ, T., GOODFELLOW, M. & MINNIKIN, D. E. 1977 Distribution of menaquinones in actinomycetes and corynebacteria. *Journal of General Microbiology* **100,** 221-230.

COLLINS, M. D., GOODFELLOW, M. & MINNIKIN, D. E. 1979 Isoprenoid quinones in the classification of coryneform and related bacteria. *Journal of General Microbiology* **110,** 127-136.

CROMBACH, W. H. J. 1978 *Caseobacter polymorphus* gen. nov., sp. nov., a coryneform bacterium from cheese. *International Journal of Systematic Bacteriology* **28,** 354-366.

CROSS, T. & GOODFELLOW, M. 1973 Taxonomy and classification of the actinomycetes. In *Actinomycetales: Characteristics and Practical Importance* ed. Sykes, G. & Skinner, F. A. pp. 11-112. London: Academic Press.

DEMARTEAU-GINSBURG, H. & LEDERER, E. 1963 Sur la structure chimique du mycoside B. *Biochimica et Biophysica Acta* **70,** 442- 451.

DHARIWAL, K. R., CHANDER, A. & VENKITASUBRAMANIAN, T. A. 1975 Phospholipids of *Mycobacterium phlei. Experimentia* **31,** 776-778.

DHARIWAL, K. R., CHANDER, A. & VENKITASUBRAMANIAN, T. A. 1976 Alteration in lipid constituents during growth of *Mycobacterium smegmatis* CDC 46 and *Mycobacterium phlei* ATCC 354. *Microbios* **16,** 169-182.

DHARIWAL, K. R., CHANDER, A. & VENKITASUBRAMANIAN, T. A. 1977 Environmental effects on lipids of *Mycobacterium phlei* ATCC 354. *Canadian Journal of Microbiology* **23,** 7-19.

DHARIWAL, K. R., CHANDER, A. & VENKITASUBRAMANIAN, T. A. 1978 Turnover of lipids in *Mycobacterium smegmatis* CDC 46 and *Mycobacterium phlei* ATCC 354. *Archives of Microbiology* **116,** 69-75.

DRAPER, P. 1974 The mycoside capsule of *Mycobacterium avium* 357. *Journal of General Microbiology* **83,** 431- 433.

DRAPER, P. & REES, R. J. W. 1973 The nature of the electron-transparent zone that surrounds *Mycobacterium lepraemurium* inside host cells. *Journal of General Microbiology* **77,** 79-87.

DRUCKER, D. B. & OWEN, I. 1973 Chemotaxonomic fatty acid finger-prints of bacteria grown with, and without, aeration. *Canadian Journal of Microbiology* **19,** 247-250.

DUNLAP, K. R. & PERRY, J. J. 1967 Effect of substrate on the fatty acid composition of hydrocarbon-utilizing microorganisms. *Journal of Bacteriology* **94,** 1919-1923.

DUNLAP, K. R. & PERRY, J. J. 1968 Effect of substrate on the fatty acid composition of hydrocarbon- and ketone-utilizing microorganisms. *Journal of Bacteriology* **96,** 318-321.

DUNPHY, P. J., PHILLIPS, P. G. & BRODIE, A. F. 1971 Separation and identification of menaquinones from microorganisms. *Journal of Lipid Research* **12,** 442-449.

ETÉMADI, A-H. 1964 Techniques microanalytiques d'étude de structure d'esters α-ramifiés β-hydroxylés. Chromatographie en phase vapeur et spectrométrie de masse. *Bulletin de la Société Chimique de France* 1537-1541.

ETÉMADI, A-H. 1967a Correlations structurales et biogénétiques des acides mycoliques en rapport avec la phylogenèse de quelques genres d'actinomycétales. *Bulletin de la Société de Chimie Biologique* **49,** 695-706.

ETÉMADI, A-H. 1967b Les acides mycoliques structure, biogenèse et intérêt phylogénétique. *Exposés Annuels de Biochimie Médicale* **28,** 77-109.

ETÉMADI, A-H 1967c Isomerisation de mycolates de methyle en milieu alcalin. *Chemistry and Physics of Lipids* **1,** 165-175.

ETÉMADI, A-H. 1967d The use of pyrolysis gas chromatography and mass spectroscopy in the study of the structure of mycolic acids. *Journal of Gas Chromatography* **5,** 447-456.

ETÉMADI, A-H. & CONVIT, J. 1974 Mycolic acids from "noncultivable" mycobacteria. *Infection and Immunity* **10,** 235-239.

ETÉMADI, A-H., PINTE, F. & MARKOVITS, J. 1967 Nouvelle analyse des acides α-mycoliques de *Mycobacterium smegmatis. Bulletin de la Société Chimique de France* 195–199.

FARSHTCHI, D. & McCLUNG, N. M. 1970 Effect of substrate on fatty acid production in *Nocardia asteroides. Canadian Journal of Microbiology* 16, 213–217.

FREGNAN, G. B., SMITH, D. W. & RANDALL, H. M. 1961 Biological and chemical studies on mycobacteria. Relationship of colony morphology to mycoside content for *Mycobacterium kansasii* and *Mycobacterium fortuitum. Journal of Bacteriology* 82, 517–527.

FREGNAN, G. B., SMITH, D. W. & RANDALL, H. M. 1962 A mutant of a scotochromogenic *Mycobacterium* detected by colony morphology and lipid studies. *Journal of Bacteriology* 83, 828–836.

FULCO, A. J., LEVY, R. & BLOCH, K. 1964 The biosynthesis of 9- and 5-monounsaturated fatty acids by bacteria. *Journal of Biological Chemistry* 239, 998–1003.

FURUCHI, A. & TOKUNAGA, T. 1972 Nature of the receptor substance of *Mycobacterium smegmatis* for D4 bacteriophage adsorption. *Journal of Bacteriology* 111, 404–411.

GALE, P. H., ARISON, B. H., TRENNER, N. R., PAGE, A. C. & FOLKERS, K. 1963 Characterization of vitamin K_9 (H) from *Mycobacterium phlei. Biochemistry* 2, 200–203.

GANGADHARAM, P. R. J., COHN, M. L. & MIDDLEBROOK, G. 1963 Infectivity, pathogenicity and sulpholipid fraction of some Indian and British strains of tubercle bacilli. *Tubercle* 44, 452–455.

GASTAMBIDE-ODIER, M. 1973 Variantes de mycosides caractérisées par des residus glycosidiques substitues par des chaines acyles–I: Spectres de masse des mycosides G^1 et A^1 peracetyles. *Organic Mass Spectrometry* 7, 845–860.

GASTAMBIDE-ODIER, M. 1974 Differentiation des espèces pathogènes de mycobactéries. *Bulletin de l'Institut Pasteur, Paris* 72, 221–237.

GASTAMBIDE-ODIER, M. & SARDA, P. 1970 Contribution à l'étude de la structure et de la biosynthèse de glycolipides spécifiques isolés de mycobacteries: les mycosides A et B. *Pneumonologie* 142, 241–255.

GASTAMBIDE-ODIER, M. & VILLÉ, C. 1970 Desoxysucres isolés du mycoside A: Identification des dérivés acétylés des méthyl 2,4-di-o-méthyl-rhamnopyranoside, 2-0-méthyl-rhamnofuranoside, 3-0-méthyl-rhamnofuranoside, 2-0-méthyl-fucopyranoside et 3-0-méthyl-fucofuranoside. *Bulletin de la Société de Chimie Biologique* 47, 2047–2067.

GASTAMBIDE-ODIER, M., SARDA, P. & LEDERER, E. 1965 Structure des aglycones des mycosides A et B. *Tetrahedron Letters* 3135–3143.

GENSLER, W. J. & MARSHALL, J. P. 1977 Structure of mycobacterial bis-cyclopropane mycolates by mass spectrometry. *Chemistry and Physics of Lipids* 19, 128–143.

GEORG, L. K. & BROWN, J. M. 1967 *Rothia,* gen. nov. an aerobic genus of the family Actinomycetaceae. *International Journal of Systematic Bacteriology* 17, 79–88.

GIEUSSE, M. & LANÉELLE, G. 1968 Sur l'analogie structurale du phthiotriol A et du phthiocérol A. *Compte Rendu Hebdomadaire Des Séances de l'Académie des Sciences* 266C, 1107–1109.

GOCHNAUER, M. B., LEPPARD, G. G., KOMARATAT, P., KATES, M., NOVITSKY, T. & KUSHNER, D. J. 1975 Isolation and characterization of *Actinopolyspora halophila* gen. et sp. nov., an extremely halophilic actinomycete. *Canadian Journal of Microbiology* 21, 1500–1511.

GOLDFINE, H. 1972 Comparative aspects of bacterial lipids. *Advances in Microbial Physiology* 8, 1–58.

GOMES, N. F., IONEDA, T. & PUDLES, J. 1966 Purification and chemical constitution of the phospholipids from *Corynebacterium diphtheriae* PW8. *Nature, London* 211, 81–82.

GOODFELLOW, M. 1971 Numerical taxonomy of some nocardioform bacteria. *Journal of General Microbiology* 69, 33–80.

GOODFELLOW, M. 1973 Characterisation of *Mycobacterium, Nocardia, Corynebacterium* and related taxa. *Annales de la Société Belge Médicine Tropicale* 53, 287–298.

GOODFELLOW, M. & ALDERSON, G. 1977 The actinomycete genus *Rhodococcus*: a

home for the 'rhodochrous' complex. *Journal of General Microbiology* **100**, 99-122.
GOODFELLOW, M. & MINNIKIN, D. E. 1977 Nocardioform bacteria. *Annual Review of Microbiology* **31**, 159-180.
GOODFELLOW, M. & MINNIKIN, D. E. 1978 Numerical and chemical methods in the classification of *Nocardia* and related taxa. *Zentralblatt für Bakteriologie, Parasitenkunde, Infektionskrankheiten und Hygiene. I. Abteilung, Supplement* **6**, 43-51.
GOODFELLOW, M. & SCHAAL, K. P. 1979 Identification methods for *Nocardia, Actinomadura* and *Rhodococcus*. In *Identification Methods for Microbiologists*, ed. Lovelock, D. W. & Skinner, F. A., pp. 261-276. Society of Applied Bacteriology Technical Series No. 14. London: Academic Press.
GOODFELLOW, M., MINNIKIN, D. E., PATEL, P. V. & MORDARSKA, H. 1973 Free nocardomycolic acids in the classification of nocardiae and strains of the *rhodochrous* complex. *Journal of General Microbiology* **14**, 185-188.
GOODFELLOW, M., COLLINS, M. D. & MINNIKIN, D. E. 1976 Thin-layer chromatographic analysis of mycolic acid and other long chain components in whole-organism methanolysates of coryneform and related taxa. *Journal of General Microbiology* **96**, 351-358.
GOODFELLOW, M., ORLEAN, P. A. B., COLLINS, M. D., ALSHAMAONY, L. & MINNIKIN, D. E. 1978 Chemical and numerical taxonomy of strains received as *Gordona aurantiaca*. *Journal of General Microbiology* **109**, 57-68.
GOODFELLOW, M., ALDERSON, G. & LACEY, J. 1979 Numerical taxonomy of *Actinomadura* and related actinomycetes. *Journal of General Microbiology* **112**, 95-111.
GOODWIN, T. W. 1972 Carotenoids in fungi and non-photosynthetic bacteria. *Progress in Industrial Microbiology* **11**, 29-88.
GORDON, R. E., MISHRA, S. K. & BARNETT, D. A. 1978 Some bits and pieces of the genus *Nocardia: N. carnea, N. vaccinii, N. transvalensis, N. orientalis* and *N. aerocolonigenes*. *Journal of General Microbiology* **109**, 69-78.
GOREN, M. B. 1970a Sulfolipid I of *Mycobacterium tuberculosis*, strain H_{37} Rv. I. Purification and properties. *Biochimica et Biophysica Acta* **210**, 116-126.
GOREN, M. B. 1970b Sulfolipid I of *Mycobacterium tuberculosis*, strain H_{37} Rv II. Structural studies. *Biochimica et Biophysica Acta* **210**, 127-138.
GOREN, M. B. 1972 Mycobacterial lipids: selected topics. *Bacteriological Reviews* **36**, 33-64.
GOREN, M. B. 1975 Cord-factor revisited: A tribute to the late Dr. Hubert Bloch. *Tubercle* **56**, 65-71.
GOREN, M. B., BROKL, O., DAS, B. C. & LEDERER, E. 1971 Sulfolipid I of *Mycobacterium tuberculosis*, strain H_{37} Rv. Nature of the acyl substituents. *Biochemistry* **10**, 72-81.
GOREN, M. B., BROKL, O. & SCHAEFER, W. B. 1974a Lipids of putative relevance to virulence in *Mycobacterium tuberculosis:* correlation of virulence and elaboration of sulfatides and strongly acidic lipids. *Infection and Immunity* **9**, 142-149.
GOREN, M. B., BROKL, O. & SCHAEFER, W. B. 1974b Lipids of putative relevance to virulence in *Mycobacterium tuberculosis:* correlation of virulence and elaboration of uation indicator lipid. *Infection and Immunity* **9**, 150-158.
GOREN, M. B., BROKL, O., ROLLER, P., FALES, H. M. & DAS, B. C. 1976 Sulfatides of *Mycobacterium tuberculosis:* The structure of the principal sulfatide (SL-1). *Biochemistry* **15**, 2728-2735.
GOREN, M. B., CERNICH, M. & BROKL, O. 1978 Some observations on mycobacterial acid-fastness. *American Review of Respiratory Diseases* **118**, 151-154.
GUINAND, M. & MICHEL, G. 1966 Structure d'un peptidolipide isolé de *Nocardia asteroides*, la peptidolipine NA. *Biochimica et Biophysica Acta* **125**, 75-91.
GUINAND, M., VACHERON, M. J. & MICHEL, G. 1970 Structure des parois cellulaires des *Nocardia*. I. Isolement et composition des parois de *Nocardia kirovani*. *FEBS Letters* **6**, 37-39.
GUZEVA, L. N., EFIMOVA, T. P., AGRE, N. S. & KRASIL'NIKOV, N. A. 1973 Fatty acids in the mycelia of actinomycetes that form catenate spores. *Mikrobiologiya* **42**, 26-31 (English translation, *Microbiology* **42**, 19-23).

HECHT, S. T. & CAUSEY, W. A. 1976 Rapid method for the detection and identification of mycolic acids in aerobic actinomycetes and related bacteria. *Journal of Clinical Microbiology* **4**, 284–287.

HOLMBERG, K. & HALLANDER, H. O. 1973 Numerical taxonomy and laborary identification of *Bacterionema matruchotii, Rothia dentocariosa, Actinomyces naeslundii, Actinomyces viscosus* and some related bacteria. *Journal of General Microbiology* **76**, 43–63.

HUNG, J. G. C. & WALKER, R. W. 1970 Unsaturated fatty acids of mycobacteria. *Lipids* **5**, 720–722.

IONEDA, T., LENZ, M. & PUDLES, J. 1963 Chemical constitution of a glycolipid from *Corynebacterium diphtheriae* PW8. *Biochemical and Biophysical Research Communications* **13**, 110–114.

IONEDA, T., LEDERER, E. & ROZANIS, J. 1970 Sur la structure des diesters de tréhalose ("cord factors") produit par *Nocardia asteroides* et *Nocardia rhodochrous. Chemistry and Physics of Lipids* **4**, 375–392.

JAVORA, J. & BACILEK, J. 1972 The quantitative analysis of mycobacterial fatty acids (C_{10}-C_{26}) by gas chromatography. *Studia Pneumologica et Phtiseologica Cechoslovaca* **32**, 241–245.

JENKINS, P. A. & MARKS, J. 1973 Thin-layer chromatography of mycobacterial lipids as an aid to classification. *Annales de la Société Belge Médicine Tropicale* **53**, 331–337.

JENKINS, P. A., MARKS, J. & SCHAEFER, W. B. 1971 Lipid chromatography and sero-agglutination in the classification of rapidly growing mycobacteria. *American Review of Respiratory Diseases* **103**, 179–187.

JENKINS, P. A., MARKS, J. & SCHAEFER, W. B. 1972 Thin layer chromatography of mycobacterial lipids as an aid to classification: The scotochromogenic mycobacteria, including *Mycobacterium scrofulaceum, M.xenopi, M.aquae, M.gordonae, M.flavescens. Tubercle* **53**, 118–127.

JENSEN, H. L. 1931 Contributions to our knowledge of the Actinomycetales. II. The definition and subdivision of the genus *Actinomyces* with a preliminary account of Australian soil actinomycetes. *Proceedings of the Linnean Society of New South Wales* **56**, 345–370.

JENSEN, H. L. 1953 The genus *Nocardia* (or *Proactinomyces*) and its separation from other Actinomycetales, with some reflections on the phylogeny of the actinomycetes. In *Symposium Actinomycetales, Proceeding VI.* pp. 69–88. International Congress of Microbiology, Rome: Emanuele Paterno.

JIČINKSÁ, E. 1973 *Nocardia*-like mutants of a soil coryneform bacterium. *Archiv für Mikrobiologie* **89**, 269–272.

JONES, D. 1975 A numerical taxonomic study of coryneform and related bacteria. *Journal of General Microbiology* **87**, 52–96.

JONES, D. 1978 An evaluation of the contributions of numerical taxonomic studies to the classification of coryneform bacteria. In *Coryneform Bacteria* ed. Bousfield, I. J. & Cally, A. G. pp. 13–46. London: Academic Press.

KANEDA, T. 1977 Fatty acids of the genus *Bacillus*: an example of branched-chain preference. *Bacteriological Reviews* **41**, 391–418.

KANETSUNA, F. & BARTOLI, A. 1972 A simple chemical method to differentiate *Mycobacterium* from *Nocardia. Journal of General Microbiology* **70**, 209–212.

KANZAKI, T., SUGIYAMA, Y., KITANO, K., ASHIDA, Y. & IMADA, I. 1974 Quinones of *Brevibacterium. Biochimica et Biophysica Acta* **348**, 162–165.

KATAOKA, T. & NOJIMA, S. 1967 The phospholipid compositions of some actinomycetes. *Biochimica et Biophysica Acta* **144**, 681–683.

KATO, M. & GOREN, M. B. 1974 Synergistic action of cord factor and mycobacterial sulfatides on mitochondria. *Infection and Immunity* **10**, 733–741.

KEDDIE, R. M. 1978 What do we mean by coryneform bacteria? In *Coryneform Bacteria* ed. Bousfield, I. J. & Cally, A. G. pp. 1–12. London: Academic Press.

KEDDIE, R. M. & CURE, G. L. 1977 The cell wall composition and distribution of free mycolic acids in named strains of coryneform bacteria and in isolates from various natural sources. *Journal of Applied Bacteriology* **42**, 229–252.

KEDDIE, R. M. & CURE, G. L. 1978 Cell wall composition of coryneform bacteria. In *Coryneform Bacteria* ed. Bousfield, I. J. & Callely, A. G. pp. 47–83. London: Academic Press.

KHULLER, G. K. 1976 The mannophosphoinositides of *Nocardia asteroides*. *Experimentia* **32**, 1371–1372.

KHULLER, G. K. 1977*a* Phospholipid composition of *Nocardia* species. *Indian Journal of Medical Research* **65**, 567–660.

KHULLER, G. K. 1977*b* Changes in phospholipid composition of *Nocardia polychromogenes* during temperature adaptation. *Experimentia* **33**, 1277–1278.

KHULLER, G. K. & BRENNAN, P. J. 1972*a* The mannophosphoinositides of the unclassified *Mycobacterium* P_6. *American Review of Respiratory Diseases* **106**, 892–896.

KHULLER, G. K. & BRENNAN, P. J. 1972*b* The polar lipids of some species of *Nocardia*. *Journal of General Microbiology* **73**, 409–412.

KHULLER, G. K. & SUBRAHMANYAM, D. 1968 On the mannophosphoinositides of *Mycobacterium* 607. *Experimentia* **24**, 851–852.

KHULLER, G. K. & SUBRAHMANYAM, D. 1970 On the ornithinyl ester of phosphatidylglycerol of *Mycobacterium* 607. *Journal of Bacteriology* **101**, 654–656.

KHULLER, G. K. & SUBRAHMANYAM, D. 1974 Fatty acid composition of phospholipids and the phosphatidyl inositomannoside-antigens of mycobacteria. *Indian Journal of Chest Diseases* **16**, 208–213.

KHULLER, G. K. & TRANA, A. K. 1977 Alterations in the phospholipid composition of *Nocardia polychromogenes* during growth. *Experimentia* **33**, 1422–1423.

KHULLER, G. K., BANERJEE, B., SHARMA, B. V. S. & SUBRAHMANYAM, D. 1972 Effect of age on the composition of major phospholipids of mycobacteria. *Indian Journal of Biochemistry and Biophysics* **9**, 274–275.

KIKUCHI, M. & NAKAO, Y. 1973*a* Relation between fatty acid composition of cellular phospholipids and the excretion of L-glutamic acid by a glycerol auxotroph of *Corynebacterium alkanolyticum*. *Agricultural and Biological Chemistry* **37**, 509–514.

KIKUCHI, M. & NAKAO, Y. 1973*b* Relation between cellular phospholipids and the excretion of L-glutamic acid by a glycerol auxotroph of *Corynebacterium alkanolyticum*. *Agricultural and Biological Chemistry* **37**, 515–519.

KING, D. H. & PERRY, J. J. 1975 Characterization of branched and unsaturated fatty acids in *Mycobacterium vaccae* strain JOB5. *Canadian Journal of Microbiology* **21**, 510–512.

KOBAYASHI, K., IKEDA, S., TAKINAMI, K., HIROSE, Y. & SHIRO, T. 1971 Production of L-glutamic acid from hydrocarbon by penicillin-resistant mutants of *Corynebacterium hydrocarboclastus*. *Agricultural and Biological Chemistry* **35**, 1241–1247.

KOMURA, I., KOMAGATA, K. & MITSUGI, K. 1973 A comparison of *Corynebacterium hydrocarboclastus* Iizuka and Komagata 1964 and *Nocardia erythropolis* (Gray & Thornton) Waksman and Henrici 1948. *Journal of General and Applied Microbiology* **19**, 161–170.

KOMURA, I., YAMADA, K., OTSUKA, S. & KOMAGATA, K. 1975 Taxonomic significance of phospholipids in coryneform and nocardioform bacteria. *Journal of General and Applied Microbiology* **21**, 251–261.

KOUL, A. K. & GASTAMBIDE-ODIER, M. 1977 Microanalyse rapide de dimycocérosate de phthiocérol, de mycosides et de glycérides dans les extraits à l'éther de pétrole de *Mycobacterium kansasii* et du BCG, souche Pasteur. *Biochimie* **59**, 535–538.

KRASIL'NIKOV, N. A. & KORONELLI, T. V. 1971 Nature of polar lipids from a paraffinoxidizing culture of *Mycobacterium rubrum*. *Mikrobiologiya* **40**, 230–235. (English translation *Microbiology* **40**, 196–200).

KROPPENSTEDT, R. M. & KUTZNER, H. J. 1976 Biochemical markers in the taxonomy of the Actinomycetales. *Experimentia* **32**, 318–319.

KROPPENSTEDT, R. M. & KUTZNER, H. J. 1978 Biochemical taxonomy of some problem actinomycetes. *Zentralblatt für Bakteriologie, Parasitenkunde, Infektionskrankheiten und Hygiene. I. Abteilung, Supplement* **6**, 125–133.

KUBICA, G. P., BAESS, I., GORDON, R. E., JENKINS, P. A., KWAPINSKI, J. B. G.,

McDURMONT, C., PATTYN, S. R., SAITO, H., SILCOX, V., STANFORD, J. L., TAKEYA, K. & TSUKAMURA, M. 1972 A co-operative numerical analysis of rapidly-growing mycobacteria. *Journal of General Microbiology* **73**, 55-70.

KUSAMRAN, K., POLGAR, N. & MINNIKIN, D. E. 1972 The mycolic acids of *Mycobacterium phlei*. *Journal of the Chemical Society Chemical Communications* 111-112.

La BELLE, Y. S. L. & WALKER, R. W. 1972 The phospholipids of *Mycobacterium smegmatis*. *American Review of Respiratory Diseases* **105**, 625-628.

LACAVE, C., ASSELINEAU, J. & TOUBIANA, R. 1967 Sur quelques constituants lipidiques de *Corynebacterium ovis*. *European Journal of Biochemistry* **2**, 37-43.

LACEY, J., GOODFELLOW, M. & ALDERSON, G. 1978 The genus *Actinomadura* Lechevalier and Lechevalier. *Zentralblatt für Bakteriologie Parasitenkunde, Infektionskrankheiten und Hygiene. I. Abteilung, Supplement* **6**, 107-117.

LAMONICA, G. & ETÉMADI, A-H. 1967a Nouvelle confirmation de la structure et de la biogenèse des acides α-avi-mycoliques. *Compte Rendu Hebdomadaire des Séances de l'Académie des Sciences* **265C**, 1197-1200.

LAMONICA, G. & ETÉMADI, A-H. 1967b Sur la présence simultanée d'acides mycoliques comportant deux cycles propaniques ou deux insaturations dans les lipides de *Mycobacterium phlei*. *Compte rendu hebdomadaire des séances de l'Academie des sciences* **264C**, 1711-1714.

LANÉELLE, G. 1963 Nature des acides mycoliques de *Mycobacterium paratuberculosis*, applications de la chromatographie sur couche mince à leur fractionnement. *Compte Rendu Hebdomadaire des Séances de l'Académie des Sciences* **257**, 781-783.

LANÉELLE, G. 1969 Mise en evidence d'une conformation stable d'un peptidolipide. *FEBS Letters* **4**, 210-212.

LANÉELLE, G. & ASSELINEAU, J. 1968 Structure d'un glycoside de peptidolipide isolé d'une mycobactérie. *European Journal of Biochemistry* **5**, 487-491.

LANÉELLE, G., ASSELINEAU, J., WOLSTENHOLME, W. A. & LEDERER, E. 1965c Détermination de séquences d'acides aminés dans des oligopeptides par la spectrométrie de masse. III. Structure d'un peptidolipide de *Mycobacterium johnei*. *Bulletin de la Société Chimique de France*, 2133-2134.

LANÉELLE, G., ASSELINEAU, J. & CHAMOISEAU, G. 1971 Présence de mycosides C (formes simplifiées de mycoside C) dans les bactéries isolées de bovins atteints du farcin. *FEBS Letters* **19**, 109-111.

LANÉELLE, M. A. & ASSELINEAU, J. 1977 Glycolipids of *Brevibacterium vitarumen*. *Biochimica et Biophysica Acta* **486**, 205-208.

LANÉELLE, M. A. & LANÉELLE, G. 1970 Structure d'acides mycoliques et d'un intermediaire dans la biosynthèse d'acides mycoliques dicarboxyliques. *European Journal of Biochemistry* **12**, 296-300.

LANÉELLE, M. A., ASSELINEAU, J. & CASTELNUOVO, G. 1965a Études sur les mycobactéries et les nocardiae. IV. Composition des lipides de *Mycobacterium rhodochrous*, *M.pellegrino* sp., et de quelques souches de nocardiae. *Annales de l'Institut Pasteur, Paris* **108**, 69-82.

I ANÉELLE, M. A., LANÉELLE, G., BENNET, P. & ASSELINEAU, J. 1965b Sur les lipides d'une souche non-photochromogène de mycobactérie. *Bulletin de la Société de Chimie Biologique* **47**, 2047-2067.

LANÉELLE, M. A., ASSELINEAU, J. & CASTELNUOVO, G. 1968 Relations chimiques et immunologiques chez les Actinomycetales. IV. Composition chimiques des lipides de quatre souches de *Streptomyces* et d'une souche *N.(Str.) gardneri*. *Annales de l'institut Pasteur, Paris* **114**, 305-312.

LARSSON, L. & MÅRDH, P-A. 1976 Gas chromatographic characterisation of mycobacteria: Analysis of fatty acids and trifluoroacetylated whole-cell methanolysates. *Journal of Clinical Microbiology* **3**, 81-85.

LECHEVALIER, H. A. & LECHEVALIER, M. P. 1970a A critical evaluation of the genera of aerobic actinomycetes. In *The Actinomycetales* ed. Prauser, H., pp. 393-405. Jena: Gustav Fischer Verlag.

LECHEVALIER, H. A., LECHEVALIER, M. P. & BECKER, B. 1966 Comparison of the chemical composition of cell walls of nocardiae with that of other aerobic actinomycetes. *International Journal of Systematic Bacteriology* **16**, 151–160.

LECHEVALIER, H. A., LECHEVALIER, M. P. & GERBER, N. N. 1971*a* Chemical composition as a criterion in the classification of actinomycetes. *Advances in Applied Microbiology* **14**, 47–72.

LECHEVALIER, M. P. 1968 Identification of aerobic actinomycetes of clinical importance. *Journal of Laboratory and Clinical Medicine* **71**, 934–944.

LECHEVALIER, M. P. 1972 Description of a new species, *Oerskovia xanthineolytica,* and emendation of *Oerskovia* Prauser et al. *International Journal of Systematic Bacteriology* **22**, 260–264.

LECHEVALIER, M. P. 1976 The taxonomy of the genus *Nocardia:* some light at the end of the tunnel? In *The Biology of the Nocardiae* ed. Goodfellow, M., Brownell, G. H. & Serrano, J. A. pp. 1–38. London: Academic Press.

LECHEVALIER, M. P. 1977 Lipids in bacterial taxonomy—a taxonomist's viewpoint. In *CRC Critical Reviews in Microbiology* pp. 109–210. Ohio: CRC Press.

LECHEVALIER, M. P. & LECHEVALIER, H. 1970*b* Chemical composition as a criterion in the classification of aerobic actinomycetes. *International Journal of Systematic Bacteriology* **20**, 435–444.

LECHEVALIER, M. P. & LECHEVALIER, H. 1974 *Nocardia amarae* sp. nov., an actinomycete common in foaming activated sludge. *International Journal of Systematic Bacteriology* **24**, 278–288.

LECHEVALIER, M. P., HORAN, A. C. & LECHEVALIER, H. 1971*b* Lipid composition in the classification of nocardiae and mycobacteria. *Journal of Bacteriology* **105**, 313–318.

LECHEVALIER, M. P., LECHEVALIER, H. & HORAN, A. C. 1973 Chemical characteristics and classification of nocardiae. *Canadian Journal of Microbiology* **19**, 965–972.

LECHEVALIER, M. P., DE BIERVE, C. & LECHEVALIER, H. 1977 Chemotaxonomy of aerobic actinomycetes: Phospholipid composition. *Biochemical Systematics and Ecology* **5**, 249–260.

LEDERER, E. 1967 Glycolipids of mycobacteria and related micro-organisms. *Chemistry and Physics of Lipids* **1**, 294–315.

LEDERER, E. 1976 Cord factor and related trehalose esters. *Chemistry and Physics of Lipids* **16**, 91–106.

LEDERER, E., ADAM, A., CIORBARU, R., PETIT, J-F. & WIETZERBIN, J. 1975 Cell walls of mycobacteria and related organisms; chemistry and immunostimulant properties. *Molecular and Cellular Biochemistry* **7**, 87–104.

LEHMANN, K. B. & NEUMANN, R. O. 1896 *Bakteriologische Diagnostik.* Munich: Lehmann Verlag.

LEHMANN, K. B. & NEUMANN, R. O. 1927 *Bakteriologische Diagnostik,* 7th edn. Munich.

LIAAEN-JENSEN, S. & ANDREWES, G. 1972 Microbial carotenoids. *Annual Review of Microbiology* **26**, 225–248.

LUCCHESI, M., CATTANEO, C. & DE RITIS, G. C. 1967 The chromatographic separation of fatty acids (C11–C20) from mycobacteria. *Bulletin of the International Union against Tuberculosis* **39**, 65–70.

McCARTHY, C. 1976 Synthesis and release of sulfolipid by *Mycobacterium avium* during growth and cell division. *Infection and Immunity* **14**, 1241–1252.

McCLUNG, N. M. 1974 Family VI. Nocardiaceae Castellani and Chalmers 1919, pp. 1040. In *Bergey's Manual of Determinative Bacteriology*, 8th edn. Buchanan, R. E. & Gibbons, N. E. pp. 726–747. Baltimore: Williams & Wilkins.

MACLENNAN, A. P. 1962 The monosaccharide units in specific glycolipids of *Mycobacterium avium.* *Biochemical Journal* **82**, 394–400.

MAGNUSSON, M. 1976 Sensitin tests as an aid in the taxonomy of *Nocardia* and its pathogenicity. In *The Biology of the Nocardiae,* ed. Goodfellow, M., Brownell, G. H. & Serrano, J. A. pp. 236–265. Academic Press: London.

MÁRA, M., GALLIOVÁ, J., ŠIR, Z., MOHELSKÁ, H., PRUCHOVA, J. & JULÁLAK, J. 1975 Biochemistry of BCG lipids and their role in antituberculous immunity and hypersensitivity. *Journal of Hygiene, Epidemiology, Microbiology and Immunology* **19**, 444–452.

MARKOVITS, J., PINTE, F.& ETÉMADI, A-H. 1966 Sur la structure des acides mycoliques dicarboxyliques insaturés de *Mycobacterium phlei*. *Compte Rendu Hebdomadaire des Séances de l'Académie des Sciences* **263**C, 960–962.

MARKS, J. & SZULGA, T. 1965 Thin layer chromatography of mycobacterial lipids as an aid to classification. *Tubercle*, **46**, 400–411.

MARKS, J., JENKINS, P. A. & SCHAEFER, W. B. 1971 Thin layer chromatography of mycobacterial lipids as an aid to classification: Technical improvements: *Mycobacterium avium*, *M. intracellulare* (Battey bacilli). *Tubercle*, **52**, 219–225.

MARKS, J., JENKINS, P. A. & TSUKAMURA, M. 1972 *Mycobacterium szulgai*–A new pathogen. *Tubercle* **53**, 210–214.

MAURICE, M. T., VACHERON, M. J. & MICHEL, G. 1971 Isolément d'acides nocardiques de plusieurs espèces de *Nocardia*. *Chemistry and Physics of Lipids* **7**, 9–18.

MEISSNER, G., SCHRODER, K. H., AMADIO, G. E., ANZ, W., CHAPARAS, S., ENGEL, H. W. B., JENKINS, P. A., KÄPPLER, W., KLEEBERG, H. H., KUBALA, E., KUBIN, M., LAUTERBACH, D., LIND, A., MAGNUSSON, M., MIKOVA, Zd., PATTYN, S. R., SCHAEFER, W. B., STANFORD, J. L., TSUKAMURA, M., WAYNE, L. G., WILLERS, I. & WOLINSKY, E. 1974 A co-operative numerical analysis of nonscoto- and nonphotochromogenic slowly growing mycobacteria. *Journal of General Microbiology* **83**, 207–235.

MEYER, J. 1976 *Nocardiopsis*, a new genus of the order Actinomycetales. *International Journal of Systematic Bacteriology* **26**, 487–493.

MICHALEC, C. & MÁRA, M. 1975 Thin-layer chromatography of mycolic acids and its application for the characterization of BCG strains. *Journal of Hygiene, Epidemiology, Microbiology and Immunology* **19**, 467–470.

MICHALEC, C., MÁRA, M. & ADAMCOVÁ, D. 1975 Qualitative and quantitative thin-layer chromatography of mycolic acids in *Mycobacterium tuberculosis* var. *bovis*–BCG. *Journal of Chromatography* **104**, 460–464.

MIDDLEBROOK, G., COLEMAN, C. M. & SCHAEFER, W. B. 1959 Sulfolipid from virulent tubercle bacilli. *Proceedings of the National Academy of Sciences, U.S.A.* **45**, 1801–1804.

MINNIKIN, D. E. 1972 Ring location in cyclopropane fatty acid esters by boron trifluoride-catalysed methoxylation followed by mass spectroscopy. *Lipids* **7**, 398–403.

MINNIKIN, D. E. 1978 Location of double bonds and cyclopropane rings in fatty acids by mass spectrometry. *Chemistry and Physics of Lipids* **21**, 313–347.

MINNIKIN, D. E. & GOODFELLOW, M. 1976 Lipid composition in the classification and identification of nocardiae and related taxa. In *The Biology of the Nocardiae* ed. Goodfellow, M., Brownell, G. H. & Serrano, J. A. pp. 160–219. London: Academic Press.

MINNIKIN, D. E. & POLGAR, N. 1966a Stereochemical studies on the mycolic acids. *Chemical Communications*, 648–649.

MINNIKIN, D. E.& POLGAR, N. 1966b Studies relating to phthiocerol. Part V. Phthiocerol A and B. *Journal of the Chemical Society C*, 2107–2112.

MINNIKIN, D. E. & POLGAR, N. 1967a The mycolic acids from human and avian tubercle bacilli. *Chemical Communications*, 916–918.

MINNIKIN, D. E. & POLGAR, N. 1967b The methoxymycolic and ketomycolic acids from human tubercle bacilli. *Chemical Communications*, 1172–1174.

MINNIKIN, D. E. & POLGAR, N. 1967c Studies relating to phthiocerol. Part VII. Phthiodiolone A. *Journal of the Chemical Society C*, 803–807.

MINNIKIN, D. E., PATEL, P. V. & GOODFELLOW, M. 1974 Mycolic acids of representative strains of *Nocardia* and the "*rhodochrous*" complex. *FEBS Letters* **39**, 322–324.

MINNIKIN, D. E., ALSHAMAONY, L. & GOODFELLOW, M. 1975 Differentiation of *Mycobacterium, Nocardia* and related taxa by thin-layer chromatographic analysis of whole-cell methanolysates. *Journal of General Microbiology* **88**, 200–204.

MINNIKIN, D. E., PIROUZ, T. & GOODFELLOW, M. 1977a Polar lipid composition in the classification of some *Actinomadura* species. *International Journal of Systematic Bacteriology* 27, 118–121.

MINNIKIN, D. E., PATEL, P. V., ALSHAMAONY, L. & GOODFELLOW, M. 1977*b* Polar lipid composition in the classification of *Nocardia* and related bacteria. *International Journal of Systematic Bacteriology* 27, 104–117.

MINNIKIN, D. E., COLLINS, M. D. & GOODFELLOW, M. 1978*a* Menaquinone patterns in the classification of nocardioform and related bacteria. *Zentralblatt für Bakteriologie, Parasitenkunde, Infektionskrankheiten und Hygiene. I. Abteilung, Supplement* 6, 85–90.

MINNIKIN, D. E., GOODFELLOW, M. & ALSHAMAONY, L. 1978*b* Mycolic acids in the classification of nocardioform bacteria. *Zentralblatt für Bakteriologie Parasitenkunde, Infektionskrankheiten und Hygiene. I. Abteilung, Supplement* 6, 63–66.

MINNIKIN, D. E., GOODFELLOW, M. & COLLINS, M. D. 1978*c* Lipid composition in the classification and identification of coryneform and related taxa. In *Coryneform Bacteria* ed. Bousfield, I. J. & Callely, A. G., pp. 85–160. London: Academic Press.

MINNIKIN, D. E., COLLINS, M. D. & GOODFELLOW, M. 1979 Fatty acid and polar lipid composition in the classification of *Cellulomonas, Oerskovia* and related taxa. *Journal of Applied Bacteriology* 47, 87–95.

MINNIKIN, D. E., HUTCHINSON, I. G., CALDICOTT, A. B. & GOODFELLOW, M. 1980 Thin-layer chromatography of methanolysates of mycolic acid-containing bacteria. *Journal of Chromatography* 188, 221–233.

MIQUEL, A-M., DAS, B. C. & ETÉMADI, A-H. 1966 Sur la structure des α-kansamycolones, cétones à haut poids moléculaire, isolées de *Mycobacterium kansasii. Bulletin de la Société Chimique de France* 2342–2345.

MOMPON, B., FEDERICI, C., TOUBIANA, R. & LEDERER, E. 1978 Isolation and structural determination of a "cord factor" (trehalose 6,6′ dimycolate) from *Mycobacterium smegmatis. Chemistry and Physics of Lipids* 21, 97–101.

MORDARSKA, H. 1968 A trial of using lipids for the classification of actinomycetes. *Archivum Immunologiae et Therapiae Experimentalis* 16, 45–50.

MORDARSKA, H. & MORDARSKI, M. 1969 Comparative studies on the occurrence of lipid A, diaminopimelic acid and arabinose in *Nocardia* cells. *Archivum Immunologiae et Therapiae Experimentalis* 17, 739–743.

MORDARSKA, H. & RÉTHY, A. 1970 Preliminary studies on the chemical character of the lipid fraction in *Nocardia. Archivum Immunologiae et Therapiae Experimentalis* 18, 455–459.

MORDARSKA, H., MORDARSKI, M. & GOODFELLOW, M. 1972 Chemotaxonomic characters and classification of some nocardioform bacteria. *Journal of General Microbiology* 71, 77–86.

MORDARSKI, M., GOODFELLOW, M., SZYBA, K., PULVERER, G. & TKACZ, A. 1978*a* Deoxyribonucleic acid base composition and homology studies on *Rhodococcus* and allied taxa. *Zentralblatt für Bakteriologie, Parasitenkunde Infektionskrankheiten und Hygiene. I. Abteilung, Supplement* 6, 99–106.

MORDARSKI, M., SCHAAL, K. P., TKACZ, A., PULVERER, G., SZYBA, K. & GOODFELLOW, M. 1978*b* Deoxyribonucleic acid base composition and homology studies on *Nocardia. Zentralblatt für Bakteriologie, Parasitenkunde, Infektionskrankheiten und Hygiene. I. Abteilung, Supplement* 6 91–97.

MORGAN, E. D. & POLGAR, N. 1957 Constituents of the lipids of tubercle bacilli. Part VIII. Studies on mycolic acid. *Journal of the Chemical Society,* 3779–3786.

MOTOMIYA, M., MAYAMA, A. FUJIMOTO, M., SATO, H. & OKA, S. 1969 Chemistry and biology of phospholipids from an unclassified mycobacteria, P6. *Chemistry and Physics of Lipids* 3, 159–167.

NACASCH, C. & VILKAS, E. 1967 Sur la présence simultanée de deux types de mycosides dans *Mycobacterium kansasii. Compte rendu hebdomadaire des séances de l'Academie des Sciences* 265C, 413–415.

NAVALKAR, R. G., WIEGESHAUS, E. H. & SMITH, D. W. 1964 Relationship of mycoside content and colony morphology for group II and group III unclassified mycobacteria. *Journal of Bacteriology* 88, 255–259.

NAVALKAR, R. G., WIEGESHAUS, E., KONDO, E., KIM, H. K. & SMITH, D. W. 1965 Mycoside G, a specific glycolipid in *Mycobacterium marinum (balnei)*. *Journal of Bacteriology* 90, 262-265.

NOCARD, M. E. 1888 Note sur le maladie des boeufs de la Guadeloupe, connue sous le nom de farcin. *Annales de l'Institut Pasteur, Paris* 2, 291-302.

NOLL, H. 1956 The chemistry of cord factor, a toxic glycolipid of *M. tuberculosis*. *Advances in Tuberculosis Research* 7, 149-183.

NOLL, H. 1957 The chemistry of some native constituents of the purified wax of *Mycobacterium tuberculosis*. *Journal of Biological Chemistry* 224, 149-164.

OHASHI, D. K., WADE, T. J. & MANDLE, R. J. 1977 Characterization of ten species of mycobacteria by reaction gas-liquid chromatography. *Journal of Clinical Microbiology* 6, 469-473.

OKA, S., FUKUSHI, K., FUJIMOTO, M., SATO, H. & MOTOMIYA, M. 1968a La distribution subcellulaire des phospholipides de la mycobactérie. *Comptes Rendu Société Franco-Japonaise de Biologie* 162, 1648-1650.

OKA, S., MAYAMA, A., FUJIMOTO, M., SATO, H. & MOTOMIYA, M. 1968b Activités biologiques des phospholipides de la souche P6 de la mycobactérie dite atypique et scotochromogene. *Comptes Rendu Société Franco-Japonaise de Biologie* 163, 253-255.

OKAZAKI, H., KANAZAKI, T. & FUKUDA, H. 1968 L-glutamic acid fermentation. Part V. Behaviour of oleic acid in an oleic acid-requiring mutant. *Agricultural and Biological Chemistry* 32, 1464-1470.

OKUYAMA, H., KANKURA, T. & NOJIMA, S. 1967 Positional distribution of fatty acids in phospholipids from mycobacteria. *Journal of Biochemistry* 61, 732-737.

PANGBORN. M. C. 1968 Structure of mycobacterial phosphatides. *Annals New York Academy of Sciences* 154, 133-139.

PANGBORN, M. C. & MCKINNEY, J. A. 1966 Purification of serologically active phosphoinositides of *Mycobacterium tuberculosis*. *Journal of Lipid Research* 7, 627-633.

POLGAR, N. 1971 Natural alkyl-branched long-chain acids. In *Topics in Lipid Chemistry*, Vol. 2. ed. Gunstone, F. D. pp. 207-246. London: Logos Press.

POMMIER, M. T. & MICHEL, G. 1972 Isolement et caractéristiques d'un nouveau glycolipide de *Nocardia caviae*. *Compte Rendu Hebdomadaire des Séances de l'Académie des Sciences* 275C, 1323-1326.

POMMIER, M. T. & MICHEL, G. 1973 Phospholipid and acid composition of *Nocardia* and nocardoid bacteria as criteria of classification. *Biochemical Systematics* 1, 3-12.

POMMIER, M. T. & MICHEL, G. 1976 Caracterisation de l'acide D (-) glycerique dans un glycolipide de *Nocardia caviae*. *Biochimica et Biophysica Acta* 441, 327-333.

PORTELANCE, V. & ASSELINEAU, J. 1970 Nonphosphorus contaminants of mycobacterial phospholipid preparations. *American Review of Respiratory Diseases* 103, 853-854.

PRAUSER, H. 1967 Contributions to the taxonomy of the Actinomycetales. *Publications of the Faculty of Science, J. E. Purkyné University, Brno*, K40, 196-199.

PRAUSER, H. 1978 Considerations on taxonomic relations among Gram-positive, branching bacteria. *Zentralblatt für Bakteriologie, Parasitenkunde, Infektionskrankheiten und Hygiene. I. Abteilung, Supplement* 6, 3-12.

PRAUSER, H. & BERGHOLZ, M. 1974 Taxonomy of actinomycetes and screening for antibiotic substances. *Postepy Higieny I Medycyny Doswiadczalnej* 28, 441-457.

PRAUSER, H., LECHEVALIER, M. P. & LECHEVALIER, H. 1970 Description of *Oerskovia* gen. n. to harbor Ørskov's motile *Nocardia*. *Applied Microbiology* 19, 534.

PROMÉ, J. C. 1968 Localisation d'un cycle propanique dans une substance aliphatique par examen du spectre de masse des cétones obtennes par oxydation. *Bulletin de la Société Chimique de France*, 655-660.

PROMÉ, J. C., ASSELINEAU, C. & ASSELINEAU, J. 1966 Acides mycobactériques nouveaux acides isolés des lipides des mycobactéries. *Compte Rendu Hebdomadaire des Séances de l'Academie des Sciences* 263C, 448-457.

PROMÉ, J. C., LACAVE, C. & LANÉELLE, M. A. 1969 Sur les structures de lipides à ornithine de *Brucella melitensis* et de *Mycobacterium bovis*. *Compte Rendu Hebdo-*

madaire des Séances de l'Académie des Sciences **269C**, 1664–1667.

PROMÉ, J. C., LÁCAVE, C., AHIBO-COFFY, A. & SAVAGNAC, A. 1976 Séparation et étude structurale des espèces moléculaires de monomycolates et de dimycolates de α-D-tréhalose présents chez *Mycobacterium phlei. European Journal of Biochemistry* **63**, 543–552.

PUDLES, J. & LEDERER, E. 1954 Sur l'isolement et la constitution chimique de l'acide coryno-mycolénique et de deux cetones des lipides du bacille diphthérique. *Bulletin de la Société de Chimie Biologique* **36**, 759–777.

PUZO, G. & PROMÉ, J. C. 1973 Fragmentation des aldehydes cyclopropaniques en spectrometry de masse. *Tetrahedron* **29**, 3619–3629.

QURESHI, N., TAKAYAMA, K., JORDI, H. C. & SCHNOES, H. K. 1978 Characterisation of the purified components of a new homologous series of α-mycolic acids from *Mycobacterium tuberculosis* H_{37} Ra. *Journal of Biological Chemistry* **253**, 5411–5417.

RANDALL, H. M. & SMITH, D. W. 1964 Characterisation of mycobacteria by infrared spectroscopic examination of their lipid fractions. *Zentralblatt für Bakteriologie, Parasitenkunde, Infektionskrankheiten und Hygiene. I. Abteilung* **194**, 190–201.

RATLEDGE, C. 1976 The physiology of the mycobacteria. *Advances in Microbial Physiology* **13**, 115–244.

RATLEDGE, C. & PATEL, P. V. 1976 Lipid-soluble, iron-binding compounds in *Nocardia* and related organisms. In *The Biology of the Nocardiae* ed. Goodfellow, M., Brownell, G. H. & Serrano, J. A. pp. 372–385. London: Academic Press.

RÖHRSCHEIDT, E. & TÁRNOK, I. 1972 Investigations on *Nocardia* pigments. Chromatographic properties of the pigments and their significance for differentiation of pigmented *Nocardia* strains. *Zentralblatt für Bakteriologie, Parasitenkunde, Infektionskrankheiten und Hygiene. I. Abteilung, Originale A* **221**, 221–233.

RUNYON, E. H. 1959 Anonymous mycobacteria in pulmonary disease. *Medical Clinics of North America* **43**, 273–290.

RUNYON, E. H. 1965 Pathogenic mycobacteria. *Advances in Tuberculosis Research* **14**, 235–287.

RUNYON, E. H., WAYNE, L. G. & KUBICA, G. P. 1974 Family II. Mycobacteriaceae Chester 1897, 63. In *Bergey's Manual of Determinative Bacteriology*, 8th edn. ed. Buchanan, R. E. & Gibbons, N. E. pp. 681–701. Baltimore: Williams & Wilkins.

SAITO, H., GORDON, R. E., JUHLIN, I., KÄPPLER, W., KWAPINSKI, J. B. G., McDURMONT, C., PATTYN, S. R., RUNYON, E. H., STANFORD, J. L., TÁRNOK, I., TASAKA, H., TSUKAMURA, M. & WEISZFEILER, J. 1977 Co-operative numerical analysis of rapidly growing mycobacteria. *International Journal of Systematic Bacteriology* **27**, 75–85.

SARDA, P. & GASTAMBIDE-ODIER, M. 1967 Structure chimique de l'aglycone du mycoside G de *Mycobacterium marinum. Chemistry and Physics of Lipids* **1**, 434–444.

SASAKI, A. 1975 Isolation and characterisation of serologically active phosphatidylinositol oligomannosides of *Mycobacterium tuberculosis. Journal of Biochemistry* **78**, 547–554.

SCHAAL, K. P. & REUTERSBERG, H. 1978 Numerical taxonomy of *Nocardia asteroides. Zentralblatt für Bakteriologie, Parasitenkunde, Infektionskrankheiten und Hygiene. I. Abteilung Supplement* **6**, 53–62.

SCHLEIFER, K. H. & KANDLER, O. 1972 Peptidoglycan types of bacterial cell walls and their taxonomic implication. *Bacteriological Reviews* **36**, 407–477.

SCHOLES, P. B. & KING, H. K. 1965 Isolation of a naphthaquinone with partly hydrogenated side chain from *Corynebacterium diphtheriae. Biochemical Journal* **97**, 766–768.

SEHRT, I., KÄPPLER, W. & LANGE, A. 1975 Differentiation and identification of mycobacteria by means of thin-layer chromatography of their lipids. I. Application for the classification of rapidly growing mycobacteria. *Zeitschrift fuer Erkrankungen der Atmungsorgane* **143**, 203–211.

SEHRT, I., KÄPPLER, W. & LANGE, A. 1975 Differentiation and identification of mycobacteria by means of thin-layer chromatography of their lipids. II. Application for the classification of slowly growing mycobacteria: *M.avium* and *M.intracellulare. Zeitschrift fuer Erkrankungen der Atmungsorgare* **144**, 146–154.

SENN, M. T., IONEDA, T., PUDLES, J. & LEDERER, E. 1967 Spectrométrie de masse de glycolipids. 1. Structure du "cord factor" de *Corynebacterium diphtheriae*. *European Journal of Biochemistry* 1, 353-356.

SHAW, N. 1974 Lipid composition as a guide to the classification of bacteria. *Advances in Applied Microbiology* 17, 63-108.

SHAW, N. 1975 Bacterial glycolipids and glycophospholipids. *Advances in Microbial Physiology* 12, 141-167.

SHIBUKAWA, M., KURIMA, M. & COHUCHI, S. 1970 L-Glutamic acid fermentation with molasses. Part XII. Relationship between the kind of phospholipids and their fatty acid composition in the mechanism of extracellular accumulation of L-glutamate. *Agricultural and Biological Chemistry* 34, 1136-1141.

SMITH, D. W., RANDALL, H. M., GASTAMBIDE-ODIER, M. M. & KOEVOET, A. L. 1957 The characterisation of mycobacterial strains by the composition of their lipid extracts. *Annals New York Academy of Sciences* 69, 145-157.

SMITH, D. W., RANDALL, H. M., MacLENNAN, A. P. & LEDERER, E. 1960a Mycosides: a new class of type specific glycolipids of mycobacteria. *Nature, London* 186, 887-888.

SMITH, D. W., RANDALL, H. M., MacLENNAN, A. P., PUTNEY, R. K. & RAO, S. V. 1960b Detection of specific lipids in mycobacteria by infrared spectroscopy. *Journal of Bacteriology* 79, 217-229.

SNEATH, P. H. A. 1976 An evaluation of numerical taxonomic techniques in the taxonomy of *Nocardia* and allied taxa. In *The Biology of the Nocardiae* ed. Goodfellow, M., Brownell, G. H. & Serrano, J. A. pp. 74-101. London: Academic Press.

SNOW, G. A. 1970 Mycobactins: iron-chelating growth factors from mycobacteria. *Bacteriological Reviews* 34, 99-125.

SOINA, V. S. & AGRE, N. S. 1976 Fine structure of vegetative and sporulating hyphae in *Micropolyspora fascifera*. *Mikrobiologiya* 45, 287-291.

STANECK, J. L. & ROBERTS, G. D. 1974 Simplified approach to identification of aerobic actinomycetes by thin-layer chromatography. *Applied Microbiology* 28, 226-231.

STECK, P. A., SCHWARTZ, B. A., ROSENDAHL, M. S. & GRAY, G. R. 1978 Mycolic acids. A reinvestigation. *Journal of Biological Chemistry* 253, 5625-5629.

STODOLA, F. H., LESUK, A. & ANDERSON, R. J. 1938 The chemistry of lipids of tubercle bacilli. LIV. The isolation and properties of mycolic acid. *Journal of Biological Chemistry* 126, 505-513.

STRAIN, S. M., TOUBIANA, R., RIBI, E. & PARKER, R. 1977 Separation of the mixture of trehalose 6,6'-dimycolates comprising the mycobacterial glycolipid fraction, "P3". *Biochemical and Biophysical Research Communications* 77, 449-456.

SUBRAHMANYAM, D. 1965 Fatty acid composition of phosphatidylethanoloamine of *Mycobacterium tuberculosis*. *Indian Journal of Biochemistry* 2, 274-275.

SUZUKI, T., TANAKA, K., MATSUBARA, I. & KINOSHITA, S. 1969 Trehalose lipid and α-branched-β-hydroxy fatty acid formed by bacteria grown on *n*-alkanes. *Agricultural and Biological Chemistry* 33, 1619-1627.

SZULGA, T., JENKINS, P. A. & MARKS, J. 1966 Thin-layer chromatography of mycobacterial lipids as an aid to classification, *Mycobacterium kansasii* and *Mycobacterium marinum (balnei)*. *Tubercle* 47, 130-136.

TADAYON, R. A. & CARROLL, K. K. 1971 Effect of growth conditions on the fatty acid composition of *Listeria monocytogenes* and comparison with fatty acids of *Erysipelothrix* and *Corynebacterium*. *Lipids* 6, 820-825.

TAKAYAMA, K., SCHNOES, H. K., ARMSTRONG, E. L. & BOYLE, R. W. 1975 Site of inhibitory action of isoniazid in the synthesis of mycolic acids in *Mycobacterium tuberculosis*. *Journal of Lipid Research* 16, 308-317.

TAKINAMI, K., YOSHII, H., YAMADA, Y., OKADA, H. & KINOSHITA, K. 1968 Control of L-glutamic acid fermentation by biotin and fatty acid. *Amino Acid Nucleic Acid* 18, 120-156.

TÁRNOK, I. & TÁRNOK, ZS. 1970 Carotenes and xanthophylls in mycobacteria. I. Technical procedures; thin-layer chromatographic patterns of mycobacterial pigments. *Tubercle* 51, 305-312.

TÁRNOK, I. & TÁRNOK, ZS. 1971 Carotenes and xanthophylls in mycobacteria. II. Lycopene, α- and β-carotene and xanthophyll in mycobacterial pigments. *Tubercle* **52**, 127-135.

THOEN, C. O., KARLSON, A. G. & ELLEFSON, R. D. 1971*a* Comparison by gas-liquid chromatography of the fatty acids of *Mycobacterium avium* and some other nonphotochromogenic mycobacteria. *Applied Microbiology* **22**, 560-563.

THOEN, C. O., KARLSON, A. G. & ELLEFSON, R. D. 1971*b* Fatty acids of *Mycobacterium kansasii. Applied Microbiology* **21**, 628-632.

THOEN, C. O., KARLSON, A. G. & ELLEFSON, R. D. 1972 Differentiation between *Mycobacterium kansasii* and *Mycobacterium marinum* by gas-liquid chromatographic analysis of cellular fatty acids. *Applied Microbiology* **24**, 1009-1010.

TIMBOL, C. R. & SHIMIZU, K. 1977 Lipid analysis of atypical mycobacteria by thin-layer chromatography. *Philippine Journal of Veterinary Medicine* **15**, 28-38.

TREVISAN, V. 1889 *I generi e le specie delle Batteriacae.* Milano: Zanaboni & Gabussi.

TSUKAMURA, M. 1969 Numerical taxonomy of the genus *Nocardia. Journal of General Microbiology* **56**, 265-287.

TSUKAMURA, M. 1971 Proposal of a new genus, *Gordona,* for slightly acid-fast organisms occurring in sputa of patients with pulmonary disease and in soil. *Journal of General Microbiology* **68**, 15-26.

TSUKAMURA, M. 1974 A further numerical taxonomic study of the *rhodochrous* group. *Japanese Journal of Microbiology* **18**, 37-44.

TSUKAMURA, M. 1975 Numerical analysis of the relationship between *Mycobacterium, rhodochrous* group and *Nocardia* by use of hypothetical median organisms. *International Journal of Systematic Bacteriology* **25**, 329-335.

TSUKAMURA, M. & MIZUNO, S. 1975 Differentiation among mycobacterial species by thin-layer chromatography. *International Journal of Systematic Bacteriology* **25**, 271-280.

TSUKAMURA, M. & MIZUNO, S. 1978 Differentiation of *Rhodococcus (Gordona)* and *Nocardia* by thin-layer chromatography after uptake of (^{35}S) methionine. *Journal of General Microbiology* **105**, 159-160.

TSUKAMURA, M., MIZUNO, S., TSUKAMURA, S. & TSUKAMURA, J. 1979 Comprehensive numerical classification of 369 strains of *Mycobacterium, Rhodococcus,* and *Nocardia. International Journal of Systematic Bacteriology* **29**, 110-129.

TSYGANOV, V. A., EFIMOVA, T. P. & SOBOLEVA, L. V. 1973 Fatty acids produced by proactinomycetes. *Mikrobiologiya* **42**, 795-799. (English translation, *Microbiology* **42**, 705-709).

TSYGANOV, V. A., NAMESTNIKOVA, V. P. & KRASILNIKOV, V. A. 1966 DNA composition in various genera of the Actinomycetales. *Mikrobiologiya* **35**, 92-95.

UCHIDA, K. & AIDA, K. 1977 Acyl type of bacterial cell wall: Its simple identification by colorimetric method. *Journal of General and Applied Microbiology* **23**, 249-260.

VACHERON, M. J., GUINAND, M., MICHEL, G. & GHUYSEN, J. M. 1972 Structural investigations on cell walls of *Nocardia* sp. The wall lipid and peptidoglycan moieties of *Nocardia kirovani. European Journal of Biochemistry* **29**, 156-166.

VILKAS, E. & LEDERER, E. 1968 N-Methylation de peptides par la methode de Hakomori. Structure du mycoside C_{b1} *Tetrahedron Letters* 3089-3092.

VILKAS, E. & ROJAS, A. 1964 Sur les lipides de *Mycobacterium fortuitum. Bulletin de la Société de Chimie Biologique* **46**, 689-701.

VILKAS, E., ADAM, A. & SENN, M. 1968*a* Isolement d'un nouveau type de diester de trehalose a partir de *Mycobacterium fortuitum. Chemistry and Physics of Lipids* **2**, 11-16.

VILKAS, E., GROS, C. & MASSOT, J. C. 1968*b* Sur la structure chimique d'un mycoside C isolé de *Mycobacterium scrofulaceum. Compte Rendu Hebdomadaire des Séances de l'Académie des sciences* **266C**, 837-840.

VILLÉ, C. & GASTAMBIDE-ODIER, M. 1970 Le 3-0-méthyl-L-rhamnose, sucre isolé du mycoside G de *Mycobacterium marinum. Carbohydrate Research* **12**, 97-107.

VOILAND, A., BRUNETEAU, M. & MICHEL, G. 1971 Étude du mycoside C_2 de *Mycobac-*

terium avium. Détermination de la structure. *European Journal of Biochemistry* **21**, 285-291.

WAKSMAN, S. A. 1961 *The Actinomycetes,* Volume II. Baltimore: Williams & Wilkins.

WAKSMAN, S. A. & HENRICI, A. T. 1943 The nomenclature and classification of the actinomycetes. *Journal of Bacteriology* **46**, 337-341.

WALKER, R. W., MALLMANN, W. L. & BRUNNER, J. R. 1967 Type-specific lipids of mycobacteria. *American Review of Respiratory Diseases* **95**, 1065-1067.

WALKER, R. W., BARAKAT, H. & HUNG, J. G. C. 1970 The positional distribution of fatty acids in the phospholipids and triglycerides of *Mycobacterium smegmatis* and *M. bovis* BCG. *Lipids* **5**, 684-691.

WAYNE, L. G., DIETZ, T. M., GERNEZ-RIEUX, C., JENKINS, P. A., KÄPPLER, W., KUBICA, G. P., KWAPINSKI, J. B. G., MEISSNER, G., PATTYN, S. R., RUNYON, E. H., SCHRÖDER, K. H., SILCOX, V. A., TACQUET, A., TSUKAMURA, M. & WOLINSKY, E. 1971 A co-operative numerical analysis of scotochromogenic slowly growing mycobacteria. *Journal of General Microbiology* **66**, 255-271.

WAYNE, L. G., ANDRADE, L., FROMAN, S., KÄPPLER, W., KUBALA, E., MEISSNER, G. & TSUKAMURA, M. 1978 A co-operative numerical analysis of *Mycobacterium gastri, Mycobacterium kansasii* and *Mycobacterium marinum. Journal of General Microbiology* **109**, 319-327.

WEIR, M. P., LANGRIDGE, W. H. R. & WALKER, R. W. 1972 Relationships between oleic acid uptake and lipid metabolism in *Mycobacterium smegmatis. American Review of Respiratory Diseases* **106**, 450-457.

WELBY-GIEUSSE, M. & TOCANNE, J. F. 1970 Configuration absolue du phthiocerol A₁ du phthiotriol A et de la phthiodiolone A. *Tetrahedron* **26**, 2875-2882.

WHITESIDE, T. L., de SIERVO, A. J. & SALTON, M. R. J. 1971 Use of antibody to membrane adenosine triphosphatase in the study of bacterial relationships. *Journal of Bacteriology* **105**, 957-967.

WILLIAMS, S. T., SHARPLES, G. P., SERRANO, J. A., SERRANO, A. A. & LACEY, J. 1976 The micromorphology and fine structure of nocardioform organisms. In *The Biology of the Nocardiae* ed. Goodfellow, M., Brownell, G. H. & Serrano, J. A. pp. 102-140. London: Academic Press.

YAMADKA, K. & KOMAGATA, K. 1972 Taxonomic studies on coryneform bacteria. V. Classification of coryneform bacteria. *Journal of General and Applied Microbiology* **18**, 417-431.

YAMADA, Y., INOUYE, G., TAHARA, Y. & KONDO, K. 1976 The menaquinone system in the classification of coryneform and nocardioform bacteria and related organisms. *Journal of General and Applied Microbiology* **22**, 203-214.

YAMADA, Y., ISHIKAWA, T., TAHARA, Y. & KONDO, K. 1977 The menaquinone system in the classification of the genus *Nocardia. Journal of General and Applied Microbiology* **23**, 207-216.

YAMAGUCHI, T. 1967 Similarity in DNA of various morphologically distinct actinomycetes. *Journal of General and Applied Microbiology, Tokyo* **13**, 63-71.

YANO, I. & SAITO, K. 1972 Gas chromatographic and mass spectrometric analysis of molecular species of corynomycolic acids from *Corynebacterium ulcerans. FEBS Letters* **23**, 352-356.

YANO, I., FURUKAWA, Y. & KUSUNOSE, M. 1968 Incorporation of radioactivity from methionine—methyl-¹⁴C into phospholipids by *Nocardia polychromogenes. Journal of Biochemistry* **63**, 133-135.

YANO, I., FURUKAWA, Y. & KUSUNOSE, M. 1969 Phospholipids of *Nocardia coeliaca. Journal of Bacteriology* **98**, 124-130.

YANO, I., FURUKAWA, Y. & KUSUNOSE, M. 1970 α-Hydroxy fatty acid-containing phospholipids of *Nocardia leishmanii. Biochimica et Biophysica Acta* **202**, 189-191.

YANO, I., FURUKAWA, Y. & KUSUNOSE, M. 1971 Occurrence of acylated trehaloses in *Nocardia. Journal of General and Applied Microbiology* **17**, 329-334.

YANO, I., SAITO, K., FURUKAWA, Y. & KUSUNOSE, M. 1972 Structural analysis of molecular species of nocardomycolic acids from *Nocardia erythropolis* by the combined system of gas chromatography and mass spectrometry. *FEBS Letters* **21**, 215-219.

YANO, I., KAGEYAMA, K., OHNO, Y., MASUI, M., KUSUNOSE, E., KUSUNOSE, M. & AKIMORI, N. 1978 Separation and analysis of molecular species of mycolic acids in *Nocardia* and related taxa by gas chromatography mass spectrometry. *Biomedical Mass Spectrometry* **5**, 14-24.

Fermentation Products (Using g.l.c.) in the Differentiation of Non-Sporing Anaerobic Bacteria

R. HAMMANN AND H. WERNER

Institut für Medizinische Mikrobiologie und Immunologie der Universität Bonn, Venusberg, Bonn, FRG

Contents

1. Introduction

IT HAS BEEN KNOWN for many years that micro-organisms are able to produce acids and this property is currently exploited in the fermentation of sauerkraut, milk and other foodstuffs. Even today the production of organic acids, e.g. citric acid, by fungi is commercially feasible.

In aerobic organisms, acid production normally occurs only under certain growth conditions, e.g. high C/N ratio, or in mutants which have lost the terminal steps of energy gain, including oxidative phosphorylation, so that an intermediate accumulates and is excreted. In fermentative organisms, however, acids are obligate end products of carbohydrate or amino acid metabolism. In anaerobes, primarily produced acids such as pyruvic acid serve as electron acceptors instead of oxygen which is used by aerobes. The metabolism and energy generation of anaerobes has been excellently reviewed by Thauer *et al.* (1977).

Although acid production in bacteria, especially in anaerobes, has long been known to be a constant property of genera or species, it was the introduction of gas chromatography into bacteriology (Moore *et al.* 1966) that revolutionized the analysis of many products which formerly required labour and time-consuming methods. Whereas many morphological and biochemical characters of anaerobes are found to be unreliable for both classification and identification, the acid end products of metabolism can generally be considered as characters of prime taxonomic importance. The fatty acids formed as end products of protein or carbohydrate metabolism are, therefore, of particular importance in the current classification of anaerobes (Holdeman *et al.* 1977). In medical microbiology, accurate and prompt identification of anaerobes, largely based on the analysis of acid fermentation products, is essential for effective treatment of infected patients.

In the present paper the contribution of gas chromatography to the classification and identification of non-sporing anaerobes is discussed.

2. Methods

Gas liquid chromatography (g.l.c.) can be carried out with a Perkin-Elmer gas chromatograph using diethylhexyl-sebacinic acid on Chromosorb WAW, which allows the separation of non-branched and iso-branched fatty acids from C_2 to C_6 and up to C_8 from ether extracts of acidified culture media (Werner 1972*a*). Strains are grown in peptone-yeast extract medium (PY) and peptone-yeast extract-glucose medium (PYG). Some alcohols can also be detected by this method (Holdeman *et al.* 1977).

The detection of non-volatile lactic and succinic acids, which are often produced by anaerobic bacteria, requires methylation. Esterification can be carried out directly using a centrifuged aliquot of the culture fluid and the methanol-sulfuric acid reagent of Holdeman *et al.* (1977). For quantitative work it is better to use a freeze-dried extract of the culture fluid which is methylated using boron trifluoride-methanol (Alcock 1969; Hammann 1977). In each case, the methyl esters are extracted using chloroform. Chloroform extracts can be separated on the same column as the volatile acids. Peaks are identified by comparing their retention times with the corresponding mixture of pure volatile acids or with a methylated mixture of lactic and succinic acids respectively.

The use of gas solid chromatography on Chromosorb 101 columns, introduced by Carlsson (1973), has the advantage that the fermentation broth can be used directly and that even lactic and succinic acids are eluted. However, this method has disadvantages, in particular the part of the column near the injector must be cleaned and refilled after several analyses, and the succinic acid is eluted very late.

The impact of g.l.c. methods can best be seen in certain fields of anaerobic bacteriology where theoretical and practical problems are involved. In medical bacteriology the Gram negative non-sporing anaerobic rods of the family Bacteroidaceae have been studied extensively using such methods.

3. Bacteroidaceae

The Gram negative obligately anaerobic bacteria classified in the family Bacteroidaceae are found in complex ecosystems in man. Thus, the saccharolytic *Bacteroides* spp., *Bcrd. fragilis*, *Bcrd. thetaiotaomicron*, *Bcrd. vulgatus*, *Bcrd.*

TABLE 1

Acid fermentation products of various anaerobic bacteria

Taxon	Description and acid end products
(1) Family Bacteroidaceae	Anaerobic, Gram negative, more or less pleomorphic rods
Genus *Bacteroides* species	*Iso*-butyric and *iso*-valeric acids are typical fermentation products
fragilis *thetaiotaomicron* *vulgatus* *distasonis* *variabilis* *eggerthii* *uniformis*	Acetic, propionic, *iso*-butyric, *iso*-valeric, lactic and succinic acids produced. Traces of butyric acid found in some strains. Strongly saccharolytic and proteolytic. Normal flora of human intestinal tract; *Bcrd fragilis* often found in anaerobic infections
melaninogenicus subsp.*melaninogenicus* subsp.*intermedius* *oralis*	Same acidic products as *fragilis*-group except propionic acid
asaccharolyticus *splanchnicus* *putredinis*	Same products as *fragilis*-group with large amounts of butyrate also produced. Succinic acid formed only by *Bcrd. splanchnicus*
Genus *Sphaerophorus*	Gram negative, usually pleomorphic rods; no *iso*-branched acids produced. Acetic, propionic and butyric acids, and varying amounts of lactate and succinate produced
Genus *Fusobacterium*	Gram negative rods, often very long cells with tapered ends. Acetic, propionic, butyric, lactic and succinic acids produced
(2) Family Peptococcaceae	Anaerobic, Gram positive cocci; found as normal flora in humans and in clinical specimens
Genus *Peptococcus*	Acetic and often butyric acids produced; other acids found variably
Genus *Peptostreptococcus*	Anaerobic cocci forming chains in culture media. Lactic acid not amongst the major fermentation products. Acetic acid formed often together with other acids. Most species contain strains which are more proteolytic than saccharolytic
Genus *Coprococcus*	Anaerobic cocci which are more saccharolytic than proteolytic. Formic, acetic, propionic and butyric acids produced
Genus *Ruminococcus*	Symbiotic inhabitants of herbivores, attacking plant fibres by fermenting cellulose. Produce acetic, formic, and succinic acids, and ethanol
(3) Miscellaneous Genus *Lactobacillus*	Gram positive rods, mostly facultatively anaerobic but sometimes obligately anaerobic with a typical lactic acid fermentation
Genus *Streptococcus*	Gram positive cocci occurring in chains, mostly facultatively anaerobic but sometimes obligately anaerobic with a typical lactic acid fermentation

TABLE 1 (*cont.*)
Acid fermentation products of various anaerobic bacteria

Taxon	Description and acid end products
Genus *Eubacterium*	Gram positive, obligately anaerobic, non-sporing rods. Products include acetic, propionic and butyric acids together with lactic and succinic acids. Members of some species exhibit very low fermentative activities. Propionic or lactic acids not the only major fermentation products
Genus *Propionibacterium*	Gram positive, pleomorphic rods. Propionic acid the major fermentation product, together with acetic, *iso*-valeric, lactic and succinic acids
Genus *Megasphaera*	Only one species, *Masf. elsdenii*, isolated from faeces of man and animals. Characteristic end products: acetic, propionic, *iso*-butyric, butyric, *iso*-valeric, valeric, caproic, lactic and succinic acids
Genus *Anaerobiospirillum* *Anbp. succinoproducens*	Spiral shaped, strict anaerobic bacteria isolated from dogs. Succinic and acetic acids formed

distasonis, Bcrd. variabilis, Bcrd. uniformis and *Bcrd. eggerthii*, are inhabitants of the intestinal tract in which they may reach populations of up to 10^{10}/g of faeces. Members of some of these species appear in clinical specimens, e.g. wound material, cervical swabs or even in cerebrospinal fluid and blood cultures. It has been reported (Werner & Hammann 1977) that *Bcrd. fragilis* accounts for more than 60% of Gram negative obligately anaerobic bacteria recovered from clinical specimens.

The results of g.l.c. analysis of members of the so-called *fragilis*-goup, which contains at least the seven species mentioned above, are shown in Table 1. The metabolic products are acetic, *iso*butyric, *iso*valeric, lactic, propionic and large amounts of succinic acids (Table 1). *Bacteroides fragilis, Bcrd. thetaiotaomicron, Bcrd. vulgatus, Bcrd. distasonis, Bcrd. variabilis, Bcrd. uniformis* and *Bcrd. eggerthii*, considered to be good species by Werner (1974) and Cato & Johnson (1976), cannot be separated solely on the basis of g.l.c. analyses. Members of these species can, however, be easily distinguished on the basis of carbohydrate fermentation patterns (Werner & Rintelen 1968; Werner 1974).

The g.l.c. is helpful in distinguishing members of the saccharolytic *fragilis*-group from weakly or non-saccharolytic *Bacteroides* spp. Most members of the genus *Bacteroides* produce relatively high amounts of *iso*butyric and *iso*valeric acids but, at best, only traces of butyric acid are formed by the *fragilis*-group, *Bcrd. oralis* or by *Bcrd. melaninogenicus*. *Bacteroides oralis* strains are weakly saccharolytic, colonize the human oral cavity and vagina and have little or no virulence. Members of the *fragilis*-group can be distinguished from *Bcrd. oralis* and *Bcrd. melaninogenicus* by the production of propionic acid which is virtually absent in strains of the latter. Typical butyric acid-producers are the oral and intestinal inhabitant *Bcrd. asaccharolyticus*, which is often associated with anaerobic infections of the upper and lower respiratory tract, and strains of two species of low virulence isolated from the human intestine, namely the weakly saccharolytic *Bcrd. splanchnicus* (Werner *et al.* 1975) and the non-saccharolytic *Bcrd. putredinis* (Table 1).

Bacteroides melaninogenicus formerly comprised the subspecies *asaccharolyticus*, *melaninogenicus*, and *intermedius* having in common black pigment formation (Holdeman & Moore 1974a). On the basis of G+C base ratios and butyric acid formation, *Bcrd. melaninogenicus* ssp. *asaccharolyticus* has been assigned to a separate species, *Bcrd. asaccharolyticus* (Finegold & Barnes 1977). *Bacteroides asaccharolyticus* strains can be distinguished from the saccharolytic pigmented strains as they are able to produce *n*-butyric acid but not succinic acid. In addition, the mucopeptide of the wall of *Bcrd. asaccharolyticus* strains contain lysine instead of the diaminopimelic acid found in the strains of the subspecies *melaninogenicus* and *intermedius* of *Bcrd. melaninogenicus* (Williams *et al.* 1975).

The g.l.c. is also of value in the classification of genera in the family Bacteroidaceae. Thus, members of the *Fusobacterium-Sphaerophorus* group can be separated from the genus *Bacteroides* by their inability to produce *iso*-branched acids. *Sphaerophorus* spp. (*Shrp. necrophorus, Shrp. varius, Shrp. freundii* var. *mortiferum*) form acetic, propionic and large amounts of butyric acids and trace amounts of some alcohols. Some strains of the *Sphaerophorus* group also produce *n*-valeric acid (Werner 1972b). In *Fusobacterium fusiforme* (*Fsob. nucleatum*) the tendency to form alcohols is much less pronounced than in *Sphaerophorus* spp. (Werner *et al.* 1971). On the basis of butyric acid production and similar DNA-base ratios, *Sphaerophorus* spp. and *Fusobacterium fusiforme* (*Fsob. nucleatum*) are currently classified in the genus *Fusobacterium* (Holdeman & Moore 1974a, Holdeman *et al.* 1977). However, restricting the genus name *Fusobacterium* to anaerobic Gram negative rods which produce butyric acid raises the problem of how to classify those *Sphaerophorus* and *Fusobacterium* spp. that are butyrate-negative but have many other characters in common with the afore-mentioned species (*Shrp. necrophorus, Shrp. varius, Fsob. nucleatum*). These problems have been discussed in detail (Barnes 1977; Langworth 1977).

The taxonomic importance of butyric acid production is further reduced by data from recent studies on the origin of fatty acid formation. The saccharolytic *Bacteroides* spp. (*Bcrd. fragilis, Bcrd. thetaiotaomicron* etc.), *Bcrd. oralis* and *Bcrd. melaninogenicus*, excluding strains now assigned to *Bcrd. asaccharolyticus*, normally do not form butyric acid when incubated in PYG medium. Without glucose, however, certain saccharolytic *Bacteroides* strains, such as the *Bcrd. vulgatus* strain shown in Fig. 1, may produce small amounts of butyric acid. Recent studies (Hammann & Werner, unpublished data) were designed to eluci-

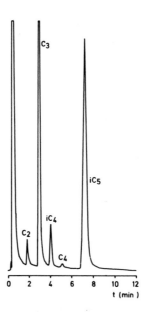

'Fig. 1. *Bacteroides vulgatus* strain 27-C20 grown in PY medium. Trace amounts of butyric acid (C_4) are produced besides the typical fermentation pattern comprising acetic (C_2), propionic (C_3), *iso*-butyric (iC_4) and *iso*-valeric (iC_5) acids.

date which amino acid yields butyric and other short chain fatty acids when incubated with members of the family Bacteroidaceae. In these experiments, PYG-grown cells were washed in sterile phosphate buffer and incubated anaerobically overnight in sterilized solutions of various L-amino acids. Assays were done in duplicate, and controls of cells in phosphate buffer without amino acid were run in parallel.

Amino acids which gave positive results with one or more of the test strains when examined for volatile acids by gas chromatography are shown in Table 2. It is evident that alanine is a very poor substrate for the organisms tested. Cysteine, the degradation of which leads to the intermediate pyruvate, yielded propionate with members of the saccharolytic *Bacteroides* spp. whereas *Bcrd. asaccharolyticus* and most of the *Sphaerophorus* spp. tested formed butyric acid. Methionine and threonine were fermented to propionic and, to a lesser extent, to butyric acid. The branched-chain amino acid leucine was degraded to the corresponding *iso*valeric acid by representatives of all the *Bacteroides* spp. while *Sphaerophorus* spp. were clearly unable to form and excrete any *iso*-branched fatty acid (Table 2).

It is evident that glutamic acid, histidine and lysine are typical "butyrogenic' amino acids. In the case of glutamic acid and lysine large amounts of butyric

TABLE 2

Degradation of amino acids to fatty acids by washed cells of strains representing the family Bacteroidaceae

Strains	Alanine	Cysteine	Methionine	Threonine	Leucine	Glutamic Acid	Histidine	Lysine
Bacteroides								
fragilis Kö 57a	–	p	p	p,b	a,P,iV	A,p,B	A,p,B	A,P,B
fragilis NCTC 9343	–	–	–	p,b	a,p,b,iV	A,p,B	A,p,B	A,p,B
thetaiotaomicron NCTC 10582	–	p	–	p,b	a,p,b,iv	p	p	A,p,B
distasonis AI 149	–	p	p	p	a,p,b,iv	a,P,b	A,p,B	A,P,B
vulgatus E 1/1	p	a,P	p	P	a,P,iV	P,b	p	a,P
vulgatus BM 137	p	p	p	A,P,b	a,p,b,iv	a,P,B	p	A,P,B
putredinis S 1/36	–	p	–	p,b	a,p,B,iV	A,P,B	–	a,p,B
putredinis S 27/35	–	–	–	p	a,iv	A,B	–	A,B
asaccharolyticus NCTC 9337	–	–	p	p	a,b	A,p,B	a,p,B	A,B
asaccharolyticus VPI 3280	b	b	P,b	P,b	a,p,b,iv	–	A,B	A,p,B
Sphaerophorus								
freundii ATCC 9817	–	B	P	P	a,b	–	a,B	B
freundii Lille 252	–	B	P,b	P	a,b	–	B	B
necrophorus Hu 2495	–	–	P,B	P,b	a,p,b	–	B	a,B
necrophorus F 2597	–	b	p,b	P,B	b	a,B	B	A,B
varius ATCC 8501	–	B	P	P	b	A,B	a,B	A,B
varius IPP 700	–	a,B	P	P	B	A,B	A,B	A,B

a, A, acetic acid; b, B, butyric acid; iv, iV, *iso*-valeric acid; p, P, propionic acid. Large amounts of the respective acids are given in capitals.

264   R. HAMMANN AND H. WERNER

acids are found in addition to acetic and propionic acids. This also holds true of members of the saccharolytic *Bacteroides* spp. (Table 2), which form at best only traces of butyric acid when grown in PYG or PY. Representative data obtained with saccharolytic *Bacteroides* cells incubated with glutamic acid and lysine, respectively, are shown in Fig. 2. High peaks of butyric acid, together with acetic and propionic acids, are evident.

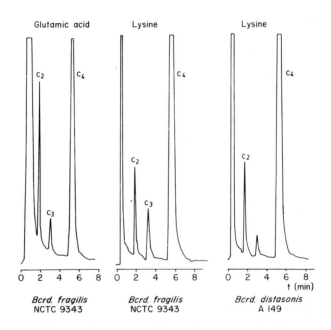

Fig. 2. Production of fatty acids from amino acids by washed cell suspensions of saccharolytic *Bacteroides* spp. C_2, acetic; C_3, propionic; C_4, butyric acids.

These results show that the saccharolytic *Bacteroides* spp. also have the enzymatic ability to form butyrate. It appears, however, that this property is fully or partially repressed when the bacteria are grown in complex media. The presence and activity of amino acid decarboxylases leading to the production of amines may also be significant and may compete with the deaminating enzymes. Again it is known that the content of certain amino acids in various kinds of peptone is different (Jennens 1954). Consequently, although butyric acid is produced by some *Bacteroides* strains, it is not a typical end product of the saccharolytic species.

In members of the butyrate-positive *Sphaerophorus-Fusobacterium* group the absence of *iso*-valeric and *iso*-butyric acids seems to be a character of prime taxonomic importance. Reproducibility studies have shown that these acids are formed neither in complex media nor in media containing single amino acids.

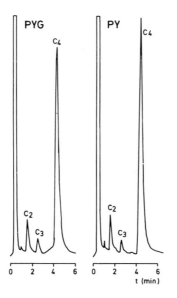

Fig. 3. Fermentation pattern of *Peptococcus asaccharolyticus*. C_2, acetic; C_3, propionic; C_4, butyric acids; PYG, peptone yeast-extract glucose medium; PY, peptone yeast-extract medium.

4. Peptococcaceae

The anaerobic Gram positive cocci also play a significant role in human and animal pathology. Rogosa (1971c) classified these bacteria in the family Peptococcaceae which includes the genera *Peptococcus, Peptostreptococcus, Ruminococcus* and *Coprococcus*. Members of all of these taxa except *Ruminococcus* are inhabitants of man and exhibit varying degrees of pathogenicity. *Peptostreptococcus anaerobius* frequently causes puerperal fever in humans, and other species and has been isolated from human genital and intestinal tracts, respiratory organs, appendicitis, blood, as well as from infections in animals.

The genus *Peptococcus* contains butyric acid-producing and butyrate-negative species. Varying amounts of lactic and propionic acids are produced but only traces of succinic acids are formed. The taxonomy of *Peptococcus prevotii, Petc. asaccharolyticus*, and *Petc. aerogenes* has already been discussed in detail by several workers (Schleifer & Nimmermann 1973; Bentley & Dawes 1974; Montague & Dawes 1974; Böhm & Werner 1975). However, the present taxonomic status of most of the *Peptococcus* spp. is not sound and further systematic studies are needed to determine the relationship between peptococci. In the meantime, however, gas-chromatographic analysis of metabolic products is of value in the identification and characterization of single strains.

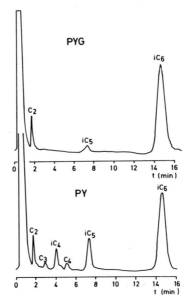

Fig. 4. Volatile fatty acids produced by *Peptostreptococcus anaerobius* in PYG and PY medium, C_2, acetic; C_3, propionic; iC_4, *iso*-butyric; C_4, butyric; iC_5, *iso*-valeric; iC_6, *iso*-caproic acids; PYG, peptone yeast-extract glucose medium; PY, peptone yeast-extract medium.

The genus *Peptostreptococcus* is heterogeneous. *Peptostreptococcus anaerobius* has a marked characteristic fermentation pattern (Fig. 4) typically producing acids from C_2 to C_6 in PY medium. Strains of some species formerly assigned to the genus *Peptostreptococcus* are probably microaerophilic or anaerobic members of the genus *Streptococcus*. Such strains produce neither fetid odours nor volatile fatty acids with more than two carbon atoms but exhibit a typical lactic acid fermentation and were consequently excluded from the genus *Peptostreptococcus* in the latest edition of *Bergey's Manual of Determinative Bacteriology* (Rogosa 1974). Given this classification it is easy to determine whether an anaerobic coccus should be assigned to the genus *Peptostreptococcus* or to the genus *Streptococcus*.

The genus *Coprococcus* was introduced (Holdeman & Moore 1974*b*) for Gram positive cocci that exhibit a greater saccharolytic ability than *Peptococcus* and *Peptostreptococcus*. Coprococci utilize carbohydrates as a source of energy and only amino acids as nitrogen sources (Holdeman & Moore 1974*b*). The fermentation patterns of these bacteria include formic, acetic, propionic and butyric acids; lactic acid may or may not be produced (Table 1).

The genus *Ruminococcus* includes cellulose-splitting strains which inhabit the rumen. These bacteria produce acetic acid, ethanol, varying amounts of lactic and succinic acids but fatty acids with more than two carbon atoms are not found (Table 1).

5. Miscellaneous

Gas chromatography can also be useful for the identification of other taxa of anaerobic bacteria. Thus the importance of gas chromatography in the identification of *Clostridium* spp. was first demonstrated by Moore *et al.* (1966). The large group of anaerobic, Gram positive, non-sporing rods provides another vast domaine for the application of g.l.c., as the classification of many of these organisms leaves a lot to be desired. Morphological differences where they exist at all are often only slight, though physiological tests may sometimes be helpful (Holdeman *et al.* 1977). DNA homology studies may help to clarify the classification of these bacteria but such studies are still too laborious for most routine laboratories (see Chapter 1).

Many strains of *Lactobacillus* behave anaerobically on primary isolation but can be easily recognized by their volatile and non-volatile acids. Lactic acid is the sole major fermentation product in lactobacilli but, while acetic and succinic acids are produced in traces, fatty acids with more than two carbon atoms are not found. These features distinguish *Lactobacillus* from strains classified in the genera *Propionibacterium, Bifidobacterium* and *Eubacterium.*

The genus *Eubacterium* contains obligately anaerobic Gram positive rods which do not produce major amounts of lactic or propionic acids but usually form fatty acids with more than two carbon atoms. Most *Eubacterium* species produce major amounts of butyric acid. However, some strains of *Eubacterium* can only be separated from *Clostridium* by their inability to form endospores. In such cases, therefore, gas chromatography is not decisive in identifying butyric acid-producing Gram positive rods growing anaerobically and appropriate tests for spore formation must be made. The volatile acid spectrum of a typical *Eubacterium* strain, *Eubc. limosum*, isolated from human faeces, is shown in Fig. 5. The difference in acidic products between strains grown in PYG and PY is, together with some other physiological and morphological properties, characteristic of this species. Other *Eubacterium* spp., e.g. *Eubc. lentum*, are only weakly saccharolytic or proteolytic and exhibit little metabolic activity.

Gram positive, rod-like, sometimes more or less pleomorphic, propionibacteria can be easily identified by gas chromatography. The fermentation products of such strains always include large amounts of propionic and acetic acids with differing amounts of *iso*-valeric, lactic and succinic acids (Table 1).

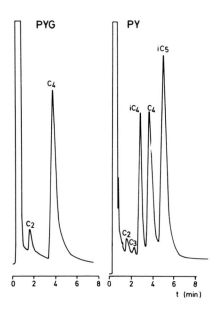

Fig. 5. Fermentation pattern of *Eubacterium limosum* strain 28-C46. C_2, acetic; C_3, propionic; iC_4, *iso*-butyric; C_4, butyric; iC_5, *iso*-valeric acids; PYG, peptone yeast-extract glucose medium; PY, peptone yeast-extract medium.

Another little known anaerobic bacterium that can be easily recognized by gas chromatographic analysis of metabolic end products is *Megasphaera elsdenii*. Strains of this taxon were isolated and described as 'organism LC' (Elsden & Lewis 1953; Elsden *et al.* 1956) and classified in the genus *Peptostreptococcus* by Gutierrez *et al.* (1959). Rogosa (1971*a*) transferred these strains from the genus *Peptostreptococcus* to the new genus *Megasphaera* on the basis of characteristic morphological and cytological properties and included the taxon in the new family Veillonellaceae (Rogosa 1971*b*). Werner (1973) was the first to isolate *Masf. elsdenii* from human faeces, an observation repeated by Sugihara *et al.* in 1974. The organism is a large coccus, is mostly Gram negative but can be Gram variable, is extremely oxygen-sensitive and has a typical volatile fatty acid pattern consisting of *iso*- as well as the non-branched butyric, valeric and *n*-caproic acids (Table 1).

Gas chromatography of metabolic end products has also been found to be of value in the classification of new bacteria. This was true of *Anaerobiospirillum succinoproducens* proposed by Davis *et al.* (1976) for isolates from dogs. The name was chosen because these bacteria produce succinic acid as the main metabolic end product.

6. Conclusions

Among the numerous possible applications of gas chromatography in the biological sciences, the analysis of volatile fatty and non-volatile acids produced during growth of anaerobic bacteria represents a method which cannot be dispensed with in the classification and identification of anaerobes. The impact of the method, however, varies according to the anaerobic taxon involved. Only few species, e.g. *Peptostreptococcus anaerobius*, *Eubacterium limosum* and *Megasphaera elsdenii*, can be unmistakably identified by their acidic products alone. Closely related species. e.g. *Bacteroides fragilis*, *Bcrd. thetaiotaomicron* and the other taxa in the saccharolytic *Bacteroides* group, may show the same fermentation pattern.

Certain features may be characteristic of a higher taxon such as a genus. Thus *iso*-butyric and *iso*-valeric acids are produced by all 'true' *Bacteroides* spp.

Some characters are shared by members of different taxa. This is particularly true of butyric acid production which is observed in members of the genera *Fusobacterium*, *Sphaerophorus*, *Peptococcus*, and *Eubacterium*, in some *Bacteroides* spp. (*Bcrd. asaccharolyticus*, *Bcrd. splanchnicus*) as well as in many *Clostridium* spp. In saccharolytic *Bacteroides* strains (e.g. *Bcrd. fragilis*, *Bcrd. thetaiotaomicron*) butyric acid formation is more pronounced after incubation with certain amino acids. Butyric acid formation can, therefore, probably not serve as a primary taxonomic character.

As the fermentation patterns of anaerobic bacteria may be assumed to be more stable than the utilization of single carbohydrates, which might be changed by mutation of single enzymes, analysis of metabolic end products by g.l.c. can generally be considered a taxonomic character of prime importance. On the other hand, an anaerobic bacterial strain cannot be properly identified without due regard to other taxonomic criteria such as biochemical, physiological and morphological properties.

The support of the Deutsche Forschungsgemeinschaft and the technical assistance of Mrs M. Clotten, Mrs A. Schmitz and Miss I. Werner are gratefully acknowledged.

7. References

ALCOCK, N. W. 1969 Separation of citric acid and related compounds by gas chromatography. In *Methods in Enzymology* Vol. XIII, ed. Colowick, S. P. & Kaplan, N. O. pp. 397–415. London: Academic Press.

BARNES, E. M. 1977 Problems in the classification of anaerobic bacteria. Metronidazole: Proceedings, Montreal, May 26-28, 1976, *Excerpta Medica* 1977.

BENTLEY, C. M. & DAWES, E. A. 1974 The energy-yielding reactions of *Peptococcus prévotii*, their behaviour on starvation and the role and regulation of threonine dehydratase. *Archives of Microbiology* 100, 363-387.

BÖHM, G. & WERNER, H. 1975 Das serologische Verhalten der pyogenen Anaerobier *Peptococcus asaccharolyticus* und *P. prevotii. Pathologia et Microbiologia* 43, 17-25.

CARLSSON, J. 1973 Simplified gas chromatographic procedure for identification of bacterial metabolic products. *Applied Microbiology* 25, 287-289.

CATO, E. P. & JOHNSON, J. L. 1976 Reinstatement of species rank for *Bacteroides fragilis, B. ovatus, B. distasonis, B. thetaiotaomicron* and *B. vulgatus*: Designation of neotype strains for *Bacteroides fragilis* (Veillon and Zuber) Castellani and Chalmers and *Bacteroides thetaiotaomicron* (Distaso) Castellani and Chalmers. *International Journal of Systematic Bacteriology* 26, 230-237.

DAVIS, C. P., CELVEN, D., BROWN, J. & BALISH, E. 1976 *Anaerobiospirillum*, a new genus of spiral-shaped bacteria. *International Journal of Systematic Bacteriology* 26, 498-504.

ELSDEN, S. R. & LEWIS, D. 1953 The production of fatty acids by a Gram-negative coccus. *Biochemical Journal* 55, 183-189.

ELSDEN, S. R., VOLCANI, B. E., GILCHRIST, F. M. C. & LEWIS, D. 1956 Properties of a fatty acid forming organism isolated from the rumen of sheep. *Journal of Bacteriology* 72, 681-689.

FINEGOLD, S. M. & BARNES, E. M. 1977 Report of the ICSB Taxonomic Sub-committee on Gram-negative Anaerobic Rods: Proposal that the saccharolytic and asaccharolytic strains at present classified in the species *Bacteroides melaninogenicus* (Oliver and Wherry) should be reclassified in two species as *Bacteroides melaninogenicus* and *Bacteroides asaccharolyticus. International Journal of Systematic Bacteriology* 27, 388-391.

GUTIERREZ, J., DAVIS, R. E., LINDAHL, L. & WARWICH, E. J. 1959 Bacterial changes in the rumen during the onset of feed-lot bloat of cattle and characteristics of *Peptostreptococcus elsdenii* n. sp. *Applied Microbiology* 7, 16-22.

HAMMANN, R. 1977 *Untersuchungen zur Physiologie der Ordnung Actinomycetales Buchanan 1917: Bildung organischer Säuren und Neutralprodukte, Oxidation von C_1-Verbindungen und Abbau von Aromaten.* Ph.D. Thesis, Technische Hochschule Darmstadt, Germany.

HOLDEMAN, L. V. & MOORE, W. E. C. 1974a Bacteroidaceae. In *Bergey's Manual of Determinative Bacteriology*, 8th edn. ed. Buchanan, R. E. & Gibbons, N. E. pp. 384-418. Baltimore: Williams & Wilkins.

HOLDEMAN, L. V. & MOORE, W. E. C. 1974b New genus, *Coprococcus*, twelve new species, and emended descriptions of four previously described species of bacteria from human feces. *International Journal of Systematic Bacteriology* 24, 260-277.

HOLDEMAN, L. V., CATO, E. P. & MOORE, W. E. C. (Ed.) 1977 *Anaerobe Laboratory Manual*, 4th edn. Blacksburg: Virginia Polytechnic Institute and State University, Anaerobe Laboratory.

JENNENS, M. G. 1954 The methyl red test in peptone media. *Journal of General Microbiology* 10, 121-126.

LANGWORTH, B. F. 1977 *Fusobacterium necrophorum*: Its characteristics and role as an animal pathogen. *Bacteriological Reviews* 41, 373-390.

MONTAGUE, M. D. & DAWES, E. A. 1974 The survival of *Peptococcus prévotii* in relation to the adenylate energy charge. *Journal of General Microbiology* 80, 291-299.

MOORE, W. E. C., CATO, E. P. & HOLDEMAN, L. V. 1966 Fermentation patterns of *Clostridium* species. *International Journal of Systematic Bacteriology* 16, 383-415.

ROGOSA, M. 1971a Transfer of *Peptostreptococcus elsdenii* Gutierrez *et al.* to a new genus. *Megasphaera* [*M. elsdenii* (Gutierrez *et al.*) comb. nov.] *International Journal of Systematic Bacteriology* 21, 187-189.

ROGOSA, M. 1971b Transfer of *Veillonella* Prévot and *Acidaminococcus* Rogosa from

Neisseriaceae to Veillonellaceae fam. nov., and the inclusion of *Megasphaera* Rogosa in Veillonellaceae. *International Journal of Systematic Bacteriology* **21**, 231–233.

ROGOSA, M. 1971c Peptococcaceae, a new family to include the Gram-positive, anaerobic cocci of the genera *Peptococcus, Peptostreptococcus,* and *Ruminococcus. International Journal of Systematic Bacteriology* **21**, 234–237.

ROGOSA, M. 1974 Peptococcaceae. In *Bergey's Manual of Determinative Bacteriology,* 8th edn. ed. Buchanan, R. E. & Gibbons, N. E. pp. 517–528. Baltimore: Williams & Wilkins.

SCHLEIFER, K. H. & NIMMERMANN, E. 1973 Peptidoglycan types of strains of the genus *Peptococcus. Archiv für Mikrobiologie* **93**, 245–258.

SUGIHARA, P. T., SUTTER, V. L., ATTEBERY, H. R., BRICKNELL, K. S. & FINEGOLD, S. M. 1974 Isolation of *Acidaminococcus fermentans* and *Megasphaera elsdenii* from normal human feces. *Applied Microbiology* **27**, 274–275.

THAUER, R. K., JUNGERMANN, K. & DECKER, K. 1977 Energy conservation in chemotrophic anaerobic bacteria. *Bacteriological Reviews* **41**, 100–180.

WERNER, H. 1972a Anaerobierdifferenzierung durch gaschromatographische Stoffwechselanalysen. *Zentralblatt für Bakteriologie, Parasitenkunde, Infektionskrankheiten und Hygiene I. Abteilung Originale A* **220**, 446–451.

WERNER, H. 1972b A comparative study of 55 *Sphaerophorus* strains. Differentiation of 3 species: *Sphaerophorus necrophorus, Sph. varius* and *Sph. freundii. Medical Microbiology and Immunology* **157**, 299–314.

WERNER, H. 1973 *Megasphaera elsdenii*–ein normaler Bewohner des menschlichen Dickdarmes? *Zentralblatt für Bakteriologie, Parasitenkunde, Infektionskrankheiten und Hygiene I. Abteilung Originale A* **223**, 343–347.

WERNER, H. 1974 Differentiation and medical importance of saccharolytic intestinal *Bacteroides. Arzneimittel-Forschung (Drug Research)* **24**, 340–343.

WERNER, H. & HAMMANN, R. 1977 Comparative studies of normal anaerobic flora and infectious processes in man as a means to establish the virulence of *Bacteroides* species. *Zentralblatt für Bakteriologie, Parasitenkunde, Infektionskrankheiten und Hygiene I. Abteilung Originale A, Supplement* 7, 193–196.

WERNER, H. & RINTELEN, G. 1968 Untersuchungen über die Konstanz der Kohlenhydratspaltung bei intestinalen *Bacteroides-* (Eggerthella-) Arten. *Zentralblatt für Bakteriologie, Parasitenkunde, Infektionskrankheiten und Hygiene I. Abteilung Originale* **208**, 521–528.

WERNER, H., NEUHAUS, F. & HUSSELS, H. 1971 A biochemical study of fusiform anaerobes. *Medical Microbiology and Immunology* **157**, 10–16.

WERNER, H., RINTELEN, G. & KUNSTEK-SANTOS, H. 1975 Eine neue buttersäurebildende *Bacteroides*-Art: *B. splanchnicus* n. sp. *Zentralblatt für Bakteriologie, Parasitenkunde, Infektionskrankheiten und Hygiene I. Abteilung Originale A* **231**, 133–144.

WILLIAMS, R. A. D., BOWDEN, G. H., HARDIE, J. M. & SHAH, H. 1975 Biochemical properties of *B. melaninogenicus* subspecies. *International Journal of Systematic Bacteriology* **25**, 298–300.

Classification and Identification of Bacteria by Electrophoresis of Their Proteins

K. KERSTERS AND J. DE LEY

Laboratorium voor Microbiologie en microbiële Genetica, Faculteit der Wetenschappen, Rijksuniversiteit Gent, Gent, Belgium

Contents

1. Introduction

A BACTERIAL STRAIN, growing in standardized conditions always produces the same set of proteins. The amino acid sequence, the molecular weight and the net electrical charge of each protein species are determined by the nucleotide sequence of the corresponding cistron in the DNA. The number of copies, i.e. the amount of each protein species is likewise genetically determined. Zone electrophoresis of the mixture of all these proteins in well-defined standardized conditions produces protein banding patterns (electrophoregrams) which can be considered as 'finger-prints' of the bacterial strains under investigation. Electrophoresis of total soluble or envelope proteins of a bacterial strain yields a complex pattern, where each protein band usually consists of a number of structurally different protein species with identical electrophoretic mobility. It should be emphasized that identical electrophoretic mobility of protein bands from different bacteria does not necessarily imply that these proteins possess an identical primary structure. Interchange among the neutral amino acids does not usually alter the electrophoretic mobility of a protein. Substitution of a lysine residue by arginine will likewise have little effect on the electrophoretic characteristics of the protein, whereas a single substitution of a basic residue by a neutral or acidic one can drastically change the electrophoretic properties. In spite of these theoretical considerations, it is our experience that proteins of genetically closely related bacterial strains display similar or almost identical electrophoregrams (Kersters & De Ley 1975; Swings et al. 1976).

Protein electrophoresis is a powerful, relatively simple and inexpensive method that has been extensively used for the classification and identification of bacteria. It has the advantage that electrophoretic mobilities of a large number of proteins, encoded by a large part of the bacterial genome, can be compared in well-defined standardized conditions. The method is usually faster than DNA:DNA hybridizations (see Chapter 1) and elaborate phenotypic comparisons between strains (see Chapter 4), but slower than pyrolysis-gas-liquid chromatography.

In this contribution the impact of gel electrophoresis of proteins on the classification and identification of several bacterial genera is discussed. Special emphasis is given to the results obtained using computer-assisted techniques for comparing and grouping the electrophoretic data.

2. Methods

Electrophoresis of bacterial proteins is usually carried out according to one of the following general principles:

(1) The proteins from a bacterial strain (e.g. soluble proteins from a cell-free extract or solubilized cell-envelope proteins) are submitted to electrophoresis (usually in polyacrylamide gels), stained with amido black or coomassie blue and the entire banding pattern examined and compared with electrophoregrams from other strains.

(2) The native proteins of a bacterial strain (e.g. a cell-free extract) are submitted to electrophoresis in a suitable stabilizing medium (e.g. starch gel, agar gel, polyacrylamide gel or polyacrylamide–agarose mixtures) and stained for certain enzymes only. This zymogram technique detects individual enzymes by specific staining and compares their electrophoretic mobilities.

The present contribution is mainly concerned with the application of comparative electrophoresis of total soluble proteins; the zymogram technique is considered in the paper by Williams & Shah (Chapter 12).

A. Preparation of sample and polyacrylamide gel electrophoresis

Two different types of polyacrylamide gel electrophoresis (PAGE) are currently used for bacterial proteins: (i) the disc electrophoretic technique in cylindrical gels pioneered by Ornstein (1964) and Davis (1964), and (ii) the vertical and horizontal slab gel techniques. The theoretical and practical aspects of PAGE are well-documented, and details can be found in a number of monographs and articles (Maizel 1969; Cooksey 1971; Maurer 1971; Ames 1974; Gordon 1975). It can be seen from Table 2 that a great number of sometimes widely divergent gel and buffer systems are used depending upon the type of proteins separated (soluble, solubilized, ribosomal and cell-envelope proteins). This diversity in methodology makes inter-laboratory comparisons almost impossible. Moreover, several authors have pointed out that small variations in the position of protein bands can occur between different electrophoretic runs of the same cell-free extract (Moses & Wild 1969; Armstrong & Yu 1970; Morris 1973; Kersters & De Ley 1975; Swings *et al.* 1976). Although the technique of PAGE is relatively simple, the entire procedure consists of a great number of experimental steps from the growth of the bacteria to the scanning of the electrophoregrams. It is for this reason that special attention has been paid to numerous aspects of the reproducibility of PAGE of soluble bacterial proteins in continuous alkaline buffer systems (Kersters & De Ley 1975). Reproducible and objective comparison of electrophoregrams from different electrophoretic runs became possible by inclusion of internal reference proteins (e.g. ovalbumin and thyroglobulin), allowing normalization of the electrophoregrams (Fig. 1).

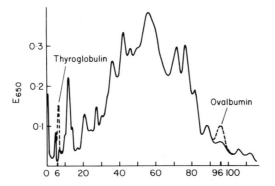

Fig. 1. Densitometric tracing of the electrophoretic protein pattern of *Alcaligenes faecalis* NCIB 8156 (——— ; reference strain) and part of the corresponding protein pattern with the reference proteins thyroglobulin and ovalbumin (- - - -). The entire tracing is divided in unit distances on the horizontal axis. A unit distance is 1/90th of the distance between both reference proteins.

The following basically different techniques have found wide application in bacterial classification and identification (see Table 2):

(1) Cell-free extracts are prepared e.g. by ultrasonic disintegrator, French pressure cell or X-press and the native soluble proteins submitted to electrophoresis in an alkaline gel and buffer system (usually Tris-HCl

buffer, pH 8.7–8.9) (Davis 1964; Maurer 1971; Gordon 1975; Kersters & De Ley 1975).

(2) Cellular proteins are solubilized by a mixture of phenol-acetic acid-water (4:2:1; w/v/v) and electrophoresis carried out in an acetic acid and urea containing gel and buffer system (Takayama et al. 1966; Razin & Rottem 1967; Cooksey 1971). Several authors used this technique for studying representatives of the Mycoplasmatales (see Table 2). This method offers a greater safety to workers preparing cell-free extracts from pathogenic bacteria.

Electrophoresis of proteins in the presence of sodium-dodecyl sulphate (SDS) (Maizel 1969; Cooksey 1971; Ames 1974) and recently introduced techniques such as iso-electric focusing (Vesterberg 1971), and gradient gel electrophoresis (Rodbard et al. 1971) have to date only found a limited application for the classification and identification of bacteria. The SDS-PAGE technique seems to be particularly suitable for comparing cell-envelope and ribosomal proteins (Table 2). Iso-electric focusing has been successfully applied by Matthew & Harris (1976) for comparison of β-lactamases from almost 250 strains belonging to 21 different bacterial genera.

B. The quantitative comparison of electrophoregrams by computer-assisted techniques

The degree of resemblance between electrophoregrams of a small number of bacterial extracts is usually estimated visually, either directly by comparing the stained gels or indirectly by visual comparison of the normalized photographs of the gels, densitometric tracings or schematic drawings. The eye and the human brain are unable, however, to compare objectively and remember large numbers of gels: computer-assistance is obviously required. A computer-assisted technique has been developed for quantitative comparison and grouping of normalized scannings of electrophoregrams (Kersters & De Ley 1975). In this technique the relative mobility, the sharpness and relative protein concentration of the peaks and the valleys between them are taken into account. The Pearson product-moment correlation coefficient is used as a measure of similarity between each pair of electrophoregrams. The main steps in the procedure are schematically illustrated in Table 1. Feltham (1975) used similar but more advanced computer-techniques for comparison and grouping of electrophoregrams of 40 representative strains of the Enterobacteriaceae.

The advantages of computer-assisted comparison of electrophoregrams are:

(1) objective measuring of the degree of similarity between each pair of electrophoregrams;

(2) grouping of the electrophoregrams by a variety of clustering techniques;

TABLE 1

Summary of the quantitative comparison of electrophoregrams by the method of Kersters & De Ley (1975)

(1)	Grow bacteria in standard conditions
(2)	Disrupt cells and centrifuge extracts in standard conditions
(3)	Prepare, stain and scan electrophoregrams in standard conditions
(4)	Scan two stained gels of each protein sample, one with and another without the reference proteins ovalbumin and thyroglobulin, in a microdensitometer at 650 nm
(5)	Superimpose both tracings and mark the positions of ovalbumin and thyroglobulin on the tracing without the reference proteins (see Fig. 1)
(6)	Normalize the X-axis of each tracing by dividing the distance between the position of both reference proteins into 90 equal parts (Fig. 1) and extend the division all over the electrophoregram
(7)	Punch the optical density values of all boundary positions (No 1–110 for Fig. 1) onto data cards
(8)	Computation of Pearson's product-moment correlation coefficient r between each pair of normalized densitometric tracings
(9)	Transformation of r-matrix into z-matrix (z is the inverse hyperbolic tangent of r)
(10)	Clustering of z-values by the unweighted pair-group method using arithmetic averages
(11)	Transformation of z-values of clustering levels in r-values
(12)	Draw dendrogram (Fig. 2) or shaded r-matrix (Fig. 6)

Items (8)–(12): Computer

(3) storage of large numbers of electrophoregrams on punched cards, tapes or disc files and retrieval of the data, and

(4) identification of new and unknown isolates by computer-assisted comparison of their electrophoretic protein patterns with known patterns of large numbers of reference organisms (stored e.g. on data files).

3. Some Applications of Gel Electrophoresis of Proteins in the Classification and Identification of Bacteria

A selective survey of the literature on the application of comparative polyacrylamide gel electrophoresis of proteins to the classification and identification in different bacterial genera is shown in Table 2. Only investigations on relatively large numbers of strains are considered and the genera are arranged according to the type of proteins investigated and the gel and buffer systems used. It is obvious that a great number of different experimental approaches have been used, and some of the more significant contributions will be discussed briefly. In order to demonstrate the possibilities and limitations of PAGE to the examination of total soluble bacterial proteins, some of our recent (mostly unpublished) results will be considered in greater detail, with special emphasis given to the grouping of electrophoregrams by numerical analysis.

TABLE 2

Bacterial genera examined by polyacrylamide gel electrophoresis of proteins

(A) Soluble proteins; alkaline gel and buffer system

Acetobacter, Gluconobacter (200)*	Kersters & De Ley, unpublished
Actinoplanaceae (31) 10 genera	Davies & Gottlieb (1973)
Agrobacterium (250)*	Kersters & De Ley (1975) and unpublished
Alcaligenes, Achromobacter (200)*	Kersters & De Ley, this chapter,
Arthrobacter (44)*	Rouatt *et al.* (1970)
Campylobacter (40)	Morris & Park (1973)
Enterobacteriaceae (40)*	Feltham (1975)
Enterobacteriaceae from beer breweries (60)*	van Vuuren, Kersters & De Ley (to be published) and this chapter
Mycobacterium (55)	Haas *et al.* (1972, 1974)
Pseudomonas, plant pathogenic (38)	Palmer & Cameron (1971)
Xanthomonas (70)	El-Sharkawy & Huisingh (1971)
Yersinia pestis (160)*	Hudson *et al.* (1973, 1976)
Zymomonas (43)*	Swings *et al.* (1976)

(B) Proteins soluble in phenol-acetic acid-water (4:2:1, w/v/v); acid and urea containing gel and buffer system

Brucella (38)	Morris (1973); Balke *et al.* (1977)
Campylobacter (40)	Morris & Park (1973)
Corynebacterium (38)	Larsen *et al.* (1971)
Enterobacteriaceae (84) (9 genera)	Sacks *et al.* (1969)
Mycoplasma (28)	Razin & Rottem (1967), Rottem & Razin (1967)
Mycoplasma, porcine (30)	Ross & Karmon (1970)
Mycoplasma, canine (53)	Rosendal (1973)
Mycoplasma, avian (45)	Müllegger *et al.* (1973); Müllegger & Gerlach (1974)
Mycoplasma, bovine, equine (34)	Dellinger & Jasper (1972)

(C) Cell-envelope proteins

Neutral, SDS-containing gel and buffer system

Bordetella pertussis (13)	Parton & Wardlaw (1975)
Neisseria (10)	Russell *et al.* (1975); Russell & McDonald (1976)
Salmonella (8)	Parton (1975)

Urea containing gel and buffer system; iso-electric focusing

Streptococcus, cariogenic (7)	Hamada & Mizuno (1974)

(D) Ribosomal proteins

Neutral and SDS-containing gel and buffer system

Bacillus (2), *Escherichia, Proteus, Salmonella* (3)	Sun *et al.* (1972)

Acid and urea containing gel and buffer system

Enterobacteriaceae (9) (*Escherichia, Enterobacter, Erwinia, Klebsiella, Salmonella*)	Schaad (1974)
Erwinia (9)	Kado *et al.* (1972)

The numbers of the strains investigated are shown in parentheses. An asterisk (*) indicates that the electrophoregrams were compared and grouped by numerical analysis.

A. The Mycoplasmatales

More than 40 publications deal with the comparative aspects of PAGE of acetic acid–phenol soluble proteins from different groups of *Mycoplasma* and *Acholeplasma* strains. The analysis and comparison of electrophoretic patterns is now an accepted identification technique for the Mycoplasmatales (Anon 1972). Recent descriptions of new species of *Mycoplasma* and *Acholeplasma* are frequently based on classical biochemical and serological tests together with comparison of protein electrophoregrams e.g. *Mypl. hyosynoviae* (Ross & Karmon 1970), *Mypl. cynos* (Rosendal 1973), *Mypl. capricolum* and *Mypl. putrefaciens* (Tully *et al*. 1974) and *Mypl. molare* (Rosendal 1974). *Mycoplasma arginini* and *Mypl. leonis* were considered as synonyms because of serological and biochemical similarities and their almost identical protein patterns (Tully *et al*. 1972). The success of PAGE for the differentiation and identification of *Mycoplasma* and *Acholeplasma* is due to the difficulties encountered in classifying these bacteria using classical microbiological methods.

The morphological similarity between *Mycoplasma* and bacterial L-forms poses a problem for their differentiation and identification. PAGE was found to be a simple and reliable technique for the generic identification of several bacterial L-forms from *Proteus, Staphylococcus* and *Streptococcus* (Razin & Shafer 1969; Theodore *et al*. 1969, 1971).

B. Yersinia pestis

Hudson and co-workers from the Center for Disease Control (USA) investigated more than 160 isolates of *Yersinia pestis* from South-East Asia, USSR, Manchuria, Java, Yemen, Iran, Kenya, Republic of South Africa, Madagascar, South and North America. Although the protein electrophoregrams of these strains were qualitatively very similar, significant reproducible quantitative differences existed for some of the protein bands. Hudson *et al*. (1976) applied numerical techniques and grouped their isolates in 11 major protein variants. Plague isolates from geographically and epidemiologically limited foci usually displayed similar electrophoregrams and can be differentiated from electrophoretypes from other areas. Isolates collected in the same area with a time interval of more than 10 years possessed identical protein patterns. Some important epidemiological conclusions could be drawn and Hudson *et al*. (1976) concluded that "classification by electrophoretypes may represent a potential method for epidemiological analysis of closely related isolates".

C. Zymomonas

Phenotypic properties, DNA base composition (% G+C) and DNA relatedness of approximately 40 *Zymomonas* strains from diverse origins have been determined (De Ley & Swings 1976; Swings & De Ley 1975). All the strains of *Zymomonas mobilis* subsp. *mobilis* isolated from fermenting plant saps from Zaire, Mexico, Brazil and Indonesia, and from British deteriorated beer samples, displayed very similar and sometimes identical protein electrophoregrams (Swings *et al.* 1976, 1977). The highly correlated protein patterns of these strains reflect the close genetic resemblance shown by the % G+C, DNA relatedness and numerical analysis of the phenotype. Strains of *Zymn. mobilis* subsp. *pomaceae* (isolated from sick cider and from apples) had a slightly difference protein pattern, indicative of the separate taxonomic position of these 'cider sickness' strains.

De Ley & Swings (1976) proposed an improved classification and nomenclature of the genus *Zymomonas* on the basis of pooled data from the numerical analysis of the phenotype, the % G+C and genome size of their DNA, the genome-DNA relatedness and the comparison of the protein electrophoregrams.

D. The identification of enterobacterial contaminants from beer breweries by the simultaneous application of phenotypic analysis and PAGE of soluble proteins

The work described below shows the possibilities of using PAGE as an identification technique when it is combined with a rapid phenotypic analysis of unknown isolates, e.g. by means of the API 20E microtube system. The combination of the two techniques allowed the authors to identify, fairly quickly, numerous Enterobacteriaceae isolates from beer breweries.

van Vuuren (1978) investigated the occurrence of enterobacterial contaminants at all stages of the brewing process in four South-African lager beer breweries. Forty-six out of the 60 bacterial isolates were identified with the aid of the API 20E microtube method—22 phenotypic characteristics—(API System, La Balme-les-Grottes, France). This technique is supported by a large data base comprising more than 10^5 Enterobacteriaceae mostly from human clinical material. Most of the 14 remaining unnamed isolates were correctly identified by comparison of the electrophoregrams of their soluble proteins with those of the other isolates.

(i) Enterobacter agglomerans

Eleven of the 60 isolated enterobacterial contaminants were identified by the API 20E system with *Enterobacter agglomerans*. Six of the 14 unidentified strains differed from the 11 *Enbc. agglomerans* strains in only one or at most two tests of the API 20E system. The differences concerned mainly acid formation from inositol, rhamnose, sucrose or arabinose. Computer-assisted comparison of the electrophoregrams revealed that the six unidentified isolates (strain numbers 82, 84, 94, 118, 120 and 123 in Fig. 2) clustered in between

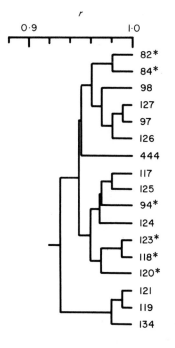

Fig. 2. Quantitative correlation of the electrophoregrams from 17 enterobacterial contaminants isolated from beer, 11 identified as *Enterobacter agglomerans* by the API system and 6 unidentified (marked by an asterisk, strain numbers 82, 84, 94, 118, 120 and 123). PAGE and clustering of the correlation coefficients *r* were carried out as described by Kersters & De Ley (1975). (From van Vuuren 1978.)

the 11 *Enbc. agglomerans* isolates (Fig. 2). The protein patterns of the 17 isolates clustered above the level of *r* = 0.93 (*r* is the Pearson product moment correlation coefficient). As the limit of reproducibility for this type of protein pattern is *r* = 0.94, it is clear that the *Enbc. agglomerans* cluster is very homogeneous. Visual comparison of the normalized protein patterns revealed that the electrophoregrams of the six unidentified isolates were almost indistinguishable from the electrophoretic patterns of the 11 *Enbc. agglomerans* strains. The close similarity between the normalized electrophoregrams of some representative *Enbc. agglomerans* strains (strain numbers 97, 126 and 117) and the unidentified isolates 84, 94 and 118 is shown in Fig. 3. Thus PAGE of the soluble proteins enabled isolates 82, 84, 94, 118, 120 and 123 to be identified as *Enbc. agglomerans*. It can also be seen from Fig. 3 that the protein electrophoregrams of the *Enbc. agglomerans* strains can be clearly distinguished from those of other enterobacterial isolates from beer such as e.g. *Klebsiella pneumoniae* (isolate 105), *Citrobacter freundii* (isolate 35) and *Hafnia protea* (isolate 90). *Enterobacter agglomerans* should be regarded as a potential beer spoilage organism in

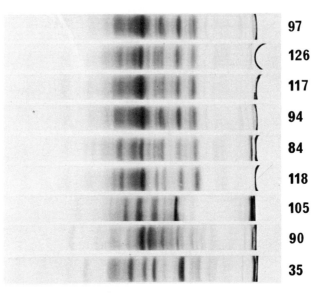

Fig. 3. Normalized protein patterns of enterobacterial contaminants isolated from beer breweries: isolates 97, 126 and 117 were identified as *Enterobacter agglomerans* by the API system; 94, 84 and 118 were unidentified (see also Fig. 2); 105, 90 and 35 were identified by the API system as *Klebsiella pneumoniae, Hafnia protea* and *Citrobacter freundii*, respectively.

lager beer breweries (at least in South Africa), because the strains were isolated from pitching yeast and fermenting wort samples in two out of four breweries investigated.

(ii) Hafnia protea

Seven of the 60 enterobacterial isolates were identified as *Hafnia protea* (*Enbc. hafniae*) with the API 20E system. Gel electrophoresis of the soluble proteins indicated that another three isolates were taxonomically related to these strains. The protein patterns of the South African and two British *Hfna. protea* strains were very similar.

Gas chromatographic analysis of volatile acids and alcohols produced from glucose in a standardized growth medium and an extensive phenotypic analysis with the aid of the API 50E system confirmed that the conclusions based on PAGE of the soluble proteins were correct (van Vuuren 1978).

(iii) The detection of misnamed strains by PAGE

Three out of the 60 bacterial isolates were tentatively identified by the API 20E system as nitrate-negative *Enbc. agglomerans* (the degree of identification given by the manufacturer was only 'good likelihood'). The protein patterns of these 3 strains were identical but totally different from the nitrate-reducing *Enbc. agglomerans* strains and from all other isolates obtained from beer. Hybridization of ^{14}C-labelled rRNA from *Escherichia coli* with DNA from one of these 3 aberrant isolates revealed that they were not even members of the family Enterobacteriaceae. They had been classified erroneously by the API 20E system. Gel electrophoresis of bacterial proteins may thus be helpful in detecting misnamed strains (see also *Alcaligenes, Achromobacter*).

E. *Classification and identification of* Alcaligenes, Achromobacter *and* Bordetella bronchiseptica

Unpublished results on the classification of *Alcaligenes* and the high % G+C *Achromobacter* strains are summarized and discussed here because they illustrate the usefulness of the computerized gel electrophoretic technique (Kersters & De Ley 1975) for grouping and identification of large numbers of bacterial strains.

Alcaligenes faecalis, Alca. odorans, Alca. denitrificans and *Achromobacter xylosoxidans* strains have been isolated frequently from various kinds of human and animal clinical specimens such as pleural and peritoneal fluids, wounds, urine, faeces, swabs of pharynx, ear and eye, etc. Yabuuchi *et al.* (1974) reported a case of fatal purulent meningitis caused by *Achr. xylosoxidans.* Members of the genus *Alcaligenes* also occur in dairy products, rotting eggs, and in fresh water, marine and terrestrial habitats. Isolates from both clinical and natural sources are usually difficult to identify because almost no specific phenotypic characteristics are known for the recognition of *Achromobacter* and *Alcaligenes* taxa. Consequently the genera *Alcaligenes* and *Achromobacter* are currently taxonomically ill-defined and contain many misnamed strains. The genus *Achromobacter* is no longer recognized in the 8th edition of Bergey's Manual (Buchanan & Gibbons 1974).

The genus *Alcaligenes* comprises strictly aerobic, Gram negative and peritrichously flagellated small rods, cocci or coccobacilli with a DNA base composition within the range 57.9–70 % G+C. A schematic representation of the DNA base composition of the four recognized *Alcaligenes* spp. (Holding & Shewan 1974), the recently described *Achr. xylosoxidans* (Yabuuchi *et al.* 1974), some marine *Alcaligenes* strains (Baumann *et al.* 1972), together with the DNA base composition of the 3 *Bordetella* spp. is shown in Fig. 4. It should be noted that *Alca. odorans* and *Alca. denitrificans* are currently synonyms of *Alca. faecalis* (Holding & Shewan 1974) although the % G+C of *Alca. denitrificans* is at least 5% higher than for *Alca. faecalis–Alca. odorans* (De Ley *et al.* 1970). *Alcaligenes eutrophus* and *Alca. paradoxus* are facultatively chemolithotrophic in an atmosphere of hydrogen, oxygen and carbon dioxide.

In order to obtain a better insight into the classification of this complex

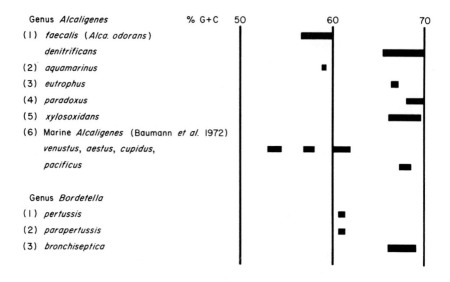

Fig. 4. Schematic representation of the DNA base composition of different *Alcaligenes* and *Bordetella* spp.

group of bacteria polyacrylamide gel electrophoregrams of more than 200 motile *Alcaligenes* and *Achromobacter* strains were prepared using the standardized conditions described previously (Kersters & De Ley 1975). The investigated strains included 15 cultures of *Bordetella bronchiseptica*. The electrophoregram of each cell-free extract was prepared in at least three independent electrophoretic runs and cell-free extracts were usually prepared twice from independently grown bacterial cultures. The electrophoregrams were compared both visually (via their normalized photographs) and by computer-assisted techniques (Kersters & De Ley 1975). Typical protein patterns of several *Alcaligenes* spp. and one *Bord. bronchiseptica* strain are represented in Fig. 5. The sorted shaded matrix of correlation coefficients (*r*) shows (Fig. 6) the similarities between the electrophoregrams of most of the strains investigated. The following groups of bacteria can be distinguished:

(1) *Alcaligenes faecalis-Alca. odorans* (56-60 % G+C). The electrophoregrams of the 50 strains are very similar; no differentiation is possible between *Alca. faecalis* and *Alca. odorans*. These findings confirm the known phenetic similarity of *Alca. faecalis* and *Alca. odorans* (Pintor & Kantór 1974; Shewan, J. and co-workers, pers. comm.).

(2) *Alcaligenes denitrificans-Achr. xylosoxidans* (65-70 % G+C). The 45 strains representing these taxa form a fairly homogeneous cluster, clearly different from both the *Alca. faecalis-Alca. odorans* cluster and the marine *Alcaligenes* strains. The electrophoregrams of *Achr. xylosoxidans*

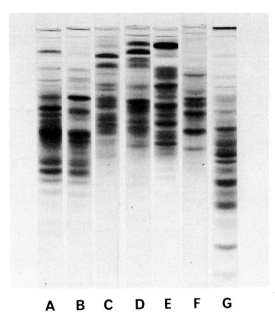

A B C D E F G

Fig. 5. Normalized protein patterns of typical *Alcaligenes* strains and one *Bordetella bronchiseptica* strain. (A) *Alcaligenes faecalis* NCIB 8156; (B) *Alcaligenes odorans* Gil79; (C) *Alcaligenes denitrificans* ATCC 15173; (D) *Achromobacter xylosoxidans* KM543; (E) *Bordetella bronchiseptica* NCTC 452; (F) *Alcaligenes eutrophus* ATCC 17697; (G) *Alcaligenes aquamarinus* NCMB 557.

and *Alca. denitrificans* are very similar. Both species are known to share many phenotypic properties (Yabuuchi *et al.* 1974). The clear-cut differentiation of the *Alca. faecalis-Alca. odorans* and *Alca. denitrificans-Achr. xylosoxidans* clusters is in complete agreement with DNA base composition data and studies on the occurrence of the modified Entner-Doudoroff pathway (De Ley *et al.* 1970). Thus, the majority of the strains in the *Alca. denitrificans* cluster (65-70 % G+C) possess D-gluconate dehydratase, the key enzyme of the modified Entner-Doudoroff pathway [(Fig. 7) (Kersters *et al.* 1971)], whereas those of the *Alca. faecalis-Alca. odorans* cluster (56-60 % G+C) lack the enzymes of both the Entner-Doudoroff and modified Entner-Doudoroff pathways (De Ley *et al.* 1970). Strains of the *Alca. denitrificans-Achr. xylosoxidans* cluster grow readily on D-gluconate as sole carbon source, whereas none of the *Alca. faecalis-Alca. odorans* strains can grow on D-gluconate or any other carbohydrate (Kersters & De Ley, unpublished).

(3) *Bordetella bronchiseptica* (67-69 % G+C). The 15 strains display almost identical electrophoregrams, all clustering values being above *r* = 0.90.

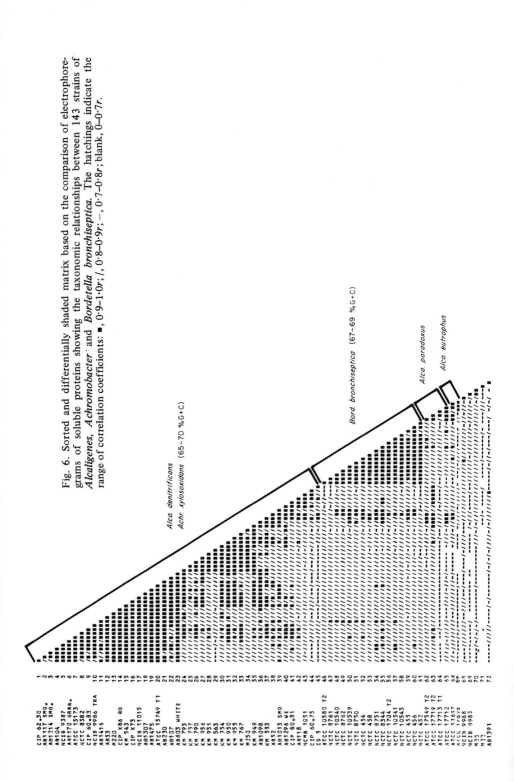

Fig. 6. Sorted and differentially shaded matrix based on the comparison of electrophoregrams of soluble proteins showing the taxonomic relationships between 143 strains of *Alcaligenes, Achromobacter* and *Bordetella bronchiseptica.* The hatchings indicate the range of correlation coefficients: ■, 0·9–1·0r; /, 0·8–0·9r; –, 0·7–0·8r; blank, 0–0·7r.

Alca. denitrificans (65–70 %G+C)
Achr. xylosoxidans (65–70 %G+C)

Bord. bronchiseptica (67–69 %G+C)

Alca. paradoxus

Alca. eutrophus

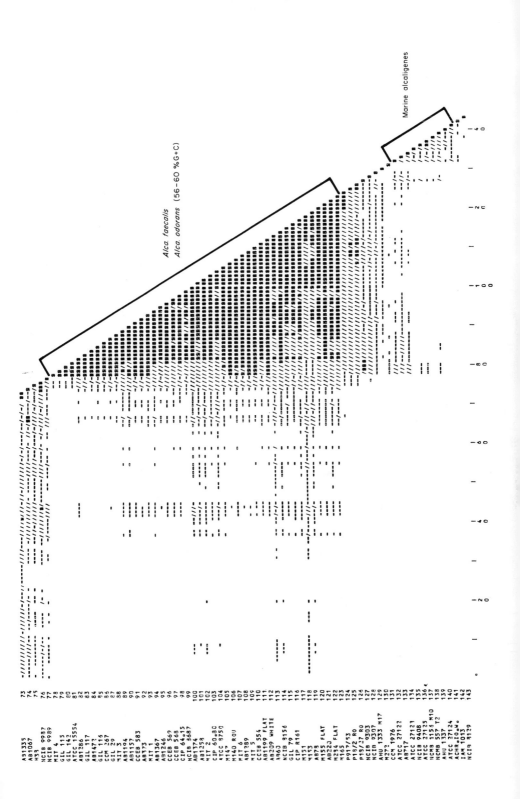

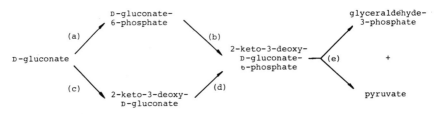

Fig. 7. Breakdown of D-gluconate by the regular (a,b,e) and modified (c,d,e) Entner-Doudoroff pathway. (a) D-gluconate kinase, (b) D-gluconate-6-phosphate dehydratase, (c) D-gluconate dehydratase, (d) 2-keto-3-deoxy-D-gluconate kinase, (e) 2-keto-3-deoxy-D-gluconate-6-phosphate aldolase.

The *Bord. bronchiseptica* strains form a tight cluster at the border of the *Alca. denitrificans-Achr. xylosoxidans* group. These data are in good agreement with the numerical phenetic study of Johnson & Sneath (1973) which highlighted the close relationship between *Alcaligenes* and *Bord. bronchiseptica.*

(4) Six strains of the hydrogen bacteria, *Alca. eutrophus* and *Alca. paradoxus*, were investigated. Although their electrophoregrams cluster at the extreme border of the *Alca. denitrificans-Achr. xylosoxidans-Bord. bronchiseptica* aggregate their banding patterns are different from all the other strains investigated (Fig. 5).

(5) The electrophoretic protein patterns of the marine *Alcaligenes* spp. (*Alca. aestus, Alca. cupidus, Alca. pacificus* and *Alca. venustus* (Baumann *et al.* 1972);*Alca. aquamarinus* are different from each other. They share an unusual feature in that several major protein bands move far towards the anode (Fig. 5), differentiating them from all the other strains investigated.

(6) The electrophoregrams of approximately 40 motile misnamed *Achromobacter* and *Alcaligenes* strains (with a % G+C varying from 57 to 68%) differ from each other and from the protein patterns of the strains in groups 1–5. Results of DNA:rRNA hybridizations (see below) indicate that these strains do not belong in the genus *Alcaligenes*. Our electrophoregrams confirm the previous recommendations of Hendrie *et al.* (1974) to remove e.g. the following strains from the genus *Alcaligenes: Achr. butyri* NCIB 9404, *Achr. halophilus* AHU 1333, *Achr. hartlebii* NCIB 8129, *Achr. viscosus* NCIB 9408 and numerous other so-called *Achromobacter* and *Alcaligenes* strains. In earlier studies (De Ley *et al.* 1970) these strains were found to possess both key enzymes of the normal Entner-Doudoroff pathway (Fig. 7), whereas the strains of the *Alca. denitrificans-Achr. xylosoxidans* cluster possess both enzymes of the modified Entner-Doudoroff pathway and those of the *Alca. faecalis-Alca. odorans* cluster lack the enzymes of both the normal and modified Entner-Doudoroff pathway.

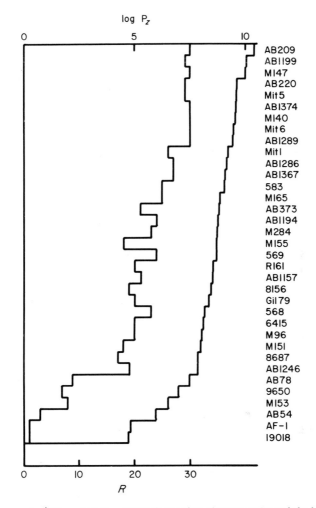

Fig. 8. An example of the calculation of the electrophoretic centrotype and the homogeneity in a cluster of strains (the *Alcaligenes faecalis–Alcaligenes odorans* cluster from Fig. 6). The hierarchical order is the product P_z of the z-values (the inverse hyperbolic tangent of the correlation coefficient r) of each strain (expressed logarithmically). The ranking order R is the number of correlation coefficients $r \geqslant 0.92$). If a cluster represents a taxon and includes as many strains as possible, the strain with the highest P_z and R values is the electrophoretic centrotype strain (AB209).

The homogeneity and the electrophoretic centrotype of large clusters were calculated. It can be seen (Fig. 8) that *Alca. odorans* AB209 is the electrophoretic centrotype of the *Alca. faecalis-Alca. odorans* cluster because it possesses both the highest hierarchical order (product Pz of the z-values) and the highest ranking order R (see also Swings *et al.* 1976).

In this study, therefore, the computerized electrophoregram technique has

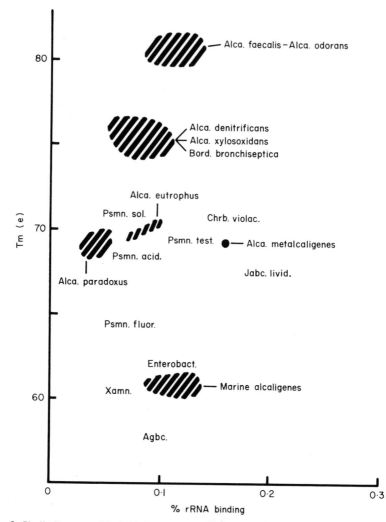

Fig. 9. Similarity map of hybrids between 23S ^{14}C-labelled rRNA from *Alcaligenes faecalis* NCIB 8156 (reference strain) and DNA from a variety of *Alcaligenes* strains and other bacteria. The position of the different *Alcaligenes* strains is indicated by the shaded areas. For the following genera or species only the approximate position is indicated on the map: Agbc. (*Agrobacterium*), Xnmn. (*Xanthomonas*), Enterobact. (Enterobacteriaceae), Psmn. fluor. (*Pseudomonas fluorescens*), Psmn. acid. (*Pseudomonas acidovorans*), Psmn. sol. (*Pseudomonas solanacearum*), Psmn. test. (*Pseudomonas testosteroni*), Chrb. violac. (*Chromobacterium violaceum*) and Jabc. livid. (*Janthinobacterium lividum*). (De Ley & Segers, unpublished results).

helped to clarify the relationships between a large number of *Alcaligenes* and *Achromobacter* strains, facilitated the identification of a number of unclassified *Alcaligenes* and *Achromobacter* strains, and detected misnamed strains. A great number of unnamed strains isolated and studied by Moore & Pickett (1960) (strain numbers starting with M in Fig. 6) and by Dr H. Lautrop (Statens Serum Institutet, Copenhagen, Denmark; strain numbers starting with AB in Fig. 6) can now be identified with *Alca. faecalis* or *Alca. denitrificans* strains. The amino acid sequence of cytochrome c' from strain NCIB 11015, tentatively named *Alcaligenes* sp. was determined by Ambler (1973). This strain was identified as *Alca. denitrificans* because its protein electrophoregram was similar to the patterns exhibited by strains of the *Alca. denitrificans-Achr. xylosoxidans* cluster.

It is impossible to deduce from the electrophoretic data the evolutionary relationship between e.g. *Alca. faecalis* and *Alca. denitrificans* or between *Alca. faecalis* and the marine *Alcaligenes* spp. It can only be concluded with certainty that the investigated strains of e.g. *Alca. faecalis* are very similar to each other and that they are different from *Alca. denitrificans* and probably even more different from the marine *Alcaligenes* spp.

Because ribosomal RNA cistrons are conservative, inter-generic relationships can be determined by hybridizations of ^{14}C-labelled rRNA and DNA of various bacteria (De Smedt & De Ley 1977). De Ley & Segers (unpublished results) prepared hybrids between ^{14}C-rRNA from *Alca. faecalis* NCIB 8156 (reference strain) and DNA from representative strains of the authentic *Alcaligenes* groups and from representatives of other genera such as *Pseudomonas, Chromobacterium, Agrobacterium*, and *Escherichia*. The number of hybridizations can be reduced drastically because good representative strains can be chosen on the basis of the computerized grouping of the protein electrophoregrams. Each DNA:rRNA hybrid is characterized by: (i) its $T_{m(e)}$, or temperature where 50% of the hybrid is denatured, and (ii) its percentage rRNA binding or the amount (μg) rRNA bound per 100 μg DNA. Both parameters were used to construct a rRNA similarity map (Fig. 9). *Alcaligenes* strains belonging to one taxon fall within one of the shaded areas on the map. The shape and the size of this area is a measure of the genetic heterogeneity of the taxon. Qualitatively the same taxa can be delineated either by DNA:rRNA hybridizations (Fig. 9) or by comparison of electrophoretic protein patterns (Fig. 6).

The following conclusions can be drawn from the gel electrophoretic and DNA:rRNA hybridization data:

(1) The genus *Alcaligenes* as defined in the eighth edition of Bergey's Manual (Holding & Shewan 1974) is extremely heterogeneous.

(2) The marine *Alcaligenes* spp. (*Alca. aquamarinus, Alca. aestus, Alca. cupidus, Alca. venustus*) do not belong to the genus *Alcaligenes*. The marine *Alcaligenes* spp. and the Enterobacteriaceae are both equally far removed from *Alca. faecalis*—in evolutionary terms (Fig. 9).

(3) The hydrogen bacteria (*Alca. paradoxus* and *Alca. eutrophus*) are very different from *Alca. faecalis, Alca. eutrophus, Alca. paradoxus, Pseudo-*

monas acidovorans, Psmn. solanacearum and *Chromobacterium* are all approximately equally far removed from *Alca. faecalis* (Fig. 9).

(4) *Alcaligenes denitrificans, Achr. xylosoxidans* and *Bord. bronchiseptica*, are related with, but different from *Alcaligenes faecalis. Alcaligenes denitrificans* should not be classified with *Alca. faecalis*; both taxa merit at least the rank of species.

(5) Strains belonging to the type species *Alca. faecalis* form a fairly homogeneous group of bacteria (Figs 6 and 8) indistinguishable from *Alca. odorans*.

(6) A thorough revision of the classification of the genus *Alcaligenes* is being prepared on the basis of these new data.

F. The sensitivity of PAGE of total soluble proteins and its use for the differentiation between colony variants and contaminants

PAGE of bacterial proteins is a refined technique which can be used to identify bacterial strains, frequently down to the individual level. Individual strain identification is often possible, because many bacterial strains possess or lack a few typical bands. In a very heterogeneous genus or species, many strains have their own typical individual protein pattern. It is, however, unlikely that mutants can be differentiated from each other, because of the usually small differences between the nature and numbers of proteins involved. The technique is surpassed in resolution by serological analysis and two dimensional PAGE (O'Farrell 1975).

Two or more colony types are sometimes detected when the bacteriological purity of a culture is checked, even when the latter is obtained from an established culture collection. It is not usually very difficult to differentiate between colony variants and contaminants but it can be very time-consuming because one has to submit each colony type to a great number of phenotypic tests. Instead of performing these lengthy procedures the authors routinely prepare cell-free extracts and compare their electrophoregrams visually. Protein patterns of two or more colony types will be identical or very similar when they are merely colony variants (e.g. smooth and rough), whereas a contaminant will yield an entirely different electrophoregram. To date the protein patterns of over 100 cases have been examined where the distinction between colony variants and contaminants was difficult (representatives of e.g. *Acetobacter, Agrobacterium, Alcaligenes, Cytophaga, Erwinia, Flavobacterium, Gluconobacter, Phyllobacterium, Pseudomonas, Vibrio* and *Zymomonas*). The colony types of more than 95% of these doubtful cases displayed almost identical electrophoregrams, indicating that they could be considered as colony variants. In a few cases a contaminant was demonstrated.

4. Conclusions

The application of polyacrylamide gel electrophoresis of bacterial proteins for classification, differentiation and identification of a great variety of bacteria has been summarized (see Table 2).

Different species and even subspecific taxa usually display a great variety of protein patterns. One exception is known so far: the genus *Zymomonas* seems to possess a genus-specific protein electrophoregram (Swings *et al.* 1976). Similarities of electrophoregrams of soluble and cell-envelope proteins are usually detected at the lowest taxonomic levels, i.e. within species, subspecies, biotypes or even individual strains.

Visual or computer-assisted comparison of electrophoretic protein patterns offers the following advantages:

(1) Large numbers of bacteria can be compared by the relatively simple inexpensive technique of polyacrylamide gel electrophoresis. Strains with a high genome DNA similarity display very similar or almost identical protein electrophoregrams.

(2) The homogeneity and heterogeneity of taxa can be determined.

(3) Once electrophoretic groups of reliable taxonomic status are delineated, unknown bacteria can be compared with them fairly quickly for eventual identification.

(4) Misnamed strains can be detected.

(5) The method can be useful to check the epidemiological spread of animal and plant pathogens.

(6) The visual comparison of electrophoretic patterns has been found to be very useful for a quick decision as to whether two or more colony types in a culture are due to variation or contamination.

(7) Computers can be used to compare, cluster, store and retrieve the electrophoretic data, provided the electrophoregrams were produced under rigorously standardized conditions.

(8) Comparison of electrophoregrams does not usually permit valid conclusions concerning the evolutionary relationships between the electrophoretic clusters. These intergroup relationships can be determined by DNA:DNA or DNA:rRNA hybridizations. The number of DNA's that needs to be prepared can, however, be drastically reduced because only a limited number of bacterial strains needs to be examined once the electrophoretic groupings are known.

The comparative electrophoregram technique suffers from a number of restrictions and drawbacks:

(1) The method is fairly rapid because large numbers of bacterial extracts can be compared in the same experiment. The technique is not applicable

for those areas of clinical bacteriology where an immediate identification is often required. Identification within a few hours after isolating a single colony is not possible by the electrophoregram technique.

(2) Comparison of electrophoregrams is only possible for bacteria grown on identical media in strictly reproducible conditions.

(3) Electrophoresis of bacterial proteins requires a relatively large number of experimental steps, an eventual source of considerable experimental error. Even when working in rigorously controlled and standardized conditions, the taxonomist needs techniques with an even greater reproducibility in order to exploit the possibilities of comparative protein electrophoresis to its extreme limits.

Electrophoretic data can be stored in binary form, and large data banks of reference protein electrophoregrams of a great variety of bacterial taxa would be very useful for computerized grouping and identification of strains. Inter-laboratory comparison and exchange of data are desirable but of limited use as long as a large number of different experimental approaches are used. In spite of these drawbacks the technique of comparative electrophoresis of bacterial proteins has already contributed considerably towards improvements and new insights in the classification and identification of, e.g. the following genera: *Agrobacterium, Alcaligenes, Brucella, Gluconobacter, Mycoplasma, Yersinia* and *Zymomonas*.

5. References

ANON. 1972 Subcommittee on the Taxonomy of Mycoplasmatales. Proposal for minimal standards for descriptions of new species of the order Mycoplasmatales. *International Journal of Systematic Bacteriology* 22, 184–188.

AMBLER, R. P. 1973 The amino acid sequence of cytochrome c' from *Alcaligenes* sp. NCIB 11015. *Biochemical Journal* 135, 751–758.

AMES, G. F. L. 1974 Resolution of bacterial proteins by polyacrylamide gel electrophoresis on slabs. *Journal of Biological Chemistry* 249, 634–644.

ARMSTRONG, D. & YU, B. 1970 Characterization of canine mycoplasmas by polyacrylamide gel electrophoresis and immunodiffusion. *Journal of Bacteriology* 104, 295–299.

BALKE, E., WEBER, A. & FRONK, B. 1977 Differenzierung von Brucellen mit Hilfe der Acrylamidgel Electrophorese. *Zentralblatt für Bakteriologie, Parasitenkunde, Infektionskrankheiten und Hygiene. I. Abteilung Originale Reihe* A238, 80–85.

BAUMANN, L., BAUMANN, P., MANDEL, M. & ALLEN, R. D. 1972 Taxonomy of aerobic marine eubacteria. *Journal of Bacteriology* 110, 402–429.

BUCHANAN, R. E. & GIBBONS, N. E. 1974 (ed.) *Bergey's Manual of Determinative Bacteriology*, 8th edn. Baltimore: Williams & Wilkins.

COOKSEY, K. E. 1971 Disc electrophoresis. In *Methods in Microbiology Vol 5B*, ed. Norris, J. R. & Ribbons, D. W. pp. 573–594, London & New York: Academic Press.

DAVIES, F. L. & GOTTLIEB, D. 1973 Polyacrylamide gel electrophoresis of soluble proteins from several genera of the family Actinoplanaceae. *International Journal of Systematic Bacteriology* 23, 43–48.

DAVIS, B. J. 1964 Disc electrophoresis–II. Method and application to human serum proteins. *Annals of the New York Academy of Sciences* **121**, 404–427.

DE LEY, J. & SWINGS, J. 1976 Phenotypic description, numerical analysis, and proposal for an improved taxonomy and nomenclature of the genus *Zymomonas* Kluyver and van Niel 1936. *International Journal of Systematic Bacteriology* **26**, 146–157.

DE LEY, J., KERSTERS, K., KHAN-MATSUBARA, J. & SHEWAN, J. M. 1970 Comparative D-gluconate metabolism and DNA base composition in *Achromobacter* and *Alcaligenes*. *Antonie van Leeuwenhoek* **36**, 193–207.

DELLINGER, J. D. & JASPER, D. E. 1972 Polyacrylamide-gel electrophoresis of cell proteins of *Mycoplasma* isolated from cattle and horses. *American Journal of Veterinary Research* **33**, 769–775.

DE SMEDT, J. & DE LEY, J. 1977 Intra- and intergeneric similarities of *Agrobacterium* ribosomal ribonucleic acid cistrons. *International Journal of Systematic Bacteriology* **27**, 222–240.

EL-SHARKAWY, T. A. & HUISINGH, D. 1971 Electrophoretic analysis of esterases and other soluble proteins from representatives of phytopathogenic bacterial genera. *Journal of General Microbiology* **68**, 149–154.

FELTHAM, R. K. A. 1975 A comparison of numerical and electrophoretic studies of the Enterobacteriaceae. Ph.D. Thesis, University of Leicester, UK.

GORDON, A. H. 1975 Electrophoresis of proteins in polyacrylamide and starch gels. In *Laboratory Techniques in Biochemistry and Molecular Biology*, ed. Work, T. S. & Work, E. pp. 7–207. Amsterdam and Oxford: North Holland Publishing Co.

HAAS, H., DAVIDSON, Y. & SACKS, T. 1972 Taxonomy of mycobacteria studied by polyacrylamide-gel electrophoresis of cell proteins. *Journal of Medical Microbiology* **5**, 31–37.

HAAS, H., MICHEL, J. & SACKS, T. 1974 Identification of *Mycobacterium fortuitum*, *Mycobacterium abscessus* and *Mycobacterium borstelense* by polyacrylamide gel electrophoresis of their cell proteins. *International Journal of Systematic Bacteriology* **24**, 366–369.

HAMADA, S. & MIZUNO, J. 1974 Isoelectric focusing in polyacrylamide gel of the membrane proteins of *Streptococcus mutans* and related streptococci. *Journal of Dental Research* **53**, 547–553.

HENDRIE, M. S., HOLDING, A. J. & SHEWAN, J. M. 1974 Emended descriptions of the genus *Alcaligenes* and of *Alcaligenes faecalis* and proposals that the generic name *Achromobacter* be rejected: status of the named species of *Alcaligenes* and *Achromobacter*. Request for an opinion. *International Journal of Systematic Bacteriology* **24**, 534–550.

HOLDING, A. J. & SHEWAN, J. M. 1974 Genus *Alcaligenes*. In *Bergey's Manual of Determinative Bacteriology*, 8th edn. ed. Buchanan, R. E. & Gibbons, N. E. pp. 273–275. Baltimore: Williams & Wilkins Co.

HUDSON, B. W., QUAN, T. J., SITES, V. R. & MARSHALL, J. D. 1973 An electrophoretic and bacteriological study of *Yersinia pestis* isolates from central Java, Asia, and the Western hemisphere. *American Journal of Tropical Medecine and Hygiene* **22**, 642–653.

HUDSON, B. W., QUAN, T. J. & BAILEY, R. E. 1976 Electrophoretic studies of the geographic distribution of *Yersinia pestis* protein variants. *International Journal of Systematic Bacteriology* **26**, 1–16.

JOHNSON, R. & SNEATH, P. H. A. 1973 Taxonomy of *Bordetella* and related organisms of the families Achromobacteriaceae, Brucellaceae, and Neisseriaceae. *International Journal of Systematic Bacteriology* **23**, 381–404.

KADO, C. I., SCHAAD, N. W. & HESKETT, M. G. 1972 Comparative gel-electrophoresis studies of ribosomal proteins of *Erwinia* species and members of the Enterobacteriaceae. *Phytopathology* **62**, 1077–1082.

KERSTERS, K. & DE LEY, J. 1975 Identification and grouping of bacteria by numerical analysis of their electrophoretic protein patterns. *Journal of General Microbiology* **87**, 333–342.

296 K. KERSTERS AND J. DE LEY

KERSTERS, K., KHAN-MATSUBARA, J., NELEN, L. & DE LEY, J. 1971 Purification and properties of D-gluconate dehydratase from *Achromobacter*. *Antonie van Leeuwenhoek* **37**, 233-246.

LARSEN, S. A., BICKHAM, S. T., BUCHANAN, T. M. & JONES, W. L. 1971 Polyacrylamide gel electrophoresis of *Corynebacterium diphtheriae*: a possible epidemiological aid. *Applied Microbiology* **22**, 885-890.

MAIZEL, J. V. 1969 Acrylamide gel electrophoresis of proteins and nucleic acids. In *Fundamental Techniques in Virology* ed. Habel, K. & Salzman, P. P. pp. 334-362. New York and London: Academic Press.

MATTHEW, M. & HARRIS, A. M. 1976 Identification of β-lactamases by analytical isoelectric focusing: correlation with bacterial taxonomy. *Journal of General Microbiology* **94**, 55-67.

MAURER, H. R. 1971 *Disc electrophoresis and related techniques of polyacrylamide gel electrophoresis*. Berlin and New York: De Gruyter W.

MOORE, H. B. & PICKETT, M. J. 1960 Organisms resembling *Alcaligenes faecalis*. *Canadian Journal of Microbiology* **6**, 43-52.

MORRIS, J. A. 1973 The use of polyacrylamide gel electrophoresis in taxonomy of *Brucella*. *Journal of General Microbiology* **76**, 231-237.

MORRIS, J. A. & PARK, R. W. A. 1973 A comparison using gel electrophoresis of cell proteins of campylobacters (vibrios) associated with infertility, abortion and swine dysentery. *Journal of General Microbiology* **78**, 165-178.

MOSES, V. & WILD, D. G. 1969 Soluble protein profiles in *Escherichia coli*. *Folia Microbiologica* **14**, 305-309.

MÜLLEGGER, P.-H. & GERLACH, H. 1974 Die Differenzierung aviärer Mykoplasmen mit Hilfe der Disc-Elektrophorese. *Zentralblatt für Bakteriologie, Parasitenkunde, Infektionskrankheiten und Hygiene. I Abteilung. Originale Reihe* **A226**, 119-136.

MÜLLEGGER, P.-H., GERLACH, H. & SCHELLNER, H.-P. 1973 Die Technik der Disc-Elektrophorese zur Differenzierung aviärer Mykoplasmen. *Zentralblatt für Bakteriologie, Parasitenkunde, Infektionskrankheiten und Hygiene. I. Abteilung. Originale Reihe* **A223**, 372-380.

O'FARRELL, P. H. 1975 High resolution two-dimensional electrophoresis of proteins. *Journal of Biological Chemistry* **250**, 4007-4021.

ORNSTEIN, L. 1964 Disc electrophoresis–I. Background and theory. *Annals of the New York Academy of Sciences* **121**, 321-349.

PALMER, B. C. & CAMERON, H. R. 1971 Comparison of plant-pathogenic pseudomonads by disc-gel electrophoresis. *Phytopathology* **61**, 984-986.

PARTON, R. 1975 Envelope proteins in *Salmonella minnesota* mutants. *Journal of General Microbiology* **89**, 113-129.

PARTON, R. & WARDLAW, A. C. 1975 Cell envelope proteins of *Bordetella pertussis*. *Journal of Medical Microbiology* **8**, 47-57.

PINTÉR, M. & KANTÓR, M. 1974 Comparison of *Alcaligenes faecalis* and *Alcaligenes odorans* var. *viridans* by carbon source utilization test. *Acta Microbiologica Academiae Scientiarum Hungaricae* **21**, 293-295.

RAZIN, S. & ROTTEM, S. 1967 Identification of *Mycoplasma* and other microorganisms by polyacrylamide-gel electrophoresis of cell proteins. *Journal of Bacteriology* **94**, 1807-1810.

RAZIN, S. & SHAFER, Z. 1969 Incorporation of cholesterol by membranes of bacterial L-phase variants. With an appendix. On the determination of the L-phase parentage by the electrophoretic patterns of cell proteins. *Journal of General Microbiology* **58**, 327-339.

RODBARD, D., KAPADIA, G. & CHRAMBACH, A. 1971 Pore gradient electrophoresis. *Analytical Biochemistry* **40**, 135-157.

ROSENDAL, S. 1973 Analysis of the electrophoretic pattern of *Mycoplasma* proteins for the identification of canine *Mycoplasma* species. *Acta Pathologica et Microbiologica Scandinavica. Section B* **81**, 273-281.

ROSENDAL, S. 1974 *Mycoplasma molare*, a new canine *Mycoplasma* species. *International Journal of Systematic Bacteriology* 24, 125-130.

ROSS, R. F. & KARMON, J. A. 1970 Heterogeneity among strains of *Mycoplasma granularum* and identification of *Mycoplasma hyosynoviae*, sp. n. *Journal of Bacteriology* 103, 707-713.

ROTTEM, S. & RAZIN, S. 1967 Electrophoretic patterns of membrane proteins of *Mycoplasma. Journal of Bacteriology* 94, 359-364.

ROUATT, J. W., SKYRING, G. W., PURKAYASTHA, V. & QUADLING, C. 1970 Soil bacteria: numerical analysis of electrophoretic protein patterns developed in acrylamide gels. *Canadian Journal of Microbiology* 16, 202-205.

RUSSELL, R. R. B. & MCDONALD, I. J. 1976 Comparison of the cell envelope proteins of *Micrococcus cryophilus*, with those of *Neisseria* and *Branhamella* species. *Canadian Journal of Microbiology* 22, 309-312.

RUSSELL, R. R. B., JOHNSON, K. G. & McDONALD, I. J. 1975 Envelope proteins in *Neisseria. Canadian Journal of Microbiology* 21, 1519-1534.

SACKS, T. G., HAAS, H. & RAZIN, S. 1969 Polyacrylamide-gel electrophoresis of cell proteins of Enterobacteriaceae. *Israel Journal of Medical Sciences* 5, 49-55.

SCHAAD, N. W. 1974 Comparative immunology of ribosomes and disc gel electrophoresis of ribosomal proteins from erwiniae, pectobacteria and other members of the family Enterobacteriaceae. *International Journal of Systematic Bacteriology* 24, 42-53.

SUN, T.-T., BICKLE, T. A. & TRAUT, R. R. 1972 Similarity in the size and number of ribosomal proteins from different prokaryotes. *Journal of Bacteriology* 111, 474-480.

SWINGS, J. & DE LEY, J. 1975 Genome deoxyribonucleic acid of the genus *Zymomonas* Kluyver and van Niel 1936: base composition, size and similarities. *International Journal of Systematic Bacteriology* 25, 324-328.

SWINGS, J., KERSTERS, K. & DE LEY, J. 1976 Numerical analysis of electrophoretic protein patterns of *Zymomonas* strains. *Journal of General Microbiology* 93, 266-271.

SWINGS, J., KERSTERS, J. & DE LEY, J. 1977 Taxonomic position of additional *Zymomonas* strains. *International Journal of Systematic Bacteriology* 27, 271-273.

TAKAYAMA, K., MACLENNAN, D. H., TZAGOLOFF, A. & STONER, C. D. 1966 Studies on the electron transfer system. LXVII. Polyacrylamide gel electrophoresis of the mitochondrial electron transfer complexes. *Archives of Biochemistry and Biophysics* 114, 223-230.

THEODORE, T. S., KING, J. R. & COLE, R. M. 1969 Identification of L-forms by polyacrylamide gel-electrophoresis. *Journal of Bacteriology* 97, 495-499.

THEODORE, T. S., TULLY, J. G. & COLE, R. M. 1971 Polyacrylamide gel identification of bacterial L-forms and *Mycoplasma* species of human origin. *Applied Microbiology* 21, 272-277.

TULLY, J. G., BARILE, M. F., EDWARD, D. G., THEODORE, T. S. & ERNØ, H. 1974 Characterization of some caprine mycoplasmas, with proposals for new species, *Mycoplasma capricolum* and *Mycoplasma putrefaciens. Journal of General Microbiology* 85, 102-120.

TULLY, J. G., DEL GIUDICE, R. A. & BARILE, M. F. 1972 Synonymy of *Mycoplasma arginini* and *Mycoplasma leonis*. *International Journal of Systematic Bacteriology* 22, 47-49.

VAN VUUREN, H. J. J. 1978 *Identification and physiology of Enterobacteriaceae isolated from South African lager beer breweries.* Doctoraatsthesis, Rijksuniversiteit Gent, België.

VESTERBERG, O. 1971 Isoelectric focusing and separation of proteins. In *Methods in Microbiology, Vol 5B*, ed. Norris, J. R. & Ribbons, D. W. pp. 595-614. London and New York: Academic Press.

YABUUCHI, E., YANO, I., GOTO, S., TANIMURA, E., ITO, T. & OHYAMA, A. 1974 Description of *Achromobacter xylosoxidans* Yabuuchi and Ohyama 1971. *International Journal of Systematic Bacteriology* 24, 470-477.

Enzyme Patterns in Bacterial Classification and Identification

R. A. D. WILLIAMS AND H. N. SHAH

*The London Hospital Medical College, University of London,
London, UK*

Contents

1. Introduction

PATTERNS OF enzymes in micro-organisms may include the presence or absence of key enzymes, or whole metabolic pathways, either under particular conditions, or under all conditions tested (De Ley 1962). Such patterns are of considerable significance as the possession of key enzymes has a bearing on the substrates that may be used, the energy yield obtainable from them and the nature of the metabolic products (Sokatch 1969). Extensive studies of this type comprise the study of comparative intermediary metabolism (De Ley 1962).

Another type of enzyme pattern is obtained by comparing the properties of one or more given enzymes. Such studies may be carried out on crude or purified preparations and include examination of their electrophoretic mobility, stability to heat or other inactivating agents, optimum conditions for activity, susceptibility to stimulation and inhibition by small molecules and immunological reactivity. The ultimate extension of this approach lies in the comparison of amino acid sequences and molecular conformations of enzymes (Hartley 1974). It is this second type of enzyme pattern which will be considered here, with electrophoretic mobility as the prime screening method. It is ten years since Norris (1968) wrote a review of this subject, and pointed out that a physiological test that depends on the presence of an enzyme has no taxonomic value within a group of organisms that all have the enzyme. The molecular properties of such an enzyme may be different, within such a group and these differences may have considerable taxonomic value.

Electrophoretic variation, both between species and between individual members of a species, has long been known to be common amongst enzymes (Shaw 1965). Although only about one in four random mutations will cause a

charge change in a protein, electrophoretic analysis has provided a major technique for biochemical genetics which has proved valuable in family and population studies, and in the identification of individuals for forensic purposes.

Molecular heterogeneity must be widespread in bacterial strains whose mean
DNA base composition may vary from 25 to 75% G+C. Micro-organisms with
significantly different DNA base ratios can have no large molecules in common,
and proteins of identical function must therefore be of different amino acid
sequence. The degree of difference in % G+C required to render identity of
protein structure impossible has been estimated as from 10% G+C (Sueoka 1961)
to 25% G+C (De Ley 1968). Electrophoretic heterogeneity is therefore to be
expected in bacteria with non-homologous DNA. Micro-organisms with a high
degree of DNA homology may have enzymes with similar or identical properties,
but point mutations that involve charge changes in protein molecules can produce marked effects on electrophoretic patterns without noticeably affecting
DNA mean base composition or nucleic acid homology. Electrophoretic differences may therefore occur in bacteria that are genetically similar. It seems,
however, that often clusters of strains that are similar by many other criteria
are also similar in their enzyme patterns. Nevertheless, the electrophoretic
mobility of enzymes cannot be considered a taxonomic criterion of primary
importance such as DNA base composition, nucleic acid homology, mucopeptide structure or even electrophoretic techniques which produce patterns of a
wide range of proteins (see Chapter 11). When electrophoretic patterns correlate
with other taxonomic data they may be used in surveying strains and perhaps in
identification because of the simplicity, cheapness and speed of the technique. It
is often possible to produce such patterns in a few hours with the cells scraped
from a plate. By contrast classical bacteriological methods such as fermentation
tests may take weeks to complete.

The occurrence of different forms of the same enzyme with similar functional
properties (isoenzymes) is a result of mutation in non-essential parts of the
enzyme molecule. Amongst the human erythrocyte enzymes examined, about
one third showed a sufficient polymorphic variation such that the population
could be categorized into two or more fairly common types (Harris 1966) by
electrophoresis. By choice of a sufficient number of polymorphic enzymes,
individuals may be compared and this technique has been used as an exclusionary
criterion in paternity testing. These variations in the enzymes of individuals
within animal species require that a cautious approach be adopted to the use of
the same method for taxonomic purposes. In animals, electrophoretic patterns
of enzymes have not been significant in classification, and where taxonomic
studies have been made, there has been a background of more fundamental
taxonomic criteria such as may not be available in bacteriological studies. It is
not possible to distinguish a chimpanzee from man by electrophoresis of haemoglobin or of many of the blood enzymes that are commonly studied (King &
Wilson 1975), and in some cases this is due to the fact that the proteins are

identical. Identity of electrophoretic mobility does not depend, however, upon molecular identity. Creatine phosphokinases have been shown to have different amino acid compositions despite identical electrophoretic patterns (Dance & Watts 1962). Nor does the degree of difference in mobility indicate the extent of molecular distinctness. Thus haemoglobin A and haemoglobin S are easily distinguished due to a mutational replacement of glu$^-$ by val at β6. Haemoglobin C has a mobility twice as far removed from A as S due to β6 glu$^-$ → lys$^+$. However, haemoglobin F, which has a totally different polypeptide chain (γ instead of β) with *ca.* 50 amino acid residues different to the β chain, has an electrophoretic mobility closer to HbA than to HbS.

Despite these cautionary remarks enzyme electrophoresis has proved useful in identification of bacteria, and also in their classification. An excellent example of the occurrence of enzyme polymorphism amongst similar bacteria, and its use to diagnose the source of an infection is provided by Norris & Burges (1963) using esterase electrophoresis to distinguish between strains of the species *Bacillus thuringiensis*.

2. Bacteroides Melaninogenicus

Bacteroides melaninogenicus strains form part of the normal flora of the human mouth, intestine and genitalia and they are common in mixed anaerobic infections (Lambe 1974). Although not frequently isolated from children, it is almost universally found in adults (Kelstrup 1966; Bailit *et al.* 1964), and is considered important in the aetiology of periodontal disease (Socransky 1970), an essentially adult disease. It has also been cited as an obligatory component of artificially induced mixed anaerobic infections (MacDonald *et al.* 1956; Socransky & Gibbons 1965). Because these organisms are anaerobic Gram negative rods which produce black to tan pigment on blood media, they are often identified presumptively on this basis alone. This produces problems because of differences in rate of formation and nature of the pigments (Shah *et al.* 1979) and the inability of some strains to form pigment under some laboratory conditions (Mochizuki *et al.* 1975). In this respect, some putative *Bacteroides oralis* strains have been identified as poorly pigmenting *Bcrd. melaninogenicus* strains.

It is clearly unsatisfactory for disease-related bacteria to be identified by trivial criteria that may be shared by a heterogeneous group of strains, some of which may be non-pathogenic. Furthermore, studies of experimental pathology to determine the involvement of *Bcrd. melaninogenicus* in disease should be carried out with appropriate strains, and the results obtained should not be attributed to fundamentally different bacteria that resemble them superficially. Taxonomically there are also objections to groups of strains that are biochemically and genetically distinct being included in the same species even if they are clearly identifiable and their pathogenic properties are known. The authors interest in *Bcrd. melaninogenicus* was due to its possible involvement in periodontal disease, and the availability of new unclassified oral isolates from the Medical Research Council Dental Epidemiology Unit survey at the London Hospital Dental Institute. The species has been subdivided into one asaccharolytic (*Bcrd. melaninogenicus* subsp. *asaccharolyticus*) and two saccharolytic subspecies (*Bcrd. melaninogenicus* subsp. *intermedius* and *Bcrd. melaninogenicus* subsp. *melaninogenicus*) (Holdeman & Moore 1970, 1973, 1974). Using the tests described by these authors, and, in the first instance, strains available from culture collections and colleagues, the electrophoretic mobility of malate dehydrogenase (MDH), the amino acid components of mucopeptide and the base composition of DNA were determined. The findings were considered sufficiently interesting to warrant a preliminary note (Williams *et al.* 1975) because each of the tests emphasized the difference between the two saccharolytic subspecies and subspecies *asaccharolyticus*. Of these three tests the electrophoresis of MDH was the most readily carried out; it can be done within a day given a blood plate of each of the pure strains. To date, MDH has been found in all the isolates of *Bacteroides* examined, including strains from the mouth and a range of clinical specimens. This enzyme is absent in strains of *Fusobacterium* and its detection therefore gives a positive criterion for distinguishing the two genera.

In a wider study of 45 strains of *Bcrd. melaninogenicus* (Shah *et al.* 1976) key tests from Holdeman & Moore (1973) such as starch hydrolysis, lactose fermentation, indole formation, terminal pH with glucose as substrate, and the nature of the acid end-products of metabolism, allowed the strains to be identi-

TABLE 1

Properties of Bacteroides melaninogenicus *subspecies**

	Bacteroides melaninogenicus subspecies		
	asaccharolyticus (13)†	melaninogenicus (10)†	intermedius (22)†
Days to become black	7–14	3–8	2–6
Starch hydrolysis	–	+	+
pH 1% (w/v) glucose	6.4–7.0	4.9–5.4	4.4–5.7
Acid end products			
Acetic	+	+	+
Propionic	+	–	+(18)–(14)
*iso*Butyric	+	–	+/tr.
n-Butyric	+	–	–
*iso*Valeric	+	+/–	+
Lactic	+/–	–	+/–
Succinic	–	+/–	+
MDH mobility	slow intermediate	fast fast	intermediate
Mucopeptide‡	lys lys	lys DAP	DAP
% G+C	45.2–46.3 49.0–51.7	38.5–39.5 36.4–39.8	39.1–43.9

*From Shah *et al.* (1976); † No. of strains tested; ‡ lys, lysine; DAP, diaminopimelic acid respectively is the dibasic amino acid of the mucopeptide.

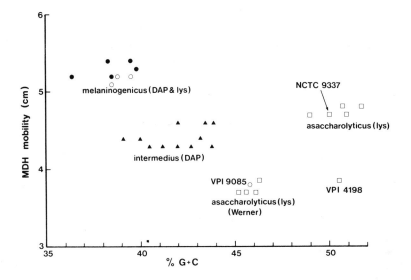

Fig. 1. Graph of DNA base composition against electrophoretic mobility of malate dehydrogenase for strains of the three subspecies of *Bacteroides melaninogenicus*. Diaminopimelic acid (DAP) mucopeptide (●, ▲), lysine (lys) mucopeptide (○, □). [Modified from Fig. 1 of Shah *et al.* (1976)].

fied with the three subspecies described at the time. It was also quite clear, however, that there was heterogeneity amongst the strains (Table 1), allocated to subspecies, *asaccharolyticus* and *melaninogenicus*, although this was not evident from the routine physiological tests. Two types of strains were found amongst subspecies *asaccharolyticus*, they differed in their MDH mobility and DNA base composition; two types of subspecies *melaninogenicus* had different dibasic amino acids in their mucopeptide. There was also a range of MDH mobility and of DNA base composition for subspecies *intermedius*, wider in the case of % G+C, but amongst which no discontinuity could be discerned. A diagrammatic representation of these properties (Fig. 1) shows how clusters of similar strains may be recognized. The strains of subspecies *intermedius* examined were also uniform as they had diaminopimelic acid in the mucopeptide.

Electrophoretically, the low mobility MDH group in subspecies *asaccharolyticus* could be distinguished from all other *Bcrd. melaninogenicus* strains. All the strains of subspecies *melaninogenicus* examined could be distinguished by the rapid mobility of MDH with the exception of VPI 9085, which is grossly atypical of the subspecies in having 45.8% G+C. This was the only strain of *Bcrd. melaninogenicus* subspecies *melaninogenicus* available during the early work, and the data of Williams *et al.* (1975), who used this strain, are not representative of the subspecies. Strain VPI 9085 falls amongst one of the two clusters of subspecies *asaccharolyticus* strains (Fig. 1) although it has physiological properties typical of subspecies *melaninogenicus*. In this case the enzyme mobility is a good indicator to identify VPI 9085 as an atypical strain. Strain 30 is identical to VPI 9085 in all tests and may be the same strain (E. M. Barnes pers. comm.).

The MDH mobilities of some strains of subspecies *intermedius* approach those of the fast mobility group of subspecies *asaccharolyticus,* and these clusters cannot be definitely distinguished from one another by this method. However, strains of these taxa can be distinguished by their wall composition. Strain VPI 4198 is typical of subspecies *asaccharolyticus* in its physiological properties and belongs to the high % G+C group although it has a low mobility MDH. As expected from its DNA base composition (Table 2) it has a low homology with the Werner strains W 83 and W 50 which it resembles electrophoretically. Its status as a member of the high G+C group is confirmed by its high homology with NCTC 9337. This provides an example of how the mutability of enzymes can cause electrophoresis to be unreliable in particular instances.

The heterogeneity amongst the asaccharolytic strains provides a problem, for while it remains desirable to remove them from *Bcrd. melaninogenicus,* any

TABLE 2

Homologies of DNA from Bacteroides melaninogenicus

Subspecies Strain (and origin)	% G+C (*)	% G+C (†)	% homology to reference strain (†)		
melaninogenicus			*VPI 4196*		
VPI 4196 (sputum)	38.5	40	100		
VPI 9085 (gingiva)	45.8	46	0		
VPI 9343 (toe infection)	38.0	35	52		
asaccharolyticus			*VPI 4198*		
VPI 4198 (empyema)	50.5	–	100		
NCTC 9337 (haemorrhoids)	50	–	73		
W 83 (clinical specimen)	45.6	46	7		
W 50 (clinical specimen)	45.2	49	7		
intermedius			*VPI 4197*		*NCTC 9336*
VPI 4197	–	44	100		29
LH 107 (dental plaque)	–	41	63		79
NCTC 9336 (gingivitis)	42.7	42	55		100
B 18 (periodontitis)	43.2	–	53		85
T 584 (plaque)	40.7	–	44		63
JP 1 (cattle horn)	43.5	43	66		78
G11ad2 (dental root)	–	42	58		81
2B (urethra)	43.9	42	44		14
LH 100 (plaque)	42.0	–	10		8
levii			*VPI 4198*	*VPI 4197*	*NCTC 9336*
JP 2 (cattle horn)	43.5	43	–	10	11
Lev 1 (rumen)	–	43	0	–	–

*Shah *et al.* (1976), and † J. L. Johnson (unpublished results).

single species proposed to accept all of them would be subdivided on chemical grounds at the outset. Despite the chemical heterogeneity of subspecies *asaccharolyticus* (Shah *et al.* 1976), no such subdivision is evident in routine physiological properties (Harding *et al.* 1976; Shah *et al.* 1976), and perhaps because of this it has been proposed that the asaccharolytic strains of *Bcrd. melaninogenicus* should be assigned to a new species, *Bacteroides asaccharolyticus* (Finegold & Barnes 1977). Whether by fortune or design, all three of the strains described by these authors belong to the high G+C group of strains (Fig. 1), but it is perhaps unlucky that the type strain (ATCC 25260 = VPI 4198) is the one strain in this group with atypical low MDH mobility. The cluster of four asaccharolytic strains with a DNA base composition of 45–46% G+C and low MDH mobility have therefore not been specifically included in *Bcrd. asaccharolyticus*. Of the four strains tested one (JMH 3A) is no longer extant. The three remaining strains, the Werner group, have a significantly different SDS-polyacrylamide pattern of proteins from the high G+C group (Swindlehurst *et al.*, 1977) and may form the nucleus of a second asaccharolytic species. The data of J. L Johnson (Table 2) for two of these Werner strains shows a DNA base composition in agreement with our own

TABLE 3

pH optimum of malate dehydrogenase of Bacteroides melaninogenicus *subspecies*

Subspecies	substrate	
	malate	oxaloacetate
melaninogenicus WAL 2728	8.0	7.3
intermedius T 588	8.0	8.4
asaccharolyticus W 83	10.2	8.4

and, not surprisingly, a negligible homology with two strains of *Bcrd. asaccharolyticus*.

As one property of MDH, its electrophoretic mobility, seemed to be of value in differentiating *Bcrd. melaninogenicus* strains, other properties of the enzyme were investigated for possible taxonomic significance. The Michaelis constants of the enzyme from strains of the three subspecies were not useful in this respect; however, the pH optima for activity were distinct for these strains (Table 3), although it was necessary to carry out the reaction in both directions in order to achieve this distinction. Therefore, further strains were not examined as the process appeared too cumbersome to be useful. The stability of MDH to incubation in alkaline buffers distinguished the asaccharolytic and the Lev strains (limited saccharolytic metabolism) from the saccharolytic types by the greater stability of their MDH; the enzyme of most strains of subspecies *intermedius* was more stable than of most subspecies *melaninogenicus* cultures (Table 4). The two subgroups within each of subspecies *asaccharolyticus* and subspecies *melani-*

TABLE 4

Time of half inactivation of malate dehydrogenase of Bacteroides melanogenicus *in alkaline buffers at 37°C*

Organisms	pH		
	9	10	11
subspecies *asaccharolyticus* and Lev strains (9 strains)	> 50	> 50	> 50
subspecies *intermedius* (10 strains)	31- > 50	24- > 50	8–40
S4 (fastest)	26	20	2.0
T 588 (slowest)	> 50	> 50	42.0
subspecies *melaninogenicus* and *Bcrd. oralis* (10 strains)	7–10	2.5–9	< 0.5–1.5
WAL 2728 (fastest)	4	2	< 0.5
G9a-c2 (slowest)	19.5	13.5	7.5

nogenicus cannot be distinguished from one another by this property of the enzyme.

It was decided to test whether such patterns of enzyme inactivation and mobility could be used to predict the identity of freshly isolated oral Gram negative anaerobic rods (from the M. R. C. Dental Epidemiology Unit) that form brown or black pigment on blood media in Petri dishes. The enzyme patterns were determined before the physiological tests. The strains were identified (Fig. 2) as five of subspecies *intermedius*, two of subspecies *asaccharolyticus* of the

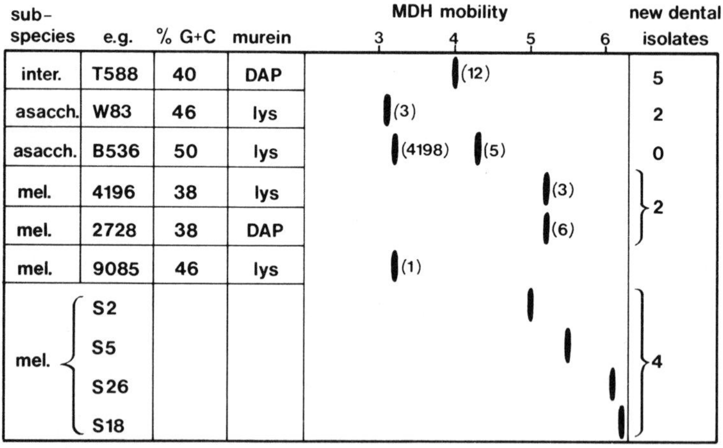

Fig. 2. DNA, mucopeptide and malate dehydrogenase mobility of *Bacteroides melaninogenicus* subgroups (Shah *et al.* 1976), and the identification of 13 fresh dental isolates by MDH electrophoresis and stability. Dibasic amino acids of murein (mucopeptide) are indicated by DAP (diaminopimelic acid) and lys (lysine). Subspecies are indicated thus: inter, subspecies *intermedius*; asacch, subspecies *asaccharolyticus* and mel, subspecies *melaninogenicus*.

low MDH mobility type, two of subspecies *melaninogenicus* with MDH like that of the strains examined previously, and four of subspecies *melaninogenicus* with a range of fast mobility of MDH. All these strains were initially identified by the stability of MDH to alkaline buffers. Subsequent electrophoresis of MDH confirmed these presumptive identifications. Despite the variability of MDH mobility in the fresh isolates of subspecies *melaninogenicus*, all of them had a faster migrating enzyme than that of the other two subspecies. The status of these strains was determined using API 20 tests for anaerobes. The identity of all strains was correctly predicted and the good correlation between alkaline stability of MDH, and these physiological tests suggests that MDH mobility and stability may be useful in the rapid identification of the three broad subgroups of pigmented *Bacteroides*.

3. Non-Pigmented *Bacteroides*

Studies with non-pigmenting (or poorly pigmenting) *Bacteroides* have not progressed as far as those with *Bacteroides melaninogenicus*. The results indicate, however, that biochemical taxonomic methods, including enzyme patterns, have a significant contribution to make. Glucose-6-phosphate dehydrogenase (G6PDH) and 6-phosphogluconate dehydrogenase (6PGDH) occur in all the strains of the *Bcrd. fragilis* group so far tested but not in any other *Bacteroides*. Where G6PDH is present, multiple electrophoretic bands are observed (Fig. 3).

Bacteroides	MDH	GDH	6PGDH	G6PDH
Bcrd. mel. VPI 4196	5·2	4·1	0	0
Bcrd. oralis VPI 8906D	5·2	4·1	0	0
Bcrd. vulgatus WBM 137	3·8	4·9	2·8	
Bcrd. oralis 7CM	3·8	4·9	2·8	

Fig. 3. Enzyme patterns of two putative strains of *Bacteroides oralis* together with standard strains of *Bacteroides melaninogenicus* subspecies *melaninogenicus* and of *Bacteroides vulgatus*. Enzymes are abbreviated thus: MDH, malate dehydrogenase; GDH, glutamate dehydrogenase; 6PGDH, 6-phosphogluconate dehydrogenase and G6PDH, glucose-6-phosphate dehydrogenase.

Strains received as *Bacteroides oralis* form a heterogeneous group which often do not show the key physiological characteristics (Table 5) described for the species (Holdeman & Moore 1973, 1974). Strain VPI 8906D closely resembles typical strains of *Bcrd. melaninogenicus* subspecies *melaninogenicus* in physiological properties, in the absence of the dehydrogenases of the pentose phosphate shunt and in the electrophoretic mobility of malate and glutamate dehydro-

TABLE 5

Reactions of putative strains of Bacteroides oralis

	*Bcrd. oralis**	S2	8906D	G9a-C2	7CM	J1	NP333	WPH 61
				Strains				
Black pigment	–	+	-†	v‡	–	–	–	–
Indole formation	–	–	–	–	–	–	–	–
Aesculin hydrolysis	+	+	+	+	–	+	+	+
Lactose	+	+	+	+	+	+	+	+
Starch	+	+	–	+	–	+	+	v
Sucrose	+	+	+	+	+	+	+	+
Xylose	–	w ‖	–	+	+	+	+	+
Arabinose	–	w	–	+	+	+	+	+
Response to bile∮	i	i	i	i	s	i	i	i

*Typical reactions of *Bcrd. oralis* (Holdeman & Moore 1974); †Occasionally forms perceptible pigment; ‡, variable; ∮ i, inhibited: s, stimulated and ‖ weak.

genases [(GDH) (Fig. 3)] and similar isoelectric points for MDH (VPI 8906D, pI = 4.3; VPI 4196, pI = 4.5). This resemblance is emphasized by the occasional formation of perceptible brown pigment by VPI 8906D. The unreliability of pigmentation as a characteristic is supported by the finding that *Bcrd. melaninogenicus* VPI 4196 does not become brown under some circumstances (Sundqvist 1976) while the supposedly non-pigmented *Bcrd. oralis* G9a-c2 (Sundqvist 1976) can form pigment intermittently (Table 5). Strain S2, which was isolated from dental plaque and originally considered to be *Bcrd. oralis*, produced brown pigment after four subcultures, contrary to the species description (Loesche *et al.* 1964), and is now considered to be *Bcrd. melaninogenicus* subspecies *melaninogenicus.* Other strains described as *Bcrd. oralis* (Table 5) ferment xylose and arabinose (7CM, J1, NP 333, WPH 61) and the growth of one (7CM) is stimulated by bile. Of these strains J1, NP 333 and WPH 61 have similar patterns of MDH and GDH to 7CM (Fig. 3) but do not contain G6PDH or 6PGDH. NP 333 has recently been identified with *Bacteroides ruminicola* subsp. *brevis* (L. V. Holdeman, pers. comm.). The biochemical comparison of these strains with reference strains of *Bcrd. ruminicola* is now in progress.

The other putative *Bcrd. oralis* (7CM) belongs to the *Bcrd. fragilis* group because its growth is stimulated by bile and it contains the dehydrogenases of the pentose phosphate shunt. Of this group it most closely resembles a strain of *Bacteroides vulgatus* (WBM 137; Fig. 3). The isoelectric points of MDH in the two strains were virtually identical.

Only a few strains of the *Bcrd. fragilis* group have been isolated from the mouth, and this, together with the fact that 7CM and J1, as originally described (Loesche *et al.* 1964) conformed to the description of *Bcrd. oralis*, must cast doubt on the authenticity of the strains presently bearing these designations.

The *Bcrd. fragilis* group is perhaps the most intensively studied of the Gram negative anaerobic rods and the five subspecies have been elevated to species rank on the basis of DNA homology (Cato & Johnson 1976). Two more species, *Bacteroides eggerthii* and *Bacteroides splanchnicus*, have been proposed recently (Holdeman *et al.* 1977). Homology subgroups were described (Cato & Johnson 1976), however, and heterogeneity of enzyme patterns is evident. Thus in five strains of *Bacteroides thetaiotaomicron*, four different patterns of superoxide dismutase were evident (Carlsson *et al.* 1977). The mobility of four dehydrogenases from representative strains of the seven species have been tested (Table 6) and where two strains of the same species were available they have proved very similar. The significance of these results awaits confirmation.

4. Lactic Acid Bacteria

One of the first papers (Lund 1965) on enzyme electrophoresis in bacterial taxonomy involved esterase and protein patterns of the enterococci. Similar findings were obtained in a study of the patterns of two dehydrogenases (Williams & Bowden 1968), namely that strains of *Streptococcus faecalis* and its varieties were quite homogeneous, and could be distinguished from *Streptococcus faecium* and the more variable *Streptococcus durans*. In both cases enzyme patterns were helpful with problematical strains.

Considerable use has been made of enzyme patterns amongst the lactobacilli, and the results have correlated well with other biochemical criteria (Williams 1975). The similarity of lactate dehydrogenase (LDH) patterns in *Lactobacillus fermentum* and *Lactobacillus cellobiosus* (Gasser 1970) and the identical DNA base composition of these species (Gasser & Mandel 1968) indicated a close relationship was possible. Meanwhile, the identity of G6PDH patterns in these species was reinforced by the findings of an exact correlation between enzyme pattern and an ornithine-containing mucopeptide; (Williams 1971; Williams & Sadler 1971). The mucopeptide of *Lact. cellobiosus* was already known to contain ornithine as the basic amino acid concerned in crosslinking (Plapp & Kandler 1967), but early reports ascribed this role to lysine in *Lact. fermentum*. In these studies *Lact. fermentum* walls were analysed by qualitative methods that would not necessarily distinguish ornithine from lysine (Cummins & Harris 1956; Ikawa & Snell 1960) and the latter workers used a strain (ATCC 9338) that does not resemble the neotype ATCC 14931. Later wall studies of putative strains of *Lact. fermentum* have also involved an atypical strain, for example ATCC 9338 which contains lysine (Wallinder & Neujahr 1971) but which has the physiological properties of *Lactobacillus brevis* (Schleifer & Kandler 1972). Strain ATCC 9338 is reputed to be the same strain as NCIB 6991 but this strain has ornithine mucopeptide and the usual enzyme pattern for *Lact. fermentum* (Williams & Sadler 1971). Further complications are the existence of a second group called *Lact. fermentum* (Reiter Type II, e.g. ATCC 23272) which differs from the neotype strain in having 40% rather than 50-53% G+C, in its fermen-

TABLE 6

Enzyme mobilities and activities amongst the Bacteroides fragilis group of species

Spicies of Bacteroides	Source	Mobility on Cellogel (cm)				Activity (nmoles/min/mg/protein)			
		MDH*	GDH	6PGDH	G6PDH	MDH	GDH	6PGDH	G6PDH
fragilis	NCTC 8560	0.0 2.9 3.8	5.9	2.1	3.3	304.6	236.9	84.6	194.6
	NCTC 9343	0.0 2.9 3.8	5.9	2.2	0.0 3.4	85.5	222.3	102.6	153.9
thetaiotaomicron	W 910	0.0 4.1 4.6	5.0	1.2	3.8	111.9	292.8	137.8	48.2
	NCTC 10582	0.0 4.1 4.6	5.0	1.2	3.7	225.4	127.7	45.1	82.9
distasonis	ATCC 8503	3.5	5.1	2.5 2.9	3.7 4.1	180.8	93.8	100.5	154.1
ovatus	ATCC 8483	3.7 4.3 5.0	5.6	0.5 1.0	1.5 2.9	65.2	36.2	57.9	18.8
vulgatus	WBM 137	3.8	4.9	2.8	0.0 4.0 4.7 5.0	301.4	562.7	26.8	57.6
eggerthii	NCTC 11155	3.7	1.6	1.4	1.7	167.1	525.3	39.8	28.6
splanchnicus	NCTC 10826	3.7	1.4	2.6 3.5	1.7	136.7	482.3	48.2	32.2

*MDH, malate dehydrogenase; GDH, glutamate dehydrogenase; 6 PGDH, 6-phosphogluconate dehydrogenase and G6PDH, glucose-6-phosphate dehydrogenase.

tation tests and enzyme mobility (Schleifer & Kandler 1972), and which therefore probably represents an undefined species; and in the apparent mislabelling
of a strain of *Lact. cellobiosus* (Williams & Sadler 1971). These problems should
not obscure the fact that the majority of *Lact. fermentum* strains (including the
neotype strain) are indistinguishable from most *Lact. cellobiosus* strains (including the type strain) in patterns of LDH and G6PDH, % G+C and in the possession of an uncommon type of mucopeptide. The two species are distinguished
by growth temperature relationships which are not as reliable taxonomically as
fermentation tests (Leifson 1966), and by the hydrolysis of cellobiose and
cleavage of aesculin (Rogosa 1974; Rogosa & Sharpe 1959). This degree of
difference is comparable to that between the varieties of other species and it
seems likely that *Lact. cellobiosus* should be regarded as a physiological variety
of *Lact. fermentum*. This is confirmed by RNA-DNA homology (Miller *et al.*
1971), and by DNA-DNA homology (Sriringanathan *et al.* 1974; Sriranganathan
et al. in press).

It has been suggested that *Lactobacillus brevis* and *Lactobacillus buchneri*
may constitute a single species (Gasser & Mandel 1968) although *Lact. brevis*
strains are heterogeneous in DNA mean base composition. They are, however,
similar in physiological properties (Rogosa 1974) and have the same group
antigen (Sharpe *et al.* 1966). The electrophoretic patterns of many strains of
both species are similar or identical (Williams & Sadler 1971), but the mobility
of LDH (Gasser 1970) and its immunological cross-reactivity (Gasser & Gasser
1971) are not the same. DNA-RNA homology between one strain of each species
was very low, but the strains used (*Lact. brevis* XI and *Lact. buchneri* BC1 =
NCDO 110 = NCIB 8007; Miller *et al.* 1971), both had unusual G6PDH patterns
(Williams & Sadler 1971). It is possible that enzyme or protein patterns have a
role in preliminary screening methods to ensure that appropriate strains are
chosen for nucleic acid homology investigations. The relationship between *Lact.
brevis* and *Lact. buchneri* remains open, but it may be that both are heterogeneous, and that many strains of each are very similar to one another.

Lactobacillus jugurti and *Lactobacillus helveticus* are very similar phenotypically (Sharpe *et al.* 1966), are of the same serotype and have identical patterns of LDH (Gasser 1970) as well as very similar electrophoretic patterns of
soluble proteins (Morichi *et al.* 1968). These species have a high nucleic acid
homology (Dellaglio *et al.* 1974) and *Lact. jugurti* has been accepted as a variety
of *Lact. helveticus* (Rogosa 1974). Despite this, the esterase (Morichi *et al.* 1968)
and G6PDH (Williams 1971) patterns are variable in strains of the combined
species. Of the other three species of subgenus *Thermobacterium* with DNA base
composition low in % G+C, *Lactobacillus salivarius* and *Lactobacillus jensenii*
are individually homogeneous in dehydrogenase patterns but distinct from one
another, while *Lactobacillus acidophilus* is a heterogeneous cluster of strains
(Gasser 1970; Williams 1971).

It has been suggested that the four species of *Thermobacterium* with DNA of approx. 50% G+C may constitute one species (Gasser & Mandel 1968). *Lactobacillus lactis* and *Lactobacillus leichmannii* both give species-specific patterns of esterase (Morichi *et al*. 1968) and G6PDH (Williams 1971) but both of these species together with *Lactobacillus delbrueckii* have LDH that is identical, both in stereospecificity and electrophoretic mobility (Gasser 1970). *Lactobacillus bulgaricus* is heterogeneous in patterns of the two former enzymes, and homogeneous but quite distinct from the other three species in lactate dehydrogenase patterns. Many strains of *Lact. lactis* and *Lact. bulgaricus* have high DNA homology with one another (Dellaglio *et al*. 1973).

One of the successes for enzyme electrophoresis has been the discovery of *Lactobacillus jensenii*, strains of which are phenotypically identical with *Lact. leichmannii* (Gasser 1970) but which have a distinctly different pattern of lactate dehydrogenases (Gasser *et al*. 1970) and DNA containing only 36.1% G+C. Gasser (1970) has also pointed out that lactate dehydrogenase electrophoresis can be used to solve specific identification problems as between *Lact. acidophilus* and *Lact. leichmannii, Lact. jugurti* and *Lact. bulgaricus* and *Lact. casei* and *Lact. plantarum.*

5. Neisseria

Except for one non-saccharolytic strain (presumably a glucokinase-deficient mutant) of *Neisseria meningitidis*, all strains of the 'true Neisseria' species contain glucokinase and G6PDH (Holten 1974*a*) whether or not they are actively saccharolytic species. These constitutive enzymes increase in activity when saccharolytic strains are grown on glucose. Glucokinase and G6PDH are absent from the 'false Neisseria' species, *Neisseria catarrhalis* (*Branhamella catarrhalis*; Reyn 1974), *Neisseria ovis* and *Neisseria caviae*. This important biochemical finding supports the removal of these species from the genus. Those species that contain glucokinase and G6PDH also possess 6PGDH and 3-keto-2-deoxy-6-phosphogluconate aldolase and all, except *Neisseria cinerea* and *Neisseria elongata*, have 6-phosphogluconate dehydrase as well (Holten 1974*b*). The electrophoretic patterns of G6PDH are not useful taxonomically as there is much variation between the strains of many of the species (Holten 1974*a*). The usefulness of patterns of a particular enzyme varies amongst different groups of organisms. This enzyme is valuable in the classification of heterofermentative lactobacilli (Williams & Sadler 1971), but not in the enterobacteria (Bowman *et al*. 1967). However, amongst the *Neisseria* the patterns of NAD- and NADP-dependent GDH are useful to some degree (Holten 1973). In each strain of the 'false neisserias' the migration of NADP- and NAD-dependent enzymes is identical and the patterns within the three species seem to be homogeneous, although different from one another. As the specific activity of this enzyme is similar with both coenzymes, and the two enzyme activities chromatograph together on DEAE cellulose columns, one protein may be active with either coenzyme in these species (Holten 1973). Furthermore, the enzymes form the 'false neisserias' do not react at all with serum against *Neis. meningitidis* glutamate dehydrogenase (Holten 1974*c*). The 'true neisserias' have two separate glutamate dehydrogenase proteins, although the low activity of the NAD-dependent enzyme does not permit its qualitative detection in some strains after electrophoresis. Except for a single strain the gonococci are similar to one another and to the meningococci, 12 strains of which give identical patterns (Holten & Jyssum 1973). Amongst strains of *Neis. sicca, Neis. mucosa*, and the *Neis. subflava* group many of the strains show similar patterns with NADP-dependent enzyme migrating about 1:10 as far as the NAD-dependent enzyme, which show the greater degree of variation. The amalgamation of *Neis. flava* and *Neis. perflava* with *Neis. subflava* (Reyn 1974) on genetic grounds receives some support as seven of nine strains have electrophoretically identical NADP specific enzyme (Holten 1974*b*), and have similar immunological distance from *Neis. meningitidis* enzyme (Holten 1974*c*). *Neisseria lactamica, Neis. cinerea*, and *Neis. flavescens* are more heterogeneous in electrophoretic patterns and immunological cross reactivity, while *Neis. elongata* is different from all the others and is relatively homogeneous.

Fox & McClain (1975) examined esterase and dehydrogenase patterns amongst eight strains from six species of *Micrococcus, Branhamella* and *Neisseria*. They concluded that *Micrococcus cryophilus* is unrelated to the 'false neisserias', but conclusions concerning the affinities of the *Neisseria* spp. are limited because of the low number of strains tested.

6. Staphylococcus

Staphylococcus aureus was one of only three named staphylococcal species (Baird-Parker 1974) until quite recently. Amongst isolates from poultry carcases, the electrophoretic mobility of the major bands amongst the complex patterns of esterases was sufficient to distinguish biotype B strains (typically of animal origin) from biotype A (typically of human origin) and intermediate strains (Gibbs *et al.* 1978). Ten new species of *Staphylococcus* have been described (Schleifer & Kloos 1975; Kloos & Schleifer 1975; Kloos *et al.* 1976) and many of these have esterase patterns that are remarkably homogeneous within the species (Zimmerman & Kloos 1976). Ten strains or more were examined for each species, and most have different enzyme bands. Some are similar in esterase patterns, however, and all of them cannot be identified by this technique.

The mobility of catalase shows similarly useful features (Zimmerman 1976) except that those species with variable and complex patterns of esterases, for example *Staf. xylosus* and *Staf.* species *3*, have more simple and constant catalase bands. Furthermore, in *Staf. cohnii* where the subspecies have different esterases there is less variability in catalase bands, and whereas the subspecies of *Staf. sciuri* and *Staf.* species 2 have very similar catalase patterns as well as similar physiological properties, only one subspecies of *Staf. sciuri* has detectable esterase bands at all. Esterase and catalase electrophoresis is particularly useful according to these authors, as species that may be mistaken for one another, such as *Staf. hominis* and *Staf. warneri*, can be distinguished by this method.

DNA hybridization has been investigated in two strains each of most of these species, including the type strains of most species (Pulverer *et al.* 1978). This study indicates that several of the species (e.g. *Staf. saprophyticus*, *Staf. xylosus* and *Staf. simulans*) form distinct homology groups, but that the strains of *Staf. warneri*, *Staf. hominis* and one of the two strains of *Staf. capitis* form a single homology group. As with many conflicting taxonomic studies a major problem is that none of the strains tested for enzyme patterns by Zimmerman & Kloos (1976) and Zimmerman (1976) were tested by Pulverer *et al.* (1978).

7. Conclusion

Electrophoretic enzyme patterns have been applied to many strains of bacteria, but some of the publications have not been useful because insufficient strains were tested, or because dissimilar bacteria (perhaps different genera) have been compared. In some groups of bacteria one enzyme may produce significant patterns, but in another group another enzyme may prove more useful. The choice of enzyme is a matter of personal preference. In general, enzymes that produce multiple electrophoretic bands provide more features for comparison and this may be one reason for the popularity of esterase electrophoresis.

Amongst *Rhizobium* spp. 3-hydroxybutyrate dehydrogenase patterns indicated a similarity between some strains of *Rhiz. leguminosarum* and *Rhiz. trifolii* (Fottrell & O'Hora 1969), which was also evident in the esterase patterns of Murphy & Masterson (1970) and those of Clark (1972) although there are some unresolved differences in the patterns of esterases in these two papers.

Goullet (1973) examined esterase patterns in *Escherichia coli* where considerable strain variation was found; in *Proteus* and *Providencia* where the patterns were species-specific with variable intraspecies difference in the mobility and type of esterase (Goullet 1975); in *Levinea* and *Citrobacter* where species-specific patterns were also observed (Goullet & Richard 1977), and in *Salmonella* where the patterns were sufficiently distinct to recognize the subgenus of most of the strains (Goullet 1977).

In general it seems that enzyme patterns are often useful when applied with caution to taxa that can be clustered by other criteria. Although the mobility of an enzyme is as likely to be changed by mutation as is the rest of a fermentation test, it seems likely that the inter-laboratory reproducibility of enzyme patterns should be higher, as is the case with other chemotaxonomic tests. Thus in fermentation tests a faint positive result may be incorrectly regarded as negative, but with enzyme electrophoresis, unless the enzyme is inactivated during preparation, differences in electrophoretic mobility can hardly be confused. The patterns of enzyme mobility may vary amongst strains of a species, or between subspecies or species. On the other hand identical patterns may be obtained with different species (or indeed with different genera). The significance of the electrophoretic results must therefore be established by reference to other techniques.

In certain instances it may prove possible to use enzyme patterns as a rapid diagnostic test in identification. Finally, there is a role of enzyme and protein electrophoresis as a screening technique to identify clusters of similar organisms prior to homology studies.

8. References

BAILIT, H. L., BALDWIN, D. C. & HUNT, E. E., Jr. 1964 Increasing prevalence of gingival *Bacteroides melaninogenicus* with age in children. *Archives of Oral Biology* 9, 435-438.

BAIRD-PARKER, A. C. 1974 *Staphylococcus*. In *Bergey's Manual of Determinative Bacteriology* 8th edn ed. Buchanan, R. E. & Gibbons, N. E. pp. 483-489. Williams & Wilkins: Baltimore.

BOWMAN, J. E., BRUBAKER, R. R., FRISCHER, H. & CARSON, P. E. 1967 Characterisation of enterobacteria by starch-gel electrophoresis of glucose-6-phosphate dehydrogenase and phosphogluconate dehydrogenase. *Journal of Bacteriology* 94, 544-551.

CARLSSON, J., WRETHEN, J. & BECKMAN, G. 1977 Superoxide dismutase in *Bacteroides fragilis* and related *Bacteroides* species. *Journal of Clinical Microbiology* 6, 280-284.

CATO, E. P. & JOHNSON, J. L. 1976 Reinstatement of species rank for *Bacteroides fragilis, B. ovatus, B. distasonis, B. thetaiotaomicron* and *B. vulgatus*: designation of neotype strains for *Bacteroides fragilis* (Veillon & Zuber) Castellani & Chalmers and *Bacteroides thetaiotaomicron* (Distaso) Castellani & Chalmers. *International Journal of Systematic Bacteriology* 26, 230-237.

CLARK, A. G. 1972 Starch-gel electrophoresis of catalase and esterase isoenzymes from some *Rhizobium* and *Agrobacterium* spp. *Journal of Applied Bacteriology* 35, 553-558.

CUMMINS, C. S. & HARRIS, H. 1956 The chemical composition of the cell wall in some Gram-positive bacteria and its possible value as a taxonomic character. *Journal of General Microbiology* 14, 583-600.

DANCE, N. & WATTS, D. C. 1962 Comparison of creatine phosphokinase from rabbit and brown hare muscle. *Biochemical Journal* 84, 114P.

DE LEY, J. 1962 Comparative biochemistry and enzymology in bacterial classification *Microbial Classification* ed. Ainsworth, G. C. & Sneath, P. H. A. pp. 164-195. Cambridge: Cambridge University Press.

DE LEY, J. 1968 Molecular biology and bacterial phylogeny. pp. 103-156 In *Evolutionary Biology* Vol. 2 Amsterdam: North Holland.

DELLAGLIO, F., BOTTAZZI, V. & TROVATELLI, L. D. 1973 Deoxyribonucleic acid homology and base composition in some thermophilic lactobacilli. *Journal of General Microbiology* 74, 289-297.

DELLAGLIO, F., VESCOVO, M. & PREMI, L. 1974 'Free' diaminopimelic acid and deoxyribonucleic acid homology of *Lactobacillus helveticus*. *International Journal of Systematic Bacteriology* 24, 235-247.

FINEGOLD, S. M. & BARNES, E. M. 1977 Report of the ICSB Taxonomic Subcomittee on Gram negative anaerobic rods. *International Journal of Systematic Bacteriology* 27, 388-391.

FOTTRELL, P. F. & O'HORA, A. 1969 Multiple forms of D (-) 3-hydroxybutyrate dehydrogenase in *Rhizobium*. *Journal of General Microbiology* 57, 287–292.

FOX, R. H. & McCLAIN, D. E. 1975 Enzyme electrophoretograms in the analysis of taxon relatedness in *Micrococcus cryophilis, Branhamella catarrhalis* and atypical neisserias. *Journal of General Microbiology* 86, 210–216.

GASSER, F. 1970 Electrophoretic characterisation of lactic dehydrogenase in the genus *Lactobacillus*. *Journal of General Microbiology* 62, 223–239.

GASSER, F. & GASSER, C. 1971 Immunological relationships among lactic dehydrogenase in the genera *Lactobacillus* and *Leuconostoc*. *Journal of Bacteriology* 106, 113–125.

GASSER, F. & MANDEL, M. 1968 Deoxyribonucleic acid base composition of the genus *Lactobacillus*. *Journal of Bacteriology* 96, 580–588.

GASSER, F., MANDEL, M. & ROGOSA, M. 1970 *Lactobacillus jensenii* sp. nov., a new representative of the subgenus *Thermobacterium*. *Journal of General Microbiology* 62, 219–222.

GIBBS, P. A., PATTERSON, J. T. & HARVEY, J. 1978 Biochemical characteristics and enterotoxigenicity of *Staphylococcus aureus* strains isolated from poultry. *Journal of Applied Bacteriology*, 44, 57–74.

GOULLET, P. L. 1973 An esterase zymogram of *Escherichia coli*. *Journal of General Microbiology* 77, 27–35.

GOULLET, P. L. 1975 Esterase zymograms of *Proteus* and *Providencia*. *Journal of General Microbiology* 87, 97–106.

GOULLET, P. L. 1977 Relationship between electrophoretic patterns of esterases from *Salmonella*. *Journal of General Microbiology* 98, 535–542.

GOULLET, P. L. & RICHARD, C. 1977 Distinctive electrophoretic patterns of esterase from *Levinea malonatica, Levinea amalonatica* and *Citrobacter*. *Journal of General Microbiology* 98, 543–549.

HARDING, G. K. M., SUTTER, V. L., FINEGOLD, S. M. & BRICKNELL, K. S. 1976 Characterisation of *Bacteroides melaninogenicus*. *Journal of Clinical Microbiology* 4, 354–359.

HARRIS, H. 1966 Enzyme polymorphisms in man. *Proceedings of the Royal Society of London, Series B*, 164, 298–310.

HARTLEY, B. S. 1974 Enzyme families. In *Evolution in the Microbial World*. ed. Carlile, N. J. Skehel, J. J. pp. 151–182. Cambridge: Cambridge University Press.

HOLDEMAN, L. V. & MOORE, W. E. C. 1970 *Outline of Clinical Methods in Anaerobic Bacteriology*. 2nd Revision. Virginia Polytechnic Institute and State University Anaerobe Laboratory, Blacksburg, Virginia.

HOLDEMAN, L. V. & MOORE, W. E. C. 1973 *Anaerobic Laboratory Manual*. Virginia Polytechnic Institute and State University Anaerobe Laboratory, Blacksburg, Virginia.

HOLDEMAN, L. V. & MOORE, W. E. C. 1974 Gram-negative anaerobic bacteria. In *Bergey's Manual of Determinative Bacteriology* 8th edn. ed. Buchanan, R. E. & Gibbons, N. E. pp. 384–404 Baltimore: Williams & Wilkins.

HOLDEMAN, L. V., CATO, E. P. & MOORE, W. E. C. 1977 *Anaerobe Laboratory Manual*. 4th Edition. V. P. I. Anaerobe Laboratory , Virginia Polytechnic Institute and State University, Blacksburg, Virginia.

HOLTEN, E. 1973 Glutamate dehydrogenases in genus *Neisseria*. *Acta Pathologica Microbiologica Scandinavica Sect. B*, 81, 49–58.

HOLTEN, E. 1974*a* Glucokinase and glucose-6-phosphate dehydrogenase in *Neisseria*. *Acta Pathologica Microbiologica Scandinavica Sect. B*, 82, 201–206.

HOLTEN, E. 1974*b* 6-phosphogluconate dehydrogenase and enzymes of the Entner-Doudoroff pathway in *Neisseria*. *Acta Pathologica Microbiologica Scandinavica Sect. B* 82, 207–213.

HOLTEN, E. 1974*c* Immunological comparison of NADP-dependent glutamate dehydrogenase and malate dehydrogenase in genus *Neisseria*. *Acta Pathologica Microbiologica Scandinavica Sect. B* 82, 849–859.

HOLTEN, E. & JYSSUM, K. 1973 Glutamate dehydrogenases in *Neisseria meningitidis*. *Acta Pathologica Microbiologica Scandinavica Sect. B* 81, 43–48.

IKAWA, M. & SNELL, E. E. 1960 Cell wall composition of lactic acid bacteria. *Journal of Biological Chemistry* 235, 1376-1382.

KELSTRUP, J. 1966 The incidence of *Bacteroides melaninogenicus* in human gingival sulci, and its prevalence in the oral cavity at different ages. *Periodontics* 4, 14-18.

KING, M. C. & WILSON, A. C. 1975 Evolution at two levels in humans and chimpanzees. *Science, New York* 188, 107-116.

KLOOS, W. E. & SCHLEIFER, K. H. 1975 Isolation and characterisation of staphylococci from human skin. II Descriptions of four new species: *Staphylococcus warneri*, *Staphylococcus capitis*, *Staphylococcus hominis* and *Staphylococcus simulans*. *International Journal of Systematic Bacteriology* 25, 62-79.

KLOOS, W. E., SCHLEIFER, K. H. & SMITH, R. F. 1976 Characterisation of *Staphylococcus sciuri* sp. nov. and its subspecies. *International Journal of Systematic Bacteriology* 26, 22-37.

LAMBE, D. W., Jr. 1974 Determination of *Bacteroides melaninogenicus* serogroups by fluorescent antibody staining. *Applied Microbiology* 28, 561-567.

LEIFSON, E. 1966 Bacterial taxonomy: a critique. *Bacteriological Reviews* 30, 257-266.

LOESCHE, W. J., SOCRANSKY, S. S. & GIBBONS, R. J. 1964 *Bacteroides oralis* proposed new species isolated from the oral cavity of man. *Journal of Bacteriology* 88, 1329-1337.

LUND, B. M. 1965 A comparison by the use of gel electrophoresis of soluble protein components and esterase enzymes of some Group D streptococci. *Journal of General Microbiology* 40, 413-419.

MACDONALD, J. B., SUTTON, R. M., KNOLL, M. L., MADLENDER, E. M. & GRAINGER, R. M. 1956 The pathogenic components of an experimental fuso-spirochaetal infection. *Journal of Infectious Diseases* 98, 15-20.

MILLER III, A., SANDINE, W. E. & ELLIKER, P. R. 1971 Deoxyribonucleic acid homology in the genus *Lactobacillus*. *Canadian Journal of Microbiology* 17, 625-634.

MOCHIZUKI, I., WETANABE, K., MIWA, T., IMAMURA, H., KOUBATA, S., NINOMIYA, K., UENO, K. & SUZUKI, S. 1975 Black colony of *Bacteroides melaninogenicus* on the laked blood media. *Igaku to Seibutsugaku (Medicine and Biology)* 91, 95-99.

MORICHI, T., SHARPE, M. E. & REITER, B. 1968 Esterases and other soluble proteins of some lactic acid bacteria. *Journal of General Microbiology* 53, 405-414.

MURPHY, P. M. & MASTERSON, C. L. 1970 Determination of multiple forms of esterase in *Rhizobium* by paper electrophoresis. *Journal of General Microbiology* 61, 121-129.

NORRIS, J. R. 1968 The application of gel electrophoresis to the classification of microorganisms. In *Chemotaxonomy and Serotaxonomy* ed. Hawkes, J. G. pp. 49-56. London: Academic Press.

NORRIS, J. R. & BURGES, H. D. 1963 Esterases of crystalliferous bacteria pathogenic for insects: epizootiological applications. *Journal of Insect Pathology* 5, 460-472.

PLAPP, R. & KANDLER, O. 1967 Die aminosäuresequenz des asparaginsäure enthaltenden mureins von *Lactobacillus coryniformis* und *Lactobacillus cellobiosus*. *Zeitschrift fur Naturforshung* 22B, 1062-1067.

PULVERER, G., MORDARSKI, M., TKACZ, A., SZYBA, K., HECZKO, P. & GOODFELLOW, M. 1978 Relationships among some coagulase negative staphylococci, based upon deoxyribonucleic acid re-association. *FEMS Microbiology Letters* 3, 51-56.

REYN, A. 1974 *Neisseria*. In *Bergey's Manual of Determinative Bacteriology*, 8th edn. ed. Buchanan, R. E. & Gibbons, N. E. pp. 428-432. Baltimore: Williams & Wilkins.

ROGOSA, M. 1974 *Lactobacillus*. In *Bergey's Manual of Determinative Bacteriology*: 8th edn. ed. Buchanan, R. E. & Gibbons, N. E. pp. 576-593. Baltimore: Willliams & Wilkins.

ROGOSA, M. & SHARPE, M. E. 1959 An approach to the classification of the lactobacilli. *Journal of Applied Bacteriology* 22, 329-340.

SCHLEIFER, K. H. & KANDLER, O. 1972 Peptidoglycan types of bacterial cell walls and their taxonomic implications. *Bacteriological Reviews* 36, 407-477.

SCHLEIFER, K. H. & KLOOS, W. E. 1975 Isolation and characterisation of *Staphylococcus epidermidis* and *Staphylococcus saprophyticus* and descriptions of three new species:

Staphylococcus cohnii, Staphylococcus haemolyticus and *Staphylococcus xylosus.*
International Journal of Systematic Bacteriology 25, 50–61.
SHAH, H. N., WILLIAMS, R. A. D., BOWDEN, G. H. & HARDIE, J. M. 1976 Comparison
of the biochemical properties of *Bacteroides melaninogenicus* from human dental
plaque and other sites. *Journal of Applied Bacteriology* 41, 473–492.
SHAH, H. N., BONNETT, R., MATEEN, B. & WILLIAMS, R. A. D. 1979 The porphyrin
pigmentation of subspecies of *Bacteroides melaninogenicus*. *Biochemical Journal*
180, 45–50.
SHARPE, M. E., FRYER, T. F. & SMITH, D. G. 1966 Identification of the lactic acid
bacteria. In *Identification Methods for Microbiologists, IA* ed. Gibbs, B. M. &
Skinner, F. A. pp. 65–79. London: Academic Press.
SHAW, C. R. 1965 Electrophoretic variation in enzymes. *Science, New York* 149, 936–942.
SOCRANSKY, S. S. 1970 Relationship of bacteria to the etiology of peridontal disease.
Journal of Dental Research 49, 203–222.
SOCRANSKY, S. S. & GIBBONS, R. 1965 Required role of *Bacteroides melaninogenicus*
in mixed anaerobic infections. *Journal of Infectious Diseases* 115, 247–253.
SOKATCH, J. R. 1969 Fermentation of sugars. In *Bacterial Physiology and Metabolism* ed.
Sokatch, J. R. Chapter 6 pp. 72–111. New York: Academic Press.
SRIRINGANATHAN, N., SEIDLER, R. J., SANDINE, W. E. & ELLIKER, P. R. 1974
Taxonomic implication of DNA homology studies of the type and neotype strains
of the species of the genus *Lactobacillus*. Abstract in *Proceedings of the Annual
Meeting of the American Society for Microbiology* pp. 30.
SRIRANGANATHAN, N., SEIDLER, R. J. & SANDINE, W. E. (1980) Nucleic acid studies
on some type and neotype strains of the genus *Lactobacillus*. *International Journal of
Systematic Bacteriology* in press.
SUEOKO, N. 1961 Variation and heterogeneity of base composition of deoxyribonucleic
acids; a compilation of old and new data. *Journal of Molecular Biology* 3, 31–40.
SUNDQVIST, G. 1976 *Bacteriological studies of neocrotic dental pulps*. Ph.D. Thesis,
University of Umea, Sweden.
SWINDLEHURST, C. A., SHAH, H. N., PARR, C. W. & WILLIAMS, R. A. D. 1977 Sodium
dodecyl sulphate-polyacrylamide gel electrophoresis of polypeptides from *Bacter-
oides melaninogenicus*. *Journal of Applied Bacteriology* 43, 319–324.
WALLINDER, I-B. & NEUJAHR, H. Y. 1971 Cell wall and peptidoglycan from *Lactobacillus
fermenti. Journal of Bacteriology* 105, 918–926.
WILLIAMS, R. A. D. 1971 Cell wall composition and enzymology of lactobacilli. *Journal of
Dental Research* 50, 1104–1117.
WILLIAMS, R. A. D. 1975 A review of biochemical techniques in the classification of lactobacilli. In *Lactic Acid Bacteria in Beverages and Food* ed. Carr, J. G., Cutting, C. V.,
& Whiting, G. C. pp. 351–367. London: Academic Press.
WILLIAMS, R. A. D. & BOWDEN, E. 1968 The starch gel electrophoresis of glucose-6-
phosphate dehydrogenase and glyceraldehyde-3-phosphate dehydrogenase of *Strep-
tococcus faecalis, S. faecium* and *S. durans*. *Journal of General Microbiology* 50,
329–336.
WILLIAMS, R. A. D. & SADLER, S. A 1971 Electrophoresis of glucose-6-phosphate dehydrogenase, cell wall composition and the taxonomy of heterofermentative lactobacilli.
Journal of General Microbiology 65, 351–358.
WILLIAMS, R. A. D., BOWDEN, G. H., HARDIE, J. M. & SHAH, H. N. 1975 Biochemical
properties of *Bacteroides melaninogenicus* subspecies. *International Journal of
Systematic Bacteriology* 25, 298–300.
ZIMMERMAN, R. J. 1976 Comparative zone electrophoresis of catalase of *Staphylococcus*
species isolated from mammalian skin. *Canadian Journal of Microbiology* 22,
1691–1698.
ZIMMERMAN, R. J. & KLOOS, W. E. 1976 Comparative zone electrophoresis of esterases
of *Staphylococcus* species isolated from mammalian skin. *Canadian Journal of
Microbiology* 22, 771–779.

Phages in the Identification of Plant Pathogenic Bacteria

E. BILLING AND C. M. E. GARRETT

East Malling Research Station, East Malling, Maidstone, Kent, UK

Contents

1. Introduction

IN THE MEDICAL and veterinary fields, phage typing has proved of great value in epidemiological studies of some bacterial pathogens, notably *Salmonella typhi*, other *Salmonella* spp. and *Staphylococcus aureus* (Adams 1959; Parker 1972). The pathogens are identified by cultural, physiological and serological methods while phages are used to differentiate epidemiological types within the species. The development of typing schemes has involved ingenuity and sophisticated approaches not used for plant pathogens where the main aim has most often been to identify a particular pathogen. A useful introduction to the principles of phage-typing, whatever approach is used, is given by Adams (1959).

With many plant pathogenic bacteria, difficulty is experienced in distinguishing them from closely related pathogens and from associated epiphytic bacteria. Conventional cultural and physiological tests sometimes suffice to identify a pathogen (e.g. an isolate of characteristic colony form from a plant with typical symptoms), but for distinguishing two very similar bacteria such tests often prove inadequate.

The reproduction of the characteristic disease symptoms by inoculation of the pathogen into its specific host plant is valuable in diagnosis, but such tests

319

are cumbersome and present practical difficulties. The ideal test should identify a specific pathogen rapidly and unambiguously.

When seeking rapid methods for distinguishing closely related bacteria, the choice often lies between serological methods and phage or bacteriocin sensitivity tests. Initially, phage and bacteriocin tests may be more easily developed. If, however, the special facilities needed for serology are available and suitable antisera are obtained, serological tests are usually quicker and more convenient. The choice of method will ultimately depend on how far the specificity of the system meets the requirements.

Although the differential value of phages has been demonstrated with some of the plant pathogens, their use has been confined largely to specialist labora-

TABLE 1

Abridged classification of some common Gram negative plant pathogenic bacteria and related epiphytes

Family: Pseudomonadaceae (G+C: 59–69 moles %)*

Pseudomonas		
fluorescens (group)†		
syringae (group)†	Section I	*campestris* (group)†,‡
cichorii		*fragariae*
caryophylli		*albilineans*
cepacia	Section II	*axonopodis*
marginata		
'olanacearum	Section III	*ampelina*
maltophilia ‡	Section IV	

Family: Enterobacteriaceae (G+C: 50–59 moles %)
   Agrobacterium
      tumefaciens
      radiobacter
      rhizogenes
      rubi

Family: Enterobacteriaceae (G+C: 50–59 moles %)
   Tribe II Klebsielleae
      Klebsiella
      Enterobacter
   Tribe V Erwinieae
      Erwinia
                 A. Amylovora group†
                 B. Herbicola group†
                 C. Carotovora group†

*(After Buchanan & Gibbons 1974.) G+C content of DNA;† Groups containing a range of pathogens (pathotypes) distinct in their host specificity or in symptoms they produce but not necessarily distinguishable by conventional laboratory tests (see text); ‡*Psmn. maltophilia* (not a plant pathogen) and *Xnmn. campestris* may be closely related (Murata & Starr 1973).

tories where they have been used as tools in studies on epidemiology and host pathogen relationships as well as for diagnosis. The main aim here is to show for which organisms phages with a useful degree of specificity (alone or in combination) have been isolated, and for what purposes they have been used.

An understanding of the classification of plant pathogenic and associated bacteria is necessary to appreciate the differential value of phage-typing. The classification in the eighth edition of *Bergey's Manual* (Buchanan & Gibbons 1974) is a useful starting point. An abridged classification of the more common Gram negative taxa is given in Table 1. Important features of the classification of each group which relate to their phage sensitivity are detailed in Sections 4-7. A fuller discussion of the classification of plant associated bacteria is given by Billing (1976).

Further information on phages of plant pathogens will be found in reviews by Okabe & Goto (1963) and Vidaver (1976) and in a discussion of interactions between phages associated with plants by Civerolo (1972).

Jones & Sneath (1970) have discussed the possible significance of shared phage sensitivity in reflecting genetic relatedness. In theory quite a low degree of relatedness may permit phage multiplication in different host bacteria, but in practice phages more often lyse strains of the taxospecies of the propagating strain than those of other taxospecies in the same genus. Cross reactions between strains in different genera appear to be rare. How much weight should be given to cross reactions, or lack of them, in phage tests is debatable.

The impact of phage tests on the classification of plant pathogens has been negligible. This paper is concerned primarily with identification but examples of shared phage sensitivities which might have some significance for classification will be noted as each taxon is discussed.

2. Methods

A. Isolation, stocks and storage

Classic methods (Adams 1959; Meynell & Meynell 1965; Billing 1969) have been used successfully for isolation, purification and preparation of high-titre stocks and for use in typing for plant pathogenic bacteria. In practice virulent phages are used almost exclusively. Most phages have been isolated from the natural habitat of the host bacterium, i.e. plant material or soil beneath infected plants, but some have come from sources remote from plants such as sewage or meat.

Chloroform can be used as a bactericide during isolation and storage of many phages but those which are sensitive to this chemical must be separated from residual bacteria by filtration. As host range and plaque mutants can arise during propagation, particularly if the propagating strain is changed, preparation of sufficient high-titre stocks for long term use is advisable in the interests of reproducibility of reaction. Goto (1969) however has reported differences in host range between *Xanthomonas oryzae* phages used soon after propagation and after prolonged storage. Many workers find storage in a nutrient broth at 0-5°C satisfactory and although the titre of plaque forming particles may fall, survival of residual phage for at least eight years has been reported (Boyd *et al.* 1965; Crosse & Garrett unpublished). With freeze-dried preparations, survival rates have often been low (P. Roberts, pers. comm.).

B. Temperate phages and bacteriocins

Many temperate phages have been isolated for pseudomonads and xanthomonads; there are also reports of lysogeny in *Erwinia herbicola* (Chatterjee & Gibbins 1971; Erskine 1973) and in *Erwn. chrysanthemi* (Paulin & Nassan 1976). There are fewer reports of their detection in other taxa, but how far this reflects the absence of lysogeny or detection difficulties is uncertain.

The incidence of lysogeny in strains of a species may be as high as 80%, e.g. *Xnmn. citri* and *Xnmn. ampelina* (Okabe 1961; Garrett unpublsihed), while in *Pseudomonas morsprunorum*, although over 50% of cherry strains are lysogenic (Garrett & Crosse 1963), temperate phages have been detected in less than 5% of plum strains (Lažar & Crosse 1969). In *Agrobacterium tumefaciens* fewer than 10% of strains have proved lysogenic (Zimmerer *et al.* 1966; Garrett 1978).

Bacteriocins affecting plant pathogens are outside the scope of this paper; most of the few studied are listed by Vidaver (1976). Both temperate phages and bacteriocins are often highly specific and this is probably the main reason why they have not been widely used at the pathogen level for identification.

C. Sensitivity tests

Where the range of phages is wide, a multipoint inoculator may be used but with routine identification of particular plant pathogens, reliance is often placed on a few phages and measured drops (*ca.* 0.02 ml) are pipetted on to lawns of the test organism. Some phages are used at high titre, because this usefully extends their host range, but it is not always clear in such cases whether or not reactions are due to phage lysis, lethal adsorption or to bacteriocin activity. For this reason, phage suspensions are normally diluted for typing purposes to the routine test dilution (RTD) which is that dilution which just gives confluent lysis with the propagating strain (Adams 1959) or 10-100 x RTD.

Unless virulent mutants can be isolated, temperate phages have the disadvantage that they may lysogenize the test organism and obscure the reaction. Lysogenic test bacteria may be resistant to some phages which can lead to typing difficulties.

3. Specificity

Some phages have a broad spectrum of activity while others, including many temperate phages, are highly specific. The art is to obtain phages with a useful degree of specificity for a particular purpose. In human and animal pathology, phages have proved most valuable for distinguishing epidemiological types of a single pathogen; in plant pathology, the main aim has been to find phages which will aid identification of individual pathogens. It has to be appreciated that absolute specificity for any purpose is unlikely to be achieved.

When isolating phages for a particular purpose a range of sources (e.g. plant material, soil, sewage) may be tried. Modification of host range of phages by propagation on different pathotypes or strains within a pathotype (Adams 1959) to increase their specificity or range of activity has rarely been attempted with phages for plant pathogens.

Infected plant material containing the pathogen in question has sometimes proved the most fruitful source of phages with a high degree of specificity (Hayward 1964b; Billing 1963; 1970a; Taylor 1970, 1972; Ritchie & Klos 1976). Some of these could well be virulent mutants of temperate phages from lysogenic strains of the pathogen or associated epiphytes. Highly specific phages have also been obtained from soil beneath infected plants (Crosse & Garrett 1963; Persley & Crosse 1978; see Table 3). For cross-reacting phages lysing both soil saprophytic and plant pathogenic pseudomonads, soil has proved a useful source (Stolp 1961) as have meat and manure for those lysing both food spoilage and plant pathogenic pseudomonads (Billing 1963, 1970a).

Before any phage or group of phages can be used with confidence for identification of bacteria, their level of specificity needs to be established. Ideally this involves testing against 10 or more authentic cultures of each of the following: cultures of the specific pathogen both from different host plants (if appropriate) and different geographical areas; other closely related pathogens, and related epiphytic bacteria commonly associated with the pathogen in question. In the *Pseudomonas syringae* group, this ideal approach could involve many phage/ bacterium test combinations. If cross-reactions with similar bacteria from other habitats or less closely related bacteria were included, in an attempt to demonstrate relationships, the number of tests could easily be doubled.

Clearly some limitation of specificity testing is inevitable and will be influenced by the main aim of the work and laboratory facilities. Claims regarding specificity need to be examined carefully; too often they are made on inadequate grounds.

Because absolute specificity can never be relied on, there are inherent risks in using single phages for identification without supporting tests. This danger can be overcome to some extent by using several phages and noting reaction patterns. Any one isolate may prove insensitive to one or other of the typing phages but the major part of the pattern usually remains.

Physiological and serological tests are often used alongside phage tests. In general, the specificity of phage tests has been greater—sometimes much greater than serological tests (Lovrekovich *et al.* 1963; Sato *et al.* 1971; Taylor 1972).

4. Pseudomonadaceae

The current classification of the genus *Pseudomonas* (Doudoroff & Palleroni 1974) is, in most cases, in good agreement with the results of RNA and DNA homology tests (Palleroni & Doudoroff 1972; Palleroni *et al.* 1973) and is concise enough to form a basis for discussion between bacteriologists concerned with different types of habitat but, for the plant pathologist, a general purpose classification is inadequate. Although the previous multiplicity of species in the genera *Pseudomonas* and *Xanthomonas* has often been deplored, the pathologist needs a special purpose classification which reflects the high degree of host specificity seen among similar bacteria in these two genera; it must also allow for the fact that some plant hosts may be infected by three or more different pseudomonads. In this paper, distinct pathogens will be referred to as pathotypes and their host of origin indicated where appropriate.

It is not surprising, therefore, that pathologists wish to retain many of the earlier specific epithets (Dye 1974; Dye *et al.* 1975; Young *et al.* 1978). There are at least 50 distinct pathogens in the *Pseudomonas syringae* group and more than 100 in the *Xanthomonas campestris* group (often named after the specific host they infect) as they are at present defined in *Bergey's Manual* (Table 1).

Members of the *Psmn. fluorescens* group are commonly found in association with plants but only a few are pathogenic. Arguments about which pathogenic or saprophytic pseudomonads merit specific rank may continue for many years but distinctions at all levels must be recognized and accommodated in any nomenclatural system.

Most of the phage studies in the family Pseudomonadaceae have concentrated on the *Psmn. syringae* and *Xnmn. campestris* groups, i.e. on the pathogens and epiphytes where distinction by other means have proved most difficult. No extensive cross-reaction tests have been made using phages lysing members of the different sections or groups in either the genus *Pseudomonas* or the genus *Xanthomonas*. There is however one case where phages isolated for *Xnmn. phaseoli* lysed four strains of *Psmn. phaseolicola* and one strain of *Psmn. syringae* (Vidaver & Schuster 1960). The taxonomic or ecological significance of these observations remains to be elucidated.

A. Pseudomonas

(i) *Section I*

All pseudomonads which produce a water-soluble green fluorescent pigment are included here but production of such a pigment is not characteristic of all members of the group. The determinative scheme of Lelliott *et al.* (1966) is commonly used to identify subgroups of plant pathogens in Section I.

In a detailed study which included many soil saprophytic pseudomonads as well as plant pathogens, Stolp (1961) observed many cases where phages lysed both saprophytes and pathogens. Neither individual phages nor patterns of sensitivity appeared to be of value in distinguishing saprophytes from pathogens or for distinguishing pathotypes from one another. Crosse & Garrett (1961) and Billing (1963, 1970*a*) also frequently observed shared phage sensitivities between pathogens, and also between pathogens, and saprophytes from plants and other sources but they succeeded in distinguishing certain—but not all—pathotypes on the basis of their phage sensitivities.

Examples of both highly specific and cross-reacting phages are shown in Table 2. The high specificity of 12B, 33, A7, 2 and 3A suggest that they should be of value in identification. Phage A7 and phages with a similar specificity to phage 2 have in fact been used as tools in special investigations (see pp. 332). Instances where patterns of sensitivity have proved useful are shown in Table 3. Both Crosse & Garrett (1961) and Billing (1970*a*) have emphasized the value of supporting physiological tests in identification rather than placing reliance on phage tests alone.

TABLE 2

Examples of specific and cross-reacting phages of Section I Pseudomonas *spp.*

Pseudomonas species (pathotype)	Sensitivity to phages*								
	12B	33	A7	2	3A	9B	12S	15	21
syringae (pear)	++	–	–	–	–	++	++	++	++
syringae (lilac)	–	++	–	–	–	++	++	++	++
morsprunorum (cherry)	–	–	++	–	–	++	++	(+)	–
phaseolicola	–	–	–	++	–	++	++	++	–
tabaci	–	–	–	–	+	++	–	–	–
cichorii	–	–	–	–	–	(+)	(+)	–	–
fluorescens	–	–	–	–	–	–	(+)	(+)	(+)

(Mainly from Billing 1970*a*).
*++, > 80% of strains sensitive; +, 50–80%; (+), < 50%; –, no isolates found sensitive.

Reasons for the apparent conflict between Stolp's findings and those of others may lie in the approach used. Crosse & Garrett (1961) used two plant pathogens from a particular ecological environment for the isolation of most of their phages whereas Stolp (1961) used mostly soil saprophytic pseudomonads. Billing's (1963, 1970*a*) approach was intermediate but weighted in favour of plant pathogens. The number of isolates of each pathotype examined by Stolp was often very low compared with the numbers studied by Crosse & Garrett and by Billing, so in some cases distinctive patterns could have been missed.

Success in distinguishing pathotypes has been only partial. The oat pathogen, *Psmn. coronafaciens*, and the ubiquitous epiphyte, *Psmn. viridiflavea*, gave very variable patterns of sensitivity (Billing 1970*a,b*). Whether or not this reflects the existence of a range of phage types, extensive lysogeny, masking of receptors or a deficiency in the range of phages used is not known.

Among stone fruit pathotypes (Crosse & Garrett 1961; Freigoun & Crosse 1975; Persley & Crosse 1978) phages have been of value in distinguishing fine differences in host specificity and differences between distinct pathogens on a single host (Table 3). Using three of the phages listed (A7, B1, A15) cherry isolates (uniquely but not invariably sensitive to A7) are distinct from plum isolates which are only lysed by A15, and Race 1 cherry isolates are distinct from Race 2 on the basis of their lysis by B1. When additional phages were used the distinctions remained but, within each group, two or more different phage types could be distinguished. The origin of four of the phages (A1, A7, A32 and B1) is reflected in the uniformity of their reactions with the homologous bacterium. The first three were isolated using Race 1 cherry strains of *Psmn. morsprunorum* and the fourth with a Race 2 strain.

TABLE 3

Phage sensitivities of Pseudomonas morsprunorum *isolates*

Bacteria	No. isolates	Typing phages (at RTD)				
		A1	A7	A32	B1	A15*
Cherry strains	{ 29	+	+	+	−	+
Race 1	{ 6	+	+	+	−	−
Cherry strains	{ 73	+	+	+	+	+
Race 2	{ 22	−	−	−	+	+
	{ 12	−	−	·+	+	+
Plum strains	{ 10	+	−	+	−	+
	{ 2	+	−	+	−	+

(Data simplified from Crosse & Garrett 1961 and Freigoun & Crosse 1975).
*A phage isolated for *Pseudomonas syringae*.

Some pathogens from different hosts were originally named *Psmn. syringae*, rather than being given a specific name reflecting their host of origin, because they resembled closely one another in physiological character. Phage tests have reflected both their host of origin and their close similarity of character depending on which phages have been used. So far, pear, citrus and lilac host types have been distinguished from one another on the basis of phage tests (Garrett *et al.* 1966; Billing 1970*a*; see also Table 2).

(ii) *Section II*

All Japanese isolates of *Psmn. caryophylli* have been reported to be susceptible to one highly specific phage but whether this phage was of value either diagnostically or epidemiologically was not indicated (Nishimura & Wakimoto 1971).

(iii) *Section III*

Among *Psmn. solanacearum* isolates from different hosts, four biotypes have been defined using physiological tests (Hayward 1964*a*). Initially, it was thought that the phage specific for biotype 2 would be valuable, in conjunction with the Gram stain, for distinguishing between *Psmn. solanacearum* and *Corynebacterium sepedonicum* which produce similar symptoms on potato tubers, thus avoiding the necessity of confirmatory tests. In practice Hayward (pers. comm.) found that some biotype 2 isolates were not lysed and reliance on a single phage was thought unwise. Bradbury (pers. comm.) prefers carbohydrate oxidation tests for distinguishing biotypes of *Psmn. solanacearum*, even though appropriate phages are available.

Thus in the Pseudomonadaceae, phages appear to have given some support to current thinking on classification, i.e. to the close relationships between the *Psmn. fluorescens* group and the *Psmn. syringae* group and within the latter but their impact has been greatest in distinguishing pathotypes. This does not mean however that species rank given to such pathotypes (Breed *et al.* 1957) should be retained.

B. Xanthomonas

Early work suggested that some *Xnmn. campestris* group phages had a high degree of specificity e.g. *Xnmn. pruni* (Eisenstark & Bernstein 1955), *Xnmn. carotae* and *Xnmn. vesicatoria* (Klement 1959) and *Xnmn. phaseoli* var. *fuscans* (Klement & Lovas 1960). These findings prompted a detailed study of phage relationships of xanthomonads by Stolp & Starr (1964) but their phages lacked a useful degree of specificity either alone or in combination.

Xanthomonas malvacearum isolates have been divided into two groups on the basis of physiological tests (Hayward 1964*b*). The same two groups were distinguishable by phage sensitivity tests. Phages were frequently detected on infected leaves and there was some correlation between the presence of members of groups 1 or 2 and the group specificity of phages isolated from the same material. Hayward suggested that detection of phage in moribund plant material might be an alternative to isolation of the pathogen in diagnosis. He successfully isolated phages for 7 other pathotypes in the *Xnmn. campestris* group in this way. When Hayward looked for reactions between *Xnmn. malvacearum* and 23 other *Xnmn. campestris* pathotypes in this group, zones of inhibition were occasionally observed but this was attributable to lethal adsorption and not to phage replication. Phages isolated for 12 other *Xnmn. campestris* pathotypes occasionally lysed group 2 isolates of *Xnmn. malvacearum*.

Thus phages isolated from plant material by Hayward appeared to be more specific than those isolated from soil by Stolp & Starr (1964). In spite of their apparent useful degree of specificity, however, Bradbury (pers. comm.) finds physiological tests adequate and more convenient for diagnosis.

Among 310 isolates of *Xnmn. phaseoli*, Sutton & Wallen (1967) distinguished 8 phage types, but for this high titre phages had to be used in some cases. There was some suggestion of a relationship between phage type and the geographical origin of isolates and also their virulence.

With *Xnmn. oryzae*, four phage types were described by Wakimoto (1960) but this number was later increased to 15 (Goto 1965). No correlation between phage type and virulence was observed. One phage, however, had a sufficiently useful degree of specificity to be used as an aid in disease forecasting.

Okabe (1961) concluded that the incidence of double or even triple lysogeny in *Xnmn. oryzae* contributed to the variation in sensitivity patterns, but variation in phage reaction between isolates from the same leaf and sometimes the same leaf lesion was more difficult to explain. Goto & Starr (1972) also concluded that phages were of little differential value for this pathogen.

So, with *Xanthomonas*, as with *Pseudomonas*, phages potentially useful in diagnosis have been found in some cases, but all too often in other cases they have proved non-specific or too specific, subdividing pathotypes into a range of phage types which are not necessarily ecologically meaningful.

5. Agrobacterium

Of the four species of *Agrobacterium* recognized, three are plant pathogens. The most common of these, *Agbc. tumefaciens*, is distinguishable from the fourth species, *Agbc. radiobacter*, most clearly in pathogenicity. Strains of the former carry a plasmid associated with their ability to induce galls on a variety of plants, to utilize specific amino acids and with their sensitivity to a specific bacteriocin.

Many phages have been isolated for agrobacteria, mostly from sewage (Roslycky *et al.* 1962; Boyd *et al.* 1970; Böttcher 1972; Garrett 1978), but no correlations between phage sensitivity and host plant, country of origin, biotype and virulence have been revealed. This may reflect the physiological heterogeneity of strains or the fact that the source of the phages was remote from the plant environment. However, cross reactions with *Rhizobium* spp., to which agrobacteria are closely related, and to which genus it has been suggested they be assigned (Heberlein *et al.* 1967), have not occurred.

Phages for *Agrobacterium* have made no contribution to either classification or to the differentiation of its species and there is no reason to assume that there will be any change in the foreseeable future.

6. Erwinia

All the plant pathogenic bacteria in the family Enterobacteriaceae are included in the genus *Erwinia* (Tribe V, Erwinieae, Table 1). Some common epiphytes are placed in this Tribe and others in Tribe II, the Klebsielleae.

In *Bergey's Manual*, the genus *Erwinia* is divided into three groups (Table 1) but this arrangement, largely one of current convenience, does not always reflect close similarity of character (cf. *Psmn. syringae* and *Xnmn. campestris* groups). Because of this heterogeneity, reassessment of the classification of the genus *Erwinia* alongside other species in the Enterobacteriaceae might allow a more rational approach where interdisciplinary communication is the main objective (Starr & Mandel 1969; White & Starr 1971).

Little is known about shared phage sensitivities between members of the tribe Erwinieae and species of the other four tribes in the family Enterobacteriaceae because comprehensive cross-reaction tests have not been made. A limited range of negative reactions is shown in Table 4 but Okabe & Goto (1963) cite a case where an *Erwn. carotovora* phage lysed strains in Tribes I and II.

TABLE 4

Sensitivity of epiphytic Erwinia herbicola *and related bacteria from apple and other pome fruits to* Erwinia amylovora *phages*

Epiphyte		No. of strains	% strains sensitive to phages*			
			L3H	4L	23	4L and/or 23
Erwinia herbicola	yellow	20	6	56	93	93
	white	38	2	52	55	63
Coli-aerogenes bacteria		21	0	0	0	0

(Billing & Hutson unpublished.)
*Phages at 10 × RTD. Reactions varied from turbid to clear confluent lysis. Some reactions were rapidly obscured by extracellular slime production on the test medium (2% glycerol yeast extract agar). Origin of phages: L3H from soil (Garrett); 4L from soil and 23 from manure (Billing).

Within the tribe Erwinieae, phages lysing the fireblight pathogen, *Erwn. amylovora*, have been studied in most detail (Billing 1960; Billing *et al.* 1961; Billing & Baker 1963; Baldwin & Goodman 1963; Hendry *et al.* 1967; Erskine 1973, Paulin 1976; Ritchie & Klos 1977). There are no reports of these phages lysing other species in the Amylovora group but members of the Herbicola group are commonly sensitive.

Some *Erwn. amylovora* phages are highly (but not absolutely) specific, e.g. phage L3H (Table 4) and P9 of Hendry *et al.* (1967) who also isolated a phage which lysed 80% of their yellow and 40% of their non-pigmented isolates from fireblight sources but not *Erwn. amylovora*. Since lysogeny has been observed in *Erwn. herbicola* but not in *Erwn. amylovora*, it has been suggested that the former provides a reservoir of phages for the latter (Erskine 1973).

The species *Erwn. herbicola* at present embraces at least two distinct epiphytes commonly found on a wide range of plants (Billing & Baker 1963). One is yellow pigmented and has sometimes been isolated from animal and plant sources (Chatterjee & Starr 1972); it grows at 37°C and may produce non-pigmented variant colonies (Chatterjee & Gibbins 1971). The other fails to grow at 37°C and is non-pigmented. Both are distinct taxonomically from *Erwn. amylovora* in a variety of characters in addition to pathogenicity (Billing & Baker 1963; Chatterjee & Starr 1972; Austin *et al.* 1978).

So far, taxonomists have not suggested that shared phage sensitivities provide good grounds to change current groupings but they are undoubtedly of ecological interest. Further information on some characteristics of *Erwn. amylovora* phages is given later (pp. 332)

Still within the Amylovora group, phages have been isolated for one walnut pathogen, *Erwn. nigrifluens*, but not for another, *Erwn. rubrifaciens* (Zeitoun & Wilson 1969). Phages for the former cleared lawns of the latter species but no plaques were observed. A limited number of other pathogens was examined, including *Erwn. amylovora*, but none was lysed.

In the Carotovora group, attempts to find phages of value for identification or for use in epidemiological studies appear to have met with little success and available phages have been little used. With *Erwn. atroseptica* (Elis-Jones & Catton pers. comm.) different isolates of the pathogen, even from the same plant, have varied in their sensitivity patterns to a range of phages isolated (cf. *Xnmn. oryzae*). Using temperate phages Paulin & Nassan (1978) had limited success in relating phage sensitivity to host of origin in *Erwn. chrysanthemi*.

In the tribe Erwinieae as a whole, serological tests have proved more specific and more convenient than phage tests for identification.

7. Coryneform Bacteria

A minority of plant pathogenic bacteria are Gram positive and most of these are classified in *Bergey's Manual* as *Corynebacterium* spp., where five subgroups are defined (Rogosa *et al.* 1974). It is recognized that these plant pathogens are a heterogeneous group and some or all will be allocated to different genera in future (Rogosa *et al.* 1974; Jones 1975; Starr *et al.* 1975; Keddie & Cure 1977; see also Bradley Chapter 1; Keddie & Bousfield Chapter 8, and Minnikin & Goodfellow Chapter 9 in this book).

Phages with an apparently high degree of specificity have been isolated for

Cnbc. nebraskense (a new species, Vidaver & Mandel 1974) and for *Cnbc. michiganense* (subgroup 4, Wakimoto *et al.* 1969; Echandi & Sun 1973) and used in diagnosis. With Echandi and Sun's phage, 4/23 isolates were only lysed by high titre preparations and the areas of lysis were turbid; 11 avirulent coryneform bacteria from tomato were not lysed. The ability of this phage to lyse other plant pathogenic coryneform bacteria was not widely tested.

A phage isolated for *Cnbc. flaccumfaciens* (subgroup 5) lysed strains of *Cnbc. poinsettiae* (Klement & Lovas 1959). This supports evidence of relationships demonstrated by other taxonomic approaches (Jones 1975; Starr *et al.* 1975). This may be a group where phages could be used more extensively because conventional identification tests have proved inadequate (P. Roberts pers. comm.).

8. Special Applications

A. Detection of low levels of infection

Highly specific phages have been used to detect plant pathogenic bacteria in material from which sampling by conventional methods would often give negative results. The presence of the pathogen is presumed when the titre of the specific phage increases some hours after its addition to the sample.

The method was first used by Katznelson & Sutton (1951) for *Pseudomonas phaseolicola* in bean seed but appears to have limitations. Taylor (pers. comm.) has observed false negative results, sometimes attributable to a deficiency in the phage/bacterium interaction or to the presence of a phage inhibitor in the seed macerates. Taylor prefers to use direct isolation followed by tube agglutination; phages (slightly more specific than antisera) are used in cases of doubt.

Attempts to isolate *Psmn. morsprunorum* from cherry buds were unsuccessful but the use of phage A7 demonstrated that the pathogen was present in about 25% of buds sampled (Shanmuganathan 1962). A phage of *Xanthomonas oryzae* has been used to detect low levels of the pathogen in rice (Wakimoto 1954).

Wherever a single phage is so used, it is essential that its specificity be properly established to guard against false positive results through phage multiplication on related bacteria. Inevitably false negative results will occur in some circumstances.

B. Epidemiology

Forecasting outbreaks of leaf blight of rice has been aided by the use of specific phages (Mizukami & Wakimoto 1969; Goto 1969). An increase of phage specific for *Xnmn. oryzae* in nursery, paddy field or irrigation water was observed prior to an outbreak of disease. Thus, it was possible to forecast disease in the paddy field by monitoring phage populations in nursery water. Forecasts were some-times upset by abnormal climatic conditions but build up of phage resistant strains did not appear to be a problem. Such disease forecasting was valuable when using chemicals for disease control.

The ability to distinguish cherry and plum strains of *Psmn. morsprunorum* rapidly and precisely by means of phage was recognized to be of great value in the field pathology of this pathogen. The unique sensitivity of cherry strains to A7, and the fact that both cherry and plum phage types were stable in passage through both hosts (Crosse & Garrett 1970), enabled cross- and mixed-infection experiments to be monitored carefully.

In leaf infection experiments, using mixed inocula of cherry and plum strains applied through stomata, phage typing of the re-isolates from lesions demon-strated that both strains were specific for the homologous host when inoculated at low, but not at high, concentrations. Autumn inoculation of cherry leaf scars with cherry strains result in severe disease but plum strains produced little infection: surprisingly, more occurred with lower than with higher inoculum doses. The anomalous result was explained by phage typing re-isolates; most low dose infections were caused by 'wild type' cherry strains; at high doses plum strains inhibited such infections. No strain specificity was observed in wound inoculation of branches (Crosse & Garrett 1970).

C. Virulence studies

Epidemiological experiments with English isolates of *Psmn. morsprunorum* suggested that there was a link between sensitivity to phage A7 (Table 3) and host specificity (Crosse & Garrett 1970). However, not all cherry strains from other countries were sensitive to A7. Furthermore plum strains, though not lysed by A7, adsorbed the phage. This suggested that A7 receptors in the cell walls were not directly involved in host specificity though cherry strain mutants resistant to A7 were attenuated in virulence. Failure of these mutants to adsorb A7 suggested a change in wall structure had occurred. So, while wall components may not act as determinants of host specificity they could well function as general virulence determinants (Garrett *et al.* 1974). Thus the use of specific phages can throw light on the complex problem of host-pathogen relationships in this pathotype.

With the fireblight pathogen, strains of *Erwn. amylovora* isolated from different hosts among the pome fruits of the family Rosaceae are not distin-guishable by available phages (cf. *Psmn. morsprunorum*). This may reflect a different level of host specificity for the two pathogens rather than a deficiency of the phages.

Erwinia amylovora phages distinguish capsulated from non-capsulated strains (Billing 1960) and, to some extent, virulent from avirulent ones (Table 5). The latter distinction is incomplete for two reasons: first, although strains which are invariably non-capsulated are invariably avirulent, capsulated strains are not invariably virulent: second, some virulent strains only form capsules on media containing a sugar or a sugar alcohol (Bennett 1978; Bennett & Billing 1978).

TABLE 5

Sensitivity of virulent and avirulent Erwinia amylovora *strains to three phages*

Strain	Sensitivity to phage*		
	L3H	4L	23
Virulent capsulated	+h	+h	(+)
Avirulent { capsulated	+h	+h	−
non-capsulated	+	−	+

* +, strong reaction; (+), weak reaction may be observed; h, halo round lysed area.

It will be seen from Table 5 that phage 4L only lysed capsulated strains whereas phage L3H has lysed nearly all strains of *Erwn. amylovora* so far tested. With both, however, the plaques produced on capsulated strains are surrounded by a halo. Evidence so far suggests that this halo results from the activity of a capsular hydrolysing enzyme in both cases (Bennett 1978). The fact that 4L only lyses capsulated strains suggests that the receptor site is either the capsule itself or a factor intimately associated with capsulation. With phage L3H, which lyses both capsulated and non-capsulated cells, it seems more likely that the receptor is associated with the cell wall and that the capsular hydrolysing enzyme plays a part in the initial stages of infection, allowing access to the receptor on capsulated cells (Lindberg 1973).

Thus, in the case of *Erwn. amylovora*, phages may be valuable tools in the study of cellular structure and virulence. The fact that these phages lyse some strains of *Erwn. herbicola* provides an added stimulus for their study. Only by awareness of such subtle interactions can the taxonomic significance or otherwise of shared phage sensitivities be assessed.

9. Discussion

The impact of phage sensitivity on the classification and identification of phyto-pathogenic bacteria has been small, in spite of intensive study of some groups in the 1960s. Why should this be so?

In classification of plant pathogens, recent trends have been towards grouping pathotypes which have much in common both genetically and phenotypically;

in the past, there was greater emphasis on differences and each pathotype was considered as a separate species. The striking contrast between the previous approach and the current one, at least as far as the pseudomonads are concerned, can be seen by comparing the seventh and eighth editions of *Bergey's Manual of Determinative Bacteriology* (Breed *et al*. 1957; Buchanan & Gibbons 1974).

That phages have contributed so little to current classification is not surprising. The major concern of most studies has been to find phages which aid distinctions or to identify specific pathotypes. This approach has favoured selection of phages which do just this. In most cases, inclusion of a comprehensive range of strains of other species which might demonstrate relationships by shared phage sensitivities would have imposed an unacceptable additional burden.

The modern taxonomist who deplores a multiplicity of species can however point to the examples of shared sensitivity amongst the fluorescent pseudomonads which support current classification. Equally, the plant pathologist can point to the fact that certain pathotypes are distinguishable both by their host specificity and by their phage sensitivity thus supporting the old divisions. The pathologist, of necessity, must continue to be primarily concerned with distinctions whilst most taxonomists will emphasize similarities. It is in nomenclature that a practical compromise must be reached if good communication between disciplines is to be sustained. But phages have no obvious role here.

In the genus *Erwinia*, shared phage sensitivities may serve to remind both taxonomists and pathologists that current groupings needs re-examination by all available means, but again it seems that phage tests will have little to offer.

In identification of plant pathogens, the lack of impact seems more surprising in view of the major role played by phages and bacteriocins in the study of human and animal disease epidemiology. Here there may be potential that could be further exploited; but the demand or the expertise or both are too often lacking.

In temperate climates, fungal and viral diseases of plants are economically far more important than those caused by bacteria and most pathology laboratories are geared to the first two. In the tropics, bacterial diseases are more prevalent and severe but fungal and viral diseases are still a major concern. In these areas, specialist bacteriologists are rare and so therefore is phage expertise. Biochemical and serological tests often provide a reasonable basis for diagnosis and there is too little incentive to develop phage typing systems.

Undoubtedly there are instances where phage tests could be exploited further for identification should the need arise. There also seems scope for further development of their use in epidemiological studies, in research on host-pathogen relationships and, perhaps, in disease control (Okabe & Goto 1963; Civerolo 1972; Vidaver 1976).

10. References

ADAMS, M. H. 1959 *Bacteriophages*. New York: Interscience.

AUSTIN, B., GOODFELLOW, M. & DICKINSON, C. H. 1978 Numerical taxonomy of phylloplane bacteria isolated from *Lolium perenne*. *Journal of General Microbiology* **104**, 139-155.

BALDWIN, C. H. & GOODMAN, R. N. 1963 Prevalence of *Erwinia amylovora* in apple buds as detected by phage typing. *Phytopathology* **53**, 1299-1303.

BENNETT, R. A. 1978 *Characteristics of the fireblight pathogen in relation to virulence*. Ph.D. Thesis, University of London, UK.

BENNETT, R. A. & BILLING, E. 1978 Capsulation and virulence in *Erwinia amylovora*. *Annals of Applied Biology* **89**, 41-45.

BILLING, E. 1960 An association between capsulation and phage sensitivity in *Erwinia amylovora*. *Nature, London* **186**, 819-820.

BILLING, E. 1963 The value of phage sensitivity tests for the identification of phytopathogenic *Pseudomonas* spp. *Journal of Applied Bacteriology* **26**, 93-210.

BILLING, E. 1969 Isolation, growth and preservation of bacteriophages. In *Methods in Microbiology*, 3B. ed. Norris, J. R. & Ribbons, D. W. London: Academic Press.

BILLING, E. 1970a Futher studies on the phage sensitivity and the determination of phytopathogenic *Pseudomonas* spp. *Journal of Applied Bacteriology* **33**, 478-491.

BILLING, E. 1970b *Pseudomonas viridiflava* (Burkholder, 1930; Clara, 1934). *Journal of Applied Bacteriology* **33**, 492-500.

BILLING, E. 1976 The taxonomy of bacteria on the aerial parts of plants. In *Microbiology of Aerial Plant Surfaces*. ed. Dickinson, C. H. & Preece, T. London: Academic Press.

BILLING, E. & BAKER, L. A. E. 1963 Characteristics of *Erwinia*-like organisms found in plant material. *Journal of Applied Bacteriology* **26**, 58-65.

BILLING, E., BAKER, L. A. E., CROSSE, J. E. & GARRETT, C. M. E. 1961 Characteristics of English isolates of *Erwinia amylovora* (Burrill) Winslow *et al. Journal of Applied Bacteriology* **24**, 195-211.

BÖTTCHER, I. 1972 Die isolation von Bakteriophagen für *Agrobacterium tumefacies* (Smith & Townsend) Conn. *Archiv für Pflanzen Schutz* **8**, 429-438.

BOYD, R. J., ALLEN, O. N. & HILDEBRANDT, A. C. 1965 Longevity of agrobacteria bacteriophages in storage. *American Journal of Botany* **52**,633 (Abs).

BOYD, R. J., HILDEBRANDT, A. C. & ALLEN, O. N. 1970 Specificity patterns of *Agrobacterium tumefaciens* phages. *Archiv für Mikrobiologie* **73**, 324-330.

BREED, R. S., MURRAY, E. G. D. & SMITH, N. R. (eds.) 1957 *Bergey's Manual of Determinative Bacteriology*, 7th edn. Baltimore: Williams & Wilkins.

BUCHANAN, R. E. & GIBBONS, N. E. 1974 *Bergey's Manual of Determinative Bacteriology*, 8th edn. Baltimore. Williams & Wilkins.

CHATTERJEE, A. K. & GIBBINS, L. N. 1971 Induction of non-pigmented variants of *Erwinia herbicola* by incubation at supra-optimal temperatures. *Journal of Bacteriology* **105**, 107-112.

CHATTERJEE, A. K. & STARR, M. P. 1972 The genus *Erwinia* : enterobacteria pathogenic to plants and animals. *Annual Review of Microbiology* **26**, 389-426.

CIVEROLO, E. L. 1972 Interaction between bacteria and bacteriophages on plant surfaces and in plant tissues. In *Proceedings of the Third International Conference on Plant Pathogenic Bacteria*, Wageningen, 14-21 April. ed. Maas Geesteranus, H. P. pp. 25-37.

CROSSE, J. E. & GARRETT, C. M. E. 1961 Relationship between phage type and host plant in *Pseudomonas mors-prunorum* Wormald. *Nature, London* **192**, 379-380.

CROSSE, J. E. & GARRETT, C. M. E. 1963 Studies on the bacteriophagy of *Pseudomonas mors-prunorum, P. syringae* and related organisms. *Journal of Applied Bacteriology* **26**, 159-177.

CROSSE, J. E. & GARRETT, C. M. E. 1970 Pathogenicity of *Pseudomonas mors-prunorum* in relation to host specificity. *Journal of General Microbiology* **62**, 315-327.

DOUDOROFF, M. & PALLERONI, N. J. 1974 The genus *Pseudomonas*. In *Bergey's Manual of Determinative Bacteriology*. 8th edn. ed. Buchanan, R. E. & Gibbons, N. E. Baltimore: Williams & Wilkins.

DYE, D. W. 1974 The problem of nomenclature of the plant pathogenic pseudomonads. *Review of Plant Pathology* 53, 953–962.

DYE, D. W., BRADBURY, J. F., DICKEY, R. S., GOTO, M., HALE, C. N., HAYWARD, A. C., KELMAN, A., LELLIOTT, R. A., PATEL, P. N., SANDS, D. C., SCHROTH, M. N., WATSON, D. R. W. & YOUNG, J. M. 1975 Proposals for a reappraisal of the status of the names of plant-pathogenic *Pseudomonas* species. *International Journal of Systematic Bacteriology* 25, 252–257.

ECHANDI, E. & SUN, M. 1973 Isolation and characterization of a bacteriophage for the identification of *Corynebacterium michiganense*. *Phytopathology* 63, 1398–1401.

EISENSTARK, A. & BERNSTEIN, L. B. 1955 Specificity of bacteriophages of *Xanthomonas pruni*. *Phytopathology* 45, 596–598.

ERSKINE, J. M. 1973 Characteristics of *Erwinia amylovora* bacteriophage and its possible role in the epidemiology of fireblight. *Canadian Journal of Microbiology* 19, 837–845.

FREIGOUN, S. O. & CROSSE, J. E. 1975 Host relations and distribution of a physiological and pathological variant of *Pseudomonas morsprunorum*. *Annals of Applied Biology* 81, 317–330.

GARRETT, C. M. E. 1978 *The epidemiology and bacteriology of* Agrobacterium tumefaciens. Ph.D. Thesis, University of London, UK.

GARRETT, C. M. E. & CROSSE, J. E. 1963 Observations on lysogeny in the plant pathogens *Pseudomonas mors-prunorum* and *P. syringae*. *Journal of Applied Bacteriology* 26, 27–34.

GARRETT, C. M. E., PANAGOPOULOS, C. G. & CROSSE, J. E. 1966 Comparison of plant pathogenic pseudomonads from fruit trees. *Journal of Applied Bacteriology* 29, 342–356.

GARRETT, C. M. E., CROSSE, J. E. & SLETTEN, A. 1974 Relations between phage sensitivity and virulence in *Pseudomonas morsprunorum*. *Journal of General Microbiology* 80, 475–483.

GOTO, M. 1965 Phage-typing of the causal bacteria of bacterial leaf blight (*Xanthomonas oryzae*) and bacterial leaf streak (*X. translucens* f. sp. *oryzae*) of rice in the tropics. *Annals of the Phytopathological Society of Japan* 30, 253–257.

GOTO, M. 1969 Ecology of phage-bacteria interaction in *Xanthomonas oryzae* (Uyeda et Ishiyama) Dowson. *Bulletin of the Faculty of Agriculture, Shizuoka University* 19, 31–67.

GOTO, M. & STARR, M. P. 1972 Phage-host relationships of *Xanthomonas citri* compared with those of other xanthomonads. *Annals of the Phytopathological Society of Japan* 38, 226–248.

HAYWARD, A. C. 1964a Characteristics of *Pseudomonas solanacearum*. *Journal of Applied Bacteriology* 27, 265–277.

HAYWARD, A. C. 1964b Bacteriophage sensitivity and biochemical group in *Xanthomonas malvacearum*. *Journal of General Microbiology* 35, 287–298.

HEBERLEIN, G. T., DE LEY, J. & TIJTGAT, R. 1967 Deoxyribonucleic acid homology and taxonomy of *Agrobacterium*, *Rhizobium* and *Chromobacterium*. *Journal of Bacteriology* 94, 116–124.

HENDRY, A. T., CARPENTER, J. A. & GARRARD, E. H. 1967 Bacteriophage studies of isolates from fireblight sources. *Canadian Journal of Microbiology* 13, 1357–1364.

JONES, D. 1975 A numerical taxonomic study of coryneform and related bacteria. *Journal of General Microbiology* 87, 52–96.

JONES, D. & SNEATH, P. H. A. 1970 Genetic transfer and bacterial taxonomy. *Bacteriological Reviews* 34, 40–81.

KATZNELSON, H. & SUTTON, M. D. 1951 A rapid phage plaque count method for detection of bacteria as applied to the demonstration of internally borne bacterial infections of seed. *Journal of Bacteriology* 61, 689–701.

KEDDIE, R. M. & CURE, G. L. 1977 The cell wall composition and distribution of free mycolic acids in named strains of coryneform bacteria and in isolates from various natural sources. *Journal of Applied Bacteriology* **42**, 229-252.

KLEMENT, Z. 1959 Some new specific bacteriophages for plant pathogenic *Xanthomonas* spp. *Nature, London* **184**, 1248-1249.

KLEMENT, Z. & LOVAS, B. 1959 Isolation and characterization of a bacteriophage for *Corynebacterium flaccumfaciens*. *Phytopathology* **49**, 107-112.

KLEMENT, Z. & LOVAS, B. 1960 Biological and morphological characterization of the phage for *Xanthomonas phaseoli* var. *fuscans*. *Phytopathologische Zeitschrift* **37**, 321-329.

LAŽAR, I. & CROSSE, J. E. 1969 Lysogeny, bacteriocinogeny and phage types in plum isolates of *Pseudomonas morsprunorum* Wormald. *Revue Roumaine de Biologie, Série de Botanique* **14**, 325-333.

LELLIOTT, R. A., BILLING, E. & HAYWARD, A. C. 1966 A determinative scheme for the fluorescent plant pathogenic pseudomonads. *Journal of Applied Bacteriology* **29**, 470-489.

LINDBERG, A. A. 1973 Bacteriophage receptors. *Annual Review of Microbiology* **27**, 205-241.

LOVREKOVICH, L., KLEMENT, Z. & DOWSON, W. J. 1963 Serological investigation of *Pseudomonas syringae* and *Pseudomonas morsprunorum* strains. *Phytopathologische Zeitschrift* **47**, 19-24.

MEYNELL, G. G. & MEYNELL, E. 1965 *Theory and Practice in Experimental Bacteriology*. Cambridge: University Press.

MIZUKAMI, T. & WAKIMOTO, S. 1969 Epidemiology and control of bacterial leaf blight of rice. *Annual Review of Phytopathology* **7**, 51-72.

MURATA, N. & STARR, M. P. 1973 A concept of the genus *Xanthomonas* and its species in the light of segmental homology of deoxyribonucleic acids. *Phytopathologische Zeitschrift* **77**, 285-323.

NISHIMURA, J. & WAKIMOTO, S. 1971 Ecological studies on bacterial wilt disease of carnation. I. Some characteristics and mode of multiplication of *Pseudomonas caryophylli* phage. *Annals of the Phytopathological Society of Japan* **37**, 301-306.

OKABE, N. 1961 Studies on the lysogenic strains of *Xanthomonas citri*. *Special Publication of the College of Agriculture National Taiwan Univerisity* **10**, 61-73.

OKABE, N. & GOTO, M. 1963 Bacteriophages of plant pathogens. *Annual Review of Phytopathology* **1**, 397-418.

PALLERONI, N. J. & DOUDOROFF, M. 1972 Some properties and taxonomic subdivisions of the genus *Pseudomonas*. *Annual Review of Phytopathology* **10**, 73-100.

PALLERONI, N. J., KUNISAWA, R., CONTOPOULOU, R. & DOUDOROFF, M. 1973 Nucleic acid homologies in the genus *Pseudomonas*. *International Journal of Systematic Bacteriology* **23**, 333-339.

PARKER, M. T. 1972 Phage-typing of *Staphylococcus aureus*. In *Methods in Microbiology*, **7B**. ed. Norris, J. R. & Ribbons, D. W. London: Academic Press.

PAULIN, J. P. 1976 Phages from *Erwinia amylovora* and *Erwinia herbicola*. *Station de Pathologie Végétale et Phytobactériologie INRA Rapport Bisannuel 1975-1976* pp. 1-2.

PAULIN, J. P. & NASSAN, A. 1976 Phage typing of *Erwinia chrysanthemi*. *Station de Pathologie Végétale et Phytobactériologie INRA Rapport Bisannuel 1975-1976* pp. 4.

PAULIN, J.P. & NASSAN, N.A. 1978 Lysogenic strains and phage-typing in *Erwinia chrysanthemi*. In *Proceedings of the IVth International Conference on Plant Pathogenic Bacteria* 539-545.

PERSLEY, G. J. & CROSSE, J. E. 1978 A bacteriophage specific to race 2 of the cherry strain of *Pseudomonas morsprunorum*. *Annals of Applied Biology* **89**, 219-222.

RITCHIE, D. F. & KLOS, E. J. 1977 Isolation of *Erwinia amylovora* bacteriophage from aerial parts of apple trees. *Phytopathology* **67**, 101-104.

ROGOSA, M., CUMMINS, C. S., LELLIOTT, R. A. & KEDDIE, R. M. 1974 The Coryne-form Group of Bacteria. In *Bergey's Manual of Determinative Bacteriology*. 8th edn. ed. Buchanan, R. E. & Gibbons, N. E. pp. 599–602. Baltimore: Williams & Wilkins.

ROSLYCKY, E. B., ALLEN, O. N. & McCOY, E. 1962 Phages of *Agrobacterium radio-bacter* with reference to host range. *Canadian Journal of Microbiology* 8, 71–78.

SATO, M., TAKAHASHI, K. & WAKIMOTO, S. 1971 Properties of the causal bacterium of bacterial blight of Mulberry, *Pseudomonas mori* (Boyer et Lambert) Stevens, and its phages. *Annals of the Phytopathological Society of Japan* 37, 128–135.

SHANMUGANATHAN, N. 1962 *Studies in bacterial canker of plum and cherry (Pseudomonas mors-prunorum Wormald)*. Ph.D. Thesis, University of London, UK.

STARR, M. P. & MANDEL, M. 1969 DNA base composition and taxonomy of phytopatho-genic and other enterobacteria. *Journal of General Microbiology* 56, 113–123.

STARR, M. P., MANDEL, M. & MURATA, N. 1975 The phytopathogenic coryneform bacteria in the light of DNA base composition and DNA-DNA segmental homology. *Journal of General and Applied Microbiology* 21, 13–26.

STOLP, H. 1961 Neue Erkenntnisse über phytopathogene Bakterien und die von ihnen verursachten Krankheiten. *Phytopathologische Zeitschrift* 42, 197–262.

STOLP, H. & STARR, M. P. 1964 Bacteriophage reactions and speciation of phytopatho-genic xanthomonads. *Phytopathologische Zeitschrift* 51, 442–478.

SUTTON, M. D. & WALLEN, V. R. 1967 Phage types of *Xanthomonas phaseoli* isolated from beans. *Canadian Journal of Botany* 45, 267–280.

TAYLOR, J. D. 1970 Bacteriophage and serological methods for the identification of *Pseudomonas phaseolicola* (Burkh.) Dowson. *Annals of Applied Biology* 66, 387–395.

TAYLOR, J. D. 1972 Specificity of bacteriophages and antiserum for *Pseudomonas pisi*. *New Zealand Journal of Agricultural Research* 15, 421–431.

VIDAVER, A. K. 1976 Prospects for control of phytopathogenic bacteria by bacteriophages and bacteriocins. *Annual Review of Phytopathology* 14, 451–465.

VIDAVER, A. K. & MANDEL, M. 1974 *Corynebacterium nebraskense*, a new orange-pigmented phytopathogenic species. *International Journal of Systematic Bacteriology* 24, 482–485.

VIDAVER, A. K. & SCHUSTER, M. L. 1960 Characterization of *Xanthomonas phaseoli* bacteriophages. *Journal of Virology* 4, 300–308.

WAKIMOTO, S. 1954 The determination of the presence of *Xanthomonas oryzae* by the phage technique. *Scientific Bulletin of the Faculty of Agriculture, Kyushu University* 14, 495–498.

WAKIMOTO, S. 1960 Classification of strains of *Xanthomonas oryzae* on the basis of their susceptibility against bacteriophages. *Annals of the Phytopathological Society of Japan* 25, 193–198.

WAKIMOTO, S., UEMATSU, T. & MIZUKAMI, T. 1969 Bacterial canker disease of tomato in Japan (2)—Properties of bacteriophages specific for *Corynebacterium michiganense* (Smith) Jensen. *Annals of the Phytopathological Society of Japan* 35, 168–173.

WHITE, J. N. & STARR, M. P. 1971 Glucose fermentation end products of *Erwinia* spp. and other enterobacteria. *Journal of Applied Bacteriology* 34, 459–475.

YOUNG, J. M., DYE, D. W., BRADBURY, J. F., PANAGOPOULOS, C. G. & ROBBS, C. F. 1978 A proposed nomenclature and classification for plant pathogenic bacteria. *New Zealand Journal of Agricultural Research* 21, 153–177.

ZEITOUN, F. M. & WILSON, E. E. 1969 The relation of bacteriophage to the walnut tree pathogens, *Erwinia nigrifluens* and *Erwinia rubrifaciens*. *Phytopathology* 59, 756–761.

ZIMMERER, R. P., HAMILTON, R. H. & POOTJES, C. 1966 Isolation and morphology of temperate *Agrobacterium tumefaciens* bacteriophage. *Journal of Bacteriology* 92, 746–750.

Immunological Detection of Food-Poisoning Toxins

J. S. CROWTHER AND R. HOLBROOK

Unilever Research, Colworth House, Sharnbrook, Bedford, UK

Contents

1. Introduction

THE FOOD-POISONING toxins produced by *Staphylococcus aureus, Clostridium botulinum, Clostridium perfringens, Bacillus cereus* and *Escherichia coli* are known to be proteins. Other than their high biological activity they do not seem to possess any chemical characteristics which can be used to differentiate them from the many other proteins present in foods or pathological specimens such as serum or faeces. Fortunately, the toxins are antigenic and can be specifically detected using immunological methods. This paper outlines the various immunological methods which can be used to identify food-poisoning organisms and their toxins and describes the impact which these methods have made or will make on our

339

knowledge of food-poisoning diseases. Detailed experimental procedures for carrying out the various tests are not included because they are to be found in the references cited. The paper also points out the methods which are now available for detecting serotypes of *Clos. perfringens* and *Bacl. cereus*. These methods lead to identification at the strain level and therefore are important to taxonomists.

2. Evolution of Immunological Tests for Detecting Food-Poisoning Toxins

The toxic nature of the various food-poisoning toxins was confirmed by the early workers by means of feeding trials in animals such as cats or monkeys. In rare instances human volunteers have been used to determine the toxic dose for man. *In vivo* assay methods were developed in which the toxin was administered to animals either by feeding or by injection; passively immunized animals served as controls. For ethical, scientific and financial reasons, animal tests have been replaced in recent years, wherever possible, by *in vitro* tests.

The *in vitro* tests can be considered for convenience to fall into two groups: those which are only moderately sensitive, and those which are relatively sensitive. The moderately sensitive tests can detect only *ca*. 1 μg toxin/ml and compose the precipitin tests in agar gels, such as simple diffusion methods on slides or in tubes, or electroimmunodiffusion methods. The more sensitive methods, the haemagglutination methods, radioimmunoassay and enzyme linked immunoabsorbent assay (ELISA) can detect approx. 1 ng/ml. All these *in vitro* tests require highly specific antisera, which in turn require highly purified toxins. Many specific antisera and pure toxins are not yet generally available and this restricts the use of these tests.

3. *Staphylococcus aureus* Food-Poisoning

A. The organism and the enterotoxins

Staphylococcus aureus forms part of the normal bacterial flora of the skin, nose and throat of man and his domestic animals. It is therefore of considerable interest and concern to the food microbiologist.

Staphylococcal enterotoxins are produced during growth of the organism and are the causative agents of staphylococcal food-poisoning. They are proteins and at least five serologically distinct types are known, designated A to E (Bergdoll 1972). Types A and D occur most frequently in food-poisoning incidents (Casman *et al.* 1967; Gilbert & Weineke 1973). Unlike the enterotoxins of *Bacl. cereus*, *Esch. coli* and *Clos. perfringens*, staphylococcal enterotoxins do not act directly on the intestinal mucosa, but probably trigger vomiting centres in the brain via the central nervous system. The minimum dose required to cause food-poisoning in man is not known precisely but it has been estimated to range from *ca*. 1 to 20 µg (Bergdoll 1972; Gilbert *et al*. 1972).

B. Immunological methods

The moderately sensitive methods for detecting enterotoxins include the micro double-diffusion slide method, the double-diffusion tube method, and electroimmunodiffusion methods. They have been described and compared in detail by Holbrook & Baird-Parker (1975). To achieve the maximum sensitivity with these methods, however, it is necessary to extract and concentrate the toxin from the food. Different extraction procedures have been published (Casman 1967; Reiser *et al*. 1974; Genigeorgis & Kuo 1976; Niskanen & Lindroth 1976) but they are time-consuming and cannot be performed easily in a routine laboratory. The more sensitive methods include haemagglutination, radioimmunoassay and enzyme linked immunoabsorbent assay (ELISA) (see Table 1). They can be

TABLE 1

Comparison of methods for detection of staphylococcal enterotoxin

Method	Sensitivity (µg/ml)	Type	Reference
Counter-current immunoelectrophoresis	1.0	A	Kimble & Anderson (1973)
Electroimmunodiffusion	0.5	A & B	Holbrook & Baird-Parker (1975)
	0.15	A	Gasper *et al.* (1973)
Crowle micro double diffusion	0.1–0.2	A to E	Casman *et al.* (1969)
Enzyme linked immunoabsorbent assay	0.001–0.003	A	Saunders & Bartlett (1977)
Reverse passive haemagglutination	0.001	A	Holbrook & Baird-Parker (1975)
		B	Shibata *et al.* (1977)
			Silverman *et al.* (1968)
Radioimmunoassay	0.001	AB & C	Orth (1977)

used to detect enterotoxin in a food using simple extraction procedures. There are technical problems, however, with both the reverse passive haemagglutination and radioimmunoassay methods (see Crowther & Holbrook 1976). The newly devised ELISA method appears to have several advantages but it has not yet been developed for all five toxin types.

C. The impact of immunological methods

Serological methods can be used to increase our knowledge of staphylococcal food-poisoning in four ways. Firstly, to detect the presence of enterotoxin in a suspect food; secondly, to demonstrate that an isolate is enterotoxigenic; thirdly, to determine in the research laboratory the conditions necessary to inhibit growth and enterotoxin production and fourthly, to study the factors affecting the destruction of enterotoxin. These will be discussed briefly in turn.

The foods most commonly implicated in staphylococcal food-poisoning incidents in the UK have been pre-cooked cold-eating meats, meat products and dairy products. Sea-foods and canned vegetables have also been involved. Enterotoxins A and D alone, or in combination, occur most frequently in food-poisoning outbreaks, accounting for *ca.* 80% of the total both in the UK and the USA. The reasons for this are probably the common occurrence of enterotoxin A and D producing strains and also the fact that enterotoxin A and possibly D are produced over a wider range of growth conditions than the other types.

In vitro serological techniques are helping to build a better picture of the proportion of *Staf. aureus* strains which are able to form enterotoxin. Casman *et al.* (1967) reported the occurrence of enterotoxin A, B, C and D producers amongst 1200 isolates from various sources in the USA (Table 2). Enterotoxins A and D

TABLE 2

Proportion of isolates of Staphylococcus aureus *found to produce enterotoxin*

Origin	No. of strains examined	% Positive	Enterotoxin types studied	Reference*
Clinical specimens	438	44	ABCD & E	
Nasal specimens	144	31	ABCD & E	
Raw milk	236	10	ABCD & E	(1)
Cows with mastitis	51	2	ABCD & E	
Chickens	42 (biotype B)	66	AB & C	(2)
Raw meat	122	20	AB & C	(3)
Frozen foods	260	30	ABCD & E	(1)
Food-poisoning episodes	80	96	ABCD & E	

*(1) Casman *et al.* (1967); (2) Gibbs *et al.* (1978) and (3) Untermann & Sinell (1970).

alone or in combination occurred most frequently from all sources, which is in agreement with the involvement of these two serological types in food poisoning incidents. Untermann & Sinell (1970) who detected enterotoxin A, B and C producers among *Staf. aureus* isolates in Germany, found a similar incident rate in the sources common to Casman's study. There is no clear cut relationship between phage type and ability to produce enterotoxin, although most strains associated with food poisoning are typed by phage group III or phage groups I and III. Casman's and Untermann's data indicate a higher percentage of enterotoxin producing strains of *Staf. aureus* amongst human compared to animal isolates and work by Hájek & Maršélák (1973) on the incidence of enterotoxinogenicity among different *Staf. aureus* biotypes substantiated this. Gibbs *et al.* (1978) found, however, as many as 27 out of 42 isolates (66%) of *Staf. aureus* (biotype B) from chickens were enterotoxin producers. This incidence rate is far higher than has been reported hitherto for domestic animals (Table 2).

TABLE 3

Limits of growth and enterotoxin production by Staphylococcus aureus†

Factor	Growth			Toxin production	
	Optimum	Range		Optimum	Range
Temperature °C	35–37	6.5–46		37–40	10–45
pH	7–7.5	4.2–9.5	Ent. A	5.3–6.8	4.0–9.8*
			Others	6–7	
NaCl (%)	0.5–4.0	0–20		0.5	0–20
Water activity	0.98–> 0.99	0.83–> 0.99		>0.99	0.86–> 0.99
Atmosphere	Aerobic	Aerobic–anaerobic	dissolved	5–20% O_2	Aerobic–anaerobic
Eh	> + 200 mv	< − 200 to > + 200 mv		> + 200 mv	< − 100 to > + 200

Data compiled from Carpenter & Silverman (1973); Genigeorgis *et al.* (1971); Reiser & Weiss (1969); Scott (1953); Scheusner & Harmon (1973) and Tatini (1973) Holbrook, unpublished.
*Using an inoculum of 10^8 cells/ml.

The range and optimal conditions of temperature, pH, NaCl concentration, water activity and Eh permitting *Staf. aureus* growth and enterotoxin production are summarized in Table 3. These data have been obtained with pure cultures of *Staf. aureus* in laboratory media rather than in naturally contaminated foods. Further details are to be found in the references cited.

It is now well established that the microbial flora of fresh foods such as meat will outgrow or inhibit the growth of *Staf. aureus*. Casman *et al.* (1963) found that only on aseptically prepared raw pork and beef slices would *Staf. aureus* grow to levels sufficient to produce detectable enterotoxin A.

In experiments using cooked minced chicken, the authors detected enterotoxin A and C in samples inoculated with 30 *Staf. aureus*/g and incubated for 1 day at 37°C, 2 at 20°C or 4 at 15°C. Enterotoxin A was detected in samples after incubation for three weeks at 10°C. Counts reached 10^8–10^9 organisms per g before enterotoxin was detected. *Staphylococcus aureus* will grow and produce toxin in cured meats only when these have been held for excessive periods of time at high ambient temperatures. In studies in this laboratory enterotoxin B was not detected in UK-produced vacuum-packed bacon inoculated with 10^2 *Staf. aureus*/g during storage at 25°C for 14 days (Baird-Parker 1971). In aerobically-stored bacon inoculated with 10^4 cells/g enterotoxin B was present after 3 days at 37°C but not after 7 days at 28°, 30° or 32°C. In fermented sausages the control of growth and enterotoxin formation by *Staf. aureus* is dependent on the rate of chemical or microbial acidulation and the oxygen tension (Barber & Deibel 1972; Daly *et al.* 1973). In laboratory cured hams inoculated with 10^3–10^6 cells/g, enterotoxin B was detected after storage at 22°C or 30°C for 3–13 days when the pH was above 5.3; toxin was detected in some hams after 2–8 weeks storage at 10°C but not if the pH was below 5.3 (Genigeorgis *et al.* 1966).

Staphylococcal enterotoxins are markedly resistant to heat and to radiation. The D values of enterotoxin B are shown in Table 4. Extrapolation of these data to thermal inactivation of enterotoxin in foods must be done with caution, as Lee *et al.* (1977) reported the presence of a low molecular weight peptide in beef broth that increased the $D_{110°}$ for crude enterotoxin B to 90 min compared to 18 min in veronal buffer at pH 7.4. The effect of pH on the D values of enterotoxins is complex but in general the D values decrease as the pH falls below neutrality (Jamlang *et al.* 1971).

4. *Clostridium botulinum* Food-Poisoning

A. The organism and its toxins

The natural habitat of *Clos. botulinum* appears to be the soil and some marine and inland waters. Recent surveys have shown that the organism is much more widespread in the UK and in Europe than had been hitherto supposed. For example, 50 (73%) out of 69 samples of mud from lakes and waterways in London were found to contain at least one toxigenic type of the organism (Smith & Moryson 1975), and farmed trout may be contaminated with *Clos. botulinum* type E (Cann *et al.* 1975). British bacon may also be frequently contaminated with *Clos. botulinum* (Roberts & Smart 1976). The results of such surveys stress the importance of maintaining *Clos. botulinum* under constant vigilance so that food-borne botulism may be prevented.

Clostridium botulinum is known to produce 7 potent neurotoxins which are serologically distinct and designated A to G. Types A, B, E and F cause food-poisoning in man, type C in birds and type D in cattle. The toxins are simple proteins. They are all antigenic and can be detected immunologically.

TABLE 4

Heat resistance of enterotoxin B in different substrates

Heating menstruum	D values* (min) at temperatures indicated			
	$D_{104.4°C}$	$D_{110°C}$	$D_{115.9°C}$	$D_{121°C}$
Pure enterotoxin B in ceronal buffer (pH 7.2)	34.4	23.5	16.6	9.9
Pure enterotoxin B in milk	46.2	26.1	16.6	9.4
Crude enterotoxin B in veronal buffer (pH 7.2)	40.5	29.7	18.8	11.4

*D value is the time required for 90% inactivation of toxin. Data from Read & Bradshaw (1966*a*, *b*).

B. Immunological methods

There are two distinct approaches to the laboratory detection of *Clos. botulinum*, namely to look for (1) soluble toxins using *in vivo* or *in vitro* tests, or (2) intact cells or spores using cultural or microscopic methods.

The mouse test is the accepted method for detecting botulinum toxins. A sample of food is extracted with buffer and injected intraperitoneally into mice. Although this method is very sensitive it has several disadvantages, such as the non-specific death of mice. Several immunological procedures have been described in the literature to overcome these difficulties, but none is yet as sensitive as the mouse test (Table 5). The methods which show the most promise are the

TABLE 5

Comparison of methods to detect botulinum toxins

Method	Sensitivity (mouse MLD/ml)	Toxin type	References
Double diffusion	370	B	Vermilyea *et al.* (1968)
Electroimmunodiffusion	140	A	Miller & Anderson (1971)
Counter-current immunoelectrophoresis	100	A	Crowther & Holbrook (1976)
Reverse passive haemagglutination	27	A	Evancho *et al.* (1973)
Enzyme linked immunoabsorbent assay	50–100	A	Notermans *et al.* (1978)
Radioimmunoassay	5–10	A	Boroff & Shu-Chen (1973)

radioimmunoassay method of Boroff & Shu-Chen (1973) and the enzyme-linked immunoabsorbent assay of Notermans *et al.* (1978).

Fluorescent antibody methods have been devised by several workers for detecting whole cells of *Clos. botulinum*. The limitations of these procedures are discussed below.

C. The potential impact of novel immunological tests

The mouse test is still the most sensitive way of detecting botulinum toxins and must be used for investigating suspected outbreaks of botulism. It has been used in all the studies reported in the literature that investigated the factors affecting toxin production by *Clos. botulinum*. However, some of the new *in vitro* tests have potential in research work as presumptive screening tests for toxins in laboratory media and model systems inoculated with pure strains of *Clos. botulinum*. In our laboratory counter current immunoelectrophoresis is used as a presumptive test. Although this saves a considerable number of mice by no means does it replace the mouse test. Hopefully, an *in vitro* method will be developed which is at least as sensitive as the mouse test.

Studies with fluorescent antibodies have shown that strains of *Clos. botulinum* can be classified into three groups based on the nature of the somatic and flagella antigens, irrespective of the serological specificity of the toxins. The first group contains proteolytic organisms of types A, B and F, the second contains types E and non-proteolytic strains of B and F, whilst the third group contains strains of types C and D (Lynt *et al.* 1971; Baillie *et al.* 1973). Some strains stain only with difficulty, however, particularly when the cultures are old. Furthermore, some strains of *Clos. sporogenes* stain in the same way as proteolytic strains of *Clos. botulinum*. Thus the fluorescent antibody method cannot be used to identify toxigenic types but it can be used as a rapid presumptive test for *Clos. botulinum*.

5. *Clostridium perfringens* Food-Poisoning

A. The organism and the enterotoxin

Major advances have been made in recent years towards a better understanding of *Clos. perfringens* food-poisoning. It is now firmly established that the disease may be caused by strains which produce heat sensitive as well as by those which produce heat resistant spores. Vegetative cells of the organism must be present on the food in counts of *ca.* 10^7/g food in order to cause food-poisoning symptoms. Once the cells have been ingested, they multiply and sporulate in the small intestine. The sporangia lyse and release free spores and enterotoxin (Labbe *et al.* 1976). The enterotoxin is quite distinct from the major and minor lethal toxins. It is a protein of molecular weight, *ca*. 36 000, and is a structural component of the spore coat. It is stable on freezing at $-21°C$ but is denatured at $60°C$ (Duncan & Strong 1969). The enterotoxin appears to act directly on the intestinal mucosa and causes fluid and electrolyte secretion in the intestine (McDonel & Asano 1975). The mode of action may involve adenylate cyclase similar to that of *Vibrio cholerae* enterotoxin (Hauschild 1973).

B. Immunological methods

Clostridium perfringens enterotoxin is highly antigenic. Specific antitoxins have been prepared and used in immunological tests such as electroimmunodiffusion (Duncan & Somers 1972), reverse passive haemagglutination (Genigeorgis *et al.* 1973), and counter immunoelectrophoresis (Skjelkvale & Uemura 1977b). These tests are more sensitive and less cumbersome that the ligated intestinal loop test or the intradermal skin test in animals (Table 6).

TABLE 6

Comparison of methods for detection of Clostridium perfringens *enterotoxin*

Method	Sensitivity (μg/ml)	Reference
Rabbit ileal loop	9.6	Genigeorgis *et al.* (1973)
Mouse lethality	6.8	
Erythemal activity in guinea pigs	3.8	Duncan & Somers (1972)
Single gel diffusion	3.0	
Electroimmunodiffusion	0.9	Genigeorgis *et al.* (1973)
	1.0	
Counterimmunoelectrophoresis	1.0	Skjelkvale & Uemura (1977a)
	0.2	Naik & Duncan (1977)
Microslide diffusion	0.5	Genigeorgis *et al.* (1973)
Reverse passive haemagglutination	0.001	

Strains of *Clos. perfringens* can be divided into serotypes by means of a scheme first devised by Hobbs *et al.* (1953) and extended by Hughes *et al.* (1976). It is based on differences in the polysaccharide antigens of the capsular layers surrounding vegetative cells.

C. Impact of immunological methods

The serotyping scheme devised at the Food Hygiene Laboratory, Colindale, has great potential in the epidemiology of *Clos. perfringens* food-poisoning. With 57 different antisera prepared against strains isolated predominantly from food-poisoning outbreaks, Hughes *et al.* (1976) were able to serotype 65% of the isolates of *Clos. perfringens* from food-poisoning outbreaks in the UK. Some isolates from human infections other than food-poisoning could also be typed. These authors are attempting to establish an international serotyping scheme for *Clos. perfringens* by integrating the Colindale sera with those from the USA and Japan.

Another way in which serological methods will help epidemiological investigations of *Clos. perfringens* food-poisoning is the introduction of a method for detecting enterotoxin in the faeces of victims. Already, counter immunoelectrophoresis and reverse passive haemagglutination have been used successfully to detect enterotoxin in faeces of some food-poisoning victims and in those of volunteers who had eaten purified enterotoxin (Skjelkvale & Uemura 1977*a, b*). These techniques have not yet been applied widely, however, because of the difficulties of obtaining specific antisera.

6. *Bacillus cereus* Food-Poisoning

A. The organism and its enterotoxins

Bacillus cereus is widely distributed in soil, in foods and in low numbers in the human and animal intestine. It is the cause of two distinct forms of food-poisoning, the 'diarrhoea' type and the 'vomiting' type. The main symptoms of the diarrhoea type of illness are diarrhoea and abdominal pain occurring between 8 and 16 h after ingestion of contaminated food. Many different kinds of foods may be involved including soups, meats, vegetables and sauces (Gilbert & Taylor 1976; Goepfert *et al.* 1972). The vomiting type of *Bacl. cereus* food-poisoning has been recognized recently in the UK and elsewhere. The main symptoms are nausea and vomiting. This type of the disease is usually associated with cooked rice (Gilbert & Taylor 1976).

Several different enterotoxins appear to be involved in *Bacl. cereus* food-poisoning. One is responsible for the vomiting symptoms and is produced by strains isolated from vomiting-type outbreaks of the disease. Another enterotoxin causes diarrhoea but not vomiting and is produced by strains isolated from diarrhoeal outbreaks (Melling *et al.* 1976). A third type of enterotoxin, termed a necrotic toxin, is produced by some pathogenic strains of *Bacl. cereus* (Turnbull 1976; Turnbull *et al.* 1977).

B. Immunological methods

A scheme for serotyping food-poisoning strains of *Bacl. cereus* by means of their flagella (H) antigens has been devised by Taylor & Gilbert (1975). It is based on the observations by Norris & Wolf (1961) that the flagella antigens show marked strain to strain specificity.

When Gilbert & Parry (1977) used the scheme to type new food-poisoning isolates of *Bacl. cereus*, they found that 5 types (types 2, 6, 9, 10 and 12) were most commonly associated with the diarrhoeal type of disease. Another five serotypes (types 1, 3, 4, 5 and 8) were commonly associated with the vomiting type of the disease, with type 1 the most common. Types 1 and 8 were also found, however, in the diarrhoeal type of disease and this shows that serotyping has a useful but limited application for differentiating isolates of the two kinds of *Bacl. cereus* food-poisoning.

Serological methods for detecting *Bacl. cereus* enterotoxins in foods have been little studied, mainly because of the difficulties in isolating pure enterotoxins and obtaining antisera. However, Gorina *et al.* (1975) have purified an enterotoxin which causes vomiting in cats and have compared the sensitivities of various serological methods for its detection. An aggregate-haemagglutination method was the most sensitive and detected 0.004 μg/ml, whilst an immunodiffusion method in gels detected 2-3 μg/ml. In contrast, toxicity tests in animals were remarkably insensitive; the minimum dose causing vomiting in cats was 75 μg/ kg and the minimum lethal dose for mice was as high as 0.3 mg per mouse. These workers have further shown that the aggregate haemagglutination method can be used to detect *Bacl. cereus* enterotoxin in foods. Their results await confirmation by others.

C. Impact of immunological methods

Our knowledge of *Bacl. cereus* food-poisoning is limited but is expanding rapidly. Already the scheme proposed by Taylor & Gilbert (1975) for serotyping isolates by means of flagella antigens has helped in understanding the epidemiology of this disease. For example, these authors have shown that certain serotypes tend to be associated with the two distinct types of food-poisoning (Gilbert & Taylor 1976; Gilbert & Parry 1977). Furthermore, whilst 84% of strains of *Bacl. cereus* from food-poisoning outbreaks were typeable (by means of 18 antisera), only 39% of isolates from routine foods were typeable. These observations suggest that not all strains of *Bacl. cereus* are capable of causing food-poisoning in man. Perhaps the food-poisoning strains are more heat resistant or grow better or produce more toxin(s) than do non food-poisoning strains. The factors which affect the production of toxins in foods are not known and will be tedious to investigate by means of animal tests, which are the only tests available. Progress in this area will be improved, however, when the toxins have been purified and when serological methods have been devised for their detection in foods.

7. *Escherichia coli* Food-Poisoning

A. *The organism and its toxins*

Escherichia coli is part of the normal intestinal flora of man and animals and occurs widely in the environment also. It is an opportunist pathogen and can be isolated from many infections. Its role as an enteric pathogen of man and young domestic animals is now well established and has been reviewed by Craig (1972) and Sack (1975). It has been firmly recognized as a food-borne pathogen since 1973, when an outbreak of *Esch. coli* gastroenteritis was traced to cheese (Marier *et al.* 1973). Human diarrhoeal diseases caused by *Esch. coli* have been classified into three types by Rowe *et al.* (1977). Firstly, an enteropathogenic type which causes severe and often fatal diarrhoea in infants but rarely in adults; enterotoxin may not be involved. Secondly, an invasive *Shigella*-type of disease, and thirdly an enterotoxigenic type of disease which involves enterotoxin(s) and causes fluid loss without mucosal damage.

Two enterotoxins have been recognized: a heat labile toxin which is destroyed by heating at 60°C for 15 min, and a heat stable toxin which survives heating at 100°C for 30 min. Both toxins are plasmid controlled and a given toxigenic strain may produce both or only one toxin (Smith & Gyles 1970). The heat labile toxin is a single protein which acts on the epithelial cells of the intestinal mucosa by stimulating adenylate cyclase activity and hence fluid accumulation in the gut in the same manner as *Vibrio cholerae* enterotoxin. It is antigenically related to cholera enterotoxin and can be neutralized by cholera antitoxin. (Finkelstein *et al.* 1976). The nature of the heat-stable enterotoxin is not fully understood but the toxin appears to cause fluid accumulation in the intestine by a different mechanism to that defined for heat-labile enterotoxin (Hughes *et al.* 1978).

B. *Immunological methods*

The heat-labile, but not the heat-stable, toxin is antigenic. It can be detected immunologically and four quantitative procedures have been described (Table 7). Two of these methods, a solid phase radioimmunoassay described by Greenberg *et al.* (1977) and an enzyme-linked immunoabsorbent assay described by Yolken *et al.* (1977), use antibody prepared against purified *Vibrio cholerae* enterotoxin. The sensitivity of the radioimmunoassay is approximately 1.0 pg enterotoxin/ml, whilst that of the enzyme linked immunoabsorbent assay is 0.1 pg/ml. Two haemolysis procedures, the lysis inhibition test (Evans & Evans

TABLE 7

Immunological methods for the detection of Escherichia coli *heat labile enterotoxin*

Method	Sensitivity	Assay time	Reference
Lysis inhibition test	?	< 1 day	Evans & Evans (1977a)
Passive immune haemolysis	< 100 mg/ml	< 1 day	Evans & Evans (1977b)
Solid-phase radioimmunoassay	ca. 1.0 pg/ml	> 1 day	Greenberg et al. (1977)
Enzyme linked immunoabsorbent assay	ca. 0.1 pg/ml	> 1 day	Yolken et al. (1977)

1977a) and the passive immune haemolysis test (Evans & Evans 1977b), use antibody against partially purified *Esch. coli* heat labile enterotoxin. They are both less sensitive than either radioimmunoassay or enzyme-linked immunoabsorbent assay (Table 7). Furthermore, the lysis inhibition test is not suitable for detecting enterotoxin in culture supernatants, presumably because of interference from haemolysins which are produced by some strains of *Esch. coli*. It is therefore unlikely that this procedure could be used in the analysis of food.

Schemes to serotype strains of *Esch. coli* by means of their somatic and capsular antigens are well established and are described in detail by Cooke (1974).

C. Potential impacts of immunological methods

Knowledge of *Esch. coli* food-poisoning is just becoming available. Immunological methods need to be applied to find the distribution of toxigenic types in the environment, particularly in water and foods, and to ascertain whether toxin is produced during growth of *Esch. coli* in food. Furthermore, immunological methods need to be devised to detect the heat-stable enterotoxin in order to investigate the importance and role of this toxin in human disease.

8. Mycotoxicoses

A. Mycotoxins and mycotoxicoses

Mycotoxins are toxic metabolites of various chemical structures which are produced by many different species of fungi. In nature they are formed during spoilage of a wide variety of foodstuffs including cereals, oilseeds, fruits and vegetables. When ingested in high doses by domestic animals they produce acute disease syndromes, whilst low doses may cause carcinogenic or teratogenic effects (Hacking & Harrison 1976). The evidence for mycotoxins causing foodborne disease in man is not as well established as it is for domestic or laboratory animals but it is increasing and has been reviewed by Austwick (1975) and Jarvis (1976). Of the many important mycotoxins known only two, aflatoxin B_1 and ochratoxin A, will be considered here because they are the only ones which so far can be assayed by immunological means. Aflatoxin B_1 is one of a group of aflatoxins produced by *Aspergillus flavus* and *Aspergillus parasiticus*. It causes hepatitis, cirrhosis and hepatoma in poultry, pigs and cattle. Ochratoxin A is produced by *Aspergillus ochraceus* and *Penicillium viridicatum* and causes nephrotoxicosis in cattle and pigs.

B. Immunological assay of ochratoxin and aflatoxin

Ochratoxin A and aflatoxin B_1 are currently assayed by chemical or biological means. Like most other mycotoxins they are both low molecular weight organic compounds and thus are devoid of any antigenicity. When ochratoxin A is conjugated with protein such as bovine serum albumin or a polypeptide such as polylysine, however, it becomes antigenic in rabbits. Chu *et al.* (1976), who used this approach to prepare specific antisera, developed a radioimmunoassay which they claim will detect as little as 1–20 ng ochratoxin ml^{-1}. Similarly, when aflatoxin B_1 is complexed with bovine serum albumin it becomes antigenic in rabbits. The antiserum is not entirely specific for aflatoxin B_1 as it will bind with aflatoxins B_2, G_1 and Q_1, albeit with low efficiency (Chu & Ueno 1977). Radioimmunoassays have been developed which will detect 0.4–4.0 ng aflatoxin B/ml (Langone & van Vunakis 1976; Chu & Ueno 1977).

C. Potential impact of immunological methods

The radioimmunoassays described here mark the very beginnings of immunological methods for detecting mycotoxins. They have not yet been used in the analysis of feedstuffs or foods. However, their potential impact is immense in this rapidly expanding area of food-borne diseases.

9. Summary

The food-poisoning toxins produced by *Staphylococcus aureus*, *Clostridium botulinum*, *Clostridium perfringens*, *Bacillus cereus* and *Escherichia coli* are briefly described and the *in vitro* immunological methods to detect them are listed and compared. Methods are well developed for staphylococcal enterotoxins and are being applied, but methods for the other toxins are not so advanced.

Immunological methods can be used to increase our knowledge of bacterial food-poisoning in four principle ways. Firstly, to determine the nature and amounts of toxin in a suspect food. Secondly, to determine the distribution of toxigenic organisms in foods and in the environment. Thirdly, to investigate in the research laboratory the conditions necessary to inhibit the formation of toxins in foods, and fourthly, to study those factors affecting the destruction of the toxins by physical and chemical means. Examples are given for the different toxins wherever possible. The review also points out the potential value of immunological methods for detecting mycotoxins and the potential value of the serotyping schemes for cells of *Clos. perfringens* and *Bacl. cereus* (as distinct from their toxins) in the epidemiology of foodborne disease.

10. References

AUSTWICK, P. K. C. 1975 Mycotoxins. *British Medical Bulletin* **31**, 222–229.
BAILLIE, A., CROWTHER, J. S. & BAIRD-PARKER, A. C. 1973 The use of fluorescent antibodies and other techniques for the detection of *Clostridium botulinum*. In *The Microbiological Safety of Food* ed. Hobbs, B. C. & Christian, J. H. B. London: Academic Press.
BAIRD-PARKER, A. C. 1971 Factors affecting the production of food poisoning toxins. *Journal of Applied Bacteriology* **34**, 181–197.
BARBER, L. & DEIBEL, R. H. 1972 Effect of pH and oxygen tension on staphylococcal growth and enterotoxin formation in fermented sausage. *Applied Microbiology* **24**, 891–898.
BERGDOLL, M. S. 1972 The enterotoxins. In *The Staphylococci* ed. Cohen, J. O. New York: Wiley Interscience.
BOROFF, D. A. & SHU-CHEN, G. 1973 Radioimmunoassay of toxin of *Clostridium botulinum*. *Federation Proceedings* **32**, 1032.
CANN, D. C., TAYLOR, L. Y. & HOBBS, G. 1975 The incidence of *Clostridium botulinum* in farmed trout raised in Great Britain. *Journal of Applied Bacteriology* **39**, 331–336.
CARPENTER, D. F. & SILVERMAN, G. J. 1973 Enterotoxin B production in a fermenter system. *Abstract from the Annual Meeting of the American Society for Microbiology*, 1.
CASMAN, E. P. 1967 Staphylococcal food poisoning. *Health Laboratory Science* **4**, 199–206.
CASMAN, E. P., McCOY, D. W., & BRANDLY, P. J. 1963 Staphylococcal growth and enterotoxin production in meat. *Applied Microbiology* **11**, 498–500.
CASMAN, E. P., BENNETT, R. W., DORSEY, A. E. & ISSA, J. A. 1967 Identification of a fourth staphylococcal enterotoxin D. *Journal of Bacteriology* **94**, 1875–1882.
CASMAN, E. P., BENNETT, R. W., DORSEY, A. E. & STONE, J. E. 1969 The microslide gel double diffusion test for the detection and assay of staphylococcal enterotoxins. *Health Laboratory Science* **6**, 185–198.
CHU, F. S. & UENO, I. 1977 Production of antibody against aflatoxin B$_1$. *Applied and Environmental Microbiology* **33**, 1125–1128.
CHU, F. S., CHANG, F. C. C. & HINSDILL, R. C. 1976 Production of antibody against ochratoxin A. *Applied and Environmental Microbiology* **31**, 831–835.
COOKE, E. M. 1974 *Escherichia coli and Man*. Edinburgh & London: Churchill Livingstone.
CRAIG, J. P. 1972 The enterotoxic enteropathies. In *Microbial Pathogenicity in Man and Animals* ed. Smith, H. & Pearce, J. H. *Society for General Microbiology, Symposium 22*. Cambridge: University Press.
CROWTHER, J. S. & HOLBROOK, R. 1976 Trends in methods for detecting food-poisoning toxins produced by *Clostridium botulinum* and *Staphylococcus aureus*. In *Microbiology in Agriculture, Fisheries and Food* ed. Skinner, F. A. & Carr, J. G. London: Academic Press.

DALY, C., LA CHANCE, M., SANDINE, W. E. & ELLIKER, P. R. 1973 Control of *Staphylococcus aureus* in sausage by starter cultures and chemical acidulation. *Journal of Food Science* **38**, 426–430.

DUNCAN, C. L. & STRONG, D. H. 1969 Ileal loop fluid accumulation and production of diarrhoea in rabbits by cell-free products of *Clostridium perfringens*. *Journal of Bacteriology* **100**, 86–94.

DUNCAN, C. L. & SOMERS, E. B. 1972 Quantitation of *Clostridium perfringens* Type A enterotoxin by electroimmunodiffusion. *Applied Microbiology* **24**, 801–804.

EVANCHO, G. M., ASHTON, D. H., BRISKEY, E. J. & SCHANTZ, E. J. 1973 A standardized reversed passive haemagglutination technique for the determination of botulinum toxin. *Journal of Food Science* **38**, 764–767.

EVANS, D. J. & EVANS, D. G. 1977a Inhibition of immune hemolysis: Serological assay for the heat-labile enterotoxin of *Escherichia coli*. *Journal of Clinical Microbiology* **5**, 100–105.

EVANS, D. J. & EVANS, D. G. 1977b Direct serological assay for the heat-labile enterotoxin of *Escherichia coli*, using passive immune hemolysis. *Infection and Immunity* **16**, 604–609.

FINKELSTEIN, R. A., LARUE, M. K., JOHNSTON, D. W., VASIL, M. L., CHO, G. J. & JONES, J. R. 1976 Isolation and properties of heat-labile enterotoxin(s) from enterotoxigenic *Escherichia coli*. *Journal of Infectious Diseases* **133**, s.120–s.137.

GASPER, E., HEIMSCH, R. C. & ANDERSON, A. W. 1973 Quantitative detection of type A staphylococcal enterotoxin by Laurell electroimmunodiffusion. *Applied Microbiology* **25**, 421–426.

GENIGEORGIS, C. & KUO, J. K. 1976 Recovery of staphylococcal enterotoxin from foods by affinity chromatography. *Applied Environmental Microbiology* **31**, 274–279.

GENIGEORGIS, C., REIMANN, H. & SADLER, W. W. 1966 Production of enterotoxin B in cured meats. *Journal of Food Science* **34**, 62–68.

GENIGEORGIS, C., FODA, M. S., MANTIS, A. & SADLER, W. A. 1971 Effect of pH and sodium chloride on enterotoxin C production. *Applied Microbiology* **21**, 862–866.

GENIGEORGIS, C., SAKAGUCHI, G. & RIEMANN, H. 1973 Assay methods for *Clostridium perfringens* Type A enterotoxin. *Applied Microbiology* **26**, 111–115.

GIBBS, P. A., PATTERSON, J. T. & HARVEY, J. 1978 Biochemical characteristics and enterotoxigenicity of *Staphylococcus aureus* strains isolated from poultry. *Journal of Applied Bacteriology* **44**, 57–74.

GILBERT, R. J., & PARRY, J. M. 1977 Serotypes of *Bacillus cereus* from outbreaks of food-poisoning and from routine foods. *Journal of Hygiene, Cambridge* **78**, 69–74.

GILBERT, R. J. & TAYLOR, A. J. 1976 *Bacillus cereus* food-poisoning. In *Microbiology in Agriculture, Fisheries and Food*, ed. Skinner, F. A. & Carr, J. G. London & New York: Academic Press.

GILBERT, R. J. & WEINEKE, A. 1973 Staphylococcal food poisoning with special reference to the detection of enterotoxin in food. In *The Microbiological Safety of Food*, ed. Hobbs, B. C. & Christian, J. H. B. London: Academic Press.

GILBERT, R. J., WEINEKE, A., LANSER, J. & SIMKOVICOVA, M. 1972 Serological detection of enterotoxin in foods implicated in staphylococcal food poisoning. *Journal of Hygiene, Cambridge* **70**, 755–762.

GOEPFERT, J. M., SPIRA, W. M. & KIM, H. U. 1972 *Bacillus cereus*: food poisoning organism. A review. *Journal of Food and Milk Technology* **35**, 213–222.

GORINA, L. G., FLUER, F. S., OLOVNIKOV, A. M. & EZEPCUK, Y. V. 1975 Use of aggregate-hemagglutination technique for determining exo-enterotoxin of *Bacillus cereus*. *Applied Microbiology* **29**, 201–204.

GREENBERG, H. B., SACK, D. A., RODRIGUEZ, W., SACK, R. B., WYATT, R. G., KALICA, A. R., HORSWOOD, L., CHANOCK, R. M. & KAPIKIAN, A. Z. 1977 Microtiter solid-phase radio-immunoassay for detection of *Escherichia coli* heat-labile enterotoxin. *Infection and Immunity* **17**, 541–545.

HACKING, A. & HARRISON, J. 1976 Mycotoxins in animal feeds. In *Microbiology in*

Agriculture Fisheries and Food ed. Skinner, F. A. & Carr, J. G. London: Academic Press.

HÁJEK, V. & MARŠÁLEK, E. 1973 The occurrence of enterotoxigenic *Staphylococcus aureus* strains in hosts of different animal species. *Zentralblatt für Bakteriologie Parasitenkunde Infektionskrankheiten und Hygiene Erste Abteilung. Originale A* **233**, 63-68.

HAUSCHILD, A. H. W. 1973 Food poisoning by *Clostridium perfringens*. *Canadian Institute of Food Science and Technology Journal* **6**, 106-110.

HOBBS, B. C., SMITH, M. E., OAKLEY, C. L., WARRACK, G. H., & CRUICKSHANK, J.C. 1953 *Clostridium welchii* food poisoning. *Journal of Hygiene, Cambridge* **51**, 75-101.

HOLBROOK, R. & BAIRD-PARKER, A. C. 1975 Serological methods for the assay of staphylococcal enterotoxins. In *Some Methods for Microbiological Assay* ed. Board, R. G. & Lovelock, D. W. *Society for Applied Bacteriology Technical Series No. 8.* London & New York: Academic Press.

HUGHES, J. A., TURNBULL, P. C. B. & STRINGER, M. 1976 Serotyping system for *Clostridium welchii (Clostridium perfringens)* Type A, and studies on type-specific antigens. *Journal of Medical Microbiology* **9**, 475-485.

HUGHES, J. M., MURAD, F., CHANG, B. & GUERRANT, R. L. 1978 Role of cyclic G.M.P. in the action of heat stable enterotoxin of *Escherichia coli*. *Nature, London* **271**, 755-756.

JAMLANG, E. M., BARTLETT, M. L. & SNYDER, H. E. 1971. The effect of pH, protein concentration and ionic strength on heat inactivation of staphylococcal enterotoxin B. *Journal of Applied Microbiology* **22**, 1034-1040.

JARVIS, B. 1976 Mycotoxins in food. In *Microbiology in Agriculture, Fisheries and Food* ed. Skinner, F. A. & Carr, J. G. London: Academic Press.

KIMBLE, C. E. & ANDERSON, A. W. 1973 Rapid, sensitive assay for staphylococcal enterotoxin A by reversed immuno-osmophoresis. *Applied Microbiology* **25**, 693-694.

LABBE, R., SOMERS, E. & DUNCAN, C. 1976 Influence of starch source on sporulation and enterotoxin production by *Clostridium perfringens* Type A. *Applied and Environmental Microbiology* **31**, 455-457.

LANGONE, J. J. & VAN VUNAKIS, H. 1976 Aflatoxin B_1: specific antibody and their use in radioimmunoassay. *Journal of the National Cancer Institute* **56**, 591-595.

LEE, I. C., STEVENSON, K. E. & HARMAN, L. G. 1977 Effect of beef broth protein on thermal inactivation of staphylococcal enterotoxin B. *Applied and Environmental Microbiology* **33**, 341-344.

LYNT, R. K., SOLOMON, H. M. & KAUTTER, D. A. 1971 Immunofluorescence among strains of *Clostridium botulinum* and other clostridia by direct and indirect methods. *Journal of Food Science* **36**, 594-599.

MCDONEL, J. L. & ASANO, T. 1975 Analysis of unidirectional fluxes of sodium during diarrhoea induced by *Clostridium perfringens* enterotoxin in the rat terminal ileum. *Infection and Immunity* **11**, 526-529.

MARIER, R., WELLS, J. G., SWANSON, R. C., CALLAHAN, W. & MEHLMAN, I. J. 1973 An outbreak of enteropathogenic *Escherichia coli* foodborne disease traced to imported French cheese. *Lancet* ii, 1376-1378.

MELLING, J., CAPEL, B. J., TURNBULL, P. C. B. & GILBERT, R. J. 1976 Identification of a novel enterotoxigenic activity associated with *Bacillus cereus*. *Journal of Clinical Pathology* **29**, 938-940.

MILLER, C. & ANDERSON, A. W. 1971 Rapid detection and quantitative estimation of type A botulinum toxin by electroimmunodiffusion. *Infection and Immunity* **4**, 126-129.

NAIK, H. S. & DUNCAN, C. L. 1977 Rapid detection and quantitation of *Clostridium perfringens* enterotoxin by counterimmunoelectrophoresis. *Applied and Environmental Microbiology* **34**, 125-128.

NISKANEN, A. & LINDROTH, S. 1976 Comparison of different purification procedures for extraction of staphylococcal enterotoxin A from foods. *Applied and Environmental Microbiology* **32**, 455-463.

NORRIS, J. R. & WOLF, J., 1961 A study of antigens of the aerobic spore-forming bacteria. *Journal of Applied Bacteriology* **24**, 42–56.

NOTERMANS, S., DUFRENNE, J. & VAN SCHOTHORST, M. 1978 Enzyme linked immunosorbent assay for detection of *Clostridium botulinum* toxin Type A. *Japanese Journal of Medical Science* **31**, 81–89.

ORTH, D. S. 1977 Statistical analysis and quality control in radioimmunoassays for staphylococcal enterotoxins A, B & C. *Applied and Environmental Microbiology* **34**, 710–714.

READ, R. B., Jr. & BRADSHAW, J. G. 1966a Staphylococcal enterotoxin B thermal inactivation in milk. *Journal of Dairy Science* **49**, 202–203.

READ, R. B., Jr. & BRADSHAW, J. G. 1966b Thermal inactivation of staphylococcal enterotoxin B in veronal buffer. *Applied Microbiology* **14**, 130–132.

REISER, R. F. & WEISS, K. F. 1969 Production of staphylococcal enterotoxins A, B and C in various media. *Applied Microbiology* **18**, 1041–1043.

REISER, R. F., CONAWAY, D. & BERGDOLL, M. S. 1974 Detection of staphylococcal enterotoxin in foods. *Applied Microbiology* **27**, 83–85.

ROBERTS, T. A. & SMART, J. L. 1976 The occurrence and growth of *Clostridium* spp. in vacuum-packed bacon with particular reference to *Cl. perfringens (welchii)* and *Cl. botulinum*. *Journal of Food Technology* **11**, 229–244.

ROWE, B., SCOTLAND, S. M. & GROSS, R. J. 1977 Enterotoxigenic *Escherichia coli* causing infantile enteritis in Britain. *Lancet* **i**, 90–91.

SACK, R. B. 1975 Human diarrhoeal disease caused by enterotoxigenic *Escherichia coli*. *Annual Review of Microbiology* **29**, 333–353.

SAUNDERS, G. C. & BARTLETT, M. L. 1977 Double-antibody solid-phase enzyme immunoassay for the detection of staphylococcal enterotoxin A. *Applied and Environmental Microbiology* **34**, 518–522.

SCHEUSNER, D. L. & HARMON, L. G. 1973 Temperature range for production of four different enterotoxins of *Staphylococcus aureus* in B.H.I. broth. *Journal of Milk and Food Technology* **36**, 249–252.

SCOTT, W. J. 1953 Water relations of *Staphylococcus aureus* at 30°C. *Australian Journal of Biological Sciences* **6**, 549–564.

SHIBATA, Y., MORITA, M., AMANO, Y. & ISHIDA, N. 1977 The passive haemagglutination inhibition test for detection of staphylococcal enterotoxin B with sensitized and lyophilised red blood cells. *Microbiology and Immunology* **21**, 45–48.

SILVERMAN, S. J., KNOTT, A. R. & HOWARD, M. 1968 Rapid sensitive assay for staphylococcal enterotoxin and a comparison of serological methods. *Applied Microbiology* **16**, 1019–1023.

SKJELKVALE, R. & UEMURA, T. 1977a Detection of enterotoxin in faeces and antienterotoxin in serum after *Clostridium perfringens* food-poisoning. *Journal of Applied Bacteriology* **42**, 355–363.

SKJELKVALE, R. & UEMURA, T. 1977b Experimental diarrhoea in human volunteers following oral administration of *Clostridium perfringens* enterotoxin. *Journal of Applied Bacteriology* **43**, 281–286.

SMITH, H. W. & GYLES, C. L. 1970 The relationship between two apparently different enterotoxins produced by enteropathogenic strains of *Escherichia coli* of porcine origin. *Journal of Medical Microbiology* **3**, 387–401.

SMITH, G. R. & MORYSON, C. J. 1975 *Clostridium botulinum* in the lakes and waterways of London. *Journal of Hygiene, Cambridge* **75**, 371–379.

TATINI, S. R. 1973 Influence of food environments on growth of *Staphylococcus aureus* and production of various enterotoxins. *Journal of Food and Milk Technology* **36**, 559–563.

TAYLOR, A. J. & GILBERT, R. J. 1975 *Bacillus cereus* food poisoning: a provisional serotyping scheme. *Journal of Medical Microbiology* **8**, 543–550.

TURNBULL, P. C. B. 1976 Studies on the production of enterotoxins by *Bacillus cereus*. *Journal of Clinical Pathology* **29**, 941–948.

TURNBULL, P. C. B., NOTTINGHAM, J. F. & GHOSH, A. C. 1977 A severe necrotic enterotoxin produced by certain food, food-poisoning and other clinical isolates of *Bacillus cereus. British Journal of Experimental Pathology* **58**, 273–280.

UNTERMANN, F. & SINELL, H. J. 1970 Beitrag zum vorkommen enterotoxin-bilender Staphylokokken. *Zentralblatt für Bakteriologie, Parasitenkunde, Infektionskrankheiten und Hygiene* **215**, 166–172.

VERMILYEA, B. L., WALKER, H. W. & AYRES, J. C. 1968 Detection of botulinum toxins by immunodiffusion. *Applied Microbiology* **16**, 21–24.

YOLKEN, R. H., GREENBERG, H. B., MERSON, M. H., SACK, R. B. & KAPIKIAN, A. Z. 1977 Enzyme linked immunosorbent assay for detection of *Escherichia coli* heat-labile enterotoxin. *Clinical Microbiology* **6**, 439–444.

Identification of Bacteria by Measurements of Enzyme Activities and its Relevance to the Clinical Diagnostic Laboratory

SHOSHANA BASCOMB

Department of Bacteriology, St. Mary's Hospital Medical School, Wright-Fleming Institute, London, UK

Contents

1. Introduction

THE CLINICAL diagnostic laboratory is involved in detection of micro-organisms in clinical specimens, identification of these organisms and determination of the antibiotic sensitivity of the pathogenic organisms found.

To appreciate the extent of operations involved it is helpful to consider the work load of a medium sized hospital laboratory such as the Clinical Diagnostic Laboratory of St. Mary's Hospital, Paddington. The daily sample load varies from 170 to 350, including urine (50%), throat swabs and sputum (25%), wound swabs (14%), stools (4%), blood (3%). Similar distribution was reported by Johnston *et al.* (1976).

Micro-organisms can be detected by methods of staining followed by microscopic examination, or by serological techniques, or by inoculation of the sample on selective media and incubation for varying lengths of time before observation of growth. The number of operations per specimen is small but, as the total number of specimens is large, these operations constitute about a third of laboratory work load.

Only 10-30% of the specimens tested actually contain a pathogenic organism and these specimens are subsequently processed further. Testing for sensitivity to antibiotics and precise identification of the pathogens involve a number of operations, as each isolate is tested against 6-10 antibiotic compounds. A similar number of tests is also used for identification. The latter tests are labour intensive, manual labour being required for preparation of sterile media in individual compartments, inoculation of each compartment with the isolated strain, reading of test results, and disposal of contaminated material at the end of testing.

Identification in a diagnostic laboratory often need not be as complete as required for specialist taxonomy, but must be rapid. Thus, in a urinary infection it is often sufficient to determine the genus of the pathogen (*Escherichia coli* versus *Pseudomonas* spp.) provided that the choice of antibiotics for treatment is validated. Occasionally a more detailed identification is needed, and techniques and skills have to be on hand.

Identification of bacteria is based on application of 'characterization tests' and assignation of the unknown organism to a defined group on the basis of the results. This characterization includes morphological, growth, biochemical and serological tests. The biochemical tests vary from crude ones where the utilization of a substrate is determined by a change in colour of a pH indicator included in the growth medium to the use of sophisticated methods for separation, identification and measurement of constituents of the bacterial cells or the growth medium. Results of identification tests are rarely available on the day the specimen arrives at the laboratory, preliminary results are available after 24 h, and final reporting is not possible until 48-72 h after receipt of specimens. By comparison results from the haematology or chemical pathology laboratories are commonly available at the end of the day of receipt of a specimen, and if necessary can be completed within 1-2 h.

Why does the clinical microbiological laboratory lag behind in providing a rapid service to the patient? This is possibly because conventional tests rely on multiplication of bacteria for amplification of the signal to a detectable level, whereas others use sophisticated equipment capable of reading minute signals.

2. Nature of Conventional Identification Methods

In identification tests which rely on growth, gross dilution of the bacterial suspension occurs during inoculation and an interval of 10-17 generation times is needed to increase the number of organisms (from 10^2 to 10^5-10^7) to produce enough metabolites to alter the pH of the relatively large volume of buffered medium. Also, some of the enzymes sought are not constitutive; they are induced by the conditions of the test. Time must elapse between induction and appearance of the enzyme, and further time is needed for the appearance of detectable products of their activity. As detection of change is by subjective observation, the change has to be substantial.

In many conventional tests the observed change is in the colour of a pH indicator included in the medium. The disadvantage of such methods is that similar observations are not necessarily caused by the same enzymes. Thus two strains may give a negative result, one because it lacked a permease system for transporting the substrate into the cell, the second because it lacked the enzymes capable of metabolizing the substrate, yet both will show the same observable results—no change in the colour of the pH indicator. Even a positive result may be produced via two different metabolic pathways, e.g. production of acid from glucose can be via the Embden-Meyerhof and mixed acid fermentation pathways as in *Esch. coli* or via the Embden-Meyerhof and butanediol fermentation pathway as in *Klebsiella*. Yet both pathways result in accumulation of organic acids in quantities sufficient to change the colour of the pH indicator and strains of these genera will give identical results. Similarly the change in colour of pH indicator of the arginine decarboxylase medium of Möller could be caused by activity of arginine decarboxylase or arginine dihydrolase (Grimont *et al.* 1977).

Interpretation of results of tests that require several days incubation is also made difficult by the possibility that selection of a rare mutant has occurred, or that the observed results were caused by a contaminating organism. The time needed for multiplication of one cell of *Psmn. aeruginosa* to 10^6 cells is less than 24 h. At the beginning of the incubation period the inoculated organism constitutes the majority of cells, and is most likely the cause of the changes observed. Six or 7 days later, however, the initially inoculated strain is in a non-active state, and even one contaminating organism may be able to grow and change the medium.

Kersters & De Ley (1971) discussed the shortcomings of tests commonly used in identification of bacteria. They advocated tests performed with resting cells which would permit detection of a single enzyme within a few hours by a procedure simple enough to apply to a large number of specimens. Sneath (1972) doubted the usefulness of enzyme tests as taxonomic tools because many enzymes are adaptive and are only produced during growth in the presence of substrate. Inconsistent results may sometimes occur with methods when little or no bacterial growth occurs. This argument may be circular in the sense that enzymes tested for by the conventional methods are detected in conditions which encourage the appearance of induced enzymes, if the same enzymes are tested for by rapid methods leaving no time for induction, results are likely to be inconsistent. If, however, the enzymes tested for are selected because they are likely to be constitutive, such as catalase or oxidase, then the results of such tests are as invariant as the best conventional tests.

3. Advantage of Enzyme Assays

The main difference between tests that depend on growth and enzyme tests is that in the former the initial conditions of the medium are designed so as to enable the growth of all organisms which are likely to be found in the specimen

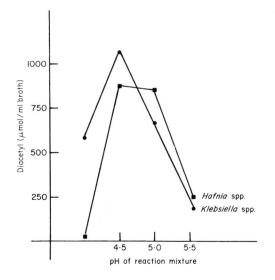

Fig. 1. Effect of pH on production of diacetyl (V-P reaction) by a strain of *Klebsiella* spp. and of *Hafnia* spp. Tested by an automated procedure (Bascomb & Spencer 1980) with 0.1 M acetate buffers (Bascomb, unpublished).

(Hedén *et al.* 1976). Conditions optimal for growth might be very different however from those optimal for the activity of the enzyme sought and growth of the organism is needed to modify the medium in order to permit the activity of the enzyme. Thus, in the conventional test for Voges-Proskauer (V-P) reaction bacteria are inoculated into a medium of pH 7.6 while the optimum pH for the reaction is 4.5 (Fig. 1). Tests with resting cells can be devised to provide optimal conditions for the assay of each enzyme and thus increase the speed of its detection. Further arguments for the use of enzyme activity tests for identification include:

(1) All constituents of bacterial cells and all changes in the growth medium are products of enzymatic activity. Theoretically, therefore, enzyme assays could substitute for the tests used in conventional classification.

(2) It is easier to magnify the signal by providing optimum conditions for enzyme activity than to increase the proportion of product in the total cell constituents. Factors which affect detection of activity and which can be regulated easily include temperature of incubation; pH, ionic strength and composition of the reaction mixture; presence of activators and inhibitors; affinity of the enzyme for the substrate; extinction coefficient of the product, and sensitivity of the sensory device.

(3) As incubation times needed for assay of enzyme activities are shorter than those needed for multiplication of bacteria, chance contamination becomes irrelevant. Maintenance of aseptic conditions during assay is

therefore unnecessary, and a substantial saving in labour and cost can be achieved.

(4) The higher sensitivity of enzyme assays enables their completion with smaller amounts of bacteria than needed for conventional tests.

(5) Sensitive equipment for study of enzyme activity developed for use in the chemical pathology laboratory (van Gemert 1973) is available.

4. Qualitative Micromethods

The first attempt to speed up identification procedures was the development of microtests (Arnold & Weaver 1948; Clarke & Cowan 1952). Two different principles were followed in the development of these methods. One method uses bacterial suspensions in a menstruum—saline or buffer—which does not support multiplication for seeding the test substrate. In the second method a heavy bacterial suspension, prepared in a medium which supports multiplication, and is added to the test substrate. A comprehensive list of micromethods is given by Cowan & Steel (1965). The relatively greater concentration of bacteria in the small volumes of media used in micromethods permits more rapid identification. With the API 20E or 50E systems (API Laboratory Products Ltd., Farnborough, Hampshire) which permit multiplication of the inoculum, identification can be achieved in 18 h compared with 24-48 h needed for conventional testing. With the Patho-Tec paper strip tests (General Diagnostics, Morris Plains, N.J.), which do not enable multiplication, identification is completed in 4-6 h. Other 4 h test schemes have been devised for identification of clinical isolates. Edberg et al. (1975) described a procedure for a 4 h identification of bacteria from blood cultures. Out of 308 specimens showing growth, 222 contained strains of enterobacteria, 40 facultative Gram positive cocci, 26 anaerobes and 20 assorted strains. All isolates were tested for the Gram staining reaction and subjected to a battery of tests accordingly. Agreement between results of 4 h and conventional methods was between 96.5-100%, depending on the type of organism. Nord et al. (1975) have described a set of 12 rapid biochemical tests for the identification of enterobacteria within 4 h and applied the system successfully to identification of 789 strains isolated from human infection (Nord et al. 1976). Agreement with identification based on conventional methods was 97%. Many of the tests used in 4 h testing are enzyme tests.

5. Qualitative Enzyme Tests

The application of enzyme tests can be considered in one or other of three categories: (1) qualitative detection of enzyme presence by observation of a colour change; (2) detection and characterization of enzymes by electrophoretic techniques (see Kersters & De Ley, Chapter 11); and (3) quantitative measurement of enzyme activity in cell suspensions or cell free extracts.

Detection of enzyme activity has been made easier in the last 30 years by the introduction initially of a variety of synthetic chromogenic substrates and latterly of fluorogenic substrates. The chromophore moieties of the former substrates include pH sensitive molecules, such as nitrophenol and phenolphthalein, the liberation of which can be followed during the reaction by adjustment of the pH of the medium. Other substrates include naphthol or naphthylamine in their molecules and the liberation of such substances is demonstrated by addition of a specific reagent which produces a colour. Dietrich & Liebermeister used the chromogenic reagents, paraphenylenediamine dihydrochloride and α-naphthol, as early as 1902 to demonstrate structure in the anthrax bacillus. Schultze (1910) used the same chromogen for differentiation of certain bacteria. Gordon & McLeod (1928) actually used this method of spotting oxidase activity for detection of neisseriae in mixed cultures. Bessey et al. (1946) described the use of nitrophenol phosphate for the rapid determination of alkaline phosphatase in serum, a method which permits detection in 5 μl samples of serum at the rate of 50-100 samples in 12 h. Huggins & Smith (1947) and Huggins & Lapides (1947) advocated the use of chromogenic substrates for assay of phenylsulphatase and esterase activities, pointing out that the advantages of such substrates lie in their simplicity, sensitivity and accuracy. Le Minor & Ben Hamida (1962) recommended the use of ortho-nitrophenol-β-D-galacto-pyranoside (ONPG) for the rapid detection of β-galactosidase in bacterial cultures. Bürger (1967a,b) and Kersters & De Ley (1971) described various methods for the detection of enzymes in suspensions of resting bacteria as well as cell free extracts.

Naphthol and naphthylamine derivatives constituted the majority of substrates used by Buissière et al. (1967) for the study of bacterial enzymes. Similar tests were used by Joubert & Buissière (1968) and Peny & Buissière (1970) for identification of staphylococci, and by Peny (1970) for the characterization of Erwiniae.

The commercially available 4 h-system, API ZYM (API Products Laboratories), was applied with encouraging results to the identification of non-haemolytic streptococci (Watkins 1976). Humble et al. (1977), who tested 81 bacterial isolates belonging to 27 taxa, found this system was easy to use, easy to read and reproducible. They suggested that the patterns of results were sufficiently different to be of use in the identification of bacteria. The most useful tests for characterization of streptococci were β-galactosidase, acid phosphatase and leucine amino-peptidase. The same system was tried by Tharagonnet et al. (1977) with regard to its suitability for identification of Gram negative anaerobic rods and cocci. The bacterial suspensions were prepared from 24, 48 and 72 h old cultures. Results obtained for each taxon were consistent regardless of the age of culture, provided that the inoculum contained sufficient organisms. The enzymatic patterns were quite distinct for all the species included, though the division to subspecies in *Bacteroides fragilis* did not agree with that obtained by conventional methods of identification. Various glycosidases were particularly useful for separation within this group of organisms.

D'Amato *et al.* (1978) tested resting cell suspensions of 172 strains belonging to 20 taxa of *Neisseria* and related genera, for activity towards 48 different chromogenic substrates. They were able to select 10 substrates (7 for assaying substrate specific amino-peptidases, β-galactosidase, acid phosphatase and esterase) which allowed the separation of *Neis. gonorrhoeae* and *Neis. meningitidis* from each other and from the remaining taxa within 4 h of availability of isolated colonies. The four most useful tests were substrate-specific amino-peptidases. The role of substrate-specific amino-peptidases in characterization of micro-organisms was reviewed by Watson (1976).

Kilian & Bülow (1976) measured the release of nitrophenol from 5 substrates for detection of glycosidases, and determination of optimal pH conditions for activity of these enzymes. They recommended a 4 h test, detection of activity being determined visually. They applied these tests to 633 strains, and found that some, β-glucuronidase and β-xylosidase, were valuable additions to the list of tests for identification of Enterobacteriaceae and Vibrionaceae at present available.

Maddocks & Greenan (1975) used 5 fluorogenic substrates for detection of glucosidases in *Esch. coli* and *Psmn. aeruginosa*. After 10 min incubation, activity of 3 enzymes was detected in *Esch. coli* cells grown on MacConkey agar, but no activity was detected in *Psmn. aeruginosa* cells grown on Direct Sensitivity agar.

6. Quantitative Enzyme Tests

Grange & Clarke (1977) used umbelliferone derivatives for the detection of β-D-glucosidase and arylsulphatases in whole cells or cell-free extracts of 56 strains of mycobacteria. By measuring the absorbance at 360 nm of the umbelliferone released, activity could be determined in 6 h compared with the 10 days needed for the conventional methods. All of the strains showed the activity of these two enzymes, but quantitative differences could be shown between the different strains. Grange (1978) also used umbelliferone derivatives, but determined the release of umbelliferone by the more sensitive measurement of its fluorescence. He measured the activity of group-specific hydrolases in small quantities of whole mycobacteria at different pH values. Levels of phosphatase of different strains varied, as well as their pH optima and heat stability. Grange suggested that such techniques would be a useful aid in the classification and identification of mycobacteria, especially slow-growing strains in which biochemical properties are very difficult to detect by other techniques. Leaback (1975) reviewed methods of fluorimetric estimation of enzyme activity.

7. Semi-Automated Enzyme Assays

The previous section has shown that quantitative measurement of enzyme activity provides further parameters for characterization of bacteria. As the number of tests necessary for identification is fairly large, however, reliable measurements of such tests are practical only with some degree of automation. Automation in the chemical pathology laboratory has followed two distinct lines, assay by continuous flow methods and by discrete sample analysis. The equipment available was reviewed by van Gemert (1973), methods for automated enzyme analysis were reviewed by Roodyn (1970) and Moss (1977).

The application of automated continuous flow methods to the study of bacterial metabolism started in the field of microbiological assay of antibiotics (Gerke *et al.* 1960). The method employs measurement of turbidity or released CO_2 for determination of bacterial growth. Further developments in this field were reviewed by Gerke & Ferrari (1968) and Ferrari & Marten (1972). Measurement of CO_2 by continuous flow methods was used by Leclerc (1967) for estimation of glutamic acid decarboxylase activity of 230 strains of Gram negative bacteria, and the presence of the enzyme in strains of *Esch. coli, Shigella, Proteus* (except *Prts. morganii*) and *Providencia* was established.

Trinel & Leclerc (1972) showed that glutamate decarboxylase was a constitutive enzyme of *Esch. coli* and that the measurement of its activity could be used for estimation of the number of *Esch. coli* cells present in a sample. The method was developed further by Moran & Witter (1976a,b) for the detection of *Esch. coli* in water and milk samples, by measuring the glutamate decarboxylase activity of samples inoculated manually into trypticase soy broth after 10 h incubation at 37°C. Trinel & Leclerc (1977) developed a completely automated method capable of detecting *Esch. coli* in water samples incubated for 12 h. The whole process was completed in 13 h, 120 samples could be processed in one day. Warren *et al.* (1978) detected *Esch. coli* in water samples by measurement of β-galactosidase activity. Detection was possible after 8–20 h compared with 22–72 h required for detection by conventional methods.

8. Identification by Semi-Automated Enzyme Assays

Identification of bacteria by continuous flow methods was first suggested by Avanzini *et al.* (1968). They described a system that measured bacterial respiratory activity by the reduction of tetrazolium red in the presence of glucose, and suggested that such a system could be used for study of respiratory activities in different species of bacteria.

Bascomb & Grantham (1973, 1975) developed an automated analytical system that measured ammonia released from different substrates by bacterial suspensions. The suspensions were prepared from bacteria grown overnight in corn steep liquor liquid medium. Each organism in suspension was tested with 41 samples comprising 15 substrates, 7 ammonia standards and 19 buffer samples in a fixed sequence. The testing of one suspension was completed within 48 min, the results being obtained within 27 h of receipt of a culture. The recorded trace of a routine assay was termed Specific Enzyme Profile (SEP). SEP was found to be reproducible and unique for each species of the tribe Proteae. Similar methods were developed for assay of dehydrogenases and nitrophenol-releasing enzymes of bacteria from urine specimens cultured on MacConkey agar (Bascomb 1976a). The eight tests most suitable for identification of bacteria from urinary infections, and a protein assay, were chosen by Bascomb & Spencer (1980) for use in a simultaneous 3-channel automated system. The tests were applied to 285 Gram negative isolates cultured from urine specimens by overnight incubation on MacConkey agar. Bacterial suspensions in saline were made from isolated colonies (1–2 colonies per ml). Each bacterial isolate was identified by comparing its SEP with patterns of activity obtained with reference strains. The results obtained with automated methods were compared with those obtained by conventional methods. Correct identification by the automated method was achieved in 99% of the cases. Individual tests were completed in 1 h but, as only 3 channels were available, results for all strains on all tests were not available until the end of one working day. Results from testing by conventional methods were available after 24–48 h.

9. Problems Inherent in Enzyme Assays

It is obvious that by measuring the activity of individual enzymes in resting cells, it is possible to speed up their detection. These methods are subject, however, to the following limitations:

(1) Only those enzymes already present in the bacteria can be measured.
(2) The active site of an enzyme must be accessible to the substrate.
(3) Lower limits of sensitivity of each technique determine minimum number of organisms needed.

The presence of enzymes in the bacterial suspension will depend on the presence of the genes coding for its molecule, and conditions that allow the phenotypic expression of the genes. In many conventional tests detection of enzymes depends on the first factor only, while rapid tests detect induced enzymes only if they were induced before the tests. The appearance of induced enzymes was studied by Dealy & Umbreit (1965) in *Esch. coli*. They noted that the ability of cells to synthesize β-galactosidase was age dependent. A 2 min incubation with the inducer was sufficient to produce enzyme detectable by

20 min incubation with ONPG. The study of Warren *et al.* (1978) suggests, however, that, when dealing with bacteria subjected to stress, periods of 2–3 h are needed. Indirect evidence could be deduced from the time needed for assay of different enzymes in the Patho-Tec system. Detection of oxidase can be completed in seconds, while detection of enzymes like urease or lysine decarboxylase needs 4–6 h. The difference in time could reflect the difference in the specific activities of these enzymes, or the time taken for their induction. Bascomb & Spencer (1980), who tested the urease activity of cells grown on MacConkeys, found that 75% and 62% of *Proteus* and *Klebsiella* strains respectively were positive in the automated (18 min incubation) urease test but all gave positive results with the conventional methods (24 h incubation). The time needed for the appearance of induced urease in cells incubated with urea only varied from 14 min to more than 3 h (Fig. 2). This evidence and the fact that such induced enzymes can be detected by the Pathotec system suggests that

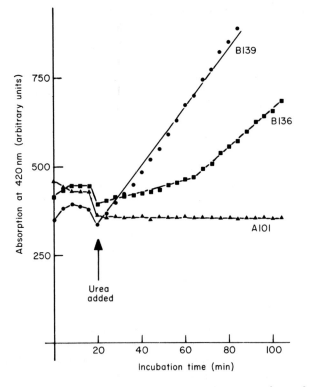

Fig. 2. Effect of period of incubation in the presence of urea on release of ammonia by *Klebsiella* spp. and *Proteus* spp. The suspension of organisms was stirred continuously at room temperature, sampled every 4 min, and its ammonia content measured by an automated Nessler method as described by Bascomb & Spencer (1980). Urea was added 16 min after beginning of sampling.

they are synthesized in resting cells, multiplication being unnecessary. It seems therefore that at least some induced enzymes can be tested for under conditions that exclude bacterial growth, so that even with longer incubation periods rigorous aseptic conditions need not be maintained in tests for enzyme activity.

On the other hand it might be possible to identify bacteria by testing basically constitutive enzymes essential for the metabolism of every organism, differentiating the enzymes on the basis of their physicochemical and biochemical characteristics. Weitzman & Jones (1968, 1975) showed that the enzyme, citrate synthase, of Gram positive differed from that of Gram negative bacteria by its response to NADH. The activity of the latter only was inhibited by addition of NADH. Moreover, the enzyme could be used for differentiation within the Gram negative bacteria as reactivation of the NADH-inhibited enzyme by AMP was typical of strict aerobes only. It was shown that the enzymes differ in their molecular weight also (Weitzman & Dunmore 1969).

Bhatti (1975) demonstrated some differences between alkaline phosphatases of *Esch. coli* and *Serratia marcescens.* The activity of the *Esch. coli* enzyme was increased more than two-fold by an increase in the ionic strength of the reaction mixture over a wide range of pH; that of the *Serratia* enzyme was increased by only 25% and the increase occurred only within a very narrow pH range. The enzymes differed also with regard to the optimal pH, the *Esch. coli* enzyme showed a high plateau at pH ranges of 8-11, the *Serratia* enzyme had a very narrow peak at pH 9.0. Bascomb (unpublished) found that alkaline phosphatase activity of *Proteus mirabilis* could be differentiated from that of *Klebsiella* spp. by addition of 1M-NaCl or cetrimide and lysozyme to the reaction mixture. Activity of the former was diminished to 38% and 5% respectively by these additions while that of the latter was increased to 212% and 240% respectively.

Accessibility of the enzyme to the substrate in rapid enzyme tests is essential. Chemical means of breaking the permeability of the barrier of the cell without adversely affecting enzyme activity would be preferable to physical ones for application to numerous samples of small volume. Dealy & Umbreit (1965) used toluene extraction, Leclerc (1967) added a few drops of toluene to the bacterial suspension, Reeves & Braithwaite (1972) used cetrimide to enable detection of β-galactosidase. Bascomb (1976b) compared the effect of various physical and chemical treatments on the activity of various enzymes. She found (unpublished data) that a combination of cetrimide and lysozyme was at least as effective as ultrasonic wave treatment for detection of NADH oxidase activity of Gram negative bacteria. It was also found that the activity of some enzymes was completely destroyed by treatments that affected the permeability barrier of the organism. It is therefore necessary to include effects on the cell membrane when the optimal conditions for assay of any enzyme are established.

The sensitivity of the assay determines the number of organisms needed for the measurement of enzyme activity. Moran & Witter (1976a) claimed that measurement of glutamate decarboxylase activity could detect 5×10^4 *Esch.*

coli. Warren *et al.* (1978) suggested that 10^6 *Esch. coli* are needed for detection by the β-galactosidase assay. Bascomb & Spencer (1980) used *ca.* 10^7 bacteria for assay of each enzyme, and bacteria from an isolated colony were sufficient for the performance of the 9 tests needed for identification. How many organisms are available in practice for identification? Clinical specimens from such sources as blood, urine or cerebro-spinal fluid (CSF) that are known to be sterile in the healthy individual yield in the majority of infections pure cultures of one strain only—infection due to more than one species can be encountered in less than a quarter of the cases. In single pathogen infections the limiting factor for detection and identification of the causative organism directly from the sample is the number of organisms present in it. Urine of patients suffering from urinary infections may contain 10^5-10^{10} organisms/ml. The total number of organisms present in the sample brought to the laboratory (20-25 ml) would vary from 2×10^6-2×10^{11}. In blood the CSF samples the number of organisms is much smaller.

Samples from sources such as stools, which are known to support a normal flora, may contain the pathogenic organism as a small proportion of the total, mainly resident. Dilution and special enrichment procedures are needed to obtain isolated colonies so that the pathogenic organisms can be distinguished from the normal flora. Under such circumstances the number of organisms available for identification is that obtained from a single colony on the selective agar medium. The number, depending on species, varies from 10^6 to 10^8 colony forming units in colonies from 24 h growth on MacConkey agar and the protein content varies from 10 to 70 μg/colony. Bascomb & Spencer (1980) showed that the measurement of enzyme activities of organisms available from a single colony enabled identification of 4 of the most important Gram negative pathogens involved in urinary infections. On the basis of these calculations it might be possible to identify bacteria from urine specimens directly in those containing more than 10^7 bacteria ml^{-1} urine. Identification of pathogens from stools could be achieved within an hour of obtaining isolated colonies.

10. Conclusions

Replacement of conventional tests by measurement of enzyme activities can reduce the time and labour involved in characterization of bacteria. The results obtained are unambiguous and can lead to a better understanding of bacterial metabolism. Availability of synthetic substrates and highly sensitive measuring devices make it possible to determine enzyme activity of 10^5-10^7 bacteria in less than 1 h. Application of such techniques directly to urine specimens could enable identification of pathogens within 1-2 h of receipt of specimen and provide better service to the patient.

11. References

ARNOLD, W. M. Jr. & WEAVER, R. H. 1948 Quick microtechniques for the identification of cultures. 1. Indole production. *Journal of Laboratory and Clinical Medicine* 33, 1334–1337.

AVANZINI, F., MAGNANELLI, D. & BOFFI, V. 1968 The automated determination of bacterial respiratory activity using tetrazolium red. In *Automation in Analytical Chemistry*, Technicon Symposia 1967, Vol. II, pp. 285–290. White Plains, New York: Mediad.

BASCOMB, S. 1976a Rapid identification of bacteria from clinical specimens by continuous flow analysis. In Proceedings of 2nd International Symposium on *Rapid Methods and Automation in Microbiology*. ed. Johnston, H. H. & Newsom, S. W. B. pp. 53. Oxford: Learned Information (Europe).

BASCOMB, S. 1976b Enzymatic activities of bacteria and their survival during treatments affecting cell integrity. *Proceedings of the Society for General Microbiology* 3, pp. 87.

BASCOMB, S. & GRANTHAM, C. 1973 Specific Enzyme Profile, an automated method for bacterial classification. In *Abstracts of International Association of Microbiological Societies*, 1st International Congress for Bacteriology Vol. II, pp. 132: Jerusalem.

BASCOMB, S. & GRANTHAM, C. A. 1975 Application of automated assay of asparaginase and other ammonia-releasing enzymes to the identification of bacteria. In *Some Methods for Microbiological Assay*, Society for Applied Bacteriology Technical Series No. 8. ed. Board, R. G. & Lovelock, D. W., pp. 29–54. London: Academic Press.

BASCOMB, S. & SPENCER, R. C. 1980 Automated methods for identification of bacteria from clinical specimens. *Journal of Clinical Pathology* 33, 36–46.

BESSEY, O. A., LOWRY, O. H. & BROCK, M. J. 1946 A method for the rapid determination of alkaline phosphatase with five cubic millimeters of serum. *Journal of Biological Chemistry* 164, 321–329.

BHATTI, A. R. 1975 Some distinctive characteristics of the alkaline phosphatase of *Serratia marcescens*. *Canadian Journal of Biochemistry and Physiology* 53, 819–822.

BUISSIÈRE, J., FOURCARD, A. & COLOBERT, L. 1967 Usage de substrats synthétiques pour l'étude de l'équipement enzymatique de micro-organismes. *Comptes Rendus Hebdomadaires des Séances de l'Académie des Sciences Série D* 264, 415–417.

BÜRGER, H. 1967a Biochemische Leistungen nichtproliferierender Mikroorganismen. II. Nachweis von Glycosid-Hydrolasen, Phosphatasen, Esterasen und Lipasen. *Zentralblatt für Bakteriologie, Parasitenkunde, Infektionskrankheiten und Hygiene. 1. Abteilung Originale* 202, 97–109.

BÜRGER, H. 1967b Biochemische Leistungen nichtproliferierender Mikroorganismen. III. Nachweis von Arylsulfatasen und Peptidasen. *Zentralblatt für Bakteriologie, Parasitenkunde, Infektionskrankheiten und Hygiene. 1. Abteilung Originale* 202, 395–401.

CLARKE, P. H. & COWAN, S. T. 1952 Biochemical methods for bacteriology. *Journal of General Microbiology* 6, 187–197.

COWAN, S. T. & STEEL, K. J. 1965 *Manual for the Identification of Medical Bacteria*. Cambridge: University Press.

D'AMATO, R. F., ERIQUEZ, L. A., TOMFOHRDE, K. A. & SINGERMAN, E. 1978 Rapid identification of *Neisseria gonorrhoeae* and *Neisseria meningitidis* by using enzymatic profiles. *Journal of Clinical Microbiology* 7, 77–81.

DEALY, J. D. & UMBREIT, W. W. 1965 The application of automated procedures for studying enzyme synthesis in *Escherichia coli*. *Annals of the New York Academy of Sciences* 130, 745–750.

DIETRICH, A. & LIEBERMEISTER, G. 1902 Sauerstoffübertragende Körnchen in Milzbranbacillen. *Centralblatt für Bakteriologie, Parasitenkunde und Infektionskrankheiten. Erste Abteilung, Originale* 32, 858–865.

EDBERG, S. C., NOVAK, M., SLATER, H. & SINGER, J. M. 1975 Direct inoculation procedure for the rapid classification of bacteria from blood culture. *Journal of Clinical Microbiology* 2, 469–473.

FERRARI, A. & MARTEN, J. 1972 Automated microbiological assay. In *Methods in Microbiology* Vol. 6B. ed. Norris, J. R. & Ribbons, D. W. pp. 331–342. London: Academic Press.

GERKE, J. R. & FERRARI, A. 1968 Review of chemical and microbiological assay of antibiotics. In *Automation in Analytical Chemistry*. Technicon Symposia (1967), Vol. 1, pp. 531–541. White Plains, New York: Mediad.

GERKE, J. R., HANEY, T. A., PAGANO, J. R. & FERRARI, A. 1960 Automation of the microbiological assay of antibiotics with an autoanalyzer instrument system. *Annals of the New York Academy of Sciences* 87, 782–791.

GORDON, J. & McLEOD, J. W. 1928 The practical application of the direct oxidase reaction in bacteriology. *Journal of Pathology and Bacteriology* 31, 185–190.

GRANGE, J. M. 1978 Fluorimetric assay of mycobacterial group-specific hydrolase enzymes. *Journal of Clinical Pathology* 31, 378–381.

GRANGE, J. M. & CLARKE, E. 1977 Use of umbelliferone derivatives in the study of enzyme activities of mycobacteria. *Journal of Clinical Pathology* 30, 151–153.

GRIMONT, P. A. D., GRIMONT, F., DULONG DE ROSNAY, H. L. C. & SNEATH, P.H.A. 1977 Taxonomy of the genus *Serratia*. *Journal of General Microbiology* 98, 39–66.

HEDÉN, C. G., ILLÉNI, T. & KÜHN, I. 1976 Mechanized identification of micro-organisms. In *Methods in Microbiology* Vol. 1. ed. Norris, J. R. & Ribbons, D. W. pp. 15–50. London: Academic Press.

HUGGINS, C. & LAPIDES, J. 1947 Chromogenic substrates IV. Acyl esters of *p*-nitrophenol as substrates for the colorimetric determination of esterases. *Journal of Biological Chemistry* 170, 467–482.

HUGGINS, C. & SMITH, D. R. 1947 Chromogenic substrates III. *p*-nitrophenol sulfate as a substrate for the assay of phenolsulfatase activity. *Journal of Biological Chemistry* 170, 391–398.

HUMBLE, M. W., KING, A. & PHILLIPS, I. 1977 API ZYM: a simple rapid system for the detection of bacterial enzymes. *Journal of Clinical Pathology* 30, 275–277.

JOHNSTON, H. H., MITCHELL, C. J. & CURTIS, G. D. W. 1976 Automation in clinical microbiology. A system for urine specimens. In Proceedings of the 2nd International Symposium on *Rapid Methods and Automation in Microbiology* ed. Johnston, H. H. & Newsom, S. W. B. pp. 210–213. Oxford: Learned Information (Europe).

JOUBERT, L. & BUISSIÈRE, J. 1968 Les chimiotypes épidémiologiques des staphylocoques animaux. *Bulletin de la Societé des Sciences Vétérinaires et de Médicine Comparée de Lyons* 70, 317–336.

KERSTERS, K. & DE LEY, J. 1971 Enzymic tests with resting cells and cell-free extracts. In *Methods in Microbiology* Vol. 6A. ed. Norris J. R. & Ribbons, D. W. pp. 33–52. London: Academic Press.

KILIAN, M. & BÜLOW, P. 1976 Rapid diagnosis of Enterobacteriaceae. I. Detection of bacterial glycosidases. *Acta Pathologica et Microbiologica Scandinavica Section B* 84, 245–251.

LEABACK, D. H. 1975 *An Introduction to the Fluorimetric Estimation of Enzyme Activities*. 2nd ed. Koch Light Laboratories Ltd.

LECLERC, H. 1967 Mise en évidence de la décarboxylase de l'acide glutamique chez les bactéries à l'aide d'une technique automatique. *Annales de l'Institut Pasteur* 112, 713–731.

LE MINOR, L. & BEN HAMIDA, F. 1962 Avantages de la recherche de la β-galactosidase sur celle de la fermentation du lactose en milieu complex dans le diagnostic bactériologique, en particulier des Enterobacteriaceae. *Annales de l'Institut Pasteur* 102, 267–277.

MADDOCKS, J. L. & GREENAN, M. J. 1975 A rapid method for identifying bacterial enzymes. *Journal of Clinical Pathology* 28, 686–687.

MORAN, J. W. & WITTER, L. D. 1976*a*. An automated rapid test for *Escherichia coli* in milk. *Journal of Food Science* 41, 165–167.

MORAN, J. W. & WITTER, L. D. 1976*b* An automated rapid method for measuring fecal pollution. *Water & Sewage Works* 123, 66–67.

MOSS, D. W. 1977 Automatic enzyme analyzers. In *Advances in Clinical Chemistry*, Vol. 19. ed. Bodansky, O. & Latner, A. L. pp. 1–54. New York: Academic Press.

NORD, C. E., LINDBERG, A. A. & DAHLBÄCK, A. 1975 Four-hour tests for the identification of Enterobacteriaceae. *Medical Microbiology and Immunology* 161, 231–238.

NORD, C. E., LINDBERG, A. A. & DAHLBÄCK, A. 1976 Rapid identification and antibiotic sensitivity testing of bacteria from clinical infections within four hours. In Proceedings of 2nd International Symposium on *Rapid Methods and Automation in Microbiology*. ed. Johnston, H. H. & Newsom, S. W. B. pp. 78. Oxford: Learned Information (Europe).

PENY, J. 1970 Caractères biochimiques des Erwiniae. Position taxonomique par rapport aux genres *Enterobacter* et *Pectobacterium*. In *Reports from the Conference on the Taxonomy of Bacteria*. University BRNO, Czechoslovakia, in *Spisy* 1–6, 37–43.

PENY, J. & BUISSIÈRE, J. 1970 Microméthode d'identification des bactéries. II. Identification du genre *Staphylococcus*. *Annales de l'Institut Pasteur* 118, 10–18.

REEVES, E. C. R. & BRAITHWAITE, J. A. 1972 The lactose system in *Klebsiella aerogenes* V9A. I. Characteristics of two Lac mutant phenotypes which revert to wild type. *Genetical Research* 20, 175–191.

ROODYN, D. B. 1970 Automated enzyme assays. In *Laboratory Techniques in Biochemistry and Molecular Biology*, Vol. 2, part 1. ed. Work, T. S. & Work, E. Amsterdam: North-Holland Publishing Co.

SCHULTZE, W. H. 1910 Über eine neue methode zum Nachwies von Reduktions und Oxydationswirkungen der Bakterien. *Centralblatt für Bakteriologie, Parasitenkunde und Infektionskrankheiten. Erste Abteilung, Originale* 56, 544–551.

SNEATH, P. H. A. 1972 Computer taxonomy. In *Methods in Microbiology* Vol. 7A. ed. Norris, J. R. & Ribbons, D. W. pp. 29–98. London: Academic Press.

THARAGONNET, D., SISSON, P. R., ROXBY, C. M., INGHAM, H. R. & SELKON, J. B. 1977 The API ZYM system in the identification of Gram-negative anaerobes. *Journal of Clinical Pathology* 30, 505–509.

TRINEL, P. A. & LECLERC, H. 1972 Automation de l'analyse bactériologique de l'eau–I. Etude d'un nouveau test spécifique de contamination fécale et de conditions optimales de se mise en évidence. *Water Research* 6, 1445–1458.

TRINEL, P. A. & LECLERC, H. 1977 Automatisation de l'analyse bactériologique de l'eau: description d'une nouvelle méthode de colimétrie. *Annales Microbiologie (Institut Pasteur)* 128A, 419–432.

VAN GEMERT, J. T. 1973 Automated wet chemical analysers and their applications. *Talanta* 20, 1045–1075.

WATKINS, S. A. 1976 Identification of streptococci using API ZYM system. In Proceedings of 2nd International Symposium on *Rapid Methods and Automation in Microbiology*. ed. Johnston, H. H. & Newsom, S. W. B. pp. 67–68. Oxford: Learned Information (Europe).

WARREN, L. S., BENOIT, R. E. & JESSEE, J. A. 1978 Rapid enumeration of fecal coliforms in water by a colorimetric-β-galactosidase assay. *Applied and Environmental Microbiology* 35, 136–141.

WATSON, R. R. 1976 Substrate specificities of aminopeptidases: a specific method for microbial differentiation. In *Methods in Microbiology* Vol. 9. ed. Norris, J. R. & Ribbons, D. W. pp. 1–14. London: Academic Press.

WEITZMAN, P. D. J. & DUNMORE, P. 1969 Citrate synthase; allosteric regulation and molecular size. *Biochimica et Biophysica Acta* 171, 198–200.

WEITZMAN, P. D. J. & JONES, D. 1968 Regulation of citrate synthase and microbial taxonomy. *Nature, London* 219, 270–272.

WEITZMAN, P. D. J. & JONES, D. 1975 The mode of regulation of bacterial citrate synthase as a taxonomic tool. *Journal of General Microbiology* 89, 187–190.

Appendix

Abstracts from Papers and Poster Session

During the *Symposium* on *The Impact of Modern Methods on Classification and Identification*, many workers active in this area presented 10-minute papers or contributed to the poster session on the morning of Wednesday, 19 July 1978. The abstracts, which have been published in the *Journal of Applied Bacteriology* (1978) **45**, I-XXIX, are reproduced in this book for the convenience of the reader.

Pyrolysis Gas-Liquid Chromatography as a Method of Characterizing Mixed Culture Systems. By M. TODD and R. S. HOLDOM (*Department of Applied Microbiology, University of Strathclyde, Glasgow G1 1XW*).

Pyrolysis gas-liquid chromatography (p.g.l.c.) is now an established technique for the characterization of large polymeric substances including biological materials. We have used a curie point pyrolysis system and temperature programmed chromatography on a low resolution packed column capable of separating the pyrolysis degradation products into *ca.* 65 peaks on the resulting chart trace. This trace is called the pyrogram.

For numerical analysis 14 peaks that were reproducible and easily identified were chosen. The measured peak heights were normalized with respect to sample weight by expressing them as a percentage of the total peak height. The coefficient of variation for individual peak heights of repeat analysis was typically 6%. Pyrograms were compared using the method described (Meuzelaar, Kistemaker & Tom 1975, in *New Approaches to the Identification of Micro-organisms*, ed. Heden, C.-G. & Illeni, T. pp. 165–178, New York: John Wiley), in which the ratio of heights for each of the 14 pairs of peaks is determined, always dividing the smaller value by the larger. The mean of these ratios is called the correspondence value. The correspondence value for repeat analyses was found to vary from 0.86–0.95.

A pure culture of *Streptococcus faecalis* was compared with mixed cultures of *Strep. faecalis* and the yeast *Saccharomyces cerevisiae*. The correspondence value decreased with increase in the percentage of yeast in the mixed culture, from 0.84 with 5% yeast to 0.71 with 50% yeast. When the pure culture of *Strep. faecalis* was compared with mixed cultures of it and *Shigella sonnei* the results were similar. From these preliminary results it is proposed that p.g.l.c. may be a useful technique for the early detection of contamination in a pure culture and as a method of characterizing more complex mixed cultures.

Analysis of Pyrolysis Gas-Liquid Chromatrography Data using Multivariate Statistical Techniques. By H. J. H. MACFIE and C. S. GUTTERIDGE (*Agricultural Research Council, Meat Research Institute, Langford, Bristol BS18 7DY*).

When bacteria are subjected to pyrolysis gas-liquid chromatography (p.g.l.c.) the immediate 'end product' is a set of complex analogue traces known as pyrograms. The aim of the multivariate analyses is to test whether a number of pyrograms, represented by a series of a peak height measurements, fall into groups which have some bacteriological significance. The effect of sample size is removed by normalizing peak heights to give a constant total peak height.

The concepts of multivariate analysis are best approached geometrically. If only two peaks are looked at, a pictorial representation of the difference can be achieved by a conventional two-dimensional graph. As more peak heights are included, there are more dimensions and principal components analysis is used to obtain co-ordinates of the pyrograms that are weighted sums of the peak heights. These are displayed relative to new orthogonal directions which display the maximum variation. Plotting the points relative to these first two (principal component) directions produces a two-dimensional representation of the scatter which most closely approximate to the multi-dimensional configuration. Principal components plots are useful for detecting outliers, assessing reproducibility and giving a pictorial representation of major clustering tendencies.

The directions of major variation among samples are not necessarily those directions that show the greatest variation between groups of pyrograms. Here we use canonical variates analysis to find successive orthogonal directions that show the maximum ratio of between group to within group variation. The coefficients of the weighted sums used to form these canonical variate scores are all scaled so that the points display a variance of 1. Thus, it is possible to place circular confidence regions around the group means on the canonical variate plots. This provides a strategy for assessing the significance of the observed differences and possibly for identifying new isolates into established data bases.

It is unusual for all the peaks to contribute to the discrimination between the groups. In fact, a high level of redundant data is the norm in p.g.l.c. studies. A further refinement of the analysis is to use stepwise discrimination which selects successively those peaks which contribute most to the analysis.

One limitation of these multivariate methods is the need to impose an *a priori* group structure on the data. Nevertheless, using these techniques, 65 strains of *Clostridium botulinum* and related organisms could be classified into three major groups which reflect known physiological and toxicological differences.

The Application of Pyrolysis Gas-Liquid Chromatography to Some Aerobic Sporeformers. By A. G.O'DONNELL (*Agricultural Research Council, Meat Research Institute, Langford, Bristol BS18 7DY*).

Pyrolysis, a process whereby complex molecules are thermally degraded in an inert atmosphere, has enhanced the use of conventional gas-liquid chromatography by enabling non-volatile compounds to be analysed. This process is therefore applicable to all types of micro-organisms and is sensitive enough to utilize small amounts of materials taken, where appropriate, directly from a culture plate.

This is a report of a study carried out on 32 cultures of aerobic sporeforming bacilli using pyrolysis gas-liquid chromatography (p.g.l.c.). Twenty-four of these strains belonged to the 'subtilis-spectrum' (*Bacillus subtilis, B. pumilus, B. licheniformis*) whilst the remaining 8 strains were *B. amyloliquefaciens*. Each culture was grown overnight at 30°C on nutrient agar containing glucose (2 g/l), harvested before sporulation into sterile distilled water and stored at −4°C.

When the pyrograms were analysed, using canonical variates and stepwise discriminant function, the 32 different strains representative of the 4 individual species clustered to form 4 distinct groups. The ability of p.g.l.c. to mimic successfully a classification derived by other means was tested by applying DNA-DNA hybridization, biochemical tests and A.P.I. systems (see pp. 394) to the strains used in the p.g.l.c. study.

Identification of Medically Important *Corynebacterium* spp. by means of Metabolic End Products detected by Gas-Liquid Chromatography. By J. E. HINE, L. R. HILL and S. P. LAPAGE (*National Collection of Type Cultures, Central Public Health Laboratory, Colindale Avenue, London, NW9 5HT*).

Metabolic end product analysis was used in an attempt to improve the identification of coryneform bacteria submitted to the NCTC for identification. Coryneform bacteria, other than *Corynebacterium diphtheriae*, comprise 20% of the miscellaneous bacteria submitted for identification. Bacteria thought to be *C. diphtheriae* are submitted to the Public Health Laboratory, Swansea. Some 172 strains of coryneforms were received in the period 1962–1975 and 35% identified by conventional means to the species level, 43% to genus or possible species and 22% unidentified.

The end products of glucose metabolism of 200 strains have been examined using the methods described for anaerobic bacteria in the Anaerobe Laboratory Manual, Virginia Polytechnic Institute. Optimal conditions for end product formation have been established and it has been shown that once present in culture fluids, end products are not metabolized further.

Qualitative analysis of the end product profiles of the coryneform bacteria held in the NCTC has revealed three broad groups: (1) *C. diphtheriae* and related species (*C. belfanti, C. ovis, C. ulcerans* and one strain of *C. xerosis*) producing large amounts of acetic, propionic, lactic and succinic acids; (2) a group including *C. minutissimum, C. renale*, some strains labelled *C. xerosis*, the opaque colony form of *C. cervicis, C. haemolyticum* and *C. pyogenes* producing acetic, lactic and succinic acid but not propionic acid; and (3) a group (comprising *C. equi, C. pseudodiphtheriticum, C. aquaticum*, the transparent colony form of *C. cervicis, Arthrobacter* spp. and *Brevibacterium* spp.) forming extremely few end products. The strains held by the NCTC as *C. xerosis* are heterogeneous and the relationships of the NCTC strains of *C. xerosis, C. minutissimum* and other skin diphtheroids are unknown.

The separation of species by a qualitative or semi-quantitative use of the results was sometimes possible, although not for those species in the *C. diphtheriae*-like group. It is possible to separate *C. minutissimum* from *C. renale* and *C. diphtheriae* from *C. pseudodiphtheriticum*. The results for *C. equi* and *C. rubrum* are quite different from those for other species, supporting the transfer of strains of these species to the genus *Rhodococcus*. The results for *C. haemolyticum* and *C. pyogenes* are similar to each other and those for strains named *C. pyogenes* var. *hominis* support the transfer of these strains to *C. haemolyticum*. Strains of *C. diphtheriae* are readily separated from some other strains producing metachromatic granules such as *C. equi*, *C. minutissimum*, strains named *C. xerosis* not producing propionic acid and strains similar to *C. bovis*.

Metabolic end product profiles were used in an attempt to identify field strains of coryneform bacteria. These results were useful, together with the results of conventional tests, where a strain appeared to belong to one of the recognized species of *Corynebacterium*. A problem exists, however, in that a significant proportion of strains of coryneform bacteria isolated from clinical specimens and which are difficult to identify are very heterogeneous and appear to belong to several as yet undefined taxa. This position will not be resolved without considerable additional research.

Quantitative results have also been recorded although they do not yield a high degree of reproducibility. Metabolic profiles are, however, readily qualitatively reproducible. The multivariate analysis methods devised by MacFie & Gutteridge (see pp. 376) have been applied in an attempt to use the quantitative data for species identification. Great problems exist in this regard both because of lack of quantitative reproducibility and as some species are poorly defined.

Capability of Pyrolysis Gas Chromatography in Differentiating Oral Bacteria at Genus, Species, and Sub-species Level. By HELEN D. DONOGHUE*, M. V. STACK and J. E. TYLER (*M.R.C. Dental Unit, University of Bristol Dental School, Bristol BS1 2LY*).

Pyrolysis gas-liquid chromatography (p.g.l.c.) has been shown to be a method capable of distinguishing between strains within a species which differ in their antigenic composition (Reiner, E. 1967 *Journal of Gas Chromatography* 5, 65–67; Huis in't Veld, J. H. J. *et al.* 1973, *Applied Microbiology* 26, 92–97). Flory and co-workers (*Journal of Dental Research* 1977, 56, Special issue B, Abstract 144) reported that serotypes of *Streptococcus mutans* may be differentiated by p.g.l.c. using capillary separating columns which give high resolution. We have demonstrated (Stack, M. V. *et al.* 1978, *Applied and Environmental Microbiology* 35, 45–50) that pyrograms resulting from packed columns, a lower resolution system, were adequate in distinguishing several species of oral streptococci. We

*Present address: School of Medical Sciences, University of Bradford, Bradford BD7 1DP, West Yorkshire.

have now shown that this method may be used to differentiate between sero-types of *Strep. mutans*.

The data comprised 15 peak heights from each pyrogram, each having six replications (three analyses of two cultures grown at different times). A hierarchy of comparisons was assembled based upon these; it ranged from a group of five sets of replicates for one strain up to a group of five organisms of different genera. For each group we calculated coefficients of variation (CV; standard deviations as percentages of means) for the 15 pyrogram peaks; attention centered on CVs of 20% or above.

There were no high CVs for the first group replicate pyrograms for one strain grown on two occasions. When five strains of *Strep. mutans* serotype *c* were examined 3/15 peaks showed high CV; 6/15 were high in the group comprising pyrograms from single strains of five *Strep. mutans* serotypes. Greater differences in the group containing single strains of the five major species of oral streptococci were expressed by 9/15 peaks having high CV. Finally, as many as 12/15 peaks showed CVs exceeding 20% in a group of pyrograms representing five different genera: *Bacterionema matruchotii*, *Rothia dentocariosa*, *Nocardia asteroides*, a *Lactobacillus* sp. and *Escherichia coli*.

The data were subjected to a cluster analysis, and both a dendrogram and minimum spanning tree were constructed. *Streptococcus mitior* and *Strep. sanguis* were confirmed as fairly homogeneous species, in contrast to the hetero-geneity of the species *Strep. mutans*. The current observations confirm the potential of p.g.l.c. in differentiating between bacteria, even at the sub-species level.

The Computer Analysis of G.L.C. Fatty Acid Fingerprints of Some Viridans Group Streptococci. By D. B. DRUCKER*, S. M. LEE, S. B. LUCAS† and V. F. HILLIER‡ (*Department of Bacteriology and Virology, †Faculty of Medicine Computation Group, ‡Department of Community Medicine, University of Manchester, Stopford Building, Oxford Road, Manchester M13 9PT).

Cellular fatty acid fingerprints were obtained for 85 strains of streptococci, mostly of the Viridans Group. The methods for growth, harvesting, preparation of methyl carboxylic esters, g.l.c. analysis and computerized data analysis are well documented (Drucker, D. B. 1976, in *Methods in Microbiology*, Vol. 9, London: Academic Press, pp. 51–126). Peaks were identified by equivalent carbon number and expressed as mean % total area. These data were subject to further computer analysis by the following methods: (i) calculation of a matrix of Pearson Coefficients of Linear Correlation; (ii) production of a matrix of Spearman Coefficients of Rank Correlation; (iii) construction of a matrix of Kendal Coefficients of Rank Correlation; (iv) calculation of a similarity matrix, treating each peak area as a quantitative attribute. Each matrix was separately subjected to cluster analysis by techniques including the following: (i) minimum spanning tree; (ii) single-linkage sorting, with production of a dendrogram; (iii) nearest neighbour analysis. As a comparative yardstick a set of physiological data on each strain was analysed in order objectively to provide names for strains, based on traditional tests.

The latter revealed that *Streptococcus milleri* fell into two distinct biotypes; *Streptococcus mitis* a catch-all species, differentiated into at least 3 biotypes; clusters representing *Streptococcus mutans, salivarius* and *sangius* were observed. Computed g.l.c. data tended to agree with these physiological clusters but further differentiated strains of *Streptococcus milleri*.

The Taxonomic Significance of the MIC of the Vibriostatic Compound 0/129 and Other Agents against Vibrionaceae. By J. V. LEE, PAMELA A. COURT, T. J. DONOVAN and A. L. FURNISS (*Public Health Laboratory, Preston Hall, Maidstone, Kent ME 20 7NH*).

Shewan, J. M., Hodgkiss, W. & Liston, J. (1954, *Nature, London* 173, 208–209) first recommended the use of 0/129 (2,4-diamino-6,7-diisopropyl-pteridine) for the identification of *Vibrio* spp. Since then it has been used widely but at different concentrations by different workers. Novobiocin and methylene blue have also been used. These 3 compounds are all said to be useful for distinguishing *Vibrio* spp. (sensitive) from *Aeromonas* and *Pseudomonas* spp. (resistant). There are no records of the MICs of these compounds for species of *Vibrio* and related genera.

We have determined the MIC of the water-soluble 0/129 phosphate for 487 strains representing the commonly encountered *Vibrio*, '*Beneckea*', *Aeromonas* and *Plesiomonas* spp. The MICs of novobiocin and methylene blue were also determined for 140 of these strains. The compounds were incorporated in Difco heart infusion agar and the medium in Petri dishes inoculated by means of a multipoint inoculator. Normal sensitivity test agars were not used because they do not contain enough electrolytes for marine strains to grow.

Each compound divided these strains into 3 groups conveniently termed sensitive (S), reduced sensitivity (RS), and resistant (R). The MICs (μg/ml) corresponding to these groups for each compound were respectively: 0/129 1–5, 10–50, $\geqslant$ 320; novobiocin 0.1–2.0, 3–20, > 20; methylene blue 2–7.5, 10–150, $\geqslant$ 200. The strains in the 3 sensitivity groups were the same for 0/129 and novobiocin except for group F (Furniss, A. L., Lee, J. V. & Donovan, T. J. 1977, *Lancet* ii, No. 8037, 565–566) which was RS with 0/129 and R with novobiocin. The groupings were not the same for methylene blue. Only *Aeromonas* spp. were 0/129 R and only *Aeromonas* and group F were novobiocin and methylene blue R. The sensitivity groups were most clearly defined with 0/129. For routine identification, discs containing 10 and 150 μg 0/129 distinguished the 3 sensitivity groups. 0/129 not only differentiated *Vibrio* from *Aeromonas* but also helped to identify species within *Vibrio* and '*Beneckea*'.

The Taxonomy of Group F Organisms: Relationships to *Vibrio* and *Aeromonas*.
By J. V. LEE, P. SHREAD and A. L. FURNISS (*Public Health Laboratory, Preston Hall, Maidstone, Kent ME20 7NH*).

Group F has characteristics intermediate between *Aeromonas* and certain *Vibrio* spp. such as *V. anguillarum* (Lee, J. V., Donovan, T. J. & Furniss, A. L. 1978, *International Journal of Systematic Bacteriology* **28**, 99–111). Similar organisms have been isolated from patients with diarrhoea (Furniss, A. L., Lee, J. V. & Donovan, T. J. 1977, *Lancet* **ii**, No. 8037, 565–566). We have carried out a numerical taxonomic study of 154 strains including 59 group F strains and representatives of possibly related species within the Vibrionaceae.

The similarities were computed using 101 characters and cluster analysis done by the nearest neighbour, furthest neighbour and group average methods. The 3 methods gave almost identical clusters. The group F strains all fell in one closely-knit cluster which contained 2 subclusters corresponding to the aerogenic and anaerogenic strains respectively. Both subclusters included strains isolated from the environment but all the strains isolated from patients with diarrhoea were anaerogenic.

Vibrio anguillarum-like strains fell into 3 clusters. One corresponds to *V. anguillarum* as it contained all but one of the working strains of this species. The second cluster included the type strain of '*Beneckea*' *nereida* and may therefore correspond to that species. The third cluster may represent a new species.

Strains of the *A. hydrophila*–*A. punctata* group formed 3 clusters broadly similar to those described by Popoff, M. & Veron, M. (1976, *Journal of General Microbiology* **94**, 11–22). The remaining strains fell into clusters corresponding to *Plesiomonas shigelloides*, *Photobacterium* and *A. salmonicida*.

As a result of this study the routine identification of these groups, which has previously been difficult, should become easier and more reliable. Group F is more closely related to the genus *Vibrio* than *Aeromonas* and we conclude that it should be placed in the former genus.

Fatty Acid Analysis of Selected Species of *Aeromonas*. By J. T. BUCKLEY and T. J. TRUST (*Department of Biochemistry and Microbiology, University of Victoria, Victoria, B.C., Canada*).

The fatty acid composition of 5 strains of *Aeromonas* was determined by gas-liquid chromatography. Cells were grown in tryptone-yeast extract broth at $10\,°C$ and $30\,°C$ aerobically, and anaerobically at $30\,°C$. The major straight-chain cellular acids were 16:1, 16:0 and 18:1, although the 14:0 and 18:0 acids were also present. The major short-chain acids detected in spent culture medium after anaerobic growth were acetic and succinic. The motile and non-motile aeromonads could not be differentiated from each other by this fatty acid analysis. The aeromonads did, however, differ from aquatic species of *Vibrio* which contained the 14:1 and 18:2 acids. The content of the 16:1 acid in the aeromonads was also significantly higher than that reported for species of *Pseudomonas* and *Alcaligenes*. Growth at $10\,°C$ increased the amount of unsaturated acids, while anaerobic growth increased the saturated acids.

Identification of *Bacillus* Species isolated from Antacid Liquids. By M. F. WILLEMSE–COLLINET, G. T. HOSPERS, Th. F. J. TROMP & T. HUISINGA (*Laboratorium voor Farmacotherapie en Receptuur, Rijkuniversiteit, Antonius Deusinglaan 2, Groningen, The Netherlands*).

In pharmaceutical practice a number of antacid liquids happen to be spoiled by micro-organisms. Most of the isolated pure cultures were *Bacillus* spp. To identify those organisms a scheme of biochemical reactions was constructed and tested against NCTC-reference cultures. A profile for each unknown isolated pure culture was compiled. This profile can be used for any additional work in the field of epidemiology of the microbial contamination of pharmaceutical preparations.

A New Approach to the Conceptual Understanding of Bacterial Classification. By D. ROY CULLIMORE (*Microbiology Unit, University of Regina, Regina, Saskatchewan, Canada*).

The conflicting needs of the different interest groups among bacteriologists has created over the years a divergence of philosophies concerning the ways by which bacteria may be classified. Of prime importance in this philosophical debate is the concern for 'condensing' or 'refining' the classification schemes. In Bergey's Manual (8th edition), both philosophies are reflected in the way different genera are now classified. The problem remains, however, of the means by which the student of microbiology may be introduced to these concepts exemplified in the Manual. Clearly, the volume of material and the depth of detail makes it very difficult for the Manual itself to be used as the teaching primer. Instead, some intermediary method for presenting the concepts in a clear and precise form needs to be developed as an introductory stage. This presentation is an attempt to provide such a device. It originates in Orla-Jensen's original concepts of sub-dividing the bacteria into nutritional groups based upon the forms of carbon and nitrogen utilized. As a direct result of developing the device while teaching systematics to third year undergraduate students over the period 1970–1977, many modifications were made. The new system consists of a series of 'generic' charts for each of the major groupings of bacteria listed in the Manual (e.g. Gram positive cocci, Gram negative bacteria). Each 'generic' chart displays the component genera by name in a configuration considered subjectively to reflect their relationship to each other. Thus, genera (or families) shown close to each other may be considered to be more closely related. These 'generic' charts give the student a rapid appreciation of the overall way in which bacterial classification is structured. A further refinement introduced is the application of shade over the 'generic' chart in order to indicate which genera possess a given feature. The shade covers completely the generic names possessing the noted feature and partially covers generic names where only some species possess the feature. For related features (such as Gram positive–Gram negative, or the various flagellar patterns), a comparison of the shaded areas for each feature allows the student to develop a comprehension of the importance of these features to the differentiation of families and genera of bacteria.

In all, a total of sixty-eight 102 × 147 mm cards have been produced including the grouping of eight major divisions of the bacteria into 'generic' charts with a special additional chart for the Enterobacteriaceae. Features incorporated into these charts total 227 for all of the charts and include mainly morphological and biochemical characteristics with a varying emphasis on cultural and environmental considerations. Students who have been taught using this concept during its development have shown an increased level of comprehension for bacterial systematics. It is to be hoped that this system will encourage more bacteriologists to identify bacteria rather than to veer around the topic (as happens so very often today).

Application of Canonical Variates to Clusters formed by Multidimensional Scaling.
By G. J. BONDE (*Institute of Hygiene, University of Aarhus, DK-8000 Aarhus, Denmark*).

Canonical variate, in this context, stands for 'authoritative, ideal coordinate', approved by long experience, and the purpose of applying canonical variates is to: (1) gain insight into the way in which clusters differ; (2) grade the tests according to significance; (3) reduce the number of tests necessary; and (4) allocate new strains to preformed groups.

A prerequisite is a preliminary grouping, e.g. by computer applying a large number of tests. The author's method of choice for clustering is Kruskal's non-metric multidimensional scaling (cf. Bonde, G. J. 1976, Kruskal's non-metric multidimensional scaling *Proceedings in Computational Statistics*, pp. 443–449, ed. J. Gordesch & P. Naeve, Wien: Physica Verlag). In the present investigation $N = 231$ *Bacillus* strains which were split into $k = 10$ groups by analysis of $m = 75$ characters in a 3-dimensional plot by this procedure.

The principle of computation is to maximize the ratio of the sum of squares between groups to the sum of squares within groups. It is attempted to replace the m-dimensional estimate of strain j in group i by test v by one single estimate, Y_{ij}:

$$Y_{ij} = \Sigma_v c_v X_{ijv},$$

i.e. the weighted sum of the individual estimates X_{ijv}, weighted by the coefficients c_v, which are chosen in such a manner as to maximize F_c:

$$F_c = \frac{\Sigma_i \Sigma_j (Y_{i.} - Y_{..})^2}{\Sigma_i \Sigma_j (Y_{ij} - Y_{i.})^2}$$

For k groups, $k - 1$ discriminators c_v (= eigenvalues) may be defined, and the contribution to each of these by the individual characters (on a relative scale) is computed (eigenvectors). The first canonical variate is the highest latent root of the matrix of groups of strains and characters, and gives the best possible discrimination from one linear function, the second is the next function which is uncorrelated with the first—and so on for $k - 1$ variates. If most of the variation between groups were given by the first two canonical variates a plot of these would give a satisfactory division. In our data, however, application of the first (or second) variate *vs* the third and fourth gives a sufficient separation of all clusters. The first four eigenvalues correspondingly explain 90% of information of the data, and the 13 characters giving a dominating contribution to these first eigenvalues, will suffice for a division into ten groups. Twenty-eight characters were of no use for classification.

The 13 necessary characters concern: breakdown of glucose, starch, arabinose and mannitol; Voges-Proskauer, ONPG, lecithinase and lysine decarboxylase tests; fat inclusions, dimension of spores, formation of gas, anaerobic growth in 10% (w/v) NaCl. Nine of the groups correspond to: *B. sphaericus*, *B. megaterium*, *B. cereus*, *B. pumilus*, *B. subtilis*, *B. licheniformis*, *B. polymyxa* and *B. circulans*. The method can also allocate unknown strains to preformed groups, but four strains (*B. similibadius* of Delaporte) in this examination formed a new (tenth) cluster.

Heterogeneity of the Ribosomal RNA Cistrons of Aerobic Nitrogen Fixing Bacteria and the Implication for their Taxonomy. By J. DE SMEDT and J. DE LEY (*Laboratorium voor Microbiologie en microbiële Genetica, Rijksuniversiteit Gent, K.L. Ledeganckstraat 35, B-9000 Gent, Belgium*).

Similarities of ribosomal RNA cistrons are valuable parameters for the classification of bacteria on the generic and suprageneric level (De Smedt, J. & De Ley, J. 1977, *International Journal of Systematic Bacteriology* 27, 222). DNA:rRNA hybrids were prepared between DNA from a great variety of bacteria and radioactively labelled rRNA from reference strains of *Azotobacter*, *Azomonas* and *Beijerinckia*. The hybrids were described by their $T_{m(e)}$ and % rRNA binding. These parameters were used to plot rRNA similarity maps. The following conclusions could be drawn: (1) Free-living nitrogen bacteria are phylogenetically very heterogeneous; they belong to different rRNA superfamilies. (2) *Azotobacter*, *Azomonas* and the *Pseudomonas* section I group are closely related; they belong in one rRNA family. (3) *Derxia* belongs to another rRNA superfamily together with *Chromobacterium*, the *Pseudomonas* section II and III groups and *Alcaligenes*. (4) *Beijerinckia* belongs to a third rRNA superfamily together with *Agrobacterium*, *Rhizobium*, *Acetobacter*, *Gluconobacter*, several nitrogen-fixing photosynthetic and nitrogen-fixing hydrogen bacteria and *Spirillum lipoferum*.

Heterogeneity of *Pseudomonas* rRNA Cistrons and its Significance for Taxonomy. By P. DE VOS and J. DE LEY (*Laboratorium voor Microbiologie en microbiële Genetica, Rijksuniversiteit Gent, K.L. Ledeganckstraat 35, B-9000 Gent, Belgium*).

The rRNA cistrons of many *Pseudomonas* spp. and of a great variety of Gram negative bacteria were compared by DNA:[14]C-rRNA hybridization. The technique of De Ley, J. & De Smedt, J. (1975, *Antonie van Leeuwenhoek Journal of Microbiology and Serology* **41**, 287–307) was used. Our results demonstrate that the genus *Pseudomonas*, as described in Bergey's Manual, 8th edn, is very heterogeneous and that it consists of five groups belonging in three rRNA superfamilies. The *Pseudomonas* spp. from Section I and *Ps. maltophilia* belong to an rRNA superfamily also containing *Azotobacter, Azomonas, Xanthomonas* (except 'X. ampelina'), and *Aplanobacter populi. Pseudomonas maltophilia* is more closely related to *Xanthomonas* than to *Pseudomonas* Section I. The *Pseudomonas* spp. from Section II and III belong in a second rRNA superfamily in which at least two different *Pseudomonas* rRNA groups can be recognized: (1) *Ps. acidovorans, Ps. testosteroni, Ps. delafieldii, Ps. facilis, Ps. palleronii*, etc. and (2) *Ps. solanacearum, Ps. cepacia, Ps. marginata*, etc. *Pseudomonas diminuta* is not a member of this genus; it belongs in an rRNA superfamily together with the acetic acid bacteria, *Agrobacterium, Rhizobium*, etc. With our technique many of the unclassified, mostly phytopathogenic *Pseudomonas* spp. could be assigned to one or another of the rRNA groups. Our results suggest that the classification of the genus *Pseudomonas* can be improved.

A Polyphasic Taxonomic Study of the Genus *Actinomadura* and Related Taxa. By GRACE ALDERSON *, M. GOODFELLOW *, D. E. MINNIKIN †, (*Departments of Microbiology * and Organic Chemistry †, The University, Newcastle upon Tyne NE1 7RU*) and J. LACEY (*Rothamsted Experimental Station, Harpenden AL5 2JQ*).

Actinomadura contains aerobic, Gram positive, non-acid fast actinomycetes which form a branched stable substrate mycelium and may exhibit an aerial mycelium carrying short chains of arthrospores. The organisms contain *meso*-diaminopimelic acid (wall chemotype III), may produce the sugar madurose (3-*o*-methyl-D-galactose) but lack mycolic acids (Goodfellow, M. & Minnikin, D. E. 1977, *Annual Review of Microbiology* **31**, 159).

The genus proposed by Lechevalier, H. A. & Lechevalier, M. P. (1970, in *The Actinomycetales*, pp. 393–405, ed. Prauser, H., Jena: Gustav Fischer Verlag) for strains described as *Nocardia dassonvillei, N. madurae* and *N. pelletieri* (Gordon, R. E. 1966, *Journal of General Microbiology* **50**, 235), was initially classified in the family Thermoactinomycetaceae but later transferred to the family Thermomonosporaceae because endospores were not produced (Cross, T. & Goodfellow,

M. 1973, in *The Actinomycetales: Characteristics and Practical Importance*, pp. 111-112, ed. Sykes, G. & Skinner, F. A., London: Academic Press). Since the erection of the genus many new species have been recognized, mainly on the basis of morphology and wall chemotype (Lacey, J., Goodfellow, M. & Alderson, G. 1978, in *Nocardia and Streptomyces*, pp. 107-117, ed. Mordarski, M., Kuryłowicz, W. & Jeljaszewicz, J., Stuttgart: Gustav Fischer Verlag) and similar isolates from mouldy hay and barley grain have been classified provisionally as *Actinomadura* strains on morphological criteria (Lacey, J. 1978, as above, pp. 161-170). Systematic studies are required to determine the relationships between the species of *Actinomadura*.

In the present study 156 *Actinomadura* and related strains were examined for 90 unit characters and the data analysed by computer using the S_{SM}, S_j and D_p coefficients and the average and single linkage algorithms (Goodfellow, M. 1977, in *CRC Handbook, 2nd Edition, Volume 1, Bacteria*, pp. 579-596, ed. Laskin, A. I. & Lechevalier, H. A., Cleveland: CRC Press). The same major and minor clusters were obtained in all of the analyses and cluster representatives were examined for the presence of isoprenoid quinones (Collins, M. D. *et al.* 1977, *Journal of General Microbiology* **100**, 221), polar lipids (Minnikin, D. E. & Abdolrahimzadeh, H. 1971, *Journal of Chromatography* **63**, 452) and fatty acids (Alshamaony, L. *et al.* 1977, *Journal of General Microbiology* **98**, 205).

Four of the six major clusters obtained were equated with the established taxa *Actinomadura dassonvillei*, *A. madurae*, *A. pelletieri* and *Streptomyces somaliensis*, the remaining ones contained grain and fodder isolates and were provisionally labelled *Actinomadura* A and *Actinomadura* B. The single representatives of *A. helvata*, *A. pusilla*, *A. roseoviolacea*, *A. spadix* and *A. verrucosospora* seemed to form nuclei of distinct clusters. It was possible to distinguish the species of *Actinomadura* on the basis of chemical, enzymic, degradative and nutritional properties.

Lipids in the Classification of Coryneform Bacteria. By M. D. COLLINS*, M. GOODFELLOW† and D. E. MINNIKIN‡ (**Department of Microbiology, University of Leicester, Leicester; Departments of Microbiology† and Organic Chemistry‡, The University, Newcastle upon Tyne NE1 7RU*).

In an attempt to clarify the classification of coryneform bacteria, mycolic acid and isoprenoid quinone composition has been studied. Mycolic acids (2-branched, 3-hydroxy long-chain acids) were investigated by thin-layer chromatography of whole-organism methanolysates followed by mass spectrometry of the purified methyl esters. Representatives of *Arthrobacter sensu stricto*, *Brevibacterium sensu stricto*, *Brochothrix*, *Cellulomonas*, *Curtobacterium*, *Erysipelothrix*, *Kurthia*, *Listeria*, *Microbacterium*, *Mycoplana*, *Oerskovia* and *Protaminobacter* did not contain mycolic acids. Small mycolic acids (22-36 carbons) were present in animal-associated corynebacteria and *Corynebacterium glutamicum* and related organisms but some strains contained mycolates similar in size (32-56 carbons) to those found in certain rhodococci.

Menaquinones, analysed by mass spectrometry, were found in all strains with the exception of *Brevibacterium leucinophagum, Corynebacterium autotrophicum, Corynebacterium nephridii, Mycobacterium flavum, Mycoplana rubra* and *Protaminobacter ruber* which contained ubiquinones. Mycolic acid-containing taxa had dihydromenaquinones with eight or nine isoprene units [MK-$8(H_2)$ or MK-$9(H_2)$]. *Arthrobacter sensu stricto* also had MK-$9(H_2)$ but *A. simplex* and *A. tumescens* had MK-$8(H_4)$. Tetrahydrogenated menaquinones [MK-$9(H_4)$] were also found in cellulomonads and oerskoviae. Fully unsaturated menaquinones were characteristic of *Kurthia* (MK–*7*), *Curtobacterium* (MK–*9*), *Corynebacterium aquaticum* (MH–*10, 11*) and *Microbacterium lacticum* (MK–*10, 11*).

Numerical and Chemical Classification of Actinomycetes with a Wall Chemotype III. By TAHEREH PIROUZ*, M. GOODFELLOW* and D. E. MINNIKIN† (*Departments of Microbiology* and Organic Chemistry†, The University, Newcastle upon Tyne NE1 7RU*).

Actinomycetes containing *meso*-diaminopimelic acids but no characteristic sugars in their walls (wall chemotype III of Lechevalier, M. P. & Lechevalier, H. A. 1970, *International Journal of Systematic Bacteriology* **20**, 435–444) can be classified into twelve genera primarily on the basis of morphological criteria (Cross, T. & Goodfellow, M. 1973, in *The Actinomycetales: Characteristics and Practical Importance*, pp. 111–112, ed. Sykes, G. & Skinner, F. A., London: Academic Press). As no systematic study has been made on representatives of these taxa little is known of their taxonomic affinities.

In the present investigation over 100 strains with a wall type III were examined, with markers representing the taxa *Micropolyspora, Micromonospora, Nocardia* and *Streptomyces somaliensis* for 108 unit characters. The data were analysed by computer using the S_{SM}, S_J and D_p coefficients and the average and single linkage algorithms (Goodfellow, M. 1977, in *CRC Handbook, 2nd edition, Volume 1, Bacteria*, pp. 579–596, ed. Laskin, A. I. & Lechevalier, H. A., Cleveland: CRC Press). Confidence can be placed in the numerical classification as good congruence was found between the results of the numerical analyses and the test error, *p*, was only 3%. Representatives of the genera *Actinomadura, Dermatophilus, Geodermatophilus, Microbispora, Microtetraspora, 'Nocardiopsis', Planomonospora, Planobispora, Spirillospora, Streptosporangium, Thermomonospora* and *Thermoactinomyces* showed little relationship either to one another or to the marker strains, indicating that actinomycetes with a wall type III form an heterogeneous assemblage. The genus *Micropolyspora* was also found to be heterogeneous with little similarity found between *M. angiospora, M. brevicatena, M. caesia* and *M. faeni*. An attempt was made to evaluate the numerical classification by examining representative strains of some of the defined clusters for diagnostic isoprenoid quinones (Collins, M. D. *et al.* 1977, *Journal of General Microbiology* **100**, 221–230), fatty acids (Alshamaony, L. *et al.* 1977, *Journal of General Microbiology* **98**, 205–213) and polar lipids (Minnikin, D. E. & Abdolrahimzadeh, H. 1971, *Journal of Chromatography* **63**, 452–454).

Preservation of Actinomycete Inoculum in Frozen Glycerol. By E. M. H.
WELLINGTON and S. T. WILLIAMS (*Botany Department, Liverpool University, Liverpool L69 3BX*).

A simple and efficient method for preservation of inoculum of actinomycetes
was tested. Dense cell or spore suspensions in 10–20% (v/v) glycerol were frozen
at $-20°C$. These served as a means of long-term preservation of cultures while at
the same time providing a convenient source of inoculum for a variety of purposes. Viability of test strains frozen in glycerol for one year compared very
favourably with that in soft agar or lyophilized preparations stored at $4°C$. The
viability of strains subject to 20 freeze-thaw cycles decreased but the final concentration of live cells was still sufficient to provide an adequate source of
inoculum. Results indicated that a concentration of 10% (v/v) glycerol was the
most satisfactory.

Fatty Acids and Classification of Actinomycetales. By R. M. KROPPENSTEDT
and H. J. KUTZNER (*Deutsche Sammlung von Mikroorganismen, Teilsammlung Darmstadt Schnittspahnstraße 9, D-6100 Darmstadt, West Germany*).

Members of the order Actinomycetales differ significantly in their pattern of
fatty acids as these can be determined by gas-liquid chromatography after transesterification. The following Table summarizes in a very simplified manner the
results of our study of more than 300 organisms representing all genera of this
order.

Considering only four types of fatty acids and neglecting quantitative differences one can recognize 4 patterns: (1) cyclopropane fatty acids were only
found in *Actinomyces*; (2) tuberculostearic acid and its precursor oleic acid are
characteristic of *Mycobacterium* and *Nocardia*; (3) *Streptomyces* contains large
quantities of *iso*- and *anteiso*-branched fatty acids, and (4) the fourth pattern
contains three of the four selected types of acids: tuberculostearic acid in
Actinomadura, and 10-methyl-heptadecanoic acid in *Nocardia autotrophica*,
Micromonospora and 4 genera of *Actinoplanaceae*.

TABLE 1
Four main fatty acid patterns occurring among
Actinomycetales

Type	Iso & anteiso branched	unsaturated	10-Methyl- branched	Cyclopropane
1	–	+	–	+
2	–	+	+	–
3	+	(v)	–	–
4	+	+	+	–

Studies on the Physiology of Actinomycetes. By R. HAMMANN and H. J. KUTZNER *(Deutsche Sammlung von Mikroorganismen, Teilsammlung Darmstadt Schnittspahnstraße 9, D-6100 Darmstadt, West Germany).*

In the course of our research for physiological criteria for the characterization of actinomycetes we investigated their ability to degrade aromatic compounds. A simple screening on agar medium showed that numerous members of *Nocardia* and *Mycobacterium* utilized benzoate (B) and *p*-hydroxybenzoate (phB). Only 7 species of *Streptomyces* of the 160 tested gave a positive reaction with B; among those were 4 producing sclerotia: 'genus *Chainia*'. However, 38% of the 160 streptomycetes degraded phB. The investigation of the mechanism of the splitting of the aromatic ring of catechol (C) and protocatechuate (P) showed that the organisms used the *beta*-ketoadipate pathway; only one strain degraded C via *meta*-cleavage, P in the *ortho* position. *Meta*-hydroxybenzoate was utilized via gentisate. Cell free extracts of 4 selected organisms showed activity of C-1,2-oxigenase (CO) and P-3,4-oxigenase (PO). In *Streptomyces* CO was only found in cells grown on B, and PO only in cells grown on phB. On the contrary, in *Nocardia opaca* B-cells oxidized C as well as P; phB-cells, however, showed only PO activity.

Thermophilic Actinomycetes Implicated in Farmer's Lung: Numerical Taxonomy of *Thermoactinomyces* species. By B. A. UNSWORTH and T. CROSS *(School of Biological Sciences, University of Bradford, Bradford BD7 1DP).*

The morphological and biochemical properties of 184 strains in the genus *Thermoactinomyces* (Tsiklinsky, P. 1899, *Annales de l'Institut Pasteur* **13**, 500–504) were determined. The strains included isolates from a wide variety of substrates and from culture collections. The test results were analysed numerically: five major clusters were formed with the Simple Matching Coefficient and Group Average Linkage of Sokal and Mitchener. Four clusters could be identified with previously described species, namely *Thermoactinomyces* (*Actinobifida*) *dichotomica* (Krassilnikov, N. A. & Agre, N. S. 1964, *Mikrobiologiya* **33**, 935–943), *Tha. sacchari* (Lacey, J. 1971, *Journal of General Microbiology* **66**, 327–338), *Tha. vulgaris* (Tsiklinsky 1899) and *Tha. thalpophilus* (Waksman, S. A. & Corke, C. T. 1953, *Journal of Bacteriology* **66**, 377–378). It is proposed that the name *Tha. vulgaris* be restricted to those strains unable to produce amylase (amyl −ve) so corresponding to Tsiklinsky's original description; such strains do not form a melanin pigment from L-tyrosine (mel -ve). *Tha. candidus* (Kurup, V. P., Barboriak, J. J., Fink, J. N. & Lechevalier, M. P. 1975, *International Journal of Systematic Bacteriology* **25**, 150–154) then becomes a junior synonym of *Tha. vulgaris*. It is also proposed that the name *Tha. thalpophilus* should be resurrected for mel +ve, amyl +ve thermoactinomycetes with sessile spores. Such strains had recently been included in the *Tha. vulgaris* complex. The species *Tha. thalpophilus* would now include strains such as A64 implicated in Farmer's Lung Disease (Corbaz, R., Gregory, P. H. & Lacey, M. E. 1964, *Journal of General*

Microbiology **32**, 449–455) and CUB 76 in which the spores were shown to be endospores (Cross, T., Walker, P. D. & Gould, G. W. 1968, *Nature, London* **220**, 352–354). A new specific name *Tha. putidus* is proposed for strains in the remaining cluster. These organisms are mel +ve, amyl +ve but have spores on sporophores and form a very vigorous, wrinkled colony, often with a yellowish brown reverse, on CYC agar (Cross, T. & Attwell, R. W. 1974, in *Spore Research* 1973, ed. Barker, A. N., Gould, G. W. & Wolf, J., pp. 11–20. London: Academic Press) used routinely for isolation and maintenance. The 5 species can be identified by a simple series of tests as outlined below:

	dichotomica	putidus	sacchari	vulgaris	thalpophilus
Aerial mycelium colour	Yellow	White	(White)	White	White
Sessile spores	–	–	–	–	+
Dichotomous branching	+	–	(+)	–	–
Amylase } production	+	+	+	–	+
Melanin } production	–	+	–	–	+
Aesculin } degradation	–	+	+	+	–
Arbutin } degradation	–	+	+	+	+

Tha. vulgaris, Tha. thalpophilus, Tha. sacchari and *Tha. putidus* are serologically distinct. There is already evidence that *Tha. vulgaris* strains (mel –ve, amyl –ve) are important in Farmer's Lung Disease (Greatorex, B. D. and Pether, J. U. S. 1975, *Journal of Clinical Pathology* **28**, 1000–1002; Kurup *et al.* 1975) and strains have been found to be common in soil and hay samples. It is suggested that representatives of each species should be included in tests for Farmer's Lung Disease and related hypersensitivity pneumonitides.

Thermophilic Actinomycetes Implicated in Farmer's Lung: Application of Serological Methods to Resolve a Problem of Nomenclature in the Genus *Micropolyspora*. By M. P. ARDEN JONES, A. J. MCCARTHY and T. CROSS (*School of Biological Sciences, University of Bradford, Bradford BD7 1DP*).

Micropolyspora faeni was the specific name proposed for the thermophilic actinomycete common in mouldy fodder (Cross, T., Maciver, A. M. & Lacey, J. 1968, *Journal of General Microbiology* **50**, 351–360) and chiefly responsible for Farmer's Lung, a hypersensitivity disease of agricultural workers. An earlier name, *Thermopolyspora polyspora*, was considered inappropriate and illegitimate because the species description (Henssen, A. 1957, *Archiv für Mikrobiologie* **26**, 373–414) was based on contaminated cultures. A thermophilic actinomycete had previously been isolated from soil and named *Thermopolyspora rectivirgula* (Krassilnikov, N. A. & Agre, N. S. 1964, *Hindustan Antibiotics Bulletin* **7**, 1–17). The organism was later suggested to be a member of the genus *Micropolyspora* (Lechavelier, H. A. & Lechevalier, M. P. 1967, *Annual Review of Microbiology* **21**, 71–100) and has never been implicated as a causative agent of respiratory disease.

Five recently acquired strains of *Mip. rectivirgula* and 6 strains of *Mip. faeni* were compared. Purified extracts of the type strains of both species were used to produce specific antisera in rabbits. Two-dimensional immunoelectrophoresis was employed to achieve maximum resolution of the antigen-antibody complexes. Approximately 40 antigens were precipitated as peaks when *Mip. faeni* and *Mip. rectivirgula* extracts were tested against their respective antisera. When the extracts were tested against heterologous antisera a very high percentage of cross-reacting antigens was detected. The same extracts were also tested against pooled Farmer's Lung sera and patterns of marked similarity were produced. This cross-reactivity indicates a very close serological relationship between *Mip. faeni* and *Mip. rectivirgula*. Further tests on the 11 strains could demonstrate no significant serological differences between the two species.

The results of 59 morphological, physiological and biochemical tests applied to the eleven strains also support the contention that they belong to a single species. The specific epithet *rectivirgula* has priority but it is our intention to propose that the name *Micropolyspora faeni* be retained as a *nomen conservandum*. Retention of the epithet *faeni* will avoid confusion, especially in the fields of medical and veterinary science. The importance of this is emphasized by the relatively large number of publications (>200) referring to *Mip. faeni* in the context of its role as a disease-producing organism.

Separation of Exponential Growth Exotoxins of *Bacillus cereus* and their Preliminary Characterization. By J. M. KRAMER, P. C. B. TURNBULL, KIRSTEN JØRGENSEN*, JENNIFER PARRY and R. J. GILBERT (*Food Hygiene Laboratory, Central Public Health Laboratory, Colindale, London, NW9 5HT, and *Institute of Hygiene and Microbiology, Royal Veterinary and Agricultural University, DK-1870 Copenhagen V, Denmark*).

During the logarithmic growth phase *Bacillus cereus* elaborates several non-dialysable exotoxins, including an enterotoxin produced to a characteristic degree by any strain, and which is thought to be a contributory factor in both the diarrhoeal-type of food poisoning and non-gastrointestinal ('clinical') infections associated with *Bacl cereus*. This report describes the use of a preparative flat-bed electrofocusing (EF) technique for separating these metabolites and attempts to define their individual roles in the pathogenesis of the organism.

By this method of analysis a single zone containing the enterotoxin could be resolved and eluted. The eluant produced fluid accumulation with necrosis in ligated rabbit ileal loops and altered vascular permeability with necrosis in the rabbit skin test. These activities were readily separated from the phospholipolytic and heat-labile haemolytic (cereolysin) factors. An adjacent heat-stable (HS) haemolysin peak, closely related in molecular size and charge, could be only partially resolved by EF from the enterotoxin, the latter retaining a residual low titre (*ca.* 1:10) of HS haemolytic activity. The enterotoxin was identified as a protein of pH 4.9 and mol. wt. (on the basis of Sephadex G-100 studies) *ca.* 50 000 daltons. Ultra-violet absorption and staining properties were characteristic. The molecule was thermolabile, sensitive to extreme pH environments, and susceptible to pronase and trypsin digestion.

The relationship of the enterotoxin to the lethal factor remains unclear. With strain 4433/73, two factors lethal to mice could be separated by EF. The first was closely associated, and possibly identical with, the enterotoxin. A second, more rapidly lethal peak was separate and distinct from this and was focused between pH 6.4 and 7.0 in a position coincident with the cereolysin and phospholipase activities. Analysis of another strain (671/78) strongly enterotoxigenic by the vascular permeability assay, showed, however, mouse lethal activity to be present only in the former, enterotoxin-associated, position. Rabbit immune serum prepared against the peak enterotoxin fraction of this second strain protected against the vascular permeability and mouse lethal activities of homologous fractions and the enterotoxigenic fractions of 4433/73, and also of the crude culture filtrates of seven other unrelated strains.

Trimethylsilyl Sugar Fingerprint of *Streptococcus mutans* NCTC 10832. By H. S. A. ALUYI and D. B. DRUCKER (*Department of Bacteriology and Virology, University of Manchester*).

Since the realization that Lancefield's serogrouping of streptococci is based on surface polysaccharide antigens, the study of bacterial polysaccharides has become increasingly important in bacterial taxonomy. Gas-liquid chromatography (g.l.c.) has proved a useful method for such studies.

In this study, the best conditions for the g.l.c. analysis of the sugar profile of *Streptococcus mutans* NCTC 10832 were investigated. The cells were grown in Brain Heart Infusion for 48 h at $37\,^{\circ}$C. The polysaccharides were extracted by methanolysis, silylated and analysed on PYE 104 Gas Chromatograph in glass columns packed with 3% S.E. 30 on Chromosorb W HP. The effects of various cultural conditions on the sugar profile were investigated.

Methanolysis in contrast to aqueous hydrolysis resulted in greater recovery of sugars and better resolution of mannose and galactose peaks. Ketose sugars were, however, not detected after methanolysis. Best resolution was achieved with temperature programming of $80\,^{\circ} \times 2\,^{\circ}$/min up to $250\,^{\circ}$C maximum, and carrier gas flow rate of 25 ml/min.

Glycerol, rhamnose, glucose and *N*-acetyl glucosamine together formed over 93% of the total peak areas in the profile. The remainder consisted mainly of xylose, galactose and *N*-acetylmuramic acid. Only quantitative differences were found in the profile of whole cells and cell walls. Cells grown at various temperatures ranging from 30 to $41\,^{\circ}$C showed no significant differences in their sugar profiles. Similar results were observed in cells grown under an atmosphere of CO_2 and under anaerobic conditions. Aeration of the culture decreased the glucose component.

Cells grown for varying lengths of time showed little variation in their sugar profiles. However, increasing the concentration of glucose, sucrose, maltose and lactose in the medium resulted in increase of the glucose component of the profile.

These results indicate that g.l.c. sugar profiles might be useful in bacterial taxonomy.

A Numerical Taxonomic Study of the Genus *Shigella*. By CHRISTINE E. R. DODD and DOROTHY JONES (*Department of Microbiology, The University, Leicester*).

One hundred and ten strains from a variety of sources representing the 4 main species of the genus *Shigella*, together with 68 strains from the rest of the Enterobacteriaceae were subjected to 200 morphological, physiological and biochemical tests and the results analysed with the aid of a computer. All the *Shigella* strains grouped into 4 main clusters: (1) *Shig. sonnei*, (2) *Shig. flexneri*, (3) *Shig. boydii*, (4) *Shig. dysenteriae*. The last 3 species showed a greater similarity to each other than any one of them did to *Shig. sonnei*.

Of the other Enterobacteriaceae, *Escherichia alkalescens* strains showed the closest relationship to the genus *Shigella*. Strains labelled *Providencia* fell into two distinct groups which could be equated with *Prov. alcalifaciens* and *Prov. stuartii*. This is in agreement with the results of McKell, J. & Jones, D. (1976, *Journal of Applied Bacteriology* 41, 143-161).

The composition of the 4 *Shigella* groups was discussed in the context of the source of strains (especially clinical isolates) and serological designations.

Gas Chromatography/Mass Spectrometry in the Diagnosis of Infections caused by *Mycobacterium* and *Nocardia* spp. By L. LARSSON *, P.-A. MÅRDH† and G. ODHAM ‡ (**Institute of Technical Analytical Chemistry, †Institute of Medical Microbiology and ‡Laboratory of Ecological Chemistry, University of Lund, Lund, Sweden*).

The detection and differential diagnosis of mycobacteria isolated from clinical specimens is both time-consuming and tedious. In order to obtain a more rapid diagnosis of mycobacterial infections than is possible by conventional diagnostic techniques, gas chromatography (GC) has been applied. Analysis of serum and cerebrospinal fluid specimens from patients with such infections have yielded characteristic chromatographic elution profiles. Chromatographic patterns, typical of given species or subspecies, have also been obtained by GC analysis of mycobacterial cellular constituents. We have applied the combinations of GC and mass spectrometry (MS) for detection and characterization of mycobacteria and *Nocardia* spp.

GC/MS was adapted for selective ion monitoring (SIM), focusing on fragments characteristic to cellular compounds present exclusively in mycobacteria, *Nocardia* spp. and a few other organisms of the order Actinomycetales, e.g. 10-methyloctadecanoic acid (tuberculostearic acid). Chromatograms obtained when analyzing methanolysates of whole mycobacterial cells contained only one peak, *viz*. that of methyl tuberculostearate. When Löwenstein–Jensen medium was used for culturing, the presence of mycobacteria could be established within a few days of incubation, i.e. even before colonies were visible to the naked eye.

For species and intra-species differentiation, lyophilized mycobacterial cells were methanolyzed and derivatized using trifluoroacetic anhydride. The areas below 6 prominent peaks, present in each chromatogram, were determined and used in cluster analysis based on euclidean distances. Each mycobacterial species studied yielded characteristic and reproducible GC patterns. In this respect, *Mycobacterium avium* proved an exception; 23 strains of this species produced 5 different elution profiles. The application of cluster analysis was found valuable as a complement to visual interpretation of chromatograms.

SIM is well suited to the detection of mycobacteria, even when they arise in mixed cultures or in amounts of only a few nanograms. Whether the exclusive use of GC analysis of trifluoroacetylated whole-cell methanolysates can differentiate all known mycobacterial species still requires further studies.

Results of an International Reproducibility Trial Using the API System Applied to the Genus *Bacillus*. By N. A. LOGAN, R. C. W. BERKELEY and J. R. NORRIS* (*Department of Bacteriology, The Medical School, University of Bristol, University Walk, Bristol BS8 1TD and *Meat Research Institute, Langford, Bristol BS18 7DY*).

Evidence from a number of sources indicates that the results of many tests classically used in *Bacillus* taxonomy are poorly reproducible between laboratories. In a number of instances the problems lie with the test medium. In an attempt to get round these difficulties, tests using standardized materials from API Ltd. have been examined in an international trial involving 6 laboratories which performed 120 tests on 60 coded, representative strains of *Bacillus*. The collected results were subjected to an analysis of variance.

Of the 120 tests, 61 of which were designed for the identification of enterobacteria, only 18 appeared to have no diagnostic value for *Bacillus*. Twelve tests of the API 20E strip were used (the remaining 8 are duplicated in the 50E gallery). The reproducibility of 6 tests was good but 2, Voges–Proskauer and nitrate reduction, which require the addition of reagents, were less consistent and further standardization and more care in the reading of results was indicated. Four tests had no diagnostic value.

In the 50E gallery, comprising 49 tests and one control, 31 tests gave good reproducibility and 2 had no diagnostic value. The results obtained with the remainder were not satisfactory due, probably, to disparities in judgment of end points.

Fifty-nine enzyme tests were studied of which 27 gave consistent results and 12 had no diagnostic value. It is probable that the difficulties encountered with the remaining enzyme tests were due to variations in inoculum size, problems in end point assessment and photolability of some of the substrates.

Many of the factors underlying test inconsistency were highlighted by this study and steps were taken to eliminate them in a smaller, intra-laboratory trial. The 120 tests were each made 3 times on 20 representative strains and the results subjected to an analysis of variance. Reproducibility was improved throughout, but the overall pattern given by the 20E strip was unchanged and the Voges–Proskauer and nitrate reduction tests still caused some difficulty. In the 50E gallery, 46 tests gave good reproducibility and 3 appeared to have no diagnostic value. Fifty-two of the 59 enzyme tests gave consistent results, 2 were poorly reproducible and 5 appeared to be of no diagnostic value. The improvement in the reproducibility of the enzyme tests was largely due to careful standardization of the inoculum.

In many cases where between-laboratory inconsistency occurred, it is clear that more detailed specification of procedures and careful attention to directions for performing tests would reduce discrepancies in results. In the light of these reproducibility trials and other work it is concluded that the API system will provide a sound basis for a taxonomy of the genus *Bacillus*.

Subject Index

Acetobacter, 131, 384
Acbc. xylinum, 117, 118
Achloeplasma, 279
Achromobacter, classification of, 288
Acho. butyri, 288
Acho. halophilus, 288
Acho. hartlebii, 288
Acho. liquefaciens, 111, 122
Acho. viscosus, 288
Acho, xylosoxidans, 283, 284, 285,286, 288, 292
Acid-fast bacteria, *see also* individual organisms
  acylated trehaloses of, 227–230
  chemotaxonomy of, 192–195
  diagnostic lipids in, 234–237
  isoprenoid quinones of, 211–212
  mycolic acids of, 195–205
  mycosides and peptidolipids of, 220–227
  numerical taxonomy of, 190–192
  non-hydroxylated fatty acids of, 205–210
  phthiocerols of, 231–232
  polar lipids of, 212–220
Acinetobacter, 117, 119, 120
  classification of, 59–65
Acnb. anitratus, 111
Acnb. calcoaceticus, 59, 111
Acnb. lwoffi, 111, 114, 120, 121
Actinomadura, 142, 143, 158, 192, 193, 385, 387, 388
Actm. dassonvillei, 386
Actm. helvata, 386
Actm. madurae, 386
Actm. pelletieri, 386
Actm. pusilla, 386
Actm. roseoviolacea, 386

Actm. spadix, 386
Actm. verrucosospora, 386
Actinomyces, 140, 158, 388
Actinomycetes, *see also* individual organisms
  cell wall composition of, 141–143
  DNA reassociation of, 154–155
  fatty acids of, 388
  fine structure of spore surface of, 158
  genetic control of morphology in, 153–154
  morphology of, 142–143
  numerical taxonomy of, 154, 387
  spore development in, 156–157
  stability and reproducibility of morphological characters in, 143–145
  spore chain morphology, 145–149
  spore surface ornamentation, 145–152
Actinoplanes, 140, 142, 157, 388
Actinopolyspora, 142, 193
Actinopolyspora halophila, 200, 210, 219, 235
Aerobacter aerogenes, see also Klebsiella aerogenes, 119
Aeromonas
  fatty acid analysis of, 381
  numerical taxonomy of, 87, 90
  sensitivity of, to 0/129, 380
Aerm. formicans, 111, 117, 119
Aerm. hydrophila, 381
Aerm. punctata, 381
Aflatoxin, immunological detection of, 352
Agrobacterium, 290, 384
  phage sensitivity of, 328
Agbc. radiobacter, 320, 328
Agbc. rhizogenes, 320
Agbc. rubi, 320
Agbc. tumefaciens, 44, 320, 328

397